Mathematical Methods for Operations Research Problems

Mathematical Methods for Operations Research Problems

Editor

Frank Werner

Basel • Beijing • Wuhan • Barcelona • Belgrade • Novi Sad • Cluj • Manchester

Editor
Frank Werner
Faculty of Mathematics
Otto-von-Guericke University
Magdeburg
Germany

Editorial Office
MDPI
Grosspeteranlage 5
4052 Basel, Switzerland

This is a reprint of articles from the Special Issue published online in the open access journal *Mathematics* (ISSN 2227-7390) (available at: www.mdpi.com/journal/mathematics/special_issues/Mathematical_Methods_Operations_Research_Problems).

For citation purposes, cite each article independently as indicated on the article page online and as indicated below:

Lastname, A.A.; Lastname, B.B. Article Title. *Journal Name* **Year**, *Volume Number*, Page Range.

ISBN 978-3-7258-1626-2 (Hbk)
ISBN 978-3-7258-1625-5 (PDF)
doi.org/10.3390/books978-3-7258-1625-5

Contents

About the Editor vii

Preface ix

Frank Werner
Special Issue "Mathematical Methods for Operations Research Problems"
Reprinted from: *Mathematics* **2021**, *9*, 2762, doi:10.3390/math9212762 1

Nancy M. Arratia-Martinez, Paulina A. Avila-Torres and Juana C. Trujillo-Reyes
Solving a University Course Timetabling Problem Based on AACSB Policies
Reprinted from: *Mathematics* **2021**, *9*, 2500, doi:10.3390/math9192500 5

Saulius Minkevičius, Igor Katin, Joana Katina and Irina Vinogradova-Zinkevič
On Little's Formula in Multiphase Queues
Reprinted from: *Mathematics* **2021**, *9*, 2282, doi:10.3390/9182282 24

Emanuel Vega, Ricardo Soto, Broderick Crawford, Javier Peña and Carlos Castro
A Learning-Based Hybrid Framework for Dynamic Balancing of Exploration-Exploitation: Combining Regression Analysis and Metaheuristics
Reprinted from: *Mathematics* **2021**, *9*, 1976, doi:10.3390/math9161976 38

Nicolás Caselli, Ricardo Soto, Broderick Crawford, Sergio Valdivia and Rodrigo Olivares
A Self-Adaptive Cuckoo Search Algorithm Using a Machine Learning Technique
Reprinted from: *Mathematics* **2021**, *9*, 1840, doi:10.3390/math9161840 61

Liang Shen, Xiaodi Wang, Qinqin Liu, Yuyan Wang, Lingxue Lv and Rongyun Tang
Carbon Trading Mechanism, Low-Carbon E-Commerce Supply Chain and Sustainable Development
Reprinted from: *Mathematics* **2021**, *9*, 1717, doi:10.3390/math9151717 89

Zdravka Aljinović, Branka Marasović and Tea Šestanović
Cryptocurrency Portfolio Selection—A Multicriteria Approach
Reprinted from: *Mathematics* **2021**, *9*, 1677, doi:10.3390/math9141677 115

R. Suganya, Lewis Nkenyereye, N. Anbazhagan, S. Amutha, M. Kameswari, Srijana Acharya and Gyanendra Prasad Joshi
Perishable Inventory System with N-Policy, MAP Arrivals, and Impatient Customers
Reprinted from: *Mathematics* **2021**, *9*, 1514, doi:10.3390/math9131514 136

Mauricio Castillo, Ricardo Soto, Broderick Crawford, Carlos Castro and Rodrigo Olivares
A Knowledge-Based Hybrid Approach on Particle Swarm Optimization Using Hidden Markov Models
Reprinted from: *Mathematics* **2021**, *9*, 1417, doi:10.3390/math9121417 151

Teeradech Laisupannawong, Boonyarit Intiyot and Chawalit Jeenanunta
Mixed-Integer Linear Programming Model and Heuristic for Short-Term Scheduling of Pressing Process in Multi-Layer Printed Circuit Board Manufacturing
Reprinted from: *Mathematics* **2021**, *9*, 653, doi:10.3390/math9060653 172

Nodari Vakhania and Frank Werner
Branch Less, Cut More and Schedule Jobs with Release and DeliveryTimes on Uniform Machines
Reprinted from: *Mathematics* **2021**, *9*, 633, doi:10.3390/math9060633 197

Jiří Mazurek, Radomír Perzina, Jaroslav Ramík and David Bartl
A Numerical Comparison of the Sensitivity of theGeometric Mean Method, Eigenvalue Method, and Best–Worst Method
Reprinted from: *Mathematics* **2021**, *9*, 554, doi:10.3390/math9050554 . . . . **215**

Frank A. Hernández Mira, Ernesto Parra Inza, José M. Sigarreta Almira and Nodari Vakhania
Properties of the Global Total k-Domination Number
Reprinted from: *Mathematics* **2021**, *9*, 480, doi:10.3390/math9050480 . . . . **228**

Nicolas Dupin, Frank Nielsen and El-Ghazali Talbi
Unified Polynomial Dynamic Programming Algorithms for P-Center Variants in a 2D Pareto Front
Reprinted from: *Mathematics* **2021**, *9*, 453, doi:10.3390/math9040453 . . . . **241**

Teddy Nurcahyadi and Christian Blum
Adding Negative Learning to Ant Colony Optimization: A Comprehensive Study
Reprinted from: *Mathematics* **2021**, *9*, 361, doi:10.3390/math9040361 . . . . **271**

Gergely Kovacs, Benedek Nagy, Gergely Stomfai, Neset Deniz Turgay and Bela Vizvari
Discrete Optimization: The Case of Generalized BCC Lattice
Reprinted from: *Mathematics* **2021**, *9*, 208, doi:10.3390/math9030208 . . . . **294**

About the Editor

Frank Werner

Frank Werner studied mathematics from 1975 to 1980 and graduated from the Technical University Magdeburg (Germany) with distinction. He received a Ph.D. degree (with summa cum laude) in Mathematics in 1984 and defended his habilitation thesis in 1989. Since then, he has worked at the Faculty of Mathematics of the Otto-von-Guericke University Magdeburg in Germany and since 1998 as an extraordinary professor. In 1992, he received a grant from the Alexander von Humboldt Foundation. He was a manager of several research projects supported by the German Research Society (DFG) and the European Union (INTAS). Since 2019, he has been the Editor-in-Chief of the journal *Algorithms*. He is also an Associate Editor of the *International Journal of Production Research* since 2012 and of the *Journal of Scheduling* since 2014, as well as a member of the editorial/advisory boards of 18 further international journals. He has been a Guest Editor of Special Issues in ten international journals and has served as a member of the program committee of more than 150 international conferences. Frank Werner is an author/editor of 14 books, among them the textbooks *'Mathematics of Economics and Business'* and *'A Refresher Course in Mathematics'*. In addition, he has co-edited four proceedings volumes of the SIMULTECH conferences and published more than 300 journal and conference papers, e.g., in the *International Journal of Production Research, Computers & Operations Research, Journal of Scheduling, Applied Mathematical Modelling*, or the *European Journal of Operational Research.* He received Best Paper Awards from the *International Journal of Production Research* (2016) and *IISE Transactions* (2021). His main research subjects are scheduling, discrete optimization, graph theory, and mathematical problems in operations research.

Preface

This is the printed edition of the Special Issue published in the journal *Mathematics* in 2021. This reprint contains an editorial and 15 research papers. Among the subjects addressed in this reprint are machine learning, scheduling, timetabling, and graph theory.

Finally, I would like to thank all people who contributed to the success of this Special Issue, i.e., authors from 16 countries, many reviewers from all over the world, and also the staff of the journal *Mathematics* for their kind support. I hope that the readers of this reprint will become inspired for their own future work in the interesting and challenging research field of Operations Research.

Frank Werner
Editor

Editorial

Special Issue "Mathematical Methods for Operations Research Problems"

Frank Werner

Faculty of Mathematics, Otto-von-Guericke University, 39016 Magdeburg, Germany; frank.werner@ovgu.de; Tel.: +49-391-675-2025

Citation: Werner, F. Special Issue "Mathematical Methods for Operations Research Problems". *Mathematics* **2021**, *9*, 2762. https://doi.org/10.3390/math9212762

Received: 11 October 2021
Accepted: 25 October 2021
Published: 30 October 2021

This Special Issue of *Mathematics* is dedicated to the application of Operations Research methods to a wide range of problems. Operations Research uses mathematical modeling and algorithms for supporting decision processes and finding optimal solutions in many fields. For this Issue, high-quality papers were solicited to address both theoretical and practical issues in the wide area of Operations Research. In particular, submissions presenting new theoretical results, models and algorithms were welcome. Some topics mentioned in the Call for Papers for this Issue were linear and nonlinear programming, optimization problems on graphs, project management, scheduling, logistics and transportation, queuing theory and simulation, to name a few.

After a careful refereeing process, 15 papers were selected for this Issue. As a rule, all submissions were reviewed by three experts in the corresponding area. The authors of the accepted papers come from 16 countries: Hungary, Turkey, Spain, France, Japan, Mexico, Czech Republic, Germany, Thailand, Chile, India, Korea, Croatia, Chile, USA and Lithuania. Subsequently, the published papers were surveyed in increasing order of their publication dates for this Special Issue.

The first accepted paper [1] deals with body-centered cubic lattices which are important grids appearing in nature. The authors formulate the shortest path problem on higher dimensional body-centered grids as an integer programming problem. Finally, a Gomory cut is applied to guarantee an integer solution, and some comments on Hilbert bases of rational polyhedral cones are given.

The second paper [2] studies an alternative mechanism for using mathematical programming to incorporate negative learning into a widely used ant colony optimization. The authors compare their approach with existing negative learning approaches from the literature on two combinatorial optimization problems: the minimum dominating set problem and the multi-dimensional knapsack problem. It is shown that the new approach outperforms the existing ant colony algorithms and negative learning mechanisms.

In the third paper [3], the authors cluster the Pareto Front for a multi-objective optimization problem in a given number of clusters and identify isolated points. In particular, K-center problems and some variants are investigated and a unified formulation is given, where both discrete and continuous variants, partial K-center problems and their min-sum K-radii on a line are considered. In the case of dimension two, a polynomial dynamic programming algorithm is given, while for a higher dimension, the associated problem is NP-hard. For some variants, including the K-center problem and min-sum K-radii variants, further improvements are discussed. In addition, parallel implementations lead to a speed-up in practice.

Paper [4] deals with a graph-theoretic subject. In particular, the authors develop lower and upper bounds on the global total k-domination number of a graph. It is the minimum cardinality of a so-called global total k-dominating set of this graph. The results were obtained by using algebraic connectivity in graphs. Moreover, the authors present an approach to obtain a global total $(k+1)$-dominating set from a global total k-dominating set.

In the fifth paper [5], three methods for deriving a priority vector in the theoretical framework of pairwise comparisons are investigated with respect to sensitivity and order violation, namely, the Geometric Mean Method, the Eigenvalue Method and the Best–Worst Method. The authors apply a One-Factor-at-a-Time sensitivity analysis via Monte Carlo simulations. The investigations show that the Best–Worse Method is statistically significantly more sensitive and, thus, less robust than the other two methods.

Paper [6] investigates a parallel machine scheduling problem on uniform machines with identical processing times as well as given release and delivery times with a minimizing makespan, which is the time when the last job is delivered to the customer. The authors present a polynomial algorithm which is based on the 'branch less cut more' framework developed earlier. This algorithm generates a tree similar to a branch and bound algorithm, but the branching and cutting criteria depend on structural properties and are not based on lower bounds. The algorithm finds an optimal solution if for any pair of jobs a specified inequality with respect to the release and delivery times is satisfied. If these conditions are not satisfied or for the case of non-identical processing times, the algorithm can be used as an approximate one.

Paper [7] considers short-term scheduling of the pressing process which occurs in the fabrication of a multi-layer printed circuit board. For this problem, a mixed-integer linear programming formulation is given for the case of minimizing the makespan. In addition, a three-phase heuristic is also given. It turns out that the MILP model can solve small and medium-sized instances. On the one hand, the MILP model could not solve most of the large-sized instances within a time limit of two hours, but the heuristic found an optimal solution for all instances, for which the MILP model could find an optimal solution in much smaller times.

Paper [8] deals with particle swarm optimization. For bio-inspired algorithms, where a proper setting of the initial parameters by an expert is required. In this paper, the authors suggest a hybrid approach allowing the adjustment of the parameters based on a state deducted by the swarm algorithm. The state deduction is reached by a classification of the observations using a hidden Markov model. Extensive tests for the set covering problem show that the presented hybrid algorithm finds better regions in the heuristic space than the original particle swarm optimization, and it shows an overall good performance.

Paper [9] investigates a perishable inventory system with an (s, Q) ordering policy together with a finite waiting hall. The single server only begins serving when N customers have arrived. This is known as N-policy. The authors investigate the impatient demands which are caused by the N-policy server to an inventory system. In particular, the steady-state vector is investigated. In addition, some measures of the performance of the system are analyzed and the expected cost rate in the steady state is given.

Paper [10] deals with cryptocurrency portfolio selection and applies a multi-criteria approach based on PROMETHEE II. The authors found that their model gave the best cryptocurrency portfolio when considering the daily return, the standard deviation, the value-at-risk, the conditional value-at-risk, the volume, the market capitalization as well as nine cryptocurrencies for the period from January 2017 to February 2020. It turned out that the proposed model won against all other models considered.

In the eleventh paper [11], a game decision-making model for a low-carbon e-commerce supply chain is derived. The paper analyzes the influence of carbon trading on the regional sustainable development. It turns out that the total carbon emission is positively related to the commission rate. The empirical analysis conducted by the authors confirms that the implementation of carbon trading is conducive to the regional sustainable development and that controlling the environmental governance intensity promotes carbon productivity. In the future, the inclusion of more factors is intended to make the model more realistic.

In the twelfth paper [12], a new cuckoo search algorithm is presented which is able to self-adapt its configuration. This is reached by means of machine learning, where a cluster analysis is used. Experimental results are presented for the set covering problem. A comparison with other hybrid bio-inspired algorithms is also performed. The authors mention some possible future works, e.g., improving the criterion of population increase and decrease by using clusterization strategies or implementing further machine learning techniques.

Paper [13] develops a novel hybrid optimization framework, entitled learning-based linear balancer. The authors also design a regression model to predict better movements for the approach and to improve the performance. The approach is based on balancing the intensification and diversification performed by the hybrid approach in an online form. To test the suggested approach, 15 benchmark functions are considered. The authors also compare their approach against a spotted hyena optimizer and a neural network approach.

In the fourteenth paper [14], a subject from queuing theory is considered, where the suggested approach consists of two stages. In the first stage, Little's law in Multiphase Systems is analyzed. In particular, Strong Law of Large Numbers-type theorems are proven. Then, the results obtained in this stage are verified by means of simulation. Here, the Python concept is used to test the results obtained in the first stage.

The last paper [15] deals with the university course timetabling problem. It presents a new integer programming model for generating a timetable of an academic department considering basic workload and course overload and, also, the profile and area of each professor. A real-world case is considered. By analyzing different strategies, the efficiency of the new model is shown.

Finally, as the guest editor, it is my pleasure to thank the editorial staff of the journal *Mathematics* for the pleasant cooperation, not only during the preparation of this, but also for the previous four Special Issues which I handled as editor for the journal *Mathematics*. I would also to thank all referees for their thorough and timely reports on the submitted works and also the authors for submitting many interesting works from a broad spectrum in the Operations Research area. For potential authors who missed the deadline for this Special Issue, I remind that there is another future Special Issue in *Mathematics* entitled 'Recent Advances of Discrete Optimization and Scheduling' edited by Alexander Lazarev, Bertrand Lin and myself, which deals with similar subjects.

Funding: This research received no external funding.

Institutional Review Board Statement: Not applicable.

Informed Consent Statement: Not applicable.

Data Availability Statement: Not applicable.

Conflicts of Interest: The author declares no conflict of interest.

References

1. Kovacs, G.; Nagy, B.; Stomfai, G.; Turgay, N.D.; Vizvari, B. Discrete Optimization: The Case of Generalized BCC Lattice. *Mathematics* **2021**, *9*, 208. [CrossRef]
2. Nurcahyadi, T.; Blum, C. Adding Negative Learning to Ant Colony Optimization: A Comprehensive Study. *Mathematics* **2021**, *9*, 361. [CrossRef]
3. Dubin, N.; Nielsen, F.; Talbi, E.-G. Unified Polynomial Dynamic Programming Algorithms for P-Center Variants in a 2D Pareto Front. *Mathematics* **2021**, *9*, 453. [CrossRef]
4. Hernandez-Mira, F.A.; Parra Inza, E.; Sigarreta Almira, J.M.; Vakhania, N. Properties of the Global Total *k*-Domination Number. *Mathematics* **2021**, *9*, 480. [CrossRef]
5. Mazurek, J.; Perzina, R.; Ramik, J.; Bartl, D. A Numerical Comparison of the Sensitivity of the Geometric Mean Method, Eigenvalue Method, and Best-Worst Method. *Mathematics* **2021**, *9*, 554. [CrossRef]
6. Vakhania, N.; Werner, F. Branch Less, Cut More and Schedule Jobs with Release and Delivery Times on Uniform Machines. *Mathematics* **2021**, *9*, 633. [CrossRef]
7. Laisupannawong, T.; Intiyot, B.; Jeenanunta, C. Mixed-Integer Linear Programming Model and Heuristic for Short Term Scheduling of Pressing Process in Multi-Layer Printed Circuit Board Manufacturing. *Mathematics* **2021**, *9*, 653. [CrossRef]
8. Castello, M.; Soto, R.; Crawford, B.; Castro, C.; Olivares, R. A Knowledge-Based Hybrid Approach on Particle Swarm Optimization Using Hidden Markov Models. *Mathematics* **2021**, *9*, 1417. [CrossRef]
9. Suganya, R.; Nkenyereye, L.; Anbazhagan, N.; Amutha, S.; Kameswari, M.; Acharya, S.; Joshi, G.P. Perishable Inventory System with N-Policy, MAP Arrivals, and Impatient Customers. *Mathematics* **2021**, *9*, 1514. [CrossRef]
10. Aljinovic, Z.; Marasovic, B.; Sestanovic, T. Cryptocurrency Portfolio Selection—A Multicriteria Approach. *Mathematics* **2021**, *9*, 1677. [CrossRef]

11. Shen, L.; Wang, X.; Liu, Q.; Wang, Y.; Lv, L.; Tang, R. Carbon Trading Mechanism, Low Carbon E-Commerce Supply Chain and Sustainable Development. *Mathematics* **2021**, *9*, 1717. [CrossRef]
12. Caselli, N.; Soto, R.; Crawford, B.; Valdivia, S.; Olivares, R. A Self-Adaptive Cuckoo Search Algorithm Using a Machine Learning Technique. *Mathematics* **2021**, *9*, 1840. [CrossRef]
13. Vega, E.; Soto, R.; Crawford, B.; Pena, J.; Castro, C. A Learning-Based Hybrid Framework for Dynamic Balancing of Exploration-Exploitation: Combining Regression Analysis and Metaheuristics. *Mathematics* **2021**, *9*, 1976. [CrossRef]
14. Minkevicius, S.; Katin, I.; Katina, J.; Vinogradova-Zinkevic, I. On Little's Formula in Multiphase Queues. *Mathematics* **2021**, *9*, 2282. [CrossRef]
15. Arratia-Martinez, N.M.; Avila-Torres, P.A.; Trujillo-Reyes, J.C. Solving the University Course Timetabling Problem Based on the AACSB Policies. *Mathematics* **2021**, *9*, 2500. [CrossRef]

 mathematics

MDPI

Article

Solving a University Course Timetabling Problem Based on AACSB Policies

Nancy M. Arratia-Martinez *,†, Paulina A. Avila-Torres † and Juana C. Trujillo-Reyes †

Department of Business Administration, Universidad de las Américas Puebla/Ex-Hacienda Santa Catarina Mártir, San Andres Cholula, Puebla 72810, Mexico; paulina.avila@udlap.mx (P.A.A.-T.); juanac.trujillo@udlap.mx (J.C.T.-R.)
* Correspondence: nancy.arratia@udlap.mx
† These authors contributed equally to this work.

Abstract: The purpose of this research is to solve the university course timetabling problem (UCTP) that consists of designing a schedule of the courses to be offered in one academic period based on students' demand, faculty composition and institutional constraints considering the policies established in the standards of the Association to Advance Collegiate Schools of Business (AACSB) accreditation. These standards involve faculty assignment with high level credentials that have to be fulfilled for business schools on the road to seek recognition and differentiation while providing exceptional learning. A new mathematical model for UCTP is proposed. The model allows the course-section-professor-time slot to be assigned for an academic department strategically using the faculty workload, course overload, and the fulfillment of the AACSB criteria. Further, the courses that will require new hires are classified according to the faculty qualifications stablished by AACSB. A real-world case is described and solved to show the efficiency of the proposed model. An analysis of different strategies derived from institutional policies that impacts the resulting timetabling is also presented. The results show the course overload could be a valuable strategy for helping mitigate the total of new hires needed. The proposed model allows to create the course at the same time the AACSB standards are met.

Keywords: timetabling problem; course university timetabling problem; AACSB standards; integer linear programming

Citation: Arratia-Martinez, N.M.; Avila-Torres, P.A.; Trujillo-Reyes, J.C. Solving a University Course Timetabling Problem Based on AACSB Policies. *Mathematics* **2021**, *9*, 2500. https://doi.org/10.3390/math9192500

Academic Editors: Frank Werner and Jay Jahangiri

Received: 30 May 2021
Accepted: 17 September 2021
Published: 6 October 2021

1. Introduction

Timetabling is the process of building a timetable while satisfying several constraints. The timetabling problem has many applications such as educational and transportation issues for employees and others [1]. This research is focused on the university course timetabling problem, a problem that has been extensively studied [2]. The University Course Timetabling Problem (UCTP) consists of supplying a schedule of the courses to be offered in one academic period based on students enrolled and constraints established by the university. A course timetabling usually involves the allocation of resources (teachers, students, classrooms, etc.) and time slots to each given meeting (lectures, seminars, etc.) while satisfying constraints [3].

The UCTP has three stages: (i) faculty course assignment optimization, (ii) faculty course scheduling optimizations and (iii) faculty room assignment optimization [4], there are many constraints to be considered and they are usually divided into two categories: (i) hard constraints, these constraints must be satisfied in order to produce a feasible timetable and (ii) soft constraints, these constraints are desirable but not absolutely essential [2].

UCTP is considered one of the most interesting problems faced by universities [5] but some of them are still constructing timetables by hand [6] with the assistance of simple office applications like spreadsheets. This is a very difficult task given the many restrictions to be satisfied [3]. The automation of timetabling problems is a task that saves a lot of work and time for institutions, it also provides optimal solutions by improving the quality of education and services [5]. The educational timetabling problem has been formulated in many different ways and has been addressed using several analytic or heuristic approaches. However, it is difficult to implement the same approach to a problem because each institution has different characteristics and constraints or limitations [7].

There are some authors that consider preferences of the faculty in different areas. Such is the case of the method proposed by Immonen and Putkonen [8] where they build a timetable satisfying pre-requisite knowledge and specific preferences for faculty. Also Tavakoli et al. [9] say one subject can be taught by many lecturers but the priority must be given to the one with higher qualifications. In [10], authors use a bee colony optimization and consider the preferences of subjects a professor can teach. Another work is the one proposed by Domenech and Lusa [11] in which they propose a mathematical model considering some preferences according to the category of the professor. Al-Yakoob and Sherali [12] developed a mathematical model for assigning faculty members to classes considering their preferences of time as much as possible and the qualifications of the faculty.

Another characteristic considered in the construction of the timetable is the workload of the faculty, for example, in [11] authors propose a mathematical model where they balance the teachers' workload. Authors in [12] present a mathematical model for assigning faculty members to classes considering teaching load and qualifications, the objective of the model is to minimize the dissatisfaction of faculty members. In some cases according to the level of the professor some institutions may establish a number of days a professor can teach in order to give them the opportunity to do work in research, such is the case of the approach made by Chen and Shih [13] where teachers of specific levels can only teach two classes per week and each teacher may not teach more hours than the limit stipulated by the academic department. Further, authors in [10] consider the maximum number of courses a professor can teach, or create a fair course timetable, balancing the interests between faculty [14].

Characteristics like, preferences of different types, workload, among others are important in the construction of timetable for business schools that are accredited by the Association to Advance Collegiate Schools of Business (AACSB) or for those institutions that would like to obtain the accreditation. For the latter, it is important to build the timetable fulfilling the standards established by AACSB. The mathematical model presented by Boronico and Kong [15] determines the full-time faculty (without any decision about the timetable) required according to accreditation guidelines of AACSB for the different campuses and disciplines, this is the reason business schools must now specify their relative emphasis on teaching, intellectual contributions, service, and make explicit commitments to particular types of intellectual contributions [16].

For business schools, it is very important to obtain the AACSB accreditation, specially for those schools outside North America and Europe [17], according to Bajada and Trayler [18], the faculty of a modern business school is expected to be academically qualified (AQ) under AACSB standards. For an institution that would like to receive AACSB accreditation, a certain percentage of the business school faculty must be AQ [17].

AACSB is an important influence on many business schools, that is the reason the accredited business schools are expected to have highly qualified faculty members to complete the course timetabling, but the number of faculty available to fill open positions is not sufficient and it is difficult for schools to recruit and retain the qualified faculty [19].

AACSB is becoming more important for business schools and fulfilling the standards is determinant to achieve the accreditation or re-accreditation. This implies universities need to accomplish the percentages of professors in every category and the construction of the timetable is directly related to that standard.

In the reviewed literature, we did not find any other paper that involves the timetabling problem with all the characteristics addressed here and the AACSB standards. This research is focused in the construction of a university course timetable for a business school considering mainly the preferences and qualifications of faculty, teaching load and the category of the professors in order to fulfill the percentages of the qualification standards established by the AACSB. The output of the model will be the assignment of professor to subject and time slot, also the number of professors that have to be hired in order to achieve the percentages indicated by the standard of the AACSB.

The papers found in the reviewed literature have some similarities with the proposed model. For example, authors like Immonen and Putkonen [8] and Al-Yakoob and Sherali [12] include in their construction of timetabling the qualification of professors. The category of professors is considered by Domenech and Lusa [11] and Chen and Shih [13]. Another similarity is the consideration of maximum number of courses, a characteristic taken into account by Ojha and Sahoo [10]. But our proposal has some differences with the papers found in literature, for example, we do not consider the balance of workload whereas Domenech and Lusa [11] and Al-Yakoob and Sherali [12] include this characteristic and the models proposed by them. One of the decision made by our proposed model is whether to give or not work overload to professors, characteristic not found in any other model. Boronico and Kong [15] take into account the standards of AACSB but they do not construct the timetable. Their model indicates the number of professors needed in each campus is order to comply the percentages of each category in the AACSB standards. To sum it up, any other paper that involves the timetabling problem with all the characteristics addressed here and the AACSB standards was not found.

The article is organized as followed, the rest of the Section 1 provides a description of the AACSB accreditation as well as the faculty standard; Section 2 describes on detail the context of the case study; Section 3 presents the proposed mathematical model; Section 4 presents the case information; Section 5 presents a discussion of the results; Section 6 presents the conclusions of the present work.

AACSB Accreditation and Standards

A challenge facing economic programs in business schools is that of aligning programs to be consistent with the assessment expectations for the AACSB accreditation [20]. Business school accreditation is a way for business schools to differentiate their brand and demonstrate the highest standard of achievement [21]. The AACSB is the most important institution responsible of accrediting business schools around the world. The AACSB was founded in 1916 and established the first standards for programs in business administration in 1919. Nowadays, there are 874 business institutions in 56 countries that have earned the AACSB accreditation [22].

A business school has to follow the next process in order to apply for the AACSB Accreditation: first, the business school must establish its membership and eligibility for accreditation. During the initial accreditation process, the school is evaluated based on the AACSB accreditation standards. After earning the AACSB accreditation, the business school is periodically evaluated to continue its accreditation [22].

The nine standards that every business school have to achieve are divided into three categories: (1) Strategic management and innovation, (2) Learner success and (3) Thought leadership, engagement and societal impact learning and teaching. Standard three declares that the school should maintain and strategically deploy a sufficient number of participating (P) and supporting (S) faculty. A participating faculty actively takes part in the activities of the school besides teaching responsibilities. A supporting faculty is more dedicated to teaching responsibilities; she/he does not normally participate in the intellectual or operational life of the school [22].

According to the AACSB, the faculty is classified as follows: Scholarly Academic (SA), Practice Academic (PA), Scholarly Practitioner (SP), or Instructional Practitioner (IP). Faculty members who do not meet the definitions of any of these categories are classified as Additional Faculty (A).

- Scholarly Academics (SA) are faculty who have normally attained a terminal degree in a field related to the area of teaching and who sustain currency and relevance through scholarship and related activities.

- Practice Academics (PA) are faculty who have normally attained a terminal degree in a field related to teaching and who sustain currency and relevance through professional engagement, interaction, and relevant activities.
- Scholarly Practitioners (SP) are faculty who have normally attained a master's degree related to the field of teaching; have professional experience and produce scholarship related to their professional background and experience.
- Instructional Practitioners (IP) are faculty who have normally attained a master's degree related to the field of teaching and who have professional experience and continue their engagement related to their professional background and experience.
- Additional Faculty (A) are faculty who do not meet the expectations of the school as SA, PA, SP, or IP because the individual faculty member's initial preparation and/or on-going engagement activities are not aligned with the school's criteria.

In the first case (SA), they should have actual and relevancy research and activities linked to the same field of teaching. In the second case (PA), the teachers should be working in relevant professional positions also related to the field of teaching. The faculty classified as SA and PA must have a doctorate degree.

On the other hand, SP and IP have a master's degree related to a teaching field and have significant professional experience at the same field they are teaching. The difference in this case is that SP sustain research associated to the area of their professional background and experience where they teach, and IP show relevancy and engagement through their professional experience related to their teaching field.

For more information on the above categories, please refer to the AACSB manual.

The standards provide guidance about the criteria the school should develop. The criteria applied to faculty is the following:

- At least 60% of faculty should be participating. Faculty sufficiency related to teaching is measured through a teaching productivity metric (a particular institutional metric, e.g., contact-hours, course-hours, courses) and the overall should be at least 60% for the participating components.
- Percentage of time devoted to mission for each faculty qualification group:
  - Scholarly Academics $\geq$ 40%.
  - Scholarly Academics + Practice Academics + Scholarly Practitioners $\geq$ 60%.
  - Scholarly Academics + Practice Academics + Scholarly Practitioners + Instructional Practitioners $\geq$ 90%.
  - Additional Faculty less than 10%.

  Normally, full-time professors spend 100% of their time devoted to the mission and an adequate and rational manner to assess the percentage of time devoted to the mission should be establish for part-time faculty [23].

2. Context of the Problem

The Business School is composed of seven departments. In a fall 2020 semester the school offered 345 courses in total for 7750 registered students. In the past, the 24% of the courses of the entire Business School corresponded to the academic department under study. This department can be considered one of the biggest in the business school. It offers 9 service courses to the other departments at school and also to other six bachelors degree of others schools.

Some years ago, the university started the process of achieving the AACSB accreditation, now the construction of the timetable has to consider the characteristics of the school and it has to fulfill the requirements of standard three of the AACSB. The current process is as follows:

1. The dean's office informs the academic department the number of students demanded by each course.

2. The academic department defines the number of groups (sections) for the same course to be offered according to the maximum number of students per course allowed by the university.
3. A first draft of the timetable is created manually, trying to satisfy the percentages of each category in the AACSB standard and other requirements.
4. If the academic department notices a lower percentage than the required, then they have to recruit a professor of specific category.
5. The academic department confirms the assignment with each professor and some changes could be made.
6. Once the timetable is confirmed, it is sent to dean's office where the assignment of classrooms is performed.
7. Finally, the academic department proceeds to register the course timetable in the university's system and to publish it on the official website.

In this case, student curricula consists of eight semesters and fifty courses that are divided into general education courses, basic/initial disciplinary education courses, and disciplinary courses. In a specific academic period there are students enrolled in each of the level (from first to eighth semesters), so it is needed to program all the courses by period but just the disciplinary courses are in charge of the academic department.

The requirements of the administration are:

- There are two schemes of time slots for the academic department courses: scheme A) three sessions of one hour on Monday, Wednesday and Friday and, scheme B) two sessions of one hour and thirty minutes for Tuesday and Thursday.
- The first class of the day starts at 7:00 a.m. and the last one starts at 7:00 p.m.
- The courses that belong to the same semester in the curricula are assigned in different time slots.
- Semester courses from 1st to 4th are assigned to start in the morning (07:00 to 14:00 h.), and from the 5th to 8th in the afternoon (13:00 to 20:00 h, and 7:00 a.m.). Some courses will need to be scheduled at additional times due to their high demand.
- The course assignment for a professor is made by considering her/his knowledge area.
- The number of courses to be assigned to full-time professors is well known and it depends of her/his profile (researcher or manager position).
- Further, overload is allowed, when a course is assigned to a professor additionally to her/his official basic workload (authorized by the university).
- It is not desirable to assign more than two courses to part-time professors.
- When the total of courses to be scheduled exceeds the actual capacity (using course load and overload), then new hires should be considered.

3. Mathematical Formulation

When a department belongs to a school which wants to be accredited by high standards such as AACSB, and plays a fundamental role scheduling a large number of courses, it requires having mechanisms that facilitate decision-making to assign courses to the right teachers. Therefore, the mathematical formulations offer opportunities not only for assigning a large number of sections, but also for accomplishing the requirements of faculty qualifications.

The mathematical model proposed allows to determine the assignation of course-section-professor-time slot using the actual capacity. In case, it is not possible to schedule all courses, the remaining courses will be assigned to one faculty member category based on AACSB faculty qualifications policies and then the new hires can be established. Also the concept of course overload is contemplated.

Sets

C	Set of courses, index $i \in C$.
P	Set of professors, index $j \in P$.
$P_{full-time}$	Set of full-time professors, index $j \in P_{full-time}$, $P_{full-time} \subset P$.
$P_{part-time}$	Set of part-time professors, index $j \in P_{part-time}$, $P_{part-time} \subset P$.
T	Set of semesters, index $t \in T$.
B	Set of time slots, index $k \in B$.
B_t	Subset of time slots that are allowed to schedule courses belonging the semester $t \in T$.
C_j	Set of courses that the professor j can teach according with her/his knowledge area, $C_j \subset C$.
S_i	Set of sections needed to be schedule for the course i.
H_1	Set of faculty qualification categories, index $p \in H_1 = \{SA, PA, SP, IP, A\}$.
H_2	Set of categories for faculty composition based on the level of professors involvement, index $q \in H_2 = \{\text{participating, supporting}\}$.
H_3	Set of profiles of new hires based on the minimum academic profiles needed in order to allocate all the courses, index $r \in H_3$.
CH_r	Set of courses that can be taught by a professor with academic profile $r \in H_3$.
P^*	Set of professors with faculty composition category: participating.

Parameters

h_i	Semester to which the course i belongs.
m_j^{max}	Maximum number of courses to be assigned to professor j.
m_j^{min}	Minimum number of courses to be assigned to professor j.
c_j^{max}	Maximum number of course overload allowed for professor j.
α_j	Qualification of the professor j, $\alpha_j \in H_1$.
u	Percentage of faculty time spent dedicated to the mission per course for supporting faculty.
v	Percentage of faculty time spent dedicated to the mission per participating faculty (in some cases it could be naturally 100%).

The decision variables in our model are denoted as follows:

$$x_{ijkl} = \begin{cases} 1 & \text{if, the course } i \text{ is assigned to professor } j \text{ in time slot } k \text{ in section } l \\ 0 & \text{otherwise.} \end{cases}$$

$\sigma_j =$ Number of courses assigned to professor j without exceeding the number of courses allowed according to her/his academic profile.

$\sigma_j^+ =$ Number of additional courses assigned to full-time professors as course overload.

$w_i =$ Quantity of sections of the course i without assignation of schedule and professor.

$y_{pqr} =$ Quantity of course sections that weren't programmed due to the lack of enough teacher staff, also to be programmed for new candidate professors with faculty qualification category $p \in H_1$ and with faculty composition category based on the level involvement $q \in H_2$ and an academic profile $r \in H_3$.

$z_{pqr} =$ Auxiliary variables denoting the quantity of minimum candidates to professors to be hired with faculty qualification category $p \in H_1$, with faculty composition category based on the level involvement $q \in H_2$ and an academic profile $r \in H_3$.

The variables y_{pqr} individually help to know how many courses without schedule require a specific professor profile (new hire) to maintain the adequate levels of the AACSB standards.

The objective function has to be established in accordance with the institution's strategy, in this paper three possible objectives are stated. Naturally, it is desirable to minimize the total of courses without schedule (that need to be assigned to new hires) (Equation (1)).

$$\text{minimize} \sum_{p \in H_1} \sum_{q \in H_2} \sum_{r \in H_3} y_{pqr} \tag{1}$$

Other objective function based on a policy is to minimize the aggregation of courses without schedule and the course overload of professors, this is described in expression (2).

$$\text{minimize} \sum_{p \in H_1} \sum_{q \in H_2} \sum_{r \in H_3} y_{pqr} + \sum_{j \in P} \sigma_j^+ \tag{2}$$

Further, in expression (3) the total number of new hires is minimized since the objective in expression (1) does not differentiate courses that can be assigned to different academic profiles. The value of z_{pqr} is estimated in (20).

$$\text{minimize} \sum_{p \in H_1} \sum_{q \in H_2} \sum_{r \in H_3} z_{pqr} \tag{3}$$

Since all sections should be assigned whenever possible, the group of constraints (4) guarantees that all course sections will be scheduled only if there is enough capacity (i.e., $w_i = 0$), otherwise, the variable w_i will take a value greater than zero ($w_i > 0$).

$$\sum_{j \in P} \sum_{k \in B} \sum_{l \in S_i} X_{ijkl} + w_i = \mid S_i \mid \quad \forall i \in C \tag{4}$$

where $\mid S_i \mid$ denotes the cardinality of the set S_i.

The constraint group (5) states that each course section should be assigned just once.

$$\sum_{j \in P} \sum_{k \in B} X_{ijkl} \leq 1 \quad \forall i \in C, \forall l \in S_i \tag{5}$$

Course sections without a schedule and teacher assignment will result in new teacher hires with an AACSB faculty qualification category $p \in H_1$, a category of faculty composition based on the level of involvement $q \in H_2$ and with an academic profile $r \in H_3$. In constraint group (6) it is stated that the number of course sections without schedule that can be assigned to a candidate professor with an academic profile $r \in H_3$ should be equal to the number of courses assigned to new professors with a profile in each set of categories (H_1 and H_2). It allows to balance and to relate course sections with faculty profiles.

$$\sum_{p \in H_1} \sum_{q \in H_2} y_{pqr} = \sum_{i \in CH_r} w_i \quad r \in H_3 \tag{6}$$

All professors have a maximum and a minimum number of course sections to be assigned, this is restricted by (7) and (8). Commonly, for full-time professors this quantity represents the mandatory number of courses to be assigned, in that case $m_j^{max} = m_j^{min}$.

$$\sigma_j \leq m_j^{max} \quad \forall j \in P \tag{7}$$

$$\sigma_j \geq m_j^{min} \quad \forall j \in P \tag{8}$$

This model contemplates the concept of restricted course overload in order to consider to teach more than the maximum courses allowed for each professor. The number of assigned course sections of each professor is equal to the sum of the course load plus the course overload assigned, as it is shown in the group of constraints (9).

$$\sum_{i \in C} \sum_{k \in B} \sum_{l \in S_i} x_{ijkl} = \sigma_j + \sigma_j^+ \quad \forall j \in P \tag{9}$$

When the course overload does not apply for all professors or for part-time professors, it can be easily restricted. The constraint group (10) states the maximum number for professor j course overload.

$$\sigma_j^+ \leq c_j^{max} \quad \forall j \in P \tag{10}$$

Regarding the schedule, in (11) it is stated that all courses that belong to a same semester must be assigned to different time slots.

$$\sum_{i \in \{C|h_i = t\}} \sum_{j \in P} \sum_{l \in S_i} X_{ijkl} \leq 1 \quad \forall t \in T, \forall k \in B \tag{11}$$

As it was mention in the previous section, the courses belonging to semester t should be scheduled just in the allowed time slots (the subset B_t), as it is shown in constraint group (12).

$$\sum_{j \in P} \sum_{k \notin B_t} X_{ijkl} \leq 0 \quad \forall t \in T, \forall i \in \{C|h_i = t\}, \forall l \in S_i \tag{12}$$

The constraint group (13) ensures that all professors will be assigned their courses in different time slots. It means that, there will not be any overlap in the schedule of professors.

$$\sum_{i \in C} \sum_{l \in S_i} X_{ijkl} \leq 1 \quad \forall j \in P, \forall k \in B \tag{13}$$

The courses assigned to the professors will be according to their credentials (14). Previously, a list of courses that a professor can teach according to her/his expertise area was created.

$$X_{ijkl} \leq 0 \quad \forall j \in P, \forall i \notin C_j, \forall l \in S_i, \forall k \in B \tag{14}$$

In some cases, full-time professors need to have a free day scheme in the afternoon allowing to teach at graduate programs as expressed in constraints (15)–(19). The set of time slots in conflict is $B' \subset B$ for days scheme A, and $B'' \subset B$ for days scheme B. The auxiliary binary variables A_j and B_j denote if a professor j is free in the time slots to teach at the graduate program in the scheme A and scheme B, respectively.

$$A_j + B_j \leq 1 \quad \forall j \in P_{full-time} \tag{15}$$

$$A_j \cdot \sigma_j \geq \sum_{i \in C} \sum_{l \in S_i} \sum_{k \in B'} X_{ijkl} \quad \forall j \in P_{full-time} \tag{16}$$

$$A_j \leq \sum_{i \in C} \sum_{l \in S_i} \sum_{k \subseteq B'} X_{ijkl} \quad \forall j \in P_{full-time} \tag{17}$$

$$B_j \cdot \sigma_i \geq \sum_{i \in C} \sum_{l \in S_i} \sum_{k \in B''} X_{ijkl} \quad \forall j \in P_{full-time} \tag{18}$$

$$B_j \leq \sum_{i \in C} \sum_{l \in S_i} \sum_{k \subseteq B''} X_{ijkl} \quad \forall j \in P_{full-time} \tag{19}$$

In (20) the number of professors needed for profile is calculated.

$$z_{pqr} \geq \frac{y_{pqr}}{M}, \quad \forall p \in H_1, \forall q \in H_2, \forall r \in H_3 \tag{20}$$

where M denotes the maximum course load for a new professor.

To establish the criteria and policies for faculty based on AACSB standards the constraints (21)–(25) were incorporated.

Based on the metric selected (number of courses), we restricted that 60% of courses have to be imparted by participating faculty, as it is shown in constraint (21).

$$\sum_{j \in P^*} \left(\sigma_j + \sigma_j^+\right) + \sum_{p \in H_1} \sum_{r \in H_3} y_{pqr} \geq 0.60 \sum_{i \in C} | S_i |, \quad q = \text{participating} \tag{21}$$

The total time dedicated to mission for each AACSB faculty qualification category is calculated in (22). The constraints are defined with $q_1 =$ participating, $q_2 =$ supporting and, $\forall p \in H_1$.

$$ded_p = u\left(\sum_{j\in P-P^*|\alpha_j=p}(\sigma_j+\sigma_j^+)+\sum_{r\in H_3} y_{pq_2r}\right)+v\left(|P^*|+\sum_{r\in H_3} z_{pq_1r}\right) \quad (22)$$

Then, the time dedication to the mission of all faculty and courses with a hiring profile is restricted in (23)–(25). Different metrics can be applied according to the best structure for the institution. Here, the percentage of total time spend dedicated to the mission differentiating the percentage of participating faculty and supporting faculty is calculated.

$$ded_{SA} \geq 0.4 \sum_{p\in H_1} ded_p \quad (23)$$

$$ded_{SA}+ded_{PA}+ded_{SP} \geq 0.6 \sum_{p\in H_1} ded_p \quad (24)$$

$$ded_{SA}+ded_{PA}+ded_{SP}+ded_{IP} \geq 0.9 \sum_{p\in H_1} ded_p \quad (25)$$

Further, additional constraints to limit the number of new hires can be added for some determined profile. In constraint group (26) the latter condition is added.

$$z_{pqr} \leq a_{pqr} \quad (26)$$

where a_{pqr} are the maximum number of hires allowed by faculty qualification category p, with faculty composition category based on the level involvement $q \in H_2$ and an academic profile $r \in H_3$.

Finally, the no-negativity constraints are included for the integer variables.

$$w_i \geq 0 \quad \forall i \in C \quad (27)$$

$$y_{pqr}, z_{pqr} \geq 0 \quad \forall p \in H_1, \forall q \in H_2, \forall r \in H_3 \quad (28)$$

$$\sigma_j, \sigma_j^+ \geq 0 \quad \forall j \in P \quad (29)$$

Remarks of the Mathematical Formulation

Here some main remarks about the mathematical formulation are exposed:

- In the proposed mathematical model the variables that allow to define the timetabling are x_{ijkl}, while w_i are the number of sections by course i that could not be programmed with the current faculty. In the Equation (4) is established the relation between these variables.
- Since in the objective functions the y_{pqr} variables are minimized directly or indirectly and the sum of all y_{pqr} and the sum of all w_i variables are equal, then the w_i variables are also minimized in an indirectly form.
- The variables y_{pqr} and z_{pqr} are related in the mathematical model in constraints (20), which allow to determine the number of new hires given the number of courses without schedule for all $p \in H_1$, $q \in H_2$ and $r \in H_3$ categories.
- The values of variables x_{ijkl} impact in the assignations of the variables y_{pqr}, since as was mentioned before, the variables x_{ijkl} and w_i are related, and the variables w_i are related with y_{pqr}, then, the variables x_{ijkl} are related indirectly with the variables y_{pqr}.
- Regarding the latter, the constraints of balancing (21)–(25) are used to ensure compliance with the standard three of AACSB while defining the values for variables y_{pqr}.

- The variables x_{ijkl} and z_{pqr} are related through the relation stablished before about the pairs of variables (x_{ijkl}, w_i) (w_i, y_{pqr}) and (y_{pqr}, z_{pqr}).

4. Case Information

In this section the main information of the case studied corresponding to the fall 2020 academic period is described.

As it is stated in Table 1, in the academic period fall 2020, the academic department was constituted by 28 teachers (17 full-time professors and 11 part-time professors) with different profiles and preferences of courses according to their qualifications. They are classified as six supporting professors and 22 participating, according to the category (explained in the Section 1) the same 28 teachers can be classified as: 12 IP, 12 SA, 2 PA, 1 SP and 1 is classified as A (categories explained in Section 1).

There were 28 courses, each one with a specific number of sections (or groups), in total 82 sections to be scheduled. The course load and the list of possible courses to assign for each professor, are presented in Table 1. To get the solution, the course overload for part-time professors was considered as zero, since this concept is applicable just for full-time faculty.

Table 1. Information of professors.

Professor ID	Max. Courses	Courses	Professor ID	Max. Courses	Courses
P1	2	23	P15	3	27, 15, 10, 11
P2	3	24, 18, 2, 1	P16	2	25
P3	2	20, 8, 16	P17	4	16, 17, 7
P4	3	8, 16	P18	4	9, 16, 4, 6
P5	1	25	P19	4	16, 12
P6	3	25, 14, 21, 11, 3, 10	P20	2	18, 2, 1
P7	2	25, 16	P21	2	8,6
P8	2	11, 15, 2, 10	P22	2	28, 1, 2
P9	2	13, 5, 16, 20	P23	3	23, 12, 8, 7
P10	2	3, 24	P24	3	27, 1, 2
P11	2	26, 24	P25	2	9, 16
P12	2	16	P26	2	15
P13	3	22, 17, 16	P27	3	19, 16
P14	2	25, 3, 10, 11	P28	3	16

The courses are distributed in the follow way:

Semester 1: 16 and 19.
Semester 2: 7 and 9.
Semester 3: 12 and 13.
Semester 4: 6, 8, 10 and 25.
Semester 5: 11, 14 and 18.
Semester 6: 2, 5, 17 and 20.
Semester 7: 1, 4, 15, 21 and 26.
Semester 8: 3, 22, 23, 24, 27 and 28.

Regarding new hires, based on the conditions established by the institution, it is possible to define limits on new hires, according to the categories and profiles established. For this, the constraint group (26) are established with the following limits: zero new hires with profile SA-supporting and for PA-supporting categories, also, zero new hires with profile SP-supporting, IP-supporting and A-supporting. The latter applies for all academic profiles ($r \in H_3$).

The academic profiles are defined through the clusters of courses:

Academic profile 1: courses 10, 11, 12, 15, 20, 26, 27,
Academic profile 2: courses 1, 2, 18, 24,
Academic profile 3: courses 7, 8, 9, 13, 17, 23,
Academic profile 4: courses 3, 14, 25, 26, 27,
Academic profile 5: courses 16, 4, 5, 6, 19, 21, 22, 28.

In this way we assume a determined professor can be related with courses in one specific academic profile by her/his knowledge area.

To calculate the percentage of time dedication to the mission, course-hours for supporting professors (i.e., 7.5% of the time per course) were applied. And, the sum of time dedication to mission is 100% for all participating professors. Finally, since a new hire will be classified as either supporting or participating, for practical purposes we consider a new hire with a participating profile to be full-time and part-time for a supporting profile.

5. Results

In this section the results obtained for the UCTP based on the AACSB policies are presented.

Four strategies were applied to analyze the effects of combining new hiring profiles and course overload for professors through the minimization criteria, whether or not course overload was allowed. The four strategies are:

1. to minimize courses without schedule allowing course overload (base case). This formulation is composed of objective function (1), and constraints (4)–(29).
2. to minimize courses without schedule and zero teaching overload. This formulation is composed of objective function (1), and constraints (4)–(29). To avoid the course overload, in constraint (10), the parameter c_j^{max} is equal zero for all $j \in P$.
3. to minimize courses without schedule and teaching overload. This formulation is composed of objective function (2), and constraints (4)–(29).
4. to minimize hiring profiles (professors). This formulation is composed of objective function (3), and constraints (4)–(29). In this case, course overload is allowed.

The four strategies were solved through the proposed integer programming model implemented with ILOG CPLEX Optimization Studio version 12.8 on a computer with an Intel i7 at 2.5 GHz and 8 Gb of RAM. The solution algorithm used was the classic Branch and bound method. The execution time of the solutions was less than a second (0.58 s. in average) and in all cases the optimality was obtained with a zero value for the solution gap. To access the output of the solver CPLEX of the first strategy, see the Section Data Availability Statement.

5.1. Minimizing the Courses without Schedule Allowing Course Overload, the Base Case

The timetable for the case studied is shown in Figures 1 and 2. First, in Figure 1 we show the assigned courses to each time slot in day scheme A: Monday, Wednesday and Friday (i.e three classes of 1 h per week). The color indicates the semester to which the course belongs. We named each assignation as C#S#P#, where the number contiguous to C is the course ID, then, the number next to the S is the number of section of that course, and finally, the number after P is the professor's ID. As an example, the blue cell in time 07:00-08:00 (C19S1P27) implies that Section 1 of the course with ID19 from the first semester is assigned to professor ID P27 at that time. Naturally, Table 1 with the professors' information shows that professor ID P27 can teach course ID19.

Time	Days			
	Monday, Wednesday, Friday			
07:00-08:00	C19S1P27	C7S1P17	C12S2P19	C8S3P3
	C14S1P6	C26S1P11	C3S6P10	
08:00-09:00	C9S1P18	C25S4P7		
09:00-10:00	C25S1P5			
10:00-11:00	C6S2P21			
11:00-12:00	C25S5P5			
12:00-13:00	C7S2P17	C8S2P23		
13:00-14:00	C16S3P27	C8S4P23	C18S1P20	C1S2P2
	C27S1P15			
14:00-15:00	C9S5P25	C6S1P21	C11S1P14	C2S1P20
	C1S1P24	C28S1P22		
15:00-16:00	C9S4P18	C12S3P19	C25S3P16	C5S1P9
	C15S3P26	C27S2P24		
16:00-17:00	C11S2P8	C2S3P24	C26S2P11	C22S1P13
17:00-18:00	C22S3P13			
18:00-19:00	C18S3P20	C3S2P14		
19:00-20:00	C18S2P2	C3S3P10		

Sem#1
Sem#2
Sem#3
Sem#4
Sem#5
Sem#6
Sem#7
Sem#8

Figure 1. Timetable for courses in day scheme A: Monday, Wednesday and Friday, three classes of 1 h per week.

Time	Days			
	Tuesday and Thursday			
07:00-08:30	C9S6P18	C12S4P23	C25S2P16	C20S2P3
	C15S1P15	C3S5P10		
08:30-10:00	C16S6P19	C8S5P23		
10:00-11:30	C16S2P17	C12S1P19	C8S6P4	
11:30-13:00	C16S4P12	C9S3P25	C13S1P9	C8S1P4
13:00-14:30	C16S5P3	C9S2P18	C10S1P8	C14S2P6
	C17S1P17	C24S2P2		
14:30-16:00	C16S1P4	C25S6P7	C2S2P8	C21S2P6
	C23S4P1			
16:00-17:30	C20S1P10	C21S1P6	C3S4P14	
17:30-19:00	C2S4P22	C15S2P26	C22S2P13	
19:00-15:00	C17S2P13	C4S4P18	C23S2P1	

Sem#1
Sem#2
Sem#3
Sem#4
Sem#5
Sem#6
Sem#7
Sem#8

Figure 2. Timetable for courses in day scheme B: Tuesday and Thursday, two classes of 1 h and a half per week.

Figure 2 shows the assigned courses in day scheme B: Tuesday and Thursday classes (i.e., two classes of 1 h and 30 m per week). The same name structure for course assignation was applied.

The timetable obtained has 75 course sections scheduled (40 in Figure 1 and 35 in Figure 2) from the total of 82 needed course sections. Sixty four courses are assigned with basic faculty workload and 11 are overload assigned to full-time professors. The remaining seven non-scheduled course sections have an assigned hiring profile. The corresponding sections are: one section from course ID3, three sections from course ID4, two sections from course ID23, one section of the course ID24.

Figure 3 shows the proportions of the courses scheduled (with basic workload and with course overload) and courses with hiring profiles. It is important to note, that the contemplated assigned professors correspond to the list of active professors in the immediate preceding period, but when the courses' demand increases, it will be necessary to contemplate hiring profiles. In these cases, the solution implies seven non-scheduled courses with hiring profiles.

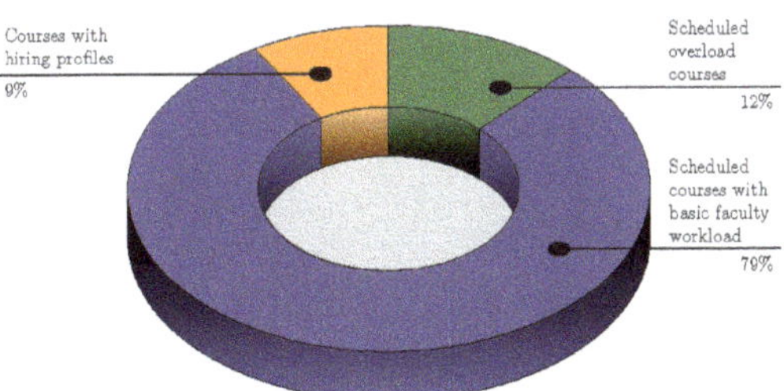

Figure 3. Proportions of courses.

As it was mentioned before, five academic profiles to new hires are defined, based on the composition of the program curricula and the creation of clusters of courses. This allows to identify the exact number of new hires needed. In this case, four professors are needed (one SA-participating, one PA-participating, two SP-supporting) each one belonging to a different academic profile. These new hires are the minimum hires to allocate the seven non-scheduled courses.

Beyond the fact that the optimal solution is obtained, it is possible to show the fulfillment of the constraints by observing the information in Figures 1 and 2. The timetabling states that 75 sections were scheduled and assigned a professor and seven sections were not scheduled but have an assigned hiring profile, giving a total of 82 sections (the grand total of sections). Here, constraint (4) is accomplished. Additionally, can be seen how each section for each course is assigned just once (constraint (5)). Constraints (7)–(10) are about the maximum numbers of courses, the course overload and the maximum overload allowed for each professor, for this, the figures (Figures 1 and 2) exhibit how this load is accomplished according to the information presented in Table 1.

Also in Figures 1 and 2, it is seen how the courses of same semester (same color) are assigned in different time (constraint (11)), the courses from first semester to forth semester are scheduled from 7 hrs. to 14 hrs. and the courses from fifth semester to eighth semester are scheduled to 13 hrs. to 20 hrs. or 7 hrs., however, there are some exceptions due to high demand (constraint (12)). As well as, it is possible to observe how the schedule of professors is at different times (constraint (13)) and the courses assigned to them are according to their credentials (constraint (14)). In addition, the timetable includes the consideration that some professors need to have a free afternoon (restriction (15)–(19)) in order to be able to teach graduate courses.

The total courses assigned to participating faculty already hired by the university (55 sections), plus the courses not scheduled but assigned to a profile (seven sections) must be greater than or equal to 60% of the total Sections (50 sections). Here constraint (21) is accomplished. For constraints related to accomplish of percentages per category (23)–(25), the results can be reviewed in Section 5.5.

To access the complete output, please refer to the data availability statement in the section.

5.2. Minimizing Courses without Schedule and Zero Teaching Overload

In the Figures 4 and 5 the timetables for the two schemes of days are presented for the application of second strategy. As is shown, when the problem presented before is solved minimizing the courses without schedule and zero teaching overload, the model can assign 69 sections with the number of available professors. The 13 remaining courses are: two sections of course ID3, four sections of course ID4, one section of course ID6, one section of course ID15, one section of the course ID20, two sections of course ID 21, and two section of the course ID24. In contrast with the first strategy, this one requires new hires to assign seven different courses, and, if the courses are not from common knowledge areas or require the same academic profile, then it will result in more hires. For this strategy, these courses are linked to academic profiles resulting in a number of six teachers needed. The distribution of this new hires, are: one PA-participating (full-time) profile and five SP-supporting (part-time) professors. The number of five SP-supporting professors needed is calculated based on the number of different courses without schedule and on the maximum load to be assigned to a full-time professor or part-time professor.

	Days			
Time	Monday, Wednesday, Friday			
07:00-08:00	C25S1P16	C11S1P15	C15S3P26	C3S4P10
08:00-09:00	C8S6P4			
09:00-10:00	C9S5P18	C12S1P19	C6S1P21	
10:00-11:00	C16S4P12	C25S6P5		
11:00-12:00	C19S1P27	C9S2P18	C12S2P19	C8S5P4
12:00-13:00	C7S2P17			
13:00-14:00	C16S1P27	C25S3P7	C22S1P13	
14:00-15:00	C9S4P18	C25S2P6	C18S2P2	C26S2P11
	C23S4P23			
15:00-16:00	C9S3P18	C25S4P16	C23S2P1	
16:00-17:00	C17S2P17	C22S2P13		
17:00-18:00	C2S1P20	C27S2P15		
18:00-19:00	C14S1P6	C5S1P9	C26S1P11	C23S1P23
19:00-20:00	C14S2P6	C17S1P17	C28S1P21	

Sem#1
Sem#2
Sem#3
Sem#4
Sem#5
Sem#6
Sem#7
Sem#8

Figure 4. Timetable of second strategy for courses in day scheme A: Monday, Wednesday and Friday, three classes of 1 h per week.

	Days			
Time	Tuesday and Thursday			
07:00-08:30	C16S5P28	C8S4P23	C18S3P2	C1S2P24
	C22S3P13			
08:30-10:00	C16S2P27	C12S3P19	C8S2P3	
10:00-11:30	C7S1P17	C13S1P9	C8S1P4	
11:30-13:00	C16S3P28	C12S4P19	C8S3P21	
13:00-14:30	C9S6P25	C25S5P7	C11S2P8	C20S2P3
	C1S1P24	C23S3P1		
14:30-16:00	C16S6P28	C9S1P25	C10S1P15	C3S5P14
16:00-17:30	C2S4P21	C27S1P24		
17:30-19:00	C18S1P2	C2S2P8	C3S6P10	
19:00-15:00	C2S3P20	C15S2P26	C3S2P14	

Sem#1
Sem#2
Sem#3
Sem#4
Sem#5
Sem#6
Sem#7
Sem#8

Figure 5. Timetable of second strategy for courses in day scheme B: Tuesday and Thursday, two classes of 1 h and a half per week.

5.3. Minimizing Courses without Schedule and Teaching Overload

In the Figures 6 and 7 the timetables for the two schemes of days are presented for the application of the third strategy. If the problem is solved considering the minimization of courses without schedule and teaching overload, the model assigns 71 sections with the number of available professors and suggests to hire five professors to teach eleven courses and only two professors will have one extra course as teaching overload. The courses to be assigned to new hires are: three sections of course ID3, three sections of course ID4, one section of course ID6, one section of course ID18, two sections of course ID 21, and one section of the course ID28. With the first strategy, seven sections of four different courses implied four new hires, while in the second strategy, thirteen sections implied five new hires, and with the application of the third strategy, eleven sections of seven different courses resulted in five new hires. This shows that to the extent that unscheduled courses are more and different, this results in a greater need for new hires.

	Days			
Time	Monday, Wednesday, Friday			
07:00-08:00	C11S2P15	C26S2P11	C3S6P10	
08:00-09:00	C9S5P18	C8S5P23		
09:00-10:00	C16S1P27	C8S1P4		
10:00-11:00	C25S5P14			
11:00-12:00	C16S2P27	C7S1P17	C12S3P19	C25S6P7
12:00-13:00	C9S4P25	C25S1P7		
13:00-14:00	C16S6P12	C7S2P17	C12S4P19	C25S3P16
	C2S3P8	C26S1P11	C27S2P24	
14:00-15:00	C13S1P9	C25S2P5	C2S4P2	C22S3P13
15:00-16:00	C16S4P28	C10S1P8	C3S4P6	
16:00-17:00	C17S1P17	C23S2P1		
17:00-18:00	C18S2P20	C20S1P3	C23S3P23	
18:00-19:00	C14S2P6	C5S1P9	C24S2P2	
19:00-20:00	C14S1P6	C17S2P17	C15S1P15	C22S1P13

Sem#1
Sem#2
Sem#3
Sem#4
Sem#5
Sem#6
Sem#7
Sem#8

Figure 6. Timetable of third strategy for courses in day scheme A: Monday, Wednesday and Friday, three classes of 1 h per week.

	Days			
Time	Tuesday and Thursday			
07:00-08:30	C9S6P25	C12S2P19	C8S2P4	C11S1P15
	C4S4P18	C23S4P1		
08:30-10:00	C19S1P27	C9S2P18	C12S1P19	C8S3P23
10:00-11:30	C16S5P28	C8S4P4		
11:30-13:00	C16S3P28	C9S3P18	C25S4P16	
13:00-14:30	C9S1P18	C6S2P21	C18S1P20	C2S1P2
	C1S1P24	C22S2P13		
14:30-16:00	C8S6P21	C20S2P3	C23S1P23	
16:00-17:30	C2S2P22	C15S3P26	C24S1P10	
17:30-19:00	C1S2P22	C27S1P24		
19:00-15:00	C15S2P26	C3S2P14		

Sem#1
Sem#2
Sem#3
Sem#4
Sem#5
Sem#6
Sem#7
Sem#8

Figure 7. Timetable of third strategy for courses in day scheme B: Tuesday and Thursday, two classes of 1 h and a half per week.

5.4. Minimizing Hiring Profiles (Professors)

In Figures 8 and 9 the timetables for the two schemes of days are presented for the application of the fourth strategy. When we solve the problem considering the minimization of hiring professors Equation (3), the model can assign 75 sections with the number of available professors and suggests to hire three professors to teach seven courses and eight professors will have teaching overload. The courses to be assigned to new hires are: three sections of the course ID3, three sections of the course ID9, and one section of the course ID24. Compared to the previous strategies, this one consists of assigning sections to new hires from only three different courses. This can result in similar academic profiles.

Time	Days			
	Monday, Wednesday, Friday			
07:00-08:00	C19S1P27	C6S1P21	C14S1P6	C2S3P20
	C4S4P18	C23S2P23		
08:00-09:00	C25S3P14			
09:00-10:00	C8S1P3			
10:00-11:00	C7S2P17	C12S2P19	C8S3P23	
11:00-12:00	C7S1P17	C12S1P19	C25S6P16	
12:00-13:00	C16S5P28	C8S5P4		
13:00-14:00	C8S4P4	C18S2P2	C26S1P11	C22S2P13
14:00-15:00	C13S1P9	C8S2P23	C2S2P24	C1S2P2
	C3S4P10			
15:00-16:00	C6S2P21	C20S2P9	C21S1P6	C27S1P15
16:00-17:00	C5S1P9	C26S2P11	C23S4P1	
17:00-18:00	C20S1P3	C4S2P18	C23S3P23	
18:00-19:00	C18S3P20	C15S2P15	C3S5P10	
19:00-20:00	C2S4P24	C23S1P1		

Sem#1 Sem#2 Sem#3 Sem#4 Sem#5 Sem#6 Sem#7 Sem#8

Figure 8. Timetable of fourth strategy for courses in day scheme A: Monday, Wednesday and Friday, three classes of 1 h per week.

Time	Days			
	Tuesday and Thursday			
07:00-08:30	C16S4P28	C9S6P25	C12S4P19	C8S6P4
	C17S1P13	C4S3P18	C27S2P24	
08:30-10:00	C16S2P17	C12S3P19	C25S4P5	
10:00-11:30	C16S1P28	C9S3P25	C10S1P15	
11:30-13:00	C16S6P7	C25S5P5		
13:00-14:30	C16S3P27	C25S2P16	C11S2P8	C17S2P17
	C21S2P6	C22S1P13		
14:30-16:00	C9S1P18	C25S1P7	C14S2P6	C1S1P22
	C22S3P13			
16:00-17:30	C18S1P20	C15S1P26	C3S3P10	
17:30-19:00	C4S1P18	C24S1P2		
19:00-15:00	C11S1P14	C2S1P8	C15S3P26	C28S1P22

Sem#1 Sem#2 Sem#3 Sem#4 Sem#5 Sem#6 Sem#7 Sem#8

Figure 9. Timetableof fourth strategy for courses in day scheme B: Tuesday and Thursday, two classes of 1 h and a half per week.

5.5. General Analysis

In the following, it is explained the behavior of each strategy in relation to the dedication of time of faculty to the mission (AACSB), also the impact of the objective function in the strategies regarding to new hires.

Table 2 presents the percentages of time dedication to the business school mission for each faculty qualification categories based on the AACSB standards. The obtained percentages are calculated for the four strategies established, strategy 1: to minimize the courses without schedule allowing course overload (base case); strategy 2: to minimize the courses without schedule and zero teaching overload; strategy 3: to minimize the courses without schedule and teaching overload; and, strategy 4: to minimize hiring profiles (professors). The percentage of dedication is obtained from the actual faculty and the new hires obtained from the application of the strategies.

Table 2. Percentage dedication of time of faculty to the Mission by strategy.

Strategy	SA	PA	SP	IP	A
Strategy 1	62%	6%	6%	21%	5%
Strategy 2	57%	6%	9%	23%	5%
Strategy 3	61%	2%	9%	23%	5%
Strategy 4	65%	1%	5%	24%	5%

In general, the percentage of time dedication to the mission for SA faculty states over 40% and the largest value (65%) was obtained with the strategy 4. Strategy 2 results in the lowest percentage (57%) for SA faculty. The actual faculty in this case has a greater number of faculty in SA and SP categories, this is why the lowest values in all faculty qualification categories are for the PA, SP and A categories in all the strategies.

Table 3 shows the distribution of new hires that are needed when one of the four proposed strategies is implemented. The first strategy minimizing the courses without schedule allowing course overload needs to hire one professor of SA profile, one professor of PA profile and two professors of SP profile, while the second strategy minimizing courses without schedule and zero teaching overload requires one professor PA and five professors of SP profile. The third strategy minimizing courses without schedule and teaching overload needs five professors of SP profile and strategy four, minimizing hiring profiles (professors) needs one professor of SA profile, one professor of SP profile and one professor of IP profile.

Table 3. Number of new hires needed by strategy.

Strategy	SA	PA	SP	IP	A
Strategy 1	1	1	2	0	0
Strategy 2	0	1	5	0	0
Strategy 3	0	0	5	0	0
Strategy 4	1	0	1	1	0

As it is seen, the first strategy suggests hiring four professors, less professors than strategy two or strategy three, due to the objective function that minimizes the courses without schedule and allows teaching overload, so with this objective, the model will try to assign more courses to current faculty and hire less professors. On the other hand, strategy two, requires six professors; in this second strategy, the objective function tries to minimize the courses without schedule, but here, the teaching overload is not allowed. This is why; this model suggests hiring more professors than any other strategy, because the teaching overload is not allowed, so the model assigns only the workload allowed to all faculty and suggests hiring more professors in order to cover the courses. In the third strategy, the model suggests hiring five professors; here, the model tries to minimize the aggregation of courses without schedule and teaching overload, it is important to notice than in this strategy all new hires are part-time (SP) given the conditions stated in Section 4. Strategy 4, suggests hiring less professors than any other strategy, due to, the model tries to minimize the hiring profiles, so, the timetable is created leaving the courses of the same profiles without schedule in order to hire less professors. The higher number of hire professors in all strategies is presented in the category SP.

6. Conclusions

This paper proposes an integer programming model to create the timetable of an academic department considering basic workload and course overload, as well as the profile and area of knowledge of each professor. The novelty in this paper is the incorporation of necessary requirements to fulfill the standards of AACSB, the most important accrediting association for business schools. It is well known that an accreditation as AACSB demands qualified faculty in each category and it is preponderant for business schools to have tools like the proposed here to support the decision making process.

The model was solved with data from a real case utilizing four different strategies that show the impact of allowing course overload and new hires. In all cases, the requirements of AACSB were met. This model will be useful to help the administration to select the best option that aligns with the objectives of the university. In this particular case, it was identified that the fourth strategy (minimize hires) is one of the best options given that it reduces the cost of new hires. This is why the course overload is a valuable resource that contributes to the last mentioned objective.

It is also common for universities that apply course overload to have a policy that establishes the maximum overload to be assigned to professors according to their hiring (the course overload implies an additional payment to a full-time professor). Therefore, it is important to explore the strategies and policies of an institution that impact the course timetabling and professor assignment.

In order to replicate this model, we suggest to classify the professors into full-time and part-time and according to the qualification categories (SA, PA, SP, IP, A), and participating or supporting (required by AACSB); to collect all the information related to the professors as the number of courses allowed to teach, as well as the courses they can teach according to their expertise. One important and necessary element is to define the time slots where the courses can take place.

The number of courses that can not be scheduled will require new hires, but identifying how many new hires are needed is important to know the knowledge area of the courses to create some categories of hiring profiles. For example, the courses about decision making methods, the courses about strategy and others. We divided the total set of disciplinary courses into disjoint sets to define these new hire profiles. In this sense, the current faculty can provide a first way to define the hire profiles with the courses that they teach as a reference.

Finally, it is important to know the strategy that the university wants to follow in order to use an appropriate objective function or criteria.

As future work, it is contemplated to add the preferences of time slots for faculty and additional necessities (course language, balancing the number of courses scheduled in the same time slot, room assignment, etc.). Further, the budget for new hires could be considered in order to not exceed the academic department budget.

Author Contributions: Conceptualization, N.M.A.-M., P.A.A.-T. and J.C.T.-R.; methodology, N.M.A.-M. and P.A.A.-T.; software, N.M.A.-M. and P.A.A.-T.; validation, N.M.A.-M. and P.A.A.-T.; formal analysis, N.M.A.-M., P.A.A.-T. and J.C.T.-R.; investigation, N.M.A.-M., P.A.A.-T. and J.C.T.-R.; resources, Not applicable; data curation, N.M.A.-M., P.A.A.-T. and J.C.T.-R.; writing—original draft preparation, N.M.A.-M. and P.A.A.-T.; writing—review and editing, N.M.A.-M., P.A.A.-T. and J.C.T.-R.; visualization, N.M.A.-M., P.A.A.-T. and J.C.T.-R.; supervision, N.M.A.-M., P.A.A.-T. and J.C.T.-R.; project administration, N.M.A.-M., P.A.A.-T. and J.C.T.-R. All authors have read and agreed to the published version of the manuscript.

Funding: The APC was funded by the institution of affiliation of the authors.

Data Availability Statement: The data corresponding fall 2020 academic period utilized in this research is reposited in Github (https://github.com/NanAMTZ/UCTP-fall2020, accessed on 20 August 2021).

Conflicts of Interest: The authors declare no conflict of interest.

Abbreviations

The following abbreviations are used in this manuscript:

AACSB	Association to Advance Collegiate Schools of Business
UCTP	University Course Timetabling Problem
AQ	Academic Qualified
P	Participating Faculty
S	Supporting Faculty
SA	Scholarly Academic Faculty
PA	Practice Academic Faculty
SP	Scholarly Practitioner Faculty
IP	Instructional Practitioner Faculty
A	Additional Faculty

References

1. Song, T.; Liu, S.; Tang, X.; Peng, X.; Chen, M. An iterated local search algorithm for the University Course Timetabling Problem. *Appl. Soft Comput.* **2018**, *68*, 597–608. [CrossRef]
2. MirHassani, S. A computational approach to enhancing course timetabling with integer programming. *Appl. Math. Comput.* **2006**, *175*, 814–822. [CrossRef]
3. Munirah Mazlan, M.M.; Khairi, A.F.K.A.; Mohamed, M.A.; Rahman, M.N.A. A study on optimization methods for solving course timetabling problem in university. *Int. J. Eng. Technol.* **2018**, *7*, 196–200. [CrossRef]
4. Gabriel, D.F.; Pangilinan, J.M.A. Faculty course scheduling optimization. *Am. Sci. Res. J. Eng. Technol. Sci.* **2018**, *44*, 170–179.
5. Perzina, R.; Ramik, J. Self-learning genetic algorithm for a timetabling problem with fuzzy constraints. *Int. J. Innov. Comput. Inf. Control* **2013**, *9*, 4565–4582.
6. Soria-Alcaraz, J.A.; Özcan, E.; Swan, J.; Kendall, G.; Carpio, M. Iterated local search using an add and delete hyper-heuristic for university course timetabling. *Appl. Soft Comput.* **2016**, *40*, 581–593. [CrossRef]
7. Junrie B. Matias, A.C.F.; Medina, R.M. A Fair Course Timetabling Using Genetic Algorithm with Guided Search Technique. *5th Int. Conf. Bus. Ind. Res.* **2018**, *1*, 77–82. [CrossRef]
8. Immonen, E.; Putkonen, A. A heuristic genetic algorithm for strategic university tuition planning and workload balancing. *Int. J. Manag. Sci. Eng. Manag.* **2018**, *12*, 118–128. [CrossRef]
9. Tavakoli, M.M.; Shirouyehzad, H.; Lotfi, F.H.; Najafi, S.E. Proposing a novel heuristic algorithm for university course timetabling problem with the quality of courses rendered approach; a case study. *Alex. Eng. J.* **2020**, *59*, 3355–3367. [CrossRef]
10. Ojha, D.; Sahoo, R.K.; Das, S. Automated timetable generation using bee colony optimization. *Int. J. Appl. Inf. Syst.* **2016**, *10*, 38–43. [CrossRef]
11. Domenech, B.; Lusa, A. A MILP model for the teacher assigment problem considering teacher preferences. *Eur. J. Oper. Res.* **2016**, *249*, 1153–1160. [CrossRef]
12. Al-Yakoob, S.M.; Sherali, H.D. Mathematical programming models and algorithms for a class faculty assignment problem. *Eur. J. Oper. Res.* **2006**, *173*, 488–507. [CrossRef]
13. Chen, R.M.; Shih, H.F. Solving university course timetabling problems using constriction particle swarm optimization with local serach. *Algorithms* **2013**, *6*, 227–244. [CrossRef]
14. Muhlenthaler, M.; Wanka, R. Fairness in academic course timetabling. *Ann. Oper. Res.* **2016**, *239*, 171–188. [CrossRef]
15. Jess Boronico, J.M.; Kong, X. Faculty Sufficiency and AACSB accreditation compliance within a global university: A mathematical modeling approach. *Am. J. Bus. Educ.* **2014**, *7*, 213–218. [CrossRef]
16. Henninger, E.A. Perceptions of the impact of the new AACSB standards on faculty qualifications. *J. Organ. Chang. Manag.* **1998**, *11*, 407–424. [CrossRef]
17. Prasad, A.; Segarra, P.; Villanueva, C.E. Acadeic life under institutional pressures for AACSB accreditation: Insights from faculty members in Mexican business schools. *Stud. High. Educ.* **2019**, *44*, 1605–1618. [CrossRef]
18. Bajada, C.; Trayler, R. Interdisciplinary business education: Curriculum through collaboration. *Educ. Train.* **2013**, *55*, 385–402. [CrossRef]
19. Koys, D.J. Judging academic qualifications, professional qualifications, and participation of faculty using AACSB guidelines. *J. Educ. Bus.* **2008**, *83*, 207–213. [CrossRef]
20. Fitzpatric, L.E.; McConnell, C. Aligning economics programs with AACSB accreditation process. *J. Econ. Educ. Res.* **2014**, *15*, 67–80.
21. Deborah, M.; Gray, V.B.; Carson, M.; Chakraborty, D. Anatomy of an MBA program capstine project assessment measure for AACSB accreditation. *Int. J. Bus. Adm.* **2015**, *6*, 1–7.
22. AACSB. *2020 Guiding Principles and Standards for Business Accreditation*; AACSB: Tampa, FL, USA, 2020; pp. 1–55.
23. AACSB. *2013 Eligibility Procedures and Accreditation Standards for Business Accreditation*; AACSB: Tampa, FL, USA, 2018; pp. 1–55.

Article

On Little's Formula in Multiphase Queues

Saulius Minkevičius [1], Igor Katin [1], Joana Katina [2,*] and Irina Vinogradova-Zinkevič [3]

[1] Institute of Data Science and Digital Technologies, Vilnius University, Akademijos st. 4, LT-08412 Vilnius, Lithuania; saulius.minkevicius@mif.vu.lt (S.M.); igor.katin@mif.vu.lt (I.K.)
[2] Institute of Computer Science, Vilnius University, Didlaukio st. 47, LT-08303 Vilnius, Lithuania
[3] Department of Information Technologies, Vilnius Gediminas Technical University, Saulėtekio al. 11, LT-10223 Vilnius, Lithuania; irina.vinogradova-zinkevic@vilniustech.lt
* Correspondence: joana.katina@mif.vu.lt

Abstract: The structure of this work in the field of queuing theory consists of two stages. The first stage presents Little's Law in Multiphase Systems (MSs). To obtain this result, the Strong Law of Large Numbers (SLLN)-type theorems for the most important MS probability characteristics (i.e., queue length of jobs and virtual waiting time of a job) are proven. The next stage of the work is to verify the result obtained in the first stage.

Keywords: multiphase systems; heavy traffic; Little's formula

Citation: Minkevičius, S.; Katin, I.; Katina, J.; Vinogradova-Zinkevič, I. On Little's Formula in Multiphase Queues. *Mathematics* **2021**, *9*, 2282. https://doi.org/10.3390/math9182282

Academic Editors: Frank Werner, Lev Klebanov and Christophe Chesneau

Received: 28 June 2021
Accepted: 13 September 2021
Published: 16 September 2021

1. Introduction

Interest in the field of multiphase queueing systems has been stimulated by the theoretical values of the results, as well as by their possible applications in information and computing systems, communication networks, and automated technological processes. The investigation methods of single phase queueing systems are provided in [1–3]. The asymptotic analysis of queueing systems in heavy traffic models are of special interest (see, for example, in [4–7]). The papers [8,9] describe the research start of diffusion approximation relative to queueing networks. Intermediate models—multiphase queueing systems—are considered rarer due to serious technical difficulties (see, for example, book [10]).

In this paper, we present a survey of articles issued between 2010 and 2021 that investigate heavy traffic networks. In [11], a multiclass queueing system was investigated—we consider a heterogeneous queueing system to consist of one large pool of identical servers. The arriving customers belong to one of several classes, which determines the service times in the distributional sense. In [12], a class of multiclass networks was analyzed—a class of stochastic processes known as semi-martingale reflecting Brownian motions is often used to approximate the dynamics of heavily loaded queuing networks. In [13], a model of approximation of resource sharing games was developed. In [14], the problem of scheduling in queueing networks was analyzed. In [15], a model of parallel multiclass queues was investigated. The model of input queued switch operation was analyzed in [16]. In [17], the stationary distribution was investigated. The authors justified the steady-state diffusion approximation of a generalized Jackson network in heavy traffic. Their approach involves the so-called limit interchange argument, which has since become a popular tool and has been employed by many others who study diffusion approximations. A survey of stochastic network analysis was presented in [18]. In [19], MapTask scheduling in heavy traffic optimality is analyzed. In [20], the authors investigate the departure process in open queuing networks. The delay process is analyzed in [21]. Motivated by the stringent requirements on delay performance in data center networks, the authors study a connection-level model for bandwidth sharing among data transfer flows, where file sizes have phase-type distributions and proportionally fair bandwidth allocation is used. In [22], universal bounds are investigated. In [23], the load balancing policy problem in heavy traffic was developed. In [24], the MaxWeight 23 scheduling algorithm is considered. Our paper on SLLN in MS is one of the first works in this area.

The study of generalized networks can be traced back to their namesake [25,26], who considered networks with inputs and exponential service times and showed that the invariant probability for the process has a simple product form. The foregoing assumptions on the arrival streams and service times were made to greatly simplify the analysis of these networks. Relaxing these assumptions was the subject of the work by Borovkov [27], where a model similar to Markovian network is considered. The finite buffer case is treated in Konstantopoulos and Walrand [28], and general point process arrival streams and general service processes are considered for networks without feedback [29].

We will next present some definitions in the theory of metric spaces (see, for example, [30]). Let C be a metric space consisting of real continuous functions in [0, 1] with a uniform metric of the following.

$$\hat{\rho}(m,n) = \sup_{0 \leq s \leq 1} |m(s) - n(s)|, \; m,n \in C.$$

Let D be a space of all real-valued right-continuous functions in [0,1] having left limits and endowed with the Skorokhod topology, induced by the metric $\hat{d}$ (under which D is complete and separable). Moreover, note that $\hat{d}(m,n) \leq \hat{\rho}(m,n)$ for $m,n \in D$. In this paper, we constantly use an analog of the theorem on converging together (see, for example, [30]).

$$\begin{aligned} &\text{Suppose } \hat{\varepsilon} > 0 \text{ and } \mathbf{M}_k,\ \mathbf{N}_k,\ \mathbf{M} \in D. \text{ If } \mathbf{Pr}\left(\lim_{k\to\infty} d(\mathbf{M}_k, \mathbf{M}) > \hat{\varepsilon}\right) = 0 \\ &\text{and } \mathbf{Pr}\left(\lim_{k\to\infty} \hat{d}(\mathbf{M}_k, \mathbf{N}_k) > \hat{\varepsilon}\right) = 0, \text{ then } \mathbf{Pr}\left(\lim_{k\to\infty} \hat{d}(\mathbf{N}_k, \mathbf{M}) > \hat{\varepsilon}\right) = 0. \end{aligned} \tag{1}$$

There is one service device in each phase of the MS; the service discipline is FCFS (i.e., first come, first served). Service time distribution and the incoming flow of jobs to the first phase of the MS are both common. We investigate here an x-phase MS (i.e., when a job is served in the ith phase of MS, it proceeds to the $i+1$ phase of MS, and it leaves MS after the job has been served in the x-phase of MS). Let us denote the time of arrival of the kth job by t_k. The service time of the kth job in the ith phase of MS is denoted by $S_k^{(i)}$; $Z_k = t_{k+1} - t_k$. Let us introduce mutually independent renewal processes $m_i(t) = \{\max_x \sum_{j=1}^{x} S_j^{(i)} \leq t\}$, $e(t) = \{\max_x \sum_{j=1}^{x} Z_j \leq t\}$ (number of jobs that arrive at MS until the time moment t).

Next, we denote the number of jobs by $\sigma_i(t)$ after service departure from the ith phase of MS until the time t; the queue length of jobs by $Q_i(t)$ in the ith phase of MS at the time moment t; $u_i(t) = \sum_{j=1}^{i} Q_j(t)$, $i = 1, 2, \ldots, x$, and $t > 0$.

Let inter arrival (Z_k) at MS and service times ($S_k^{(i)}$) in each phase of MS for $i = 1, 2, \ldots, x$ be mutually independent and identically distributed random variables.

Define $\alpha_i = (ES_k^{(i)})^{-1}$, $\alpha_0 = (EZ_k)^{-1}$, $\beta_i = \alpha_0 - \alpha_i$, $\beta_0 = 0$, $\hat{m}_i(t) = e(t) - m_i(t)$, $i = 1, 2, \ldots, x$, $t > 0$.

Suppose the following condition to be satisfied $\alpha_0 > \alpha_1 > \cdots > \alpha_x > 0$. Then, the following is the case.

$$\beta_x > \beta_{x-1} > \cdots \beta_1 > 0. \tag{2}$$

2. SLLN for the Queue Length of Jobs in MS

One of the main results of this paper is a theorem on SLLN for the summary length of jobs in MS.

Theorem 1 (SLLN for the summary length of jobs in MS). *If conditions (2) are fulfilled, then the following is the case.*

$$\left(\frac{V_1(s)}{s};\frac{V_2(s)}{s};\ldots;\frac{V_x(s)}{s}\right)\Rightarrow(\beta_1;\beta_2;\ldots;\beta_x).$$

Proof. The relations of the following:

$$Q_i(s)=\sigma_{i-1}(s)-\sigma_i(s), \tag{3}$$

$$Q_i(s)=f_s(\sigma_{i-1}(\cdot)-m_i(\cdot)), \tag{4}$$

$$Q_i(s)=f_s(\hat{m}_i(\cdot)-\sum_{j=1}^{i-1}Q_j(\cdot)) \tag{5}$$

are obtained for $i=1,2,\ldots,x$, $s>0$, and $f_s(m(\cdot))=m(s)-\inf_{0\le p\le s}m(p)$ (see [31]).

In view of (3)–(5), we find that the following is the case:

$$v_i(s)=\hat{m}_i(s)-\inf_{0\le p\le s}(\hat{m}(p)-v_{i-1}(p)), \tag{6}$$

for $i=1,2,\ldots,x$, $s>0$.

Next, using (6) for $n_i(t)=v_i(t)-\hat{m}_i(t)$, we obtain the following:

$$\begin{aligned}
n_i(t)&\le\sup_{0\le p\le t}n(p)\le\sup_{0\le m\le t}(-\inf_{0\le n\le m}(\hat{m}_i(n)-v_{i-1}(n)))\\
&=\sup_{0\le p\le t}(\sup_{0\le p\le m}(v_{i-1}(p)-\hat{m}_i(p)))\le\sup_{0\le p\le t}(v_{i-1}(p)-\hat{m}_i(p))\\
&=\sup_{0\le p\le t}(v_{i-1}(p)-\hat{m}_{i-1}(p)+\hat{m}_{i-1}(p)-\hat{m}_i(p))\\
&=\sup_{0\le p\le t}(n_{i-1}(p)+\hat{m}_{i-1}(p)-\hat{m}_i(p))\le\sup_{0\le p\le t}n_{i-1}(p)+\sup_{0\le p\le t}(\hat{m}_{i-1}(p)-\hat{m}_i(p))\\
&\le\cdots\le\sum_{j=1}^{i}\sup_{0\le p\le t}(\hat{m}_{j-1}(p)-\hat{m}_j(p))\le\sum_{j=1}^{x}\sup_{0\le p\le t}(\hat{m}_{j-1}(p)-\hat{m}_j(p)),
\end{aligned} \tag{7}$$

where $i=1,2,\ldots,x$, $t>0$.

Hence, we obtain the following:

$$v_i(s)<\hat{m}_i(s)+\sum_{j=1}^{x}\sup_{0\le p\le s}(\hat{m}_{j-1}(p)-\hat{m}_j(p)), \tag{8}$$

for $i=1,\ldots,x$, $s>0$.

Thus, for any i $(i=1,2,\ldots,x)$, we obtain the following.

$$v_i(p)=\sum_{l=1}^{i}Q_l(p)=\sum_{l=1}^{i}[\sigma_{l-1}(p)-\sigma_l(p)]\ge e(p)-\sigma_l(p)\ge e(p)-m_i(p)=\hat{m}_i(p), \tag{9}$$

From (8) and (9), we obtain the following:

$$|v_i(s)-\hat{m}_i(s)|\le\sum_{j=1}^{x}\sup_{0\le p\le s}(\hat{m}_{j-1}(p)-\hat{m}_j(p)), \tag{10}$$

for $i=1,\ldots,x$, $s>0$.

For $\hat{\varepsilon}>0$, we derive the following:

$$\begin{aligned}
&Pr\left(\left|\frac{v_i(s)}{s}-\hat{\beta}_i\right|>\hat{\varepsilon}\right)\le Pr\left(\left|\frac{\hat{m}_i(t)}{s}-\hat{\beta}_i\right|>\frac{\hat{\varepsilon}}{2}\right)+Pr\left(\left|\frac{v_i(s)-\hat{m}_i(s)}{s}\right|>\frac{\hat{\varepsilon}}{2}\right)\\
&\le Pr\left(\left|\frac{\hat{m}_i(s)}{s}-\hat{\beta}_i\right|>\frac{\hat{\varepsilon}}{2}\right)+\sum_{j=1}^{x}Pr\left(\frac{\sup_{0\le m\le t}(\hat{m}_{j-1}(m)-\hat{m}_j(m))}{s}>\frac{\hat{\varepsilon}}{2\cdot x}\right),
\end{aligned} \tag{11}$$

where $i = 1, \ldots, x, s > 0$.

Note that $\sup_{0 \le m \le s} (\hat{m}_{i-1}(m) - \hat{m}_i(m))/s \ge 0$ for $i = 1, \ldots, x$. In addition, note that the following is the case:

$$\lim_{s \to \infty} \frac{\hat{m}_{i-1}(s) - \hat{m}_i(s)}{s} = \hat{\beta}_{i-1} - \hat{\beta}_i < 0 \tag{12}$$

almost everywhere for $i = 1, \ldots, x$ (see [31]). Thus, similarly as in [31], we prove that the second item in (11) also tends to zero.

Thus, we obtain that for $\hat{\varepsilon} > 0$, the following is the case.

$$\lim_{s \to \infty} Pr\left(\left| \frac{v_i(s)}{s} - \hat{\beta}_i \right| > \hat{\varepsilon} \right) = 0, \quad i = 1, \ldots, x. \tag{13}$$

Using the convergence together theorem (see, for example, [30] and (13)), we complete the proof of the theorem. □

The theorem on SLLN for the queue length of jobs is proved similarly as Theorem 1.

Theorem 2 (SLLN for the queue length of jobs in MS). *If conditions (2) are fulfilled, then the following is the case.*

$$\left(\frac{Q_1(s)}{s}; \frac{Q_2(s)}{s}; \ldots; \frac{Q_x(s)}{s} \right) \Rightarrow (\hat{\beta}_1; \hat{\beta}_2 - \hat{\beta}_1; \ldots; \hat{\beta}_x - \hat{\beta}_{x-1}).$$

Proof. Using (13), we derive the following:

$$\left| \frac{Q_i(s)}{s} - (\hat{\beta}_i - \hat{\beta}_{i-1}) \right| \le \left| \frac{v_i(s)}{s} - \hat{\beta}_i \right| + \left| \frac{v_{i-1}(s)}{s} - \hat{\beta}_{i-1} \right|, \tag{14}$$

where $i = 1, \ldots, x, s > 0$.

Using the convergence together theorem (see, for example, [30] and (14)), we complete the proof of the theorem. □

3. SLLN for the Virtual Waiting Time of a Job in MS

In this section, we present the proof of Little's formula in MS. The main tools in proving this fact are SLLN for the queue length of jobs and the virtual length of a job in MS.

Definitions of the random variables t_k, Z_k, $S_k^{(i)}$, $e(t)$, and $m_i(t)$ for $i = 1, 2, \ldots, x$ are the same as in the proof of Theorems 1 and 2. Let us define $\bar{\beta}_i = ES_k^{(i)}$, $\bar{\beta}_0 = EZ_k$, and $\bar{\alpha}_i = \frac{\bar{\beta}_i}{\bar{\beta}_{i-1}} - 1$ for $i = 1, 2, \ldots, x$. Assume that condition (2) is fulfilled. Therefore, $\bar{\alpha}_i > 0$ for $i = 1, \ldots, x$.

In addition, let us define $V_i(t)$ as a virtual waiting time of a job in the ith phase of MS at time t. Denote $S_i(s)$ as the time, that is, the summary service of jobs arriving at the ith phase of MS until time t for $i = 1, \ldots, x$ and $s > 0$.

Note that $S_i(s) = \sum_{j=1}^{v_{i-1}(t)} S_j^{(l)}$ for $i = 1, \ldots, x$ and $s > 0$.

Moreover, let $n_i(s) = S_i(s) - s$, $f_s(n(\cdot)) = n(s) - \inf_{0 \le m \le s} n(m)$, $\hat{n}_i(s) = \sum_{j=1}^{m_{i-1}(s)} S_j^{(i)} - s$, $m_0(s) = e(s)$ for $i = 1, \ldots, x, s > 0$.

If $S_i(0) = V_i(0) = 0$, then the following is the case (see [1], p. 41).

$$V_i(s) = f_s(n_i(\cdot)) \text{ for } i = 1, \ldots, x \text{ and } s > 0$$

Thus, we prove a theorem about SLLN for the virtual waiting time of a job in MS.

Theorem 3 (SLLN for the virtual waiting time of a job in MS). *If conditions (2) are fulfilled, then the following is the case.*

$$\left(\frac{V_1(s)}{s};\frac{V_2(s)}{s};\ldots;\frac{V_x(s)}{s}\right)\Rightarrow(\bar{\alpha}_1;\bar{\alpha}_2;\ldots;\bar{\alpha}_x).$$

Proof. Using the estimation, we obtain for each fixed $\hat{\eta}>0$ that the following is the case:

$$\begin{aligned}
&Pr\left(\left|\frac{V_i(s)}{s}-\bar{\alpha}_i\right|>\hat{\varepsilon}\right)=Pr\left(\left|\frac{f_s(n_i(\cdot))}{s}-\bar{\alpha}_i\right|>\hat{\varepsilon}\right)\\
&\le P\left(\left|\frac{f_s(n_i(\cdot))}{s}-\frac{f_s(\hat{n}_i(\cdot))}{s}\right|>\frac{\hat{\varepsilon}}{2}\right)+Pr\left(\left|\frac{f_s(\hat{n}_i(\cdot))}{s}-\mu_i\right|>\frac{\hat{\varepsilon}}{2}\right)\\
&\le Pr\left(\left|\frac{n_i(s)-\hat{n}_i(s)}{s}\right|>\frac{\hat{\varepsilon}}{4}\right)+Pr\left(\left|\frac{f_s(\hat{n}_i(\cdot))-\hat{n}_i(s)}{s}\right|>\frac{\hat{\varepsilon}}{4}\right)\\
&+Pr\left(\left|\frac{\hat{n}_i(s)}{s}-\bar{\alpha}_i\right|>\frac{\hat{\varepsilon}}{4}\right)\le Pr\left(\left|\frac{n_i(s)-\hat{n}_i(s)}{\sqrt{s}}\right|>\frac{\hat{\varepsilon}}{4}\right)\\
&+Pr\left(\frac{|\sup\limits_{0\le m\le s}(-\hat{n}_i(m))|}{s}>\frac{\hat{\varepsilon}}{4}\right)+Pr\left(\left|\frac{\hat{n}_i(s)}{s}-\bar{\alpha}_i\right|>\frac{\hat{\varepsilon}}{4}\right)\\
&\le Pr\left(\left|\frac{n_i(s)-\hat{n}_i(s)}{\sqrt{s}}\right|>\frac{\hat{\varepsilon}}{4}\right)+Pr\left(\frac{\left|\sup\limits_{0\le m\le s}(-\hat{n}_i(m))\right|}{s}>\frac{\hat{\varepsilon}}{4}\right)\\
&+Pr\left(\left|\frac{\hat{n}_i(s)}{s}-\bar{\alpha}_i\right|>\frac{\hat{\varepsilon}}{4}\right),
\end{aligned}\tag{15}$$

for $i=1,2,\ldots,x$ and $s>0$.

Thus, we achieve that for each $\hat{\varepsilon}>0$:

$$\begin{aligned}
&Pr\left(\left|\frac{V_i(s)}{s}-\bar{\alpha}_i\right|>\hat{\varepsilon}\right)\le Pr\left(\left|\frac{n_i(s)-\hat{n}_i(s)}{\sqrt{s}}\right|>\frac{\hat{\varepsilon}}{4}\right)\\
&+Pr\left(\frac{\left|\sup\limits_{0\le m\le s}(-\hat{n}_i(m))\right|}{s}>\frac{\hat{\varepsilon}}{4}\right)+Pr\left(\left|\frac{\hat{n}_i(s)}{s}-\bar{\alpha}_i\right|>\frac{\hat{\varepsilon}}{4}\right),
\end{aligned}\tag{16}$$

$i=1,\ldots,x$ and $s>0$.

Since it is proven ((12)), the following is the case.

$$Pr\left(\lim_{s\to\infty}\left|\frac{n_i(s)-\hat{n}_i(s)}{\sqrt{s}}\right|>\hat{\varepsilon}\right)=0,\ i=1,\ldots,x.\tag{17}$$

Thus, the first item in inequality (16) tends to zero (see (17)). In addition, we prove that the second item in inequality (16) also tends to zero (see, for example, [4]) (if conditions (2) are fulfilled). Therefore, we apply the limit theorem for a complex renewal process (see, for example, [5]). Thus, the third item in inequality (16) also tends to zero.

We have proven that all of the items on the inequality (16) converge to zero. Thus, we achieve that for each fixed $\hat{\varepsilon}>0$, the following is the case.

$$Pr\left(\lim_{t\to\infty}\left|\frac{V_i(s)}{s}-\bar{\alpha}_i\right|>\hat{\varepsilon}\right)=0,\ i=1,\ldots,x.\tag{18}$$

Using the convergence together theorem (see, for example, [30] and (18)), we complete the proof of the theorem. □

Finally, we derive the corollary of proved theorems (Little's formula). The formula L = λW (Little's law) expresses the fundamental principle of queueing theory: Under very general conditions, the time-average or expected time-stationary number of customers in a system, L (e.g., the average queue length), is equal to the product of the arrival rate A and the customer-average or expected customer-stationary time each customer spends in the system, W (e.g., the average waiting time). The relation L = λW is very useful because the assumptions are minimal; it applies to other stochastic models in addition to queues; it applies to queueing networks and subnetworks as well as individual queues; it applies to subclasses as well as the entire customer population; and it is remarkably independent of modelling details, such as service disciplines and underlying probability distributions. Moreover, there are extensions of L = λW - the continuous, distributional, ordinal and Central Limit Theorem versions, that enable us to analyze many seemingly unrelated problems.

Corollary 1 (Little's formula in MS). *If conditions (2) are fulfilled, then the following is the case.*

$$\lim_{s\to\infty}\frac{Q_i(s)}{V_i(s)} \Rightarrow \frac{(\hat{\beta}_i-\hat{\beta}_{i-1})}{\tilde{\alpha}_i},\ i=1,\ldots,x.$$

Proof. At first, we used Theorems 2 and 3 on SLLN for the queue length of jobs and virtual waiting time of a job in MS.

Thus, the following is the case.

$$\lim_{s\to\infty}\frac{Q_i(s)}{V_i(s)} = \lim_{s\to\infty}\frac{\frac{Q_i(s)}{s}}{\frac{V_i(s)}{s}} \Rightarrow \frac{(\hat{\beta}_i-\hat{\beta}_{i-1})}{\tilde{\alpha}_i},\ i=1,\ldots,x. \quad (19)$$

The proof is complete. □

4. Simulation

4.1. Overview of Similar Simulations

We have investigated many articles in order to find a similar simulation. Although we found many articles on the same topic (MS), only a few of them described the precise simulation. While investigating a similar simulation in many articles, model descriptions have been found but most of them only described the theoretical model using formulas and algorithm block schemes.

Nevertheless, a few software models or simulations have been found. The simulations that we found can be divided into two groups: (1) simulations made with particular software packages; and (2) simulations created as programs using programming languages and/or other programming tools. We can observe that these two directions of research are developing successfully (here, we can mention the recent work on modeling retrial queue systems [32–34], etc.).

In [35], the intelligent management system and the expert system are described. The authors describe the architecture of these systems, but no code or other details were provided. In [36], the authors apply SimEvents MATLAB-Simulink and describe the block scheme of their model. However, no programming code or details were provided.

In [37], the authors reviewed the popular simulation software, such as GPSS World, AnyLogic, and Arena environments. In the authors' opinion, these software packages could make any simulation process very long and expensive because they are not optimal for such a simulation and are mostly used for business process simulations. In addition, most of them are commercial. Therefore, the authors decided to use their own neural network model, but no details were provided.

In [38], a real simulation model created using the Python programming language was described. The authors also provided the programming code. This simulation is described in detail, and all of the Python libraries that were used are provided.

After this review, some conclusions can be drawn:

1. Commercial models do not meet the requirement to make MS simulations. In addition, they are usually expensive.
2. None of the provided models meets the requirement to make all of the necessary simulations and experiments.

Consequently, we decided to create our own model that fits all of the requirements, can run on one computer and any operating system, and works in multi-threading mode.

4.2. Simulator

To implement the experiment, a Multiphase Queueing System (MS) simulator was created. The Python programming language (version 3.6.9) and its multiprocessing programming library were used. The simulator runs on any operating system that supports the Python programming language. The main new features of this simulator are the following:

- Real asynchronous processes that are not dependent on each other;
- Possibility to stop simulation at a particular time and not only when all clients are served (as [38]);
- Possibility to measure any moment of time:
  - Client enters MS t_k (and also $Z_k = t_{k+1} - t_k$ – time between two clients' arrival);
  - Waiting time V_i of each client in every queue;
  - Service time $S_{(i)k}$ of client k in the ith service of the queue;
  - Amount of clients $Q_i(t)$ in each queue after time t (when the process stopped after time t);
- Client proceeds to the consumer process not only after it pass all services (as [38]) but also immediately if MS stopped after the specified time t;
- All of the values to be measured are stored in synchronized variables for eliminating their undefined states (some issues could appear because of the operating system, the Python programming language, or hardware errors);
- Each service really stops after all of the clients pass away or after the specified time t;
- Clients enters the consumer process from any place of MS if it stops after the specified time t. For example, a client cannot pass all of the services or could even be in a waiting queue for one of the services;
- All of the calculations are performed, and only then is MS stopped or when all the clients pass through or after the specified time ends.

As shown in Figure 1, the simulator has input (producer process) and output (consumers process) storage and I (configurable) phases in the queue between them. K (configurable) clients created by producer process with random (configurable) time interval between them proceed to the first phase and then, after they have been served there, they proceed to the next phase. Each phase has its own serving time and waiting time before serve. The process continues until the last queue is stopped, after which the client comes to the consumer process.

The main difference in this model is the possibility to stop the simulation after a particular time t (configurable) interval. Imagine that somebody wants to stop the simulation after time t. Then, after the specified time, all of the phases are stopped in the state that they are in and the consumer process collects all of the clients from all of the phases. Here, all the calculations could be performed, including client number in the phase at the moment t.

In this simulation, when it finishes after the specified time t, all clients have their own states and other information, such as the following:

- All times between jobs' arrivals to MS;
- All service times of all jobs in all queues (where they had time to gain);
- All jobs waiting times in all queues and the number of jobs in all queues at time t (when the simulation stopped).

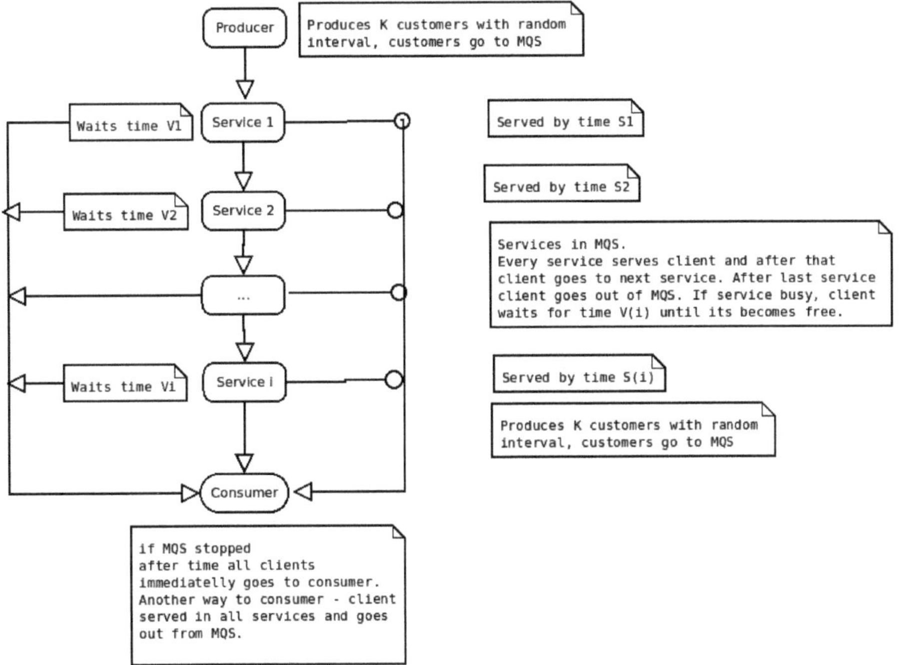

Figure 1. Algorithm of the simulation model.

All of the measurable parameters of the model are listed in Table 1.

Table 1. Model parameters.

Parameter	Description
t_k	Client's k arrival to MS time
V_i	Client's waiting time before service in the ith phase
S_i	Client's service time in the ith phase
Q_i	Clients amount in the ith phase at the moment t

After the model stops, the consumer process performs all of the computations for the values listed in Table 2.

Table 2. Computational parameters of the model.

Parameter	Description
$E(V_i)$	Estimated waiting time before service in the ith phase
$E(S_i)$	Estimated service time in the ith phase
Q_i/t	Clients amount in the queue i at time t, divided by t
Z_{k+1}	Estimated time between two clients entering MS
$E(Z)$	Estimated time between clients on entering MS

4.3. Experiment

The experiment results were obtained using this simulation with the following parameters:

- Time interval of 15 s (measurements were made each second from 1 to 15);
- Five phases in MS;

- 100,000 clients received from the producer.

During the experiment, the MS system was stopped at each second, and all of the calculations were made. In each phase, the following is the case:

- Q_1–Q_5—numbers of clients in all five phases divided by t;
- $E(V_1)$–$E(V_5)$—estimated waiting times in each phase are calculated and divided by t;
- Q_1/V_1–Q_5/V_5 ratios for each phase are calculated.

The list of hardware and software used in the experiment is provided in Table 3.

Table 3. Hardware and software used in the experiment.

OS	Linux Mint, Linux PC 5.4.0-62-generic #70-Ubuntu SMP Tue Jan 12 12:45:47 UTC 2021 ×86_64 ×86_64 ×86_64 GNU/Linux
Python programming language	Python 3.6.9 (default, 18 April 2020, 01:56:04) [GCC 8.4.0] on Linux
Python libraries	multiprocessing, time, random, numpy, sys, asyncio, matplotlib, pylab, matplotlib.pyplot, math, ctypes
IDE	Visual Studio Code
Processor	Intel(R) Core(TM) i7-8700 CPU @ 3.20 GHz
Memory	32747MB (17387MB used)
Storage	ATA Samsung SSD 860
Video adapter	NVIDIA GeForce GTX 1080 Ti/PCIe/SSE2

The results of calculation ratios Q_1/V_1, Q_2/V_2, Q_3/V_3, Q_4/V_4, and Q_5/V_5 in 1–15 s simulations are provided in Table 4.

Table 4. Results of ratio calculation.

t, s	Q_1/V_1	Q_2/V_2	Q_3/V_3	Q_4/V_4	Q_5/V_5
1	0	0	0	0	0
2	15,783.208889	7713.886687	10,173.894957	8327.816417	4849.444553
3	16,158.595911	24,539.397682	3903.148301	12,644.343116	7150.621767
4	7150.621767	14,994.150418	4764.470185	9563.342329	9489.530662
5	16,618.375117	11,304.161697	24,609.465659	11,133.676329	7467.462789
6	16,349.095476	16,200.135701	10,906.525734	21,445.316714	7687.643707
7	16,540.717478	10,413.616874	17,795.190157	11,758.146436	9688.644638
8	17,363.635526	11,608.885913	13,812.022670	14,145.828118	8471.549574
9	19,027.493286	15,717.070406	19,310.638918	13,772.830493	14,585.097103
10	17,339.523710	11,679.801550	14,368.463078	12,014.946867	10,045.985695
11	20,742.679577	23,891.867419	15,796.962250	17,233.303617	18,584.620158
12	17,426.702838	84,383.182448	12,677.306401	11,978.191479	8637.416323
13	19,019.020194	16,494.458220	21,366.420617	20,901.589822	26,116.328410
14	0	0	0	0	0
15	0	0	0	0	0

The ratios mentioned in Table 4 are provided in Figures 2–6, respectively.

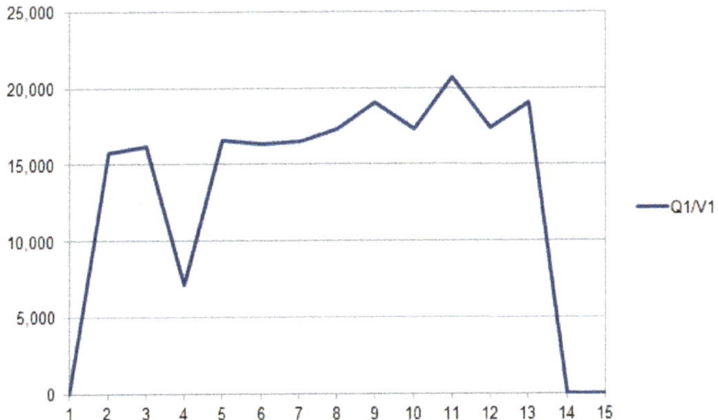

Figure 2. Ratio of Q_1/V_1.

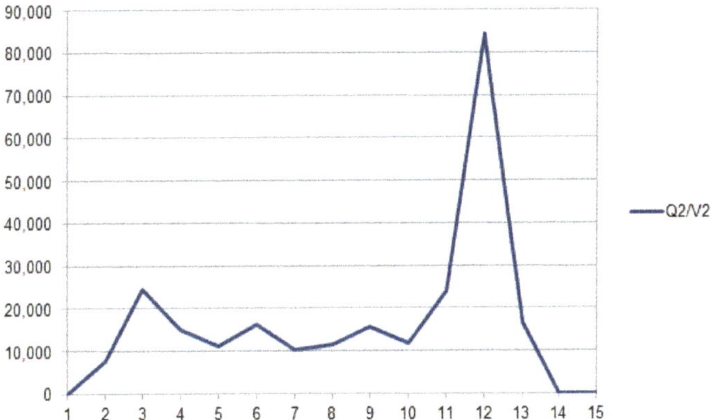

Figure 3. Ratio of Q_2/V_2.

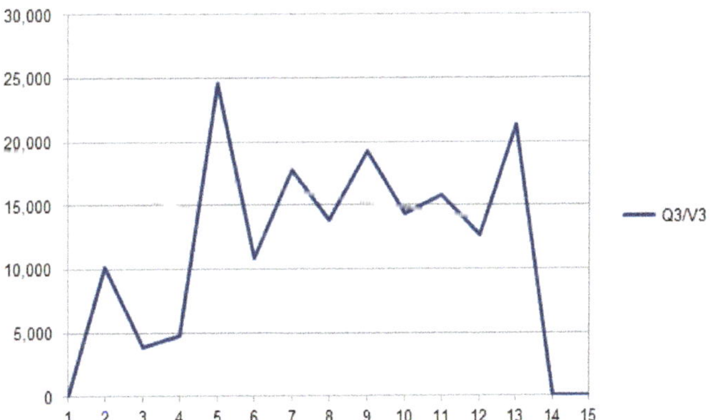

Figure 4. Ratio of Q_3/V_3.

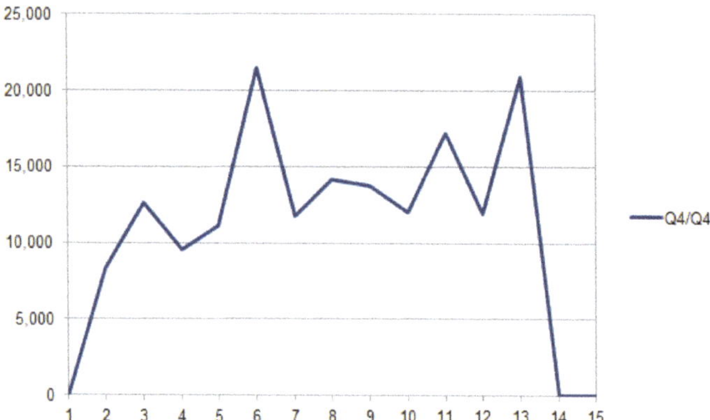

Figure 5. Ratio of Q_4/V_4.

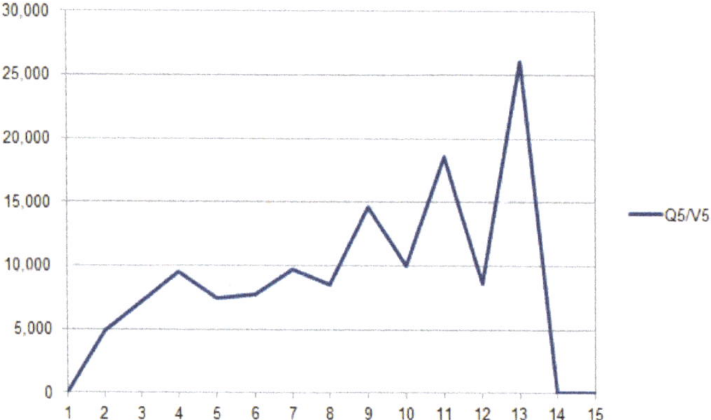

Figure 6. Ratio of Q_5/V_5.

4.4. Description of the Results

As described in the theory, each $Q_i/E(V_i)$ ratio should converge into a constant. This was an expected result—each $Q_i/E(V_i)$ ratio after some period of time converges into a constant value or becomes almost stable after some period of time and remains fairly stable for a certain period of time. In the theory, we have an infinite flow of clients, really independent phases, and infinitely large computational resources. In other words, we really have the ideal conditions and no faults or errors.

In reality, this model runs on one computer where there is one processor with real and virtual cores, memory, and storage limits. Under these conditions, some unstable work of the system is present because it is impossible to eliminate all of the faults and errors. A more important point is that the computer's operating system shares resources between processes using its own algorithms, and it can be difficult to allocate the required amount of hardware resources to a particular process.

Another significant problem is the lack of resources, which produces some limitation for any model work (e.g., to have an infinite flow of jobs). Each phase could also be "clogged" with clients when another one is waiting for them. In addition, during the experiment, other restrictions appeared in the current configuration. For example, in theory we need to use the infinite number of clients, but in reality using even 1,000,000 is difficult.

In the real condition, there are some faults and errors in multi-threading libraries when using the Python programming language, and there are some undefined states. Some other mistakes could appear due to the calculation accuracy and time measurement because approximations are used. The lack of resources especially affects measurements of time because the system could lag.

The most expected result is to find a stable interval for every phase where the equilibrium of model load and theoretical conditions are satisfied and to create the experiment where all listed theorems could be checked. A stable time interval should be found for each ratio, in this interval the ratio of $Q_j/E(V_j)$ should be the same or very similar.

All of the calculation results for each ratio are listed above. In every chart of each ratio, we can observe the state and the time when the ratio becomes stable or changes a little. Each phase becomes a stable state at a different time period. This could happen because of different times of the critical load and also because of no load for each phase.

For example, the first ratio $Q_1/E(V_1)$ becomes stable after 2 s (4 s value is an issue) and remains stable until 13 s. The second ratio is more stable from 3 to 11 s, the third from 6 to 12, the fourth from 7 to 12, and the fifth is more stable at the beginning but less stable at the end. It is difficult to find a stable period for the fifth (or last) phase because it becomes clogged after all of the other phases are empty or almost empty. When a large flow of clients arrives, the period of stability should begin, but the phase becomes empty very soon. In addition, for this configuration, all clients pass MS at the 14th second, and all phases become empty.

This experiment result shows that each phase has its own stability period; thus, it could be considered that the ratio of $Q_j/E(V_j)$ converges to a constant, as proved in the theory.

Corollary 2 (Validation of Little's formula in MS). *At this stage of the work, we confirm that the results of the first stage are correct.*

5. Concluding Remarks

- We observe that the heavy traffic condition used in the proof of the theorems on SLLN is fundamental. Abandoning this condition makes the proof of the theorems very complicated. In the future, it would be interesting to examine the situation under light traffic.
- By using another method to prove theorems and normalizing boundary processes differently than compared to SLLN (e.g., probability limit theorems or the law of the iterated logarithm—but this is implicated only in the single-phase case, and there is no research in the multiphase case), Little's law becomes the successful process or becomes its law of the iterated logarithm analog. With SLLN, the boundary process is constant—this process can be modeled.
- The theoretical results cannot be directly verified by modeling them, which is evidenced by the modeling block diagram. Modeling has its own explicit specifics; thus, a comprehensive review of the literature in this area was required.
- The ideas of the modeling part are related to the often cited work [38]. In this work, for the first time, the modern possibilities of the Python programming language were applied in order to model the queuing system. Continuing this topic, the Python concept was used to test the theoretical results of the first stage.
- The first and second parts of this paper deal with similar but completely separate subject areas. As much as possible, efforts were made to bring them closer together and to present them as a single study.

Author Contributions: Conceptualization, S.M. and I.K.; methodology, S.M.; software, I.K.; validation, J.K. and I.V.-Z.; formal analysis, S.M. and I.K.; investigation, I.K.; resources, I.K.; data curation, J.K. and I.V.-Z.; writing—original draft preparation, S.M.; writing—review and editing, J.K. and I.V.-Z.; visualization, J.K.; supervision, S.M. and I.K.; project administration, J.K. and I.V.-Z. All authors have read and agreed to the published version of the manuscript.

Funding: This research received no external funding.

Institutional Review Board Statement: Not applicable.

Informed Consent Statement: Not applicable.

Conflicts of Interest: The authors declare no conflict of interest.

References

1. Borovkov, A. *Stochastic Processes in Queueing Theory*; Nauka: Moscow, Russia, 1972. (In Russian)
2. Borovkov, A. *Asymptotic Methods in Theory of Queues*; Nauka: Moscow, Russia, 1980. (In Russian)
3. Saati, T.; Kerns, K. *Analytic Planning. Organization of Systems*; Mir: Moscow, Russia, 1971. (In Russian)
4. Iglehart, D.L.; Whitt, W. Multiple channel queues in heavy traffic. I. *Adv. Appl. Probab.* **1970**, *2*, 150–177. [CrossRef]
5. Iglehart, D.L.; Whitt, W. Multiple channel queues in heavy traffic. II: Sequences, networks and batches. *Adv. Appl. Probab.* **1970**, *2*, 355–369. [CrossRef]
6. Kingman, J. On queues in heavy traffic. *J. R. Stat. Soc. Ser. (Methodol.)* **1962**, *24*, 383–392. [CrossRef]
7. Kingman, J. The single server queue in heavy traffic. *Math. Proc. Camb. Philos. Soc.* **1961**, *57*, 902–904. [CrossRef]
8. Kobyashi, H. Application of the diffusion approximation to queueing networks I: Equilibrium queue distributions. *J. ACM* **1974**, *21*, 316–328. [CrossRef]
9. Reiman, M.I. Open queueing networks in heavy traffic. *Math. Oper. Res.* **1984**, *9*, 441–458. [CrossRef]
10. Karpelevich, F.I.; Kreinin, A.I. Heavy traffic limits for multiphase queues. *Transl. Math. Monogr.* **1994**, *137*. [CrossRef]
11. Gamarnik, D.; Stolyar, A.L. Multiclass multi-server queueing system in the Halfin-Whitt heavy traffic regime: Asymptotics of the stationary distribution. *Queueing Syst.* **2012**, *71*, 25–51. [CrossRef]
12. Gurvich, I. Validity of heavy-traffic steady-state approximations in multiclass queueing networks: The case of queue-ratio disciplines. *Math. Oper. Res.* **2014**, *39*, 121–162. [CrossRef]
13. Wu, Y.; Bui, L.; Johari, R. Heavy traffic approximation of equilibria in resource sharing games. *Internet Netw. Econ.* **2011**, 351–362. [CrossRef]
14. Markakis, M.G.; Modiano, E.; Tsitsiklis, J.N. Max-weight scheduling in queueing networks with heavy-tailed traffic. *IEEE/ACM Trans. Netw.* **2014**, *22*, 257–270. [CrossRef]
15. Anselmi, J.; Casale, G. Heavy-traffic revenue maximization in parallel multiclass queues. *Perform. Eval.* **2013**, *70*, 806–821. [CrossRef]
16. Maguluri, S.T.; Burle, S.K.; Srikant, R. Optimal heavy-traffic queue length scaling in an incompletely saturated switch. *Queueing Syst.* **2018**, *88*, 279–309. [CrossRef]
17. Braverman, A.; Dai, J.G.; Miyazawa, M. Heavy traffic approximation for the stationary distribution of a generalized Jackson network: The BAR approach. *Stochastic Syst.* **2017**, *7*, 143–196. [CrossRef]
18. Butler, R.W.; Huzurbazar, A.V. Stochastic Network Models for Survival Analysis. *J. Am. Stat. Assoc.* **1997**, *92*, 246–257. [CrossRef]
19. Wang, W.; Zhu, K.; Ying, L.; Tan, J.; Zhang, L. MapTask scheduling in mapreduce with data locality: Throughput and heavy-traffic optimality. *IEEE/ACM Trans. Netw.* **2016**, *24*, 190–203. [CrossRef]
20. Whitt, W. Some useful functions for functional limit theorems. *Math. Oper. Res.* **1980**, *5*, 67–85. [CrossRef]
21. Wang, W.; Maguluri, S.T.; Srikant, R.; Ying, L. Heavy-traffic delay insensitivity in connection-level models of data transfer with proportionally fair bandwidth sharing. *ACM Sigmetrics Perform. Eval. Rev.* **2017**, *45*, 232–245. [CrossRef]
22. Huang, J.; Gurvich, I. Beyond heavy-traffic regimes: Universal bounds and controls for the single-server queue. *Oper. Res.* **2018**, *66*, 1168–1188. [CrossRef]
23. Zhou, X.; Tan, J.; Shroff, N. Flexible load balancing with multidimensional state-space collapse: Throughput and heavy-traffic delay optimality. *Perform. Eval.* **2018**, *127–128*, 176–193. [CrossRef]
24. Maguluri, S.T.; Srikant, R. Heavy traffic queue length behavior in a switch under the MaxWeight algorithm. *Stoch. Syst.* **2016**, *6*, 211–250. [CrossRef]
25. Jackson, J.R. Jobshop-like queueing systems. *Manag. Sci.* **1963**, *10*, 131–142. [CrossRef]
26. Kelly, F.P. *Reversibility and Stochastic Networks*; Wiley: New York, NY, USA, 1987.
27. Borovkov, A.A. Limit theorems for queueing networks. I. *Theory Probab. Its Appl.* **1987**, *31*, 413–427. [CrossRef]
28. Konstantopoulps, P.; Walrand, J. On the ergodicity of networks of $\cdot/GI/1/N$ queues. *Adv. Appl. Probab.* **1990**, *22*, 263–267. [CrossRef]
29. Konstantopoulps, P.; Walrand, J. Stationarity and stability of fork-join networks. *J. Appl. Probab.* **1989**, *26*, 604–614. [CrossRef]
30. Billingsley, P. *Convergence of Probability Measures*; Wiley: New York, NY, USA, 1968.
31. Minkevičius, S. Weak convergence in multiphase queues. *Lith. Math. J.* **1986**, *26*, 347–351. [CrossRef]
32. Melikov, A.; Aliyeva, S.; Sztrik, J. Analysis of Instantaneous Feedback Queue with Heterogeneous Servers. *Mathematics* **2020**, *8*, 2186. [CrossRef]
33. Sztrick, J.; Toth, A.; Pinter, A.; Bacs, Z. Reliability Analysis of Finite-Source Retrial Queueing Systems with Two-Way Communications to the Orbit and Blocking Using Simulation. In Proceedings of the 23rd International Scientific Conference on Distributed Computer and Communication Networks: Control, Computation, Communications (DCCN-2020), Moscow, Russia, 14–18 September 2020; pp. 260–267.
34. Sztrik, J.; Tóth, Á.; Pintér, Á.; Bács, Z. The simulation of finite-source retrial queueing systems with two-way communications to the orbit and blocking. In *International Conference on Distributed Computer and Communication Networks*; Springer: Cham, Switzerland, 2020; pp. 171–182. [CrossRef]

35. Bychkov, I.V.; Kazakov, A.L.; Lempert, A.A.; Bukharov, D.S.; Stolbov, A.B. An intelligent management system for the development of a regional transport logistics infrastructure. *Autom. Remote Control* **2016**, *77*, 332–343. [CrossRef]
36. Harahap, E.; Darmawan, D.; Fajar, Y.; Ceha, R.; Rachmiatie, A. Modeling and simulation of queue waiting time at traffic light intersection. *J. Phys. Conf. Ser.* **2019**, *1188*, 012001. [CrossRef]
37. Gorbunova, A.V.; Vishnevsky, V.M.; Larionov, A.A. Evaluation of the end-to-end delay of a multiphase queuing system using artificial neural networks. In *International Conference on Distributed Computer and Communication Networks;* Lecture Notes in Computer Science; Springer: Cham, Switzerland, 2020; pp. 631–642. [CrossRef]
38. Dolgopolovas, V.; Dagienė, V.; Minkevičius, S.; Sakalauskas, L. Python for scientific computing education: Modeling of queueing systems. *Sci. Program.* **2014**, *22*, 37–51. [CrossRef]

 *mathematics*

Article

A Learning-Based Hybrid Framework for Dynamic Balancing of Exploration-Exploitation: Combining Regression Analysis and Metaheuristics

Emanuel Vega [1,*], Ricardo Soto [1], Broderick Crawford [1], Javier Peña [1] and Carlos Castro [2]

[1] Escuela de Ingeniería Informática, Pontificia Universidad Católica de Valparaíso, Valparaíso 2362807, Chile; ricardo.soto@pucv.cl (R.S.); broderick.crawford@pucv.cl (B.C.); javier.pena.r@mail.pucv.cl (J.P.)
[2] Departamento de Informática, Universidad Técnica Federico Santa María, Valparaíso 2390123, Chile; Carlos.Castro@inf.utfsm.cl
* Correspondence: emanuel.vega.m@mail.pucv.cl

Abstract: The idea of hybrid approaches have become a powerful strategy for tackling several complex optimisation problems. In this regard, the present work is concerned with contributing with a novel optimisation framework, named learning-based linear balancer (LB^2). A regression model is designed, with the objective to predict better movements for the approach and improve the performance. The main idea is to balance the intensification and diversification performed by the hybrid model in an online-fashion. In this paper, we employ movement operators of a spotted hyena optimiser, a modern algorithm which has proved to yield good results in the literature. In order to test the performance of our hybrid approach, we solve 15 benchmark functions, composed of unimodal, multimodal, and mutimodal functions with fixed dimension. Additionally, regarding the competitiveness, we carry out a comparison against state-of-the-art algorithms, and the sequential parameter optimisation procedure, which is part of multiple successful tuning methods proposed in the literature. Finally, we compare against the traditional implementation of a spotted hyena optimiser and a neural network approach, the respective statistical analysis is carried out. We illustrate experimental results, where we obtain interesting performance and robustness, which allows us to conclude that our hybrid approach is a competitive alternative in the optimisation field.

Keywords: metaheuristics; machine learning; hybrid approach; optimisation

Citation: Vega, E.; Soto, R.; Crawford, B.; Peña, J.; Castro, C. A Learning-Based Hybrid Framework for Dynamic Balancing of Exploration-Exploitation: Combining Regression Analysis and Metaheuristics. *Mathematics* **2021**, *9*, 1976. https://doi.org/10.3390/math9161976

Academic Editors: Frank Werner and Alfredo Milani

Received: 21 April 2021
Accepted: 11 August 2021
Published: 18 August 2021

1. Introduction

In recent years, the constant increase in complexity of the problems to be solved in the industry and academy have raised the necessity to further improve and evolve new techniques. In this context, hybrid approaches have been a standard and focus of multiple works. They have proved to be the most successful strategy in terms of solving capacity tackling hard optimisation problems [1]. In modern approaches, the use of randomised optimisation methods have been the focus of work by the scientist community, a well-known example are the metaheuristics. They have been successfully used to solve large instance of complex and difficult optimisation problems, being useful when exact methods are unable to provide solutions in a reasonable amount of time [2]. Usually, in the design behind an algorithm, we find multiple complex items which are in charge of carrying out the work in order to solve optimisation problems [3]. Inherent features, like intensification and diversification [4–6], which are in control on how the approach can exploit and explore the search space, respectively. Additionally, parameters and search components, such as population, probabilities, search operators, initial solutions, and so on, comprehend important family items in the work of an approach. In order to be intelligent, an agent which works in a changing environment should have the ability to learn [7]. If the approach can learn and adapt, we do not need to foresee and provide solutions for all possible situations which may appear on run-time. Machine learning, being part of artificial intelligence, encircle a number of algorithms with the aim to optimise a performance criterion using example data or past experience [8–10]. A well-known style of learning is the supervised learning, which is mainly composed by learning functions with the aim of predict values, and some of his classical objectives are regression and classification [11].

In this paper, we examine whether a formal relationship between an effective balance of intensification and diversification, influenced by a regression model, and a classic configuration of a metaheuristic exists, and whether it is sufficiently strong to be exploited for an automated framework. Most metaheuristics operate in a sequential, iterative, and in a previously designed manner, but the environment where they operate usually has a dynamic nature. Additionally, they are stochastic algorithms, which comprehends deterministic and random components. The stochastic components can take many forms, such as simple randomisation by randomly sampling the search space or by random walks. Thus, the randomness brings certain degree of uncertainness in the search. For instance, if an agent just finished performing an intensification movement, and the next step in the process performs a diversification movement, it has no certainty on reaching a better neighbourhood. In Figure 1, we illustrate a graphic example of a situation where a white agent needs to make a move; we aim to help the agent to have higher possibility to reach a green dot (possible best solution) or a yellow dot (less possible best solution) than a red dot (bad solution). The objective in the design of this framework is to let the approach learn how to orchestrate the work performed by the agents in every iteration, hence, we enforce the decision making of the approach and make him learn from previous iterations on run-time. In this regard, two components are designed: movements operators; in this work we employ movements from the spotted hyena optimiser (SHO) algorithm [12], and a regression model. First, SHO is an interesting modern metaheuristic, which has proved to yield good results in solving optimisation problems [13,14]. It is mainly based in the grouping behaviours of a special type of Hyena, where the strong point in the algorithm is the clustering features of the agents searching in the solution space. On the other hand, the learning model, is where the central axis of the work is completed by linear regression analysis. The work is completed as follows: dynamic data generated by the agents through iterations will be managed by the learning model. In this context, each time a threshold amount of iteration is met, a regression analysis is carried out by the learning model. Thus, the search will be influenced by the resulting knowledge from the previous learning process.

The efficiency of LB^2 proposed in this research is evaluated in three phases by solving 15 well-known mathematical optimisation problems. The employed benchmark concerns unimodal, multimodal, and mutimodal functions with fixed dimension. Additionally, these continuous functions comprehends multiple features, such as being convex, non-differentiable, unconstrained, and so on. Regarding the experimentation phases, we compare our results with state-of-the-art optimisation methods, such as particle swarm optimisation (PSO) [15], gravitational search algorithm (GSA) [16], differential evolution (DE) [17], whale optimisation algorithm (WOA) [18], vapour–liquid equilibrium (VLE) [19], and an hybrid between Nelder–Mead algorithm and dragonfly algorithm (INMDA) [20]. In the second phase, we compare against sequential parameter optimisation (SPO) [21]. The key work in SPO is performed by a prediction model, bringing improvements in the parameters values and algorithm performance in an iterative scheme. Third, we carry out a statistical evaluation of the results obtained by the traditional implementation of SHO, neural network (NN) [22], sine cosine algorithm (SCA) [23], and our proposed hybrid approach. Finally, we illustrate interesting experimental results, where the proposed hybrid approach achieves good performance proving to be a good and competitive option to tackle continuous optimisation problems.

The rest of this paper is organised as follows. The related work is introduced in the next section. The proposed hybrid approach is explained in Section 3. Section 4 illustrates the experimental results. Finally, we conclude and suggest some lines of future research.

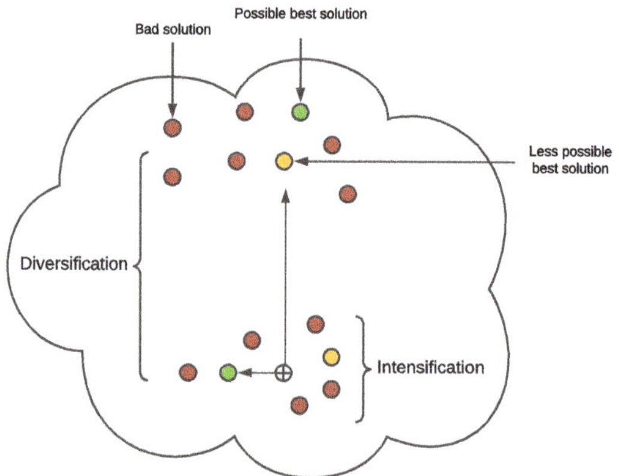

Figure 1. Graphic example of the search space illustrating green dots possible best solutions, yellow dots less possible best solutions, and red dots bad solutions.

2. Related Work

This work proposes a learning-based hybrid approach, where the main feature is the capability to influence the search performed by the agents ruled by the movements of SHO. Therefore, following the taxonomy illustrated by Talbi in [24], our proposed work can be described as a low-level teamwork hybridisation. Concerning the works reported in the literature between machine learning and metaheuristics [8,25], it is well-known that this relationship is not a one-way street, we do not have only approaches were machine learning techniques assist and enhance metaheuristics, but also the other way around: machine learning models improved by metaheuristics, is a much consolidated group in the hybridisation field [26–30]. This paper is concerned with the first group, where novel approaches have been proposed, such as [31], where a diversification-based learning (DBL) framework is proposed. DBL is designed under families of components introduced in the field of metaheuristics and machine learning that have broad applications in optimisation. Additionally, a novel approach based on two well-known components is presented in [32], an hybrid between intelligent guided adaptive search (IGAS) and path-relinking algorithm, named IGASPR. The main learning phase is ruled by the means of growing neural gas (GNG), the objective is to influence the construction of solutions controlling the features of the best solutions in each iteration. Concerning proposed works under the influence of regression analysis, [33] illustrates a data mining based approach for PSO. The main ideas behind that contribution is that the parameter selection task can appropriately be addressed by a data mining-based approach. The designed model employs a regression analysis by means of non-linear regression models, the main objective is to learn suitable parameters values from previous moves for PSO on run-time. In this field, this type of scheme is also known as specifically-located hybridisation and it is concerned with the parameter control strategies. In the literature, [34] also employ this type of hybridisation. The authors propose an hybrid employing Tabu search (TS) and support vector machine (SVM). The proposed approach is designed to tackle on hard combinatorial optimisation problems, such as knapsack problem, set covering problem, and the travelling salesman problem. The main task concerns the selection of decision rules from a corpus of solutions generated in a randomly fashion, which are used to predict high quality solutions for a given instance and it is used to fine-tune and guide the search performed by TS. However, it is stated by the authors that the complexity of the approach is a key factor, they highlight the time consumed and knowledge necessarily needed to implement, the process to build the corpus, and the extraction of the classification rule. On the other hand, regarding hybrid specialising in intensification and diversification, to the best of our knowledge there was none under the influence of a regression model. However, in [25], the feasible options on intensification employing clustering [35] and frequent itemsets using association rules [36,37] are illustrated. Regarding diversification, the use of clustering [35], self-organising maps (SOM) [38], and binary space-partitioning (BSP) trees [39] have proved to be good options balancing this issue in different approaches.

The LB^2 proposed in this work draws inspiration by the following arguments. Firstly, the scarce literature concerning machine learning mainly associated to regression model assisting metaheuristics. Second, most approaches are problem-dependant, for instance, in [32], the problem to be tackled by the regression model is the selection of best fitted parameters for PSO in order to improve the performance. It is a good exploratory and pioneer approach considering this attempt to be on run-time. However, the uncertainty in extrapolating this specifically-located implementation to other approaches is high, especially taking into account the "no free lunch" theorem. Therefore, our proposition focus in two major issues when designing a global search method, diversification, and intensification. Thus, if we analyse the metaheuristic field, they are general features who are always present. Third, the technique selected is a highly relevant issue. It is stated and explained by authors, in [33], the level of complexity is an issue to take into account in the design of the hybrid. Thus, we think this issue may have an impact replicating the results and extrapolating the implementation to an unknown environment. In this context, we employ classic techniques, where the novel mechanism are the clear advantages provided by our proposed hybrid approach.

3. Proposed Hybrid Approach

In this section we present the proposed LB^2 framework, we discuss the main ideas in the design, motivations, and inspiration behind the proposed approach. Firstly, in order to carry out the search in the solution space, the strategy employed is inspired by population-based metaheuristics. The main idea is to perform using a set of agents who evolve under the influence of multiple equations, known as movement of operators. In this regard, they are usually classified as intensification and diversification concerning the work performed, exploitation or exploration of the search space. In this work we employ SHO and his four movement operators, where each hyena is currently an agent in the framework.

The second answer proposed is concerned with the component in charge of the regression analysis. In this first attempt to design LB^2 the main concern was the complexity of the employed technique [40,41]. In this regard, multiple techniques and methods to carry out a regression analysis [42], such as linear models, SVM, and decision-tree-based models. Thus, a linear model was selected because it is the most commonly used, and all other regression methods build upon an understanding on how linear regression works [43,44]. Nevertheless, the regression model can potentially evolve in a more complex component, a more detailed explanation is presented in Section 5.

The global conceptualisation of the proposed hybrid is illustrated in Figure 2. It is based in the behaviour of multiple agents with the same attributes, also known as population. They are controlled by the movements of SHO, influenced and balanced by the learning-based model. A general description is presented in Section 3.1. The methodology and detailed explanation of the proposed approach is explained in Section 3.2. In Section 3.3, the population-based metaheuristic is presented, and the proposed algorithm is illustrated in Section 3.4.

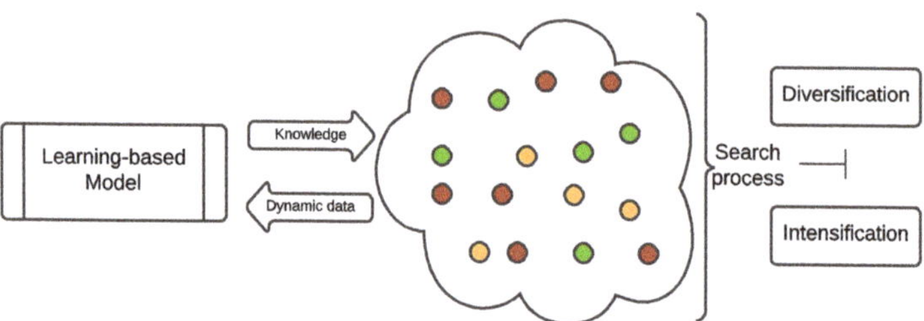

Figure 2. Graphic example of the search process.

3.1. General Description

The proposed LB^2 follows a population-based strategy, which concerns multiple agents evolving in the solution space, intensification and diversification are performed, and the process is terminated when a threshold amount of iteration is met. Dynamically adjusting the configuration and behaviour is an important topic that continues to be of growing interest. This work, in order to carry out the search, proposes two components: scheme and β. Firstly, the scheme is concerned with the amount of intensification and diversification to be performed in each iteration by the population. Regarding β, it is a parameter employed as the threshold where the learning model needs to carry out the regression analysis. The knowledge generated will be used to influence the selection mechanism, which manages the scheme that needs to be performed. In this regard, the selection will dynamically rule over the work of each agent, indicating the amount of exploration and exploitation to carry out in the search space. The proposed steps of the proposed LB^2 are described as follows:

Step 1: Set parameters concerning the population-based algorithm: B,E,h, termination criteria for the search.
Step 1.1: Set termination criteria for the search: set amount of iterations to perform LB^2.
Step 2: Set parameters concerning the learning model: scheme, probabilities, β.
Step 2.1: Set schemes for intensification and diversification.
Step 2.2: Set the probabilities for each scheme to be selected by the selection mechanism.
Step 2.3: Set the value for threshold β.
Step 3: Generation of the initial population size to perform in the search.
Step 4: while the termination criteria is not met.
Step 5: For each agent:
Step 5.1: Selection mechanism on intensification: the scheme is selected and the exploitation is carried out.
Step 5.2: Management of dynamic data generated.
Step 5.3: Selection mechanism on diversification: the scheme is selected and the exploration is carried out.
Step 5.4: Management of dynamic data generated.
Step 6: Update parameters concerning the population-based algorithm: B,E,h.
Step 7: Check if the threshold β has been met.
Step 7.1: Perform regression analysis.
Step 7.2: Management of the knowledge generated: update scheme probabilities.

3.2. Methodology

Firstly, we need to define the schemes to perform through the search, in this first attempt designing LB^2, three levels where proposed and illustrated in Table 1. They define the amount of work that needs to be performed in each iteration, the selection issue is tackled by the means of probabilities, and they are defined as follows.

$$\text{for intensification:} \quad p_i = \frac{1}{\text{IS}_{\text{soft}}} + \frac{1}{\text{IS}_{\text{medium}}} + \frac{1}{\text{IS}_{\text{hard}}} = 1$$

$$\text{for diversification:} \quad p_d = \frac{1}{\text{DS}_{\text{soft}}} + \frac{1}{\text{DS}_{\text{medium}}} + \frac{1}{\text{DS}_{\text{hard}}} = 1$$

where the probability p_i and p_d will be modified by the learning model every β amount of iterations. This model is in charge of the regression analysis ruled by the means of linear regression, where the fitted function is of the form:

$$\text{y} = \text{w}x + b$$

where y corresponds to the dependant variable, which is the fitness and the value we want to predict. x represent the independent variable, which correspond to the scheme performed. In this simple linear regression model proposed, we present the close relationship between the fitness and his convergence with the balance of intensification and diversification performed. Regarding our proposed learning-model, we define three fitted functions for each scheme on intensification and three for each scheme on diversification. They are represented as follows:

For intensification:

$$\begin{aligned} \text{y}_{i-\text{soft}} &= \text{w}_i x_{i-\text{soft}} + b_i \\ \text{y}_{i-\text{medium}} &= \text{w}_i x_{i-\text{medium}} + b_i \\ \text{y}_{i-\text{hard}} &= \text{w}_i x_{i-\text{hard}} + b_i \end{aligned}$$

For diversification:

$$\begin{aligned} \text{y}_{d-\text{soft}} &= \text{w}_i x_{d-\text{soft}} + b_d \\ \text{y}_{d-\text{medium}} &= \text{w}_i x_{d-\text{medium}} + b_d \\ \text{y}_{d-\text{hard}} &= \text{w}_i x_{d-\text{hard}} + b_d \end{aligned}$$

In order to carry out the analysis, we employ the least squares method which is a well-known approach used. We evaluate the grade of relationship between the works performed by the agents in the amount of intensification and diversification with the best fitness values reached. The model will make the decision based as follows:

$$\begin{aligned} W(x_i) &= \text{MIN}(\text{y}_{i-\text{soft}}, \text{y}_{i-\text{medium}}, \text{y}_{i-\text{hard}}) \text{ and} \\ W(x_d) &= \text{MIN}(\text{y}_{d-\text{soft}}, \text{y}_{d-\text{medium}}, \text{y}_{d-\text{hard}}) \end{aligned}$$

where $W(x_i)$ and $W(x_d)$ represent the schemes with the highest possibilities to achieve better performance in the next β iterations. The regression model will modify the probabilities of selection for each scheme. Thus, when the threshold is met, the process of selection, carried out in a Monte Carlo roulette fashion, will be influenced. Additionally, we highlight that all benchmark functions are minimisation problems which are aligned with our proposed function MIN.

Regarding the threshold β, important issues need to be considered, such as amount of total iterations, computing capacity, number of agents as population, and number of schemes in the approach. In this work, small test were carried out with β values 200, 500, and 1000. However, we concluded that the best performance was achieved with a value of 1000.

A practical example can be described as follows: At the beginning, in each iteration, the approach will select a scheme using a probabilistic roulette for the intensification and diversification. Thus, for a three way scheme, as displayed in Table 1, the initial probabilities for each scheme to be selected was in a 33.3%–33.3%–33.3% ratio. Additionally, the regression model is always storing and sorting the fitness values and agents on run-time. When the threshold β is met, the model performs a computing process corresponding to the regression analysis. Thus, it is decided which scheme have the highest chance to achieve a high performance over the next β amount of iterations. To do so, the probabilities values of each scheme for intensification and diversification are updated, giving the winning scheme a higher probability to be chosen. For instance, we designed a ratio of 60%–20%–20% ratio, a graphic example is illustrated in Figure 3. Here, the scheme assigned with a blue color had a minimum value in the resulting regression analysis compared with the other two schemes, in this case, the winner is assigned a higher value of probability to be selected, and so on.

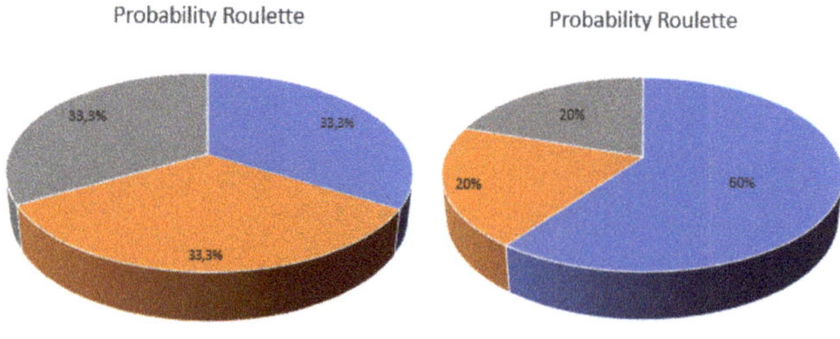

(**a**) Roulette at the beginning (**b**) Roulette after model intervention

Figure 3. Graphic example of the modification of probabilities by the model.

3.3. Spotted Hyena Optimiser

In this paper, we instantiate SHO as a means to carry out the search of solutions in order to solve optimisation problems. The movement operators are organised as illustrated in Table 2 with the aim to be employed by LB^2. The main feature of SHO is the cohesive clustering in his population [12]. The mathematical model concerns diversification methods: encircling prey, hunting, and search for prey. Additionally, intensification method: attacking prey. Additionally, they are described as follows:

1. Encircling prey: Each hyena takes the current best candidate solution as the target prey. They will try to move towards the best position defined.

$$D_h = \left|B \cdot P_p(x) - P(x)\right| \quad (1)$$

$$P(x+1) = P_p(x) - E \cdot D_h \quad (2)$$

where D_h is the distance between the current spotted hyena and the prey, x indicates the current iteration, B and E are coefficient vectors, P_p is the position of the prey, and P is the position of the spotted hyena. The vectors B and E are defined as follows:

$$B = 2 \cdot \text{rnd}_1 \quad (3)$$

$$E = 2h \cdot \text{rnd}_2 - h \quad (4)$$

$$h = 5 - (\text{Iteration} * (5/\text{Max}_{\text{iteration}})) \quad (5)$$

where $Iteration = 1, 2, 3, \ldots, \text{Max}_{\text{iteration}}$, rnd_1 and rnd_2 are random vectors in [0, 1].

2. Hunting: The hyenas make a cluster towards the best agent so far to update their positions. The equations are proposed as follows:

$$D_h = |B \cdot P_h - P_k| \tag{6}$$

$$P_k = P_h - E \cdot D_h \tag{7}$$

$$C_h = P_k + P_{k+1} + ... + P_{k+N} \tag{8}$$

where P_h is the best spotted hyena in the population, and P_k indicates the position of other spotted hyenas. Here, N is the number of spotted hyenas, which is computed as follows:

$$N = \text{count}_{\text{nos}}(P_h, P_{h+1}, P_{h+2}, ..., (P_h + M)) \tag{9}$$

Here, M is a random vector [0.5, 1], nos defines the number of solutions and count all candidate solutions plus M, and C_h is a cluster of N number of optimal solutions.

3. Attacking Prey: SHO works around the cluster forcing the spotted hyenas to assault towards the prey. The following equation was proposed:

$$P(x+1) = C_h/N \tag{10}$$

Here, $P(x+1)$ updates the positions of each spotted hyenas according to the position of the best search agent and save the best solution.

4. Search for Prey: The agents mostly search the prey based on the position of the cluster of spotted hyenas, which reside in vector C_h. SHO makes use of the coefficient vector E and B with random values to force the search agents to move far away from the prey. This mechanism allows the algorithm to search globally.

Table 1. Example of the standard work to be completed by the approach.

Scheme	Amount of Intensification	Amount of Diversification
Soft	1	1
Medium	2	2
Hard	3	3

Table 2. Organisation example of the pool of movement operators from metaheuristics.

Pool of Operators	
Intensification	**Diversification**
Exploitation movement 1	Exploration movement 1
Exploitation movement 2	Exploration movement 2
⋮	⋮

3.4. Proposed Algorithm

In this subsection, we illustrate the designed algorithm. Algorithm 1 depicts the general framework our proposed approach, where the operators of SHO performs intensification and diversification under the influence and balance of our regression model. Finally, Algorithm 2 presents the work in charge of the regression model. The regression analysis is performed and the vectors with controls values are modified.

Algorithm 1 *Proposed LB^2*

```
Set initial parameters for SHO
Set initial parameters for regression model
Generate initial population
while (i ≤ MaximumIteration) do
   for each agent in the population do
      StandardIntensification = Select-scheme-by-Roulette
      while (StandardIntensification) do
         Perform intensification operators
      end while
      if check if a best value was reached using StandardIntensification then
         Update data structures with best values reached
      end if
      StandardDiversification = Select-scheme-by-Roulette
      while (StandardDiversification) do
         Perform diversification operators
      end while
      if Check if a best value was reached using StandardDiversification then
         Update data structures with best values reached
      end if
   end for
   if Check threshold β then
      Call to Algorithm 2: Regression Model
   end if
end while
```

Algorithm 2 *Regression Model*

```
while review of dynamic-data for all x_{i-soft} do
   Management of dataframe with dynamic-data
end while
Compute statistical modelling method: y_{i-soft}
while review of dynamic-data for all x_{i-medium} do
   Management of dataframe with dynamic-data
end while
Compute statistical modelling method: y_{i-medium}
while review of dynamic-data for all x_{i-hard} do
   Management of dataframe with dynamic-data
end while
Compute statistical modelling method: y_{i-hard}
while review of dynamic-data for all x_{d-soft} do
   Management of dataframe with dynamic-data
end while
Compute statistical modelling method: y_{d-soft}
while review of dynamic-data for all x_{d-medium} do
   Management of dataframe with dynamic-data
end while
Compute statistical modelling method: y_{d-medium}
while review of dynamic-data for all x_{d-hard} do
   Management of dataframe with dynamic-data
end while
Compute statistical modelling method: y_{d-hard}
Data structures with regression analysis are updated
Check MIN(y_{i-soft}, y_{i-medium}, y_{i-hard})
Check MIN(y_{d-soft}, y_{d-medium}, y_{d-hard})
Update probabilities for intensification scheme
Update probabilities for diversification scheme
```

4. Experimental Results

This section describes the experimentation process to evaluate the performance of our proposed LB^2. In this work we make use of 15 standard benchmark test functions. These benchmark are described in Section 4.1, and the experimental setup is described in Section 4.2 along the respective analysis in Section 4.3.

4.1. Benchmark Test Functions

In order to test the performance and demonstrate the efficiency of our proposed hybrid approach, we applied 15 well-known benchmark function, Table 3. These function are divided into three main categories, such as unimodal [45] represented in Equations (11)–(14) and Figures 4 and 5, multimodal [46] represented in Equations (15)–(19) and Figures 6 and 7, and fixed-dimension multimodal [45,46], Equations (20)–(25) and Figures 8 and 9. Regarding the features of these functions, f_1 to f_9 are high-dimensional problems. On the other hand, f_{10} to f_{15} comprehends low-dimensional problems. Additionally, all test functions reflect different degrees of complexity, f_1 to f_4 are convex, f_7, f_{11}, and f_{13} are non-convex, f_5, f_6, and f_8 are non-linear functions. Regarding the justification behind the selection of this set of functions, f_1 to f_4 have only one global optimum and has no local optima, which makes this first group of functions highly appropriate to study the convergence rate and intensification ability of our proposed approach. Additionally, f_5 to f_{15} concerns large search space and multiple local solutions besides the global optimum. Thus, they are useful evaluating how efficient the approach is avoiding local optima and the diversification abilities. Additionally, it is well-known that functions from the second group, f_5 to f_9, correspond to a group of very difficult problems to solve for optimisation algorithms, where there is an exponentially increase in number of dimensions [47]. Finally, all these functions are minimisation problems.

Table 3. Optimum values reported for the benchmark functions in the literature, with their corresponding solutions and search subsets.

Function	Search Subsets	Opt	Sol
$f_1(x)$	$[-100, 100]^{30}$	0	$[0]^{30}$
$f_2(x)$	$[-10, 10]^{30}$	0	$[0]^{30}$
$f_3(x)$	$[-100, 100]^{30}$	0	$[0]^{30}$
$f_4(x)$	$[-30, 30]^{30}$	0	$[1]^{30}$
$f_5(x)$	$[-500, 500]^{30}$	−12,596.487	$[420.9687]^{30}$
$f_6(x)$	$[-5.12, 5.12]^{30}$	0	$[0]^{30}$
$f_7(x)$	$[-32, 32]^{30}$	0	$[0]^{30}$
$f_8(x)$	$[-600, 600]^{30}$	0	$[0]^{30}$
$f_9(x)$	$[-50, 50]^{30}$	0	$[1]^{30}$
$f_{10}(x)$	$[-65.536, 65.536]^2$	1	$[-32]^2$
$f_{11}(x)$	$[-5, 5]^2$	−1.0316285	$(0.08983, -0.7126)$ and $(-0.08983, 0.7126)$
$f_{12}(x)$	$[-5, 10]$ for x_1 and $[0, 15]$ for x_2	0.397887	$(-3.142, 12.275)$, $(3.142, 2.275)$, and $(9.425, 2.425)$
$f_{13}(x)$	$[-2, 2]^2$	3	$(0, -1)$
$f_{14}(x)$	$[0, 1]^3$	−3.86	$(0.114, 0.556, 0.852)$
$f_{15}(x)$	$[0, 1]^6$	−3.32	$(0.201, 0.150, 0.477, 0.275, 0.275, 0.377, 0.657)$

Unimodal functions:

Sphere Function

$$f_1(x) = f(x_1, x_2, ..., x_n) = \sum_{i=1}^{n} x_i^2 \tag{11}$$

Schwefel's Function No. 2.22

$$f_2(x) = \sum_{i=1}^{n} |x_i| + \prod_{i=1}^{n} |x_i| \tag{12}$$

Schwefel's Function No. 1.2

$$f_3(x) = \sum_{i=1}^{n} \left(\sum_{j=1}^{i} x_j \right)^2 \tag{13}$$

Generalised Rosenbrock's Function

$$f_4(x) = \sum_{i=1}^{n-1} \left[100(x_i^2 - x_{i+1})^2 + (1 - x_i)^2 \right] \tag{14}$$

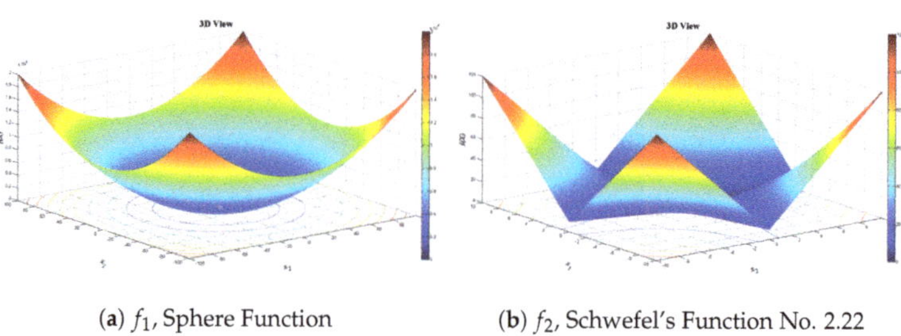

(**a**) f_1, Sphere Function (**b**) f_2, Schwefel's Function No. 2.22

Figure 4. Unimodal benchmark mathematical functions f_1 and f_2 in a 3D view.

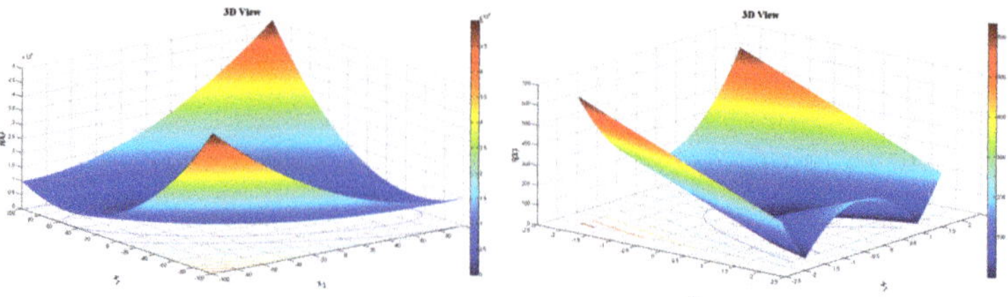

(**a**) f_3, Schwefel's Function No. 1.2 (**b**) f_4, Generalised Rosenbrock's Function

Figure 5. Unimodal benchmark mathematical functions f_3 and f_4 in a 3D view.

Multimodal functions:

Generalised Schwefel's Function No. 2.26

$$f_5(x) = -\sum_{i=1}^{n} x_i \sin(\sqrt{|x_i|}) \tag{15}$$

Generalised Rastrigin's Function

$$f_6(x) = 10n + \sum_{i=1}^{n} (x_i^2 - 10\cos(2\pi x_i)) \tag{16}$$

Ackley's Function

$$f_7(x) = -20\exp(-0.2\sqrt{\frac{1}{n}\sum_{i=1}^{n} x_i^2}) - \exp(\frac{1}{n}\sum_{i=1}^{n}\cos(2\pi x_i)) + 20 + \exp(1) \tag{17}$$

Generalised Griewank's Function

$$f_8(x) = 1 + \sum_{i=1}^{n} \frac{x_i^2}{4000} - \prod_{i=1}^{n} \cos(\frac{x_i}{\sqrt{i}}) \tag{18}$$

Generalised Penalised Function

$$f_9(x) = \frac{\pi}{n} \times \left\{ 10\sin^2(\pi y_1) + \sum_{i=1}^{n-1} (y_i - 1)^2 \left[1 + 10\sin^2(\pi y_{i+1}) \right] + (y_n - 1)^2 \right\} + \sum_{i=1}^{n} u(x_i, 10, 100, 4) \tag{19}$$

where $u(x_i, a, k, m)$ is equal to

1. $k(x_i - a)^m$ if $x_i > a$
2. 0 if $-a \leq x_i \leq a$
3. $k(-x_i - a)^m$ if $x_i < -a$

and

1. $y_i = 1 + \frac{1}{4}(x_i + 1)$

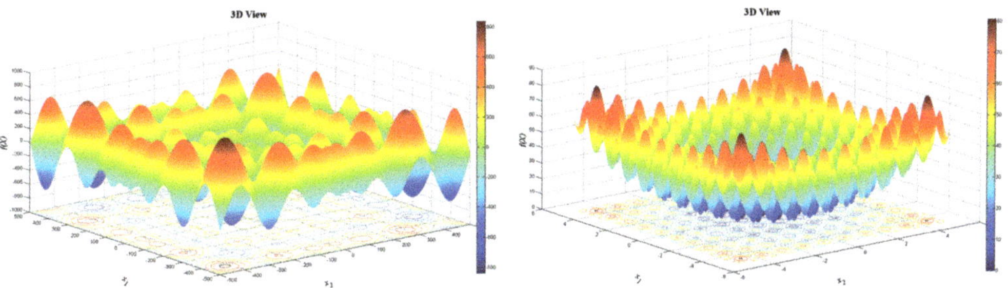

(**a**) f_5, Generalised Schwefel's Function No. 2.26 (**b**) f_6, Generalised Rastrigin's Function

Figure 6. Multimodal benchmark mathematical functions f_5 and f_6 in a 3D view.

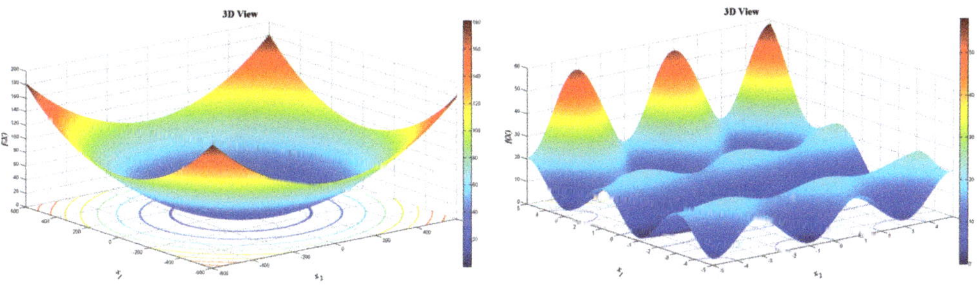

(**a**) f_8, Generalised Griewank's Function (**b**) f_9, Generalised Penalised Function No. 01

Figure 7. Multimodal benchmark mathematical functions f_8 and f_9 in a 3D view.

Multimodal functions with fixed dimensions:

Shekel's Foxholes Function

$$f_{10}(x) = \left[\frac{1}{500} + \sum_{j=1}^{25} \frac{1}{j + \sum_{i=1}^{2} (x_i - a_{i,j})^6}\right]^{-1} \tag{20}$$

where:

$$a_{i,j} = \begin{bmatrix} -32 & -16 & 0 & 16 & 32 & -32 & \ldots & 0 & 16 & 32 \\ -32 & -32 & -32 & -32 & -32 & -16 & \ldots & 32 & 32 & 32 \end{bmatrix}$$

Six-hump Camel Back Function

$$f_{11}(x) = 4x_1^2 - 2.1x_1^4 + \frac{1}{3}x_1^6 + x_1x_2 - 4x_2^2 + 4x_2^4 \tag{21}$$

Branin's Function

$$f_{12}(x) = \left(x_2 - \frac{5.1x_1^2}{4\pi^2} + \frac{5x_1}{\pi} - 6\right)^2 + 10\left(1 - \frac{1}{8\pi}\right)\cos(x_1) + 10 \tag{22}$$

Goldstein-Price Function

$$\begin{aligned} f_{13}(x) = &\left[1 + (x_1 + x_2 + 1)^2\left(19 - 14x_1 + 3x_1^2 - 14x_2 + 6x_1x_2 + 3x_2^2\right)\right] \\ &\left[30 + (2x_1 - 3x_2)^2\left(18 - 32x_1 + 12x_1^2 + 48x_2 - 36x_1x_2 + 27x_2^2\right)\right] \end{aligned} \tag{23}$$

Hartman's Function No.1

$$f_{14}(x) = -\sum_{i=1}^{4} c_i\, e^{\left[-\sum_{j=1}^{3} a_{i,j}(x_j - p_{i,j})^2\right]} \tag{24}$$

where the values of a, c, and p are tabulated in Table. 4

Table 4. Values of a_{ij}, c_i, and p_{ij} for function $f_14(x)$; $n = 3$ and $j = 1, 2, 3$.

i	a_{ij}			c_i	p_{ij}		
1	3	10	30	1	0.3689	0.1170	0.2673
2	0.1	10	35	1.2	0.4699	0.4387	0.7470
3	3	10	30	3	0.1091	0.8732	0.5547
4	0.1	10	30	3.2	0.03815	0.5743	0.8828

Hartman's Function No.2

$$f_{15}(x) = -\sum_{i=1}^{4} c_i\, e^{\left[-\sum_{j=1}^{6} a_{i,j}(x_j - p_{i,j})^2\right]} \tag{25}$$

where the values of a, c and p are tabulated in Table 5.

Table 5. Values of a_{ij}, c_i, and p_{ij} for function $f_15(x)$; $n = 6$ and $j = 1, 2, \ldots, 6$.

i	a_{ij}						c_i	p_{ij}					
1	10	3	17	3.5	1.7	8	1	0.131	0.169	0.556	0.012	0.828	0.588
2	0.05	10	17	0.1	8	14	1.2	0.232	0.413	0.830	0.373	0.100	0.999
3	3	3.5	1.7	10	17	8	3	0.234	0.141	0.352	0.288	0.304	0.665
4	17	8	0.05	10	0.1	14	3.2	0.404	0.882	0.873	0.574	0.109	0.038

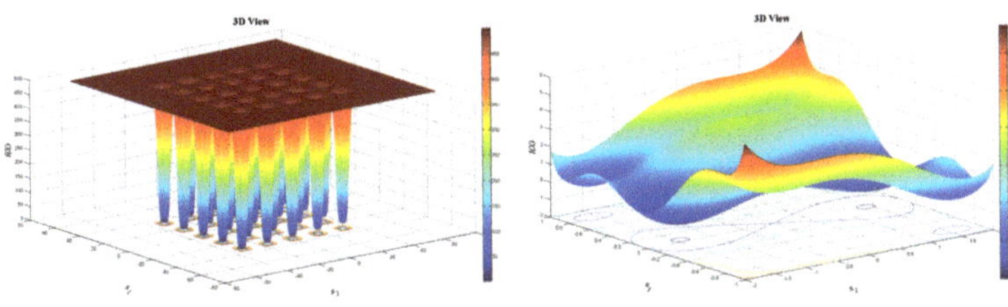

(**a**) f_{10}, Shekel's Foxholes Function (**b**) f_{11}, Six-hump Camel Back Function

Figure 8. Multimodal functions with fixed dimensions f_{10} and f_{11} in a 3D view.

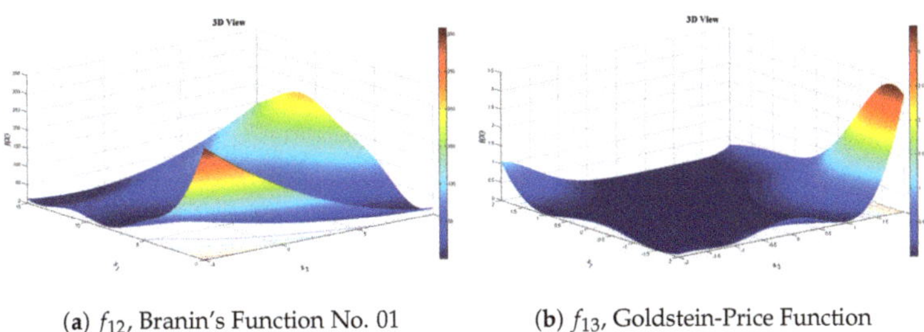

(**a**) f_{12}, Branin's Function No. 01 (**b**) f_{13}, Goldstein-Price Function

Figure 9. Multimodal functions with fixed dimensions f_{12} and f_{13} in a 3D view.

4.2. Algorithms Used for Comparison and Experimental Setup

In order to compare the results obtained, we designed this step in three phases. First phase, we compare against state-of-the-art optimisation methods reported in [18, 19,48], such as particle swarm optimisation (PSO) [15], gravitational search algorithm (GSA) [16], differential evolution (DE) [17], whale optimisation algorithm (WOA) [18], vapour–liquid equilibrium (VLE) [19], and an hybrid between Nelder–Mead algorithm and dragonfly algorithm (INMDA) [20]. In the second phase we compare against SPO [21], which is a heuristic that combines classical and modern statistical techniques to improve the performance of search algorithms. Finally, we take a closer look at the performance achieved by the traditional SHO, a neural network (NN) [22], and a sine cosine algorithm (SCA) [23] approach solving the benchmark functions in comparison with our proposed approach. Regarding the implementation of traditional SHO, the number of search agents was set to 30, control parameter h with values in range of [5, 0], the constant M in the range of [0.5, 1], and the value for number of generations was 10,000. Regarding the neural network, the design was defined as follows: For each benchmark function, a total of one million randomly generated solutions were created. On the other hand, we designed a multi-layer perceptron. The main components comprehend an input node, 7 hidden layers of 50 nodes, and an output equal to the number of dimensions for each function. The training was carried out employing the gradient descent method [49] over 1000 iterations for each randomly generated solution. The main objective behind the NN proposed is the prediction of better function values on run-time. The implementation was performed in python 3.7 and run in an environment windows 10 with 64 bits on Core i-5 processor with 2.40 GHz and 8 GB memory. Finally, regarding the experimentation phase, for each benchmark function the algorithm utilises 30 independent runs.

4.3. Performance Comparison

In this subsection, we illustrate and demonstrate the performance of our proposed LB^2 tackling the benchmark functions described in the Section 4.1.

4.3.1. First Experimentation Phase

First, all results obtained by SHO and LB^2 were rounded to four decimals. The results published for PSO, GSA, DE, WOA, and VLE, were rounded in Tables 6–8 to four decimals, using scientific notation, only for presentation purposes. However, all computations were carried out using the reported decimals by their respective authors. Regarding the performance on unimodal functions, Table 6 illustrates the results and comparison in f_1 to f_4, the average (Avg) and standard deviation (StdDev) are presented and compared, and we highlight in bold the best values reached. Additionally, it is well-known that unimodal functions can help us to measure the exploitation capabilities of our proposed approach. In this regard, it is surprising how good LB^2 performed. It was the second most efficient algorithm tackling this set of benchmark function just behind INMDA. Additionally, the small values reached corresponding to the StdDev shows that it is a very solid algorithm. Concerning the multimodal functions and multimodal functions with fixed dimensions, both sets can help us to evaluate the potential of our algorithm in carrying out the exploration. Tables 7 and 8 illustrates the results and comparison on functions f_5 to f_9 and f_{10} to f_{15} correspondingly, the average (Avg) and standard deviation (StdDev) are presented, and we highlight in bold the best values reached. Surprisingly, the LB^2 attained really good results and small Avg and StdDev values once again, proving to be a competitive approach able to tackle continuous problems. Moreover, having a relatively good performance in these three previous set of benchmark test functions, we can conclude that this first attempt corresponding to the LB^2 has the potential to be a competitive approach.

4.3.2. Second Experimentation Phase

In this subsection, we compare against SPO, which has proved to be a good and competitive option in the field of parameter tuning. In this work, we compare the results obtained by our LB^2 against the works reported and implemented in [21]. They implemented a PSO and a PSO + SPO approach, they solve 4 benchmark test functions. Table 9 illustrates the best values reached, where the first column, named problem, represent the 4 functions solved by the approach reported (2 unimodal and 2 multimodal). Column 2, 3, and 4, represent the best values achieved by PSO and PSO + SPO (both implemented by the authors), and our proposed LB^2. It is clear the superiority of our approach reaching all 4 optimum values. However, in future work, in order to improve the hybrid methodology proposed in this work, we have as an objective the implementation and comparison of SPO and F-Race approaches in order to bring a more detailed and larger competition between multiple optimisation and tuning tools.

Table 6. Results comparison in unimodal benchmark functions.

F	SHO		LB^2		WOA		DE		GSA		PSO		VLE		INMDA	
	Avg	StdDev	Avg	StdDev	Avg	StdDev	Avg	StdDev	Avg	StdDev	Avg	StdDev	Avg	StdDev	Avg	StdDev
f_1	0.0006	0.0005	**0.0000**	**0.0000**	**0.0000**	**0.0000**	8.2000×10^{-14}	5.9000×10^{-14}	2.5300×10^{-16}	0.0000	1.3600×10^{-4}	2.0200×10^{-4}	4.4989×10^{-7}	1.413×10^{-6}	**0.0000**	**0.0000**
f_2	**0.0000**	**0.0000**	**0.0000**	**0.0000**	**0.0000**	**0.0000**	1.5000×10^{-9}	9.9000×10^{-10}	5.5655×10^{-2}	0.1941	4.2144×10^{-2}	4.5421×10^{-2}	3.0840×10^{-6}	6.0498×10^{-6}	**0.0000**	**0.0000**
f_3	0.0007	0.0005	**0.0000**	**0.0000**	5.3900×10^{-7}	2.9300×10^{-6}	6.8000×10^{-11}	7.4000×10^{-11}	8.9353×10^{2}	3.1896×10^{2}	70.126	22.119	5.2020	0.7986	**0.0000**	**0.0000**
f_4	2.7511	0.0502	6.7549×10^{-7}	5.4204×10^{-7}	27.866	0.7636	**0.0000**	**0.0000**	67.543	62.225	96.718	60.116	79.199	37.400	**0.0000**	**0.0000**

Table 7. Results comparison in multimodal benchmark functions.

F	SHO		LB^2		WOA		DE		GSA		PSO		VLE		INMDA	
	Avg	StdDev	Avg	StdDev	Avg	StdDev	Avg	StdDev	Avg	StdDev	Avg	StdDev	Avg	StdDev	Avg	StdDev
f_5	-1.0867×10^{4}	0.5059	$\mathbf{-1.2569 \times 10^{4}}$	0.0014	-5.0808×10^{3}	6.9580×10^{2}	-1.1080×10^{4}	5.7470×10^{2}	-2.8211×10^{3}	4.9304×10^{2}	-4.8413×10^{3}	1.1528×10^{3}	-1.2566×10^{4}	68.705	-2245.1500	2.8400
f_6	**0.0000**	**0.0000**	**0.0000**	**0.0000**	**0.0000**	**0.0000**	69.200	38.800	25.968	7.4701	46.704	11.629	34.5830	17.8860	**0.0000**	**0.0000**
f_7	4.4408×10^{-16}	0.0000	4.4409×10^{-16}	0.0000	7.4043	9.8976	9.7000×10^{-8}	4.2000×10^{-8}	6.2087×10^{-2}	0.23628	0.27602	0.50901	3.1704	3.9211	0.0000	1.6200×10^{-16}
f_8	**0.0000**	**0.0000**	**0.0000**	**0.0000**	2.8900×10^{-4}	1.5860×10^{-3}	**0.0000**	**0.0000**	27.702	5.0403	9.2150×10^{-3}	7.7240×10^{-3}	0.5074	0.5041	**0.0000**	**0.0000**
f_9	1.906	0.0865	1.8286	1.5[illegible]85 $\times 10^{-9}$	0.3397	0.2149	7.9000×10^{-15}	8.0000×10^{-15}	1.7996	0.95114	6.9170×10^{-3}	2.6301×10^{-2}	0.2369	0.2877	0.0000	0.0000

Table 8. Results comparison in multimodal benchmark functions with fixed-dimension.

F	SHO		LB^2		WOA		DE		GSA		PSO		VLE		INMDA	
	Avg	StdDev	Avg	StdDev	Avg	StdDev	Avg	StdDev	Avg	StdDev	Avg	StdDev	Avg	StdDev	Avg	StdDev
f_{10}	2.1326×10^{-8}	5.0161×10^{-10}	**1.0000**	**0.0000**	2.1120	2.4986	0.99800	3.3000×10^{-16}	5.8598	3.8313	3.6272	2.5608	0.99800	2.5294×10^{-7}	N/A	N/A
f_{11}	0.0000	0.0000	0.0000	0.0000	-1.0316	4.2000×10^{-7}	$\mathbf{-1.0316}$	3.1000×10^{-13}	$\mathbf{-1.0316}$	4.8800×10^{-16}	$\mathbf{-1.0316}$	6.2500×10^{-16}	-1.0315	1.8408×10^{-4}	N/A	N/A
f_{12}	0.8718	0.0502	1.5436	0.4223	0.39791	2.7000×10^{-5}	0.39789	9.9000×10^{-9}	0.39789	0.0000	0.39789	0.0000	0.39815	4.5697×10^{-4}	N/A	N/A
f_{13}	36.0716	4.1607	32.6845	1.4854×10^{-8}	**3.0000**	4.2200×10^{-15}	**3.0000**	2.0000×10^{-15}	**3.0000**	4.1700×10^{-15}	**3.0000**	1.3300×10^{-15}	3.0097	1.6256×10^{-2}	N/A	N/A
f_{14}	-2.1211	0.1284	-2.0081	5.0800×10^{-10}	-3.8562	2.7060×10^{-3}	N/A	N/A	-3.8628	2.2900×10^{-15}	-3.8628	2.5800×10^{-15}	-3.8628	6.6880×10^{-5}	N/A	N/A
f_{15}	-0.8515	0.3541	-1.5870	0.5016	-2.9811	0.37665	N/A	N/A	-3.3178	2.3081×10^{-2}	-3.2663	6.0516×10^{-2}	-3.3179	2.1311×10^{-2}	N/A	N/A

Table 9. Comparison results of SPO against LB^2 proposed.

Problem	PSO	PSO + SPO	LB^2
Sphere	2.82×10^{-9}	1.66×10^{-21}	0
Rosenbrock	148.84	4.20	0
Rastrigin	10.43	0.98	0
Griewangk	0.12	0.07	0

4.3.3. Third Experimentation Phase

In this subsection, we present a comparison of LB^2 against a traditional SHO and a neural network approach, both implementation made by us. Tables 10–12 illustrates a summary of the values achieved in the experimentation phase. Column F corresponds to the test function solved, and Opt depicts the global optimum for the given function. Column best, worst, and Avg are the given values for best value reached, worst value reached, the mean value, and the Avg time achieved in 30 executions.

Regarding Table 10, LB^2 achieved 7 optimum values in f_1–f_3, f_5, f_6, f_8, and f_{10}, in comparison of SHO which achieved 5 optimum values in f_1–f_3, f_6, f_8. Additionally, regarding non-optimum values reached, our proposed approach is superior in 5 values in functions f_4, f_5, f_9,f_{13}, and f_{15}. However, the same goes for SHO in functions f_{11}, f_{12}, and f_{14}. Regarding Table 9, LB^2 achieve better values than the NN approach implemented. However, in functions f_{13} and f_{14} the performance of our proposed approach falls behind considerably. Regarding Table 12, small differences can be observed, LB^2 reached 1 more optimum value. However, the biggest difference concerns the robustness in the overall performance illustrated on columns Avg and StdDev. This can be observed in functions f_4, f_5, and f_{10}.

Regarding the average time achieved in the three illustrated tables, significant difference can be observed between NN, SCA, and LB^2. In the hardest test function, multimodal, and multimodal with fixed-dimension, NN falls significantly behind against SCA and LB^2 in solving time, which is the strong point on these types of algorithms. Additionally, we highlight the drawback behind a NN approach, the costly process of training and tuning of the model. Nevertheless, the objective behind this test, presented in Table 11, concerns the future incorporation of new learning methods to LB^2.

Regarding the room for improvements observed in the performance, values achieved in column StdDev for f_5, f_{11}, f_{14}, and f_{15} can be interpreted as the approach being trapped in local optima. The discussion follows two possible issues: the value employed as β and the scheme values for the diversification process. Firstly, the proposed value for threshold β is static through the search, the consequence can be interpreted as the approach expecting a more balanced and timely feedback from the learning model. Thus, when a local optima is detected, a proper answer can be delivered and carried out on run-time. Nevertheless, the incorporation of a learning-based component managing a dynamic β value on run-time will be proposed in order to tackle this issue. On the other hand, regarding the scheme values for diversification, the employment of static values through the search can be a critical issue. The amount and frequency on which diversification is carried out will be our next focus as a balanced exploration in the search space needs to be performed.

In order to further analyse and demonstrate the improvement in the performance of the hybridisation in optimisation tools, a statistical analysis is carried out. To this end we compare convergence and we analyse the 30 executions performed for each function through the Kolmogorov Smirnov Lilliefors (Lilliefors 1967) [50] and Wilcoxon's signed rank (Mann and Whitney 1947) [51] statistical tests. Additionally, in order to carry the statistical analysis of this phase, we make use of the RStudio software to conduct both tests.

Table 10. Results comparison of SHO vs. LB^2.

F	Opt	SHO					LB^2				
		Best	Worst	Avg	StdDev	Avg Time(s)	Best	Worst	Avg	StdDev	Avg Time(s)
f_1	0	**0**	0.0021	0.0006	0.0005	51.6432	**0**	0	0	0	50.2377
f_2	0	**0**	0	0	0	78.9275	**0**	0	0	0	80.7524
f_3	0	**0**	0.0009	0.0007	0.0005	95.5684	**0**	0	0	0	96.3627
f_4	0	2.7091	2.9351	2.7511	0.0502	75.8810	1.59197×10^{-7}	1.2262×10^{-6}	6.7549×10^{-7}	5.4204×10^{-7}	71.0024
f_5	−12,569.487	-1.1318×10^4	-0.9653×10^4	-1.0867×10^4	0.5059	121.3511	-1.2570×10^4	-1.2567×10^4	-1.2569×10^4	0.0014	110.3354
f_6	0	**0**	0	0	0	172.9312	**0**	0	0	0	60.6482
f_7	0	4.4408×10^{-16}	4.4408×10^{-16}	4.4408×10^{-16}	0	256.8700	4.4408×10^{-16}	4.4409×10^{-16}	4.4409×10^{-16}	0	24.9122
f_8	0	**0**	0	0	0	198.8546	**0**	0	0	0	21.7758
f_9	0	1.8290	2.5642	1.906	0.0865	256.8707	1.8285	1.8286	1.8286	1.5985×10^{-9}	24.9172
f_{10}	1	2.1745×10^{-8}	2.0745×10^{-8}	2.1326×10^{-8}	5.0161×10^{-10}	130.3552	**1**	1	1	0	17.5661
f_{11}	−1.0316	0	0	0	0	29.1582	0	0	0	0	7.5244
f_{12}	0.3979	0.8298	0.9523	0.8718	0.0502	22.5778	1.1905	2.0325	1.5436	0.4223	4.5528
f_{13}	3	32.6845	44.4562	36.0716	4.1607	35.7789	32.6845	32.6845	32.6845	1.4854×10^{-8}	3.6846
f_{14}	−3.86	−2.4301	−2.0081	−2.211	0.1284	53.2235	−2.0081	−2.0080	−2.0081	5.0800×10^{-10}	7.1120
f_{15}	−3.32	−1.1676	−0.4676	−0.8515	0.3541	80.4755	−2.1676	−2.1676	−2.1676	0	8.1145

Table 11. Results comparison of NN vs. LB^2.

F	Opt	NN					LB^2				
		Best	Worst	Avg	StdDev	Avg Time(s)	Best	Worst	Avg	StdDev	Avg Time(s)
f_1	0	0.0639	0.2223	0.1068	0.0435	347.4073	**0**	0	0	0	50.2377
f_2	0	1.2426	5.8827	4.4004	0.8284	375.2112	**0**	0	0	0	80.7524
f_3	0	0.0001	0.0379	0.0103	0.0108	377.0420	**0**	0	0	0	96.3627
f_4	0	211.4253	3037.6363	1376.6472	1041.5627	375.6543	1.59197×10^{-7}	1.2262×10^{-6}	6.7549×10^{-7}	5.4204×10^{-7}	71.0024
f_5	−12,569.487	-1.2557×10^4	-1.7363×10^4	-1.2057×10^4	2056.9973	357.8577	-1.2570×10^4	-1.2567×10^4	-1.2569×10^4	0.0014	110.3354
f_6	0	1.8672	7.8028	4.2664	1.5939	357.7703	**0**	0	0	0	60.6482
f_7	0	0.2687	0.5169	0.3905	0.0689	350.7136	4.4408×10^{-16}	4.4409×10^{-16}	4.4409×10^{-16}	0	24.9122
f_8	0	0.0416	1.1238	0.8017	0.1986	354.9282	**0**	0	0	0	21.7758
f_9	0	29,752,063.66	29,800,464.52	29,792,019.73	9855.6445	357.1411	1.8285	1.8286	1.8286	1.5985×10^{-9}	24.9172
f_{10}	1	0.0160	495.8931	214.7364	200.8041	343.9354	**1**	1	1	0	17.5661
f_{11}	−1.0316	−0.0079	0.0103	0.0005	0.0041	344.4559	0	0	0	0	7.5244
f_{12}	0.3979	10.0004	16.3393	12.2766	1.3941	369.1712	1.1905	2.0325	1.5436	0.4223	4.5528
f_{13}	3	3.0227	5.4062	5.3044	1.5586	369.8705	32.6845	32.6845	32.6845	1.4854×10^{-8}	3.6846
f_{14}	−3.86	−3.8417	−3.5163	−3.7177	0.0913	376.6538	−2.0081	−2.0080	−2.0081	5.0800×10^{-10}	7.1120
f_{15}	−3.32	−1.4809	−0.7560	−1.0409	0.2019	342.6122	−2.1676	−2.1676	−2.1676	0	8.1145

Table 12. Results comparison of SCA vs. LB^2.

F	Opt	SCA					LB^2				
		Best	**Worst**	**Avg**	**StdDev**	**Avg Time(s)**	**Best**	**Worst**	**Avg**	**StdDev**	**Avg time(s)**
f_1	0	**0**	**0**	**0**	**0**	10.4656	**0**	0	0	0	50.2377
f_2	0	**0**	**0**	**0**	**0**	17.3875	**0**	0	0	0	80.7524
f_3	0	**0**	**0**	**0**	**0**	74.3906	**0**	0	0	0	96.3627
f_4	0	1.33×10^{-10}	29	17.4000	14.9755	13.4296	1.59197×10^{-7}	1.2262×10^{-6}	6.7549×10^{-7}	5.4204×10^{-7}	71.0024
f_5	−12,569.487	2.51×10^{-7}	0.0696	0.0181	0.0251	9.5828	-1.2570×10^4	-1.2567×10^4	-1.2569×10^4	0.0014	110.3354
f_6	0	**0**	**0**	**0**	**0**	11.2296	**0**	0	0	0	60.6482
f_7	0	4.44×10^{-16}	4.44×10^{-16}	4.44×10^{-16}	0	15.1140	4.4408×10^{-16}	4.4409×10^{-16}	4.4409×10^{-16}	0	24.9122
f_8	0	**0**	**0**	**0**	**0**	13.3500	**0**	0	0	0	21.7758
f_9	0	1.8285	1.8416	1.8304	0.0042	78.2046	1.8285	1.8286	1.8286	1.5985×10^{-9}	24.9172
f_{10}	1	0.0003	4.9301	1.0010	2.0705	28.9609	**1**	1	1	0	17.5661
f_{11}	−1.0316	1.0316	1.0316	1.0316	2.3406×10^{-16}	2.7093	0	0	0	0	7.5244
f_{12}	0.3979	0.1555	3.9503	1.1009	1.3759	2.9765	1.1905	2.0325	1.5436	0.4223	4.5528
f_{13}	3	29.6845	30.0547	29.7444	0.1229	4.1765	32.6845	32.6845	32.6845	1.4854×10^{-8}	3.6846
f_{14}	−3.86	1.8519	1.8519	1.8519	0	7.0265	−2.0081	−2.0080	−2.0081	5.0800×10^{-10}	7.1120
f_{15}	−3.32	2.1523	2.1524	2.1524	0	9.5578	−2.1676	−2.1676	−2.1676	0	8.1145

The process is as follows, samples were tested for normality using Kolmogorov Smirnov Lilliefors test, having failed it (p-values > 0.05). Therefore, the non-parametric Mann-Whitney test subsequently used to compare the quality of SHO and LB^2 results. We need to take in consideration the next two hypothesis:

$$H_0: \; \mu_{SHO} = \mu_{LB^2}$$

$$H_1: \; \mu_{LB^2} \neq \mu_{SHO}$$

where μ_{SHO} and μ_{LB^2} are the arithmetic median of fitness values achieved corresponding to SHO and our proposed LB^2. Again, at this next step we take into consideration that the significance level is also established to 0.05, thus, smaller values that 0.05 defines that H_0 cannot be assumed. In this regard, Table 13 illustrate the comparison between the two implementations, we highlight in bold the values where there is a statistically significant winner.

Table 13. Exact p values obtained on the benchmark test functions.

F		SHO	LB^2
f_1	SHO	-	>0.05
	LB^2	>0.05	-
f_2	SHO	-	>0.05
	LB^2	>0.05	-
f_3	SHO	-	>0.05
	LB^2	>0.05	-
f_4	SHO	-	$\mathbf{2.35 \times 10^{-18}}$
	LB^2	>0.05	-
f_5	SHO	-	$\mathbf{6.611 \times 10^{-7}}$
	LB^2	>0.05	-
f_6	SHO	-	>0.05
	LB^2	>0.05	-
f_7	SHO	-	>0.05
	LB^2	>0.05	-
f_8	SHO	-	>0.05
	LB^2	>0.05	-
f_9	SHO	-	$\mathbf{7.01 \times 10^{-7}}$
	LB^2	>0.05	-
f_{10}	SHO	-	$\mathbf{1.1 \times 10^{-7}}$
	LB^2	>0.05	-
f_{11}	SHO	-	**0.02067**
	LB^2	>0.05	-
f_{12}	SHO	-	>0.05
	LB^2	**0.04**	-
f_{13}	SHO	-	>0.05
	LB^2	>0.05	-
f_{14}	SHO	-	$\mathbf{1.395 \times 10^{-6}}$
	LB^2	>0.05	-
f_{15}	SHO	-	$\mathbf{1.863 \times 10^{-9}}$
	LB^2	>0.05	-

5. Conclusions and Future Work

In this paper, a novel learning-based framework was proposed. Well-known methods and techniques are employed to design a competitive hybrid approach capable to tackle on optimisation problems. The proposed framework performs under a population-based strategy, multiple agents explore, learn, and evolve in the search space. In this regard, two main components were employed: a population-based algorithm, named spotted hyena optimiser, and a learning model which is based in a statistical modelling method.

Regarding the results achieved solving the benchmark functions, LB^2 demonstrated to be a competitive method and a promising alternative to tackle optimisation problems. However, some issues remains and improvements can be proposed. Firstly, LB^2 needs to be tested tackling benchmark functions with higher difficulty. In this regard, we are considering more complex functions with higher dimensionality, such as CEC 2021's composite functions. Additionally, the incorporation of hard optimisation problems, such as set covering problem (SCP), manufacturing cell design problem (MCDP) are being considered as future testing objectives. On another hand, results illustrated in the third experimentation phase can be interpreted as LB^2 being trapped in local optima for certain functions. Nevertheless, improvements can be carried out in order to tackle this issue. In this regard, new learning-based components will be proposed. The main objective is to dynamically adjust parameters, such as threshold β and the scheme for diversification and intensification. The idea is to keep the balance in the feedback of dynamic data and knowledge generated between the population and the learning model on run-time. Finally, new learning methods will be implemented, the objective concerns the viability, certainty, and confidence in the generated knowledge. Thus, a more complex component will be designed in order to measure the profit behind the knowledge for a better decision making through the search.

Author Contributions: Formal analysis, E.V., J.P., and R.S.; investigation, E.V., J.P., R.S., B.C., and C.C.; resource, R.S.; software, E.V. and J.P.; validation, B.C. and C.C.; writing–original draft, E.V., R.S., and B.C.; writing–review and editing, E.V. and R.S. All authors have read and agreed to the published version of the manuscript.

Funding: Ricardo Soto is supported by Grant CONICYT/FONDECYT/REGULAR/1190129. Broderick Crawford is supported by Grant ANID/FONDECYT/REGULAR/1210810, and Emanuel Vega is supported by National Agency for Research and Development ANID/Scholarship Program/DOCTORADO NACIONAL/2020-21202527.

Institutional Review Board Statement: Not applicable.

Informed Consent Statement: Not applicable.

Data Availability Statement: No new data were created or analysed in this study. Data sharing is not applicable to this article.

Conflicts of Interest: The authors declare no conflict of interest. The founding sponsors had no role in the design of the study; in the collection, analyses, or interpretation of data; in the writing of the manuscript, and in the decision to publish the results.

References

1. Talbi, E.G. Combining metaheuristics with mathematical programming, constraint programming and machine learning. *Ann. Oper. Res.* **2016**, *240*, 171–215. [CrossRef]
2. Gendreau, M.; Potvin, J.Y. *Handbook of Metaheuristics*, 2nd ed.; Springer: Berlin/Heidelberg, Germany, 2010.
3. Hussain, K.; Salleh, M.N.M.; Cheng, S.; Shi, Y. On the exploration and exploitation in popular swarm-based metaheuristic algorithms. *Neural Comput. Appl.* **2019**, *31*, 7665–7683. [CrossRef]
4. Chu, X.; Wu, T.; Weir, J.D.; Shi, Y.; Niu, B.; Li, L. Learning–interaction–diversification framework for swarm intelligence optimizers: A unified perspective. *Neural Comput. Appl.* **2020**, *32*, 1789–1809. [CrossRef]
5. Boussaïd, I.; Lepagnot, J.; Siarry, P. A survey on optimization metaheuristics. *Inf. Sci.* **2013**, *237*, 82–117. [CrossRef]
6. Tapia, D.; Crawford, B.; Soto, R.; Cisternas-Caneo, F.; Lemus-Romani, J.; Castillo, M.; García, J.; Palma, W.; Paredes, F.; Misra, S. A Q-Learning Hyperheuristic Binarization Framework to Balance Exploration and Exploitation. In *International Conference on Applied Informatics*; Springer: Cham, Switzerland, 2020; pp. 14–28.
7. Parsons, S. Introduction to Machine Learning by Ethem Alpaydin. In *The Knowledge Engineering Review*; MIT Press: Cambridge, MA, USA, 2005; Volume 20, pp. 432–433.
8. Song, H.; Triguero, I.; Özcan, E. A review on the self and dual interactions between machine learning and optimisation. *Prog. Artif. Intell.* **2019**, *8*, 143–165. [CrossRef]
9. Barber, D. *Bayesian Reasoning and Machine Learning*; Cambridge University Press: New York, NY, USA, 2012.

10. Lantz, B. *Machine Learning with R*; Packt Publishing: Birmingham, UK, 2013.
11. Dietterich, T. Machine Learning. *ACM Comput. Surv.* **1996**, *28*, 3. [CrossRef]
12. Dhiman, G.; Kumar, V. Spotted hyena optimizer: A novel bio-inspired based metaheuristic technique for engineering applications. *Adv. Eng. Softw.* **2017**, *114*, 48–70. [CrossRef]
13. Soto, R.; Crawford, B.; Vega, E.; Gómez, A.; Gómez-Pulido, J.A. Solving the Set Covering Problem Using Spotted Hyena Optimizer and Autonomous Search. Advances and Trends in Artificial Intelligence. From Theory to Practice. In *IEA/AIE 2019*; Springer: Cham, Swizterland, 2019; Volume 11606.
14. Luo, Q.; Li, J.; Zhou, Y.; Liao, L. Using spotted hyena optimizer for training feedforward neural networks. *Cogn. Syst. Res.* **2021**, *65*, 1–16. [CrossRef]
15. Kennedy, J.; Eberhart, R. Particle swarm optimization. In Proceedings of the IEEE International Conference on Neural Networks, Perth, Australia, 27 November–1 December 1995; pp. 1942–1948.
16. Rashedi, E.; Nezamabadi-Pour, H.; Saryazdi, S. GSA: A Gravitational Search Algorithm. *Inf. Sci.* **2009**, *179*, 2232–2248. [CrossRef]
17. Storn, R.; Price, K. Differential Evolution—A Simple and Efficient Heuristic for global Optimization over Continuous Spaces. *J. Glob. Optim.* **1997**, *11*, 341–359. [CrossRef]
18. Mirjalili, S.; Lewis, A. The Whale Optimization Algorithm. *Adv. Eng. Softw.* **2016**, *95*, 51–67. [CrossRef]
19. Cortés-Toro, E.M.; Crawford, B.; Gómez-Pulido, J.A.; Soto, R.; Lanza-Gutiérrez, J.M. A New Metaheuristic Inspired by the Vapour-Liquid Equilibrium for Continuous Optimization. *Appl. Sci.* **2018**, *8*, 2080. [CrossRef]
20. Xu, J.; Yan, F. Hybrid Nelder–Mead algorithm and dragonfly algorithm for function optimization and the training of a multilayer perceptron. *Arab. J. Sci. Eng.* **2019**, *44*, 3473–3487. [CrossRef]
21. Bartz-Beielstein, T.; Lasarczyk, C.W.G.; Preuss, M. Sequential parameter optimization. In Proceedings of the 2005 IEEE Congress on Evolutionary Computation, Edinburgh, UK, 2–5 September 2005; Volume 1, pp. 773–780.
22. Wang, R.L.; Tang, Z.; Cao, Q.P. A learning method in Hopfield neural network for combinatorial optimization problem. *Neurocomputing* **2002**, *48*, 1021–1024. [CrossRef]
23. Mirjalili, S. SCA: A sine cosine algorithm for solving optimization problems. *Knowl. Based Syst.* **2016**, *96*, 120–133. [CrossRef]
24. Talbi, E.G. Combining metaheuristics with mathematical programming, constraint programming and machine learning. *4OR Q. J. Belg. Fr. Ital. Oper. Res. Soc.* **2013**, *11*, 101–150. [CrossRef]
25. Talbi, E.G. *Machine Learning into Metaheuristics: A Survey and Taxonomy of Data-Driven Metaheuristics*; Working Paper or Preprint, June 2020.
26. Escalante, H.J.; Ponce-López, V.; Escalera, S.; Baró, X.; Morales-Reyes, A.; Martínez-Carranza, J. Evolving weighting schemes for the bag of visual words. *Neural Comput. Appl.* **2016**, *28*, 925–939. [CrossRef]
27. Stein, G.; Chen, B.; Wu, A.S.; Hua, K.A. Decision tree classifier for network intrusion detection with GA-based feature selection. In Proceedings of the 43rd Annual Southeast Regional Conference, Kennesaw, GA, USA, 18 March 200 ; Volume 2, pp. 136–141.
28. Sörensen, K.; Janssens, G.K. Data mining with genetic algorithms on binary trees. *Eur. J. Oper. Res.* **2003**, *151*, 253–264. [CrossRef]
29. Fernández Caballero, J.C.; Martinez, F.J.; Hervas, C.; Gutierrez, P.A. Sensitivity versus accuracy in multiclass problems using memetic pareto evolutionary neural networks. *IEEE Trans. Neural Netw.* **2010**, *21*, 750–770. [CrossRef]
30. Huang, C.L.; Wang, C.J. A GA-based feature selection and parameters optimization for support vector machines. *Expert Syst. Appl.* **2006**, *31*, 231–240. [CrossRef]
31. Glover, F.; Hao, J.K. Diversification-based learning in computing and optimization. *J. Heuristics* **2019**, *25*, 521–537. [CrossRef]
32. Máximo, V.R.; Nascimento, M.C. Intensification, learning and diversification in a hybrid metaheuristic: An efficient unification. *J. Heuristics* **2019**, *25*, 539–564. [CrossRef]
33. Lessmann, S.; Caserta, M.; Arango, I.M. Tuning metaheuristics: A data mining based approach for particle swarm optimization. *Expert Syst. Appl.* **2011**, *38*, 12826–12838. [CrossRef]
34. Zennaki, M.; Ech-Cherif, A. A new machine learning based approach for tuning metaheuristics for the solution of hard combinatorial optimization problems. *J. Appl. Sci.* **2010**, *10*,1991–2000. [CrossRef]
35. Porumbel, D.C.; Hao, J.K.; Kuntz, P. A search space "cartography" for guiding graph coloring heuristics. *Comput. Oper. Res.* **2010**, *37*, 769–778. [CrossRef]
36. Ribeiro, M.H.; Plastino, A.; Martins, S.L. Hybridization of GRASP metaheuristic with data mining techniques. *J. Math. Model. Algorithms* **2006**, *5*, 23–41. [CrossRef]
37. Dalboni, F.L.; Ochi, L.S.; Drummond, L.M.A. On improving evolutionary algorithms by using data mining for the oil collector vehicle routing problem. In Proceedings of the International Network Optimization Conference, Rio de Janeiro, Brazil, 22 April 2003; pp. 182–188.
38. Amor, H.B.; Rettinger, A. Intelligent exploration for genetic algorithms: Using self-organizing maps in evolutionary computation. In Proceedings of the 7th Annual Conference on Genetic and Evolutionary Computation, Washington DC, USA, 25–29 June 2005; pp. 1531–1538.
39. Yuen, S.Y.; Chow, C.K. A genetic algorithm that adaptively mutates and never revisits. *IEEE Trans. Evol. Comput.* **2008**, *13*, 454–472. [CrossRef]
40. Dhaenens, C.; Jourdan, L. *Metaheuristics for Big Data*; Wiley: Hoboken, NJ, USA, 2016; ISBN 9781119347606.
41. Yang, L.; Shami, A. On Hyperparameter Optimization of Machine Learning Algorithms: Theory and Practice. *Neurocomputing* **2020**, *415*, 295–316. [CrossRef]

42. Caruana, R.; Niculescu-Mizil, A. An empirical comparison of supervised learning algorithms. *ACM Int. Conf. Proc. Ser.* **2006**, *148*, 161–168.
43. Article, R. Linear Regression Analysis. *Dtsch. äRzteblatt Int.* **2010**, *107*, 776–782.
44. Almeida, A.M.D.; Castel-Branco, M.M.; Falcao, A.C. Linear regression for calibration lines revisited: Weighting schemes for bioanalytical methods. *J. Chromatogr. B* **2002**, *774*, 215–222. [CrossRef]
45. Digalakis, J.; Margaritis, K. On benchmarking functions for genetic algorithms. *Int. J. Comput. Math* **2001**, *77*, 481–506. [CrossRef]
46. Yang, X. Firefly algorithm, stochastic test functions and design optimisation. *Int. J. Bio-Inspired Comput.* **2010**, *2*, 78–84. [CrossRef]
47. Yao, X.; Liu, Y.; Lin, G. Evolutionary programming made faster. *IEEE Trans. Evol. Comput.* **1999**, *3*, 82–102.
48. Mirjalili, S.; Mirjalili, S.M.; Lewis, A. Grey Wolf Optimizer. *Adv. Eng. Softw.* **2014**, *69*, 46–61. [CrossRef]
49. Kingma, D.P.; Ba, J. Adam: A method for stochastic optimization. *arXiv* **2014**, arXiv:1412.6980.
50. Lilliefors, H. On the kolmogorov–smirnov test for normality with mean and variance unknown. *J. Am. Stat. Assoc.* **1967**, *62*, 399–402. [CrossRef]
51. Mann, H.; Whitney, D. On a test of whether one of two random variables is stochastically larger than the other. *Ann. Math. Stat.* **1947**, *18*, 50–60. [CrossRef]

 *mathematics*

Article

A Self-Adaptive Cuckoo Search Algorithm Using a Machine Learning Technique

Nicolás Caselli [1,*], Ricardo Soto [1], Broderick Crawford [1], Sergio Valdivia [2] and Rodrigo Olivares [3,*]

1 Escuela de Ingeniería Informática, Pontificia Universidad Católica de Valparaíso, Valparaíso 2362807, Chile; ricardo.soto@pucv.cl (R.S.); broderick.crawford@pucv.cl (B.C.)
2 Dirección de Tecnologías de Información y Comunicación, Universidad de Valparaíso, Valparaíso 2361864, Chile; sergio.valdivia@uv.cl
3 Escuela de Ingeniería Informática, Universidad de Valparaíso, Valparaíso 2362905, Chile
* Correspondence: nicolas.caselli.b@mail.pucv.cl (N.C.); rodrigo.olivares@uv.cl (R.O.)

Abstract: Metaheuristics are intelligent problem-solvers that have been very efficient in solving huge optimization problems for more than two decades. However, the main drawback of these solvers is the need for problem-dependent and complex parameter setting in order to reach good results. This paper presents a new cuckoo search algorithm able to self-adapt its configuration, particularly its population and the abandon probability. The self-tuning process is governed by using machine learning, where cluster analysis is employed to autonomously and properly compute the number of agents needed at each step of the solving process. The goal is to efficiently explore the space of possible solutions while alleviating human effort in parameter configuration. We illustrate interesting experimental results on the well-known set covering problem, where the proposed approach is able to compete against various state-of-the-art algorithms, achieving better results in one single run versus 20 different configurations. In addition, the result obtained is compared with similar hybrid bio-inspired algorithms illustrating interesting results for this proposal.

Keywords: clustering techniques; metaheuristics; machine learning; self-adaptive, parameter setting; exploration; exploitation

Citation: Caselli, N.; Soto, R.; Crawford, B.; Valdivia, S.; Olivares, R. A Self-Adaptive Cuckoo Search Algorithm Using a Machine Learning Technique. *Mathematics* **2021**, *9*, 1840. https://doi.org/10.3390/math9161840

Academic Editor: Frank Werner

Received: 27 June 2021
Accepted: 30 July 2021
Published: 4 August 2021

1. Introduction

Recent studies about bio-inspired procedures to solve complex optimization problems have demonstrated that finding good results and the best performance are laborious tasks, so it is necessary to apply an off-line parameter adjustment on metaheuristics [1–5]. This adjustment is considered an optimization problem itself, and several studies are proposing some solutions to solve that, but it always depends on his static therms [6]. Many of these studies use mathematical ways to change the values of one of each parameter during their execution. In this context, parameters like the population size of metaheuristics are initially set without considering their variation or behaviors. We consider using a machine learning (ML) technique that lets us analyze the population and determine the values of the number of solutions and the abandon probability.

As we can see in [7], machine learning and optimization are two topics of artificial intelligence that are rapidly expanding, having a wide range of computer science applications. Due to the rapid progress in the performance of computing and communication techniques, these two research areas have proliferated and drawn widespread attention in a wide variety of applications [8]. For example, in [9], the authors present different clustering techniques to evaluate the credit risk in a determinate population of Europe. Although both fields belong to different communities, they are fundamentally based on artificial intelligence, and the techniques from ML and optimization frequently interact with each other and themselves to improve their learning and/or search capabilities.

On the other hand, advances in operations research and computer science have brought forward new solution approaches in optimization theory, such as heuristics and metaheuristics. While the former are experience-based procedures, which usually provide good solutions in short computing times, metaheuristics are general templates that can easily be tailored to address a wide range of problems. They have been shown to provide near-optimal solutions in reasonable computing times to problems for which traditional methods are not applicable [10]. Moreover, as we can see in [11], the tendency to use a hybrid method to solve some type of recent problem, such as those that COVID-19 has brought with it, has recently proven its effectiveness. In one of the important studies of these two main topics, we are interested in exploring the integration to use ML into metaheuristics, in terms to enhance any of the characteristics or attributes of those algorithms: the solutions, performance, or time to get the results. In this way, we propose to use an unsupervised machine learning technique that let us learning in the search space of the metaheuristic, exploiting the characteristic of their attributes that let us use to enhance the metaheuristic parameters in an online way: spatial clustering based on noise application density (DBSCAN) is one of those techniques that gathers those characteristics. We propose to use noise clustering to associate with the abandon probability and the solutions clustering result for determining the number of nests of the cuckoo search algorithm (CSA). They are studies that include some hybrid propose to enhance the CSA whit an ML technique, as in [12], when the authors present a Kmean technique to determinate the discrete parameters to improve the CSA. In [13], the contribution is a hybrid method with a sin-cousin algorithm to enhance the search space to be used on the CSA. Other different scenarios, but similar cases are in [14], they present a CuckooVina, a combination of cuckoo search and differential evolution in a conformational search method. Thus, some studies propose a hybridization to enhance CSA, but they do not respond to our search to find some ML technique that allows enriching the self-adaptation capabilities of CSA in the way we present.

To illustrate and prove our approach, we apply an improvement of the cuckoo search algorithm with a self-adaptive capability (SACSADBSCAN). This approach was tested on the set covering problem, whose goal is to cover a range of needs at the lowest cost and is a widely used problem to demonstrate research.

The remainder of the paper is structured as follows: in Section 2, we introduce the theoretical background on the optimization and clustering algorithms that have been integrated into the proposed approach, then, in Section 3, we present the original CSA, and his parameters that need to be set to run, moreover we introduce to the DBSCAN algorithm, how is the characteristic operation of this algorithm, and how we propose to exploit them in the metaheuristics. Following this, in Section 4 we present the integration of DBSCAN on CSA, how parameter values are configured in the execution of our approach. Then, in Section 5 experimental results and discussions are shown. Finally, we conclude and suggest some lines of future research in Section 6.

2. Related Work

As previously mentioned, parameter adjustment of metaheuristics is a complex task and is considered an optimization problem itself, in many cases, this depends on the try-error test to find a good combination of parameters. In this scenario, some studies use different ways, many of those with a mathematical formula to vary the values of the parameters of the CSA.

For example, to determine the step size α and Pa parameter of the Cuckoo Search, Ref. [15] uses a mathematical formula that depends on the range of those minimal and maximal values of those parameters and the number of iterations of the algorithm. With that idea, the target is to use a bigger target area, also those neighboring areas. In [16], authors propose a self-adaptive step size in his studies to vary the α parameter according to the fitness in each iteration, applying a mathematical formula to get the value to set. In addition, Ref. [17] vary the α and Pa values according to their propose, α changes his value according to the algorithm iterations, meanwhile *Pa* vary randomly depending on the dimension map of the problem. Similar studies are in [18] where Q-Learning is used to set the step size value of the cuckoo search algorithm in a self-adaptive way.

Another study case of adaptive CSA is [19], where the author varies the step size and the abandon probability considering the historical values of them, including their best and worst values to determine the new one. Following with the vary in population, in [20] the authors divide the cuckoo population between making an analysis to them and improve the results.

In another case, such as in [21] use a hybrid algorithm to set the values of their parameters and reduce the population, when the diversity of population decreases, the population is reduced, allowing the algorithm to diversify in subsequent iterations.

As we can see, the machine learning techniques that we have cited up to this point are used to determine a static value in the metaheuristic parameters, which is used as it is the initial value. However, it is used to vary the value of the metaheuristic parameters. During its execution, it is seen minimally in the literature.

In other terms, machine learning has also been applied to the metaheuristic to compact the heuristic space through a clustering procedure [22] and artificial networks [23]. These manuscripts use forecasting and classification. In deep, machine learning integrates with metaheuristics is proposed to forecast different classes of problems, such as electric load [24], economic recessions [25], optimise the low-carbon flexible job shop scheduling problem [26] and other industrial problems [27]. As we mentioned, several dataset groups can be classified by employing machine learning. In this context, we can found image classification problem [28], electronic noise classification of bacterial food bones pathogens [29], urban management [30] and to mention some recently studies. Finally, in [31] the authors use the DBSCAN algorithm to make a binarization strategy and transform the continuous search space to a binary one.

Although machine learning has a big presence in the metaheuristic discipline, its use to set values is quite limited. The use to vary, in an online way, the value of the parameters of the metaheuristic is an area to dig deeper. In this way, we believe that the self-adaptive with DBSCAN, beyond the enhancement of metaheuristics, there seems to be a lot of scope for exploration, using a machine learning technique like DBSCAN with his characteristic we can make an analysis, and thanks to that, change the values of the parameters, is the aim pursued in this work.

3. Theoretical Background

3.1. Cuckoo Search Algorithm

Metaheuristics belong to a class of approximation methods. These are types of higher-level general-purpose algorithms that can be used for a wide range of different types of problems. They have been widely used to solve large-scale combinatorial optimization problems within acceptable computational time, but they do not guarantee optimum solutions [32].

There are a lot of metaheuristics. A full set of all nature-inspired algorithms can be found in [33], and one of them is CSA, which has several study cases. CSA [34] is inspired by the obligate brood parasitism of some cuckoo species by laying their eggs in the nests of other bird species. The steps of CSA are described below:

1. Each cuckoo lays an egg at a time and drops it into a randomly selected nest.
2. The best nests with high-quality eggs will be carried over to the next generations.
3. The number of available host nests is fixed, and the egg laid by a cuckoo is discovered by the host bird with a probability $Pa \in [0,1]$. In this case, a new random solution is generated.

Every new generation is determinate by Lévy flight [34], that is given by the Equation (2)

$$x_i^d(t+1) = x_i^d + \alpha \otimes Levy(\beta)$$
$$\forall i \in \{1,...,n\} \vee \forall d \in \{1,...,m\} \tag{1}$$

where x_i^d is the element d of a solution i at iteration t. $x_i^d(t+1)$ is a solution in the iteration $t+1$. $\alpha > 0$ is the step size which should be related to the scales of the problem of interest, the upper (Ub) and lower bounds (Lb) that the problem need to be determinated, in this scenario values between 0 and 1.

$$Levy \sim u = t^{\beta}, (0 < \beta < 3) \tag{2}$$

The Lévy flight represents a random walk while the random step length is drawn from a Lévy distribution which has an infinite variance with an infinite mean.

3.2. Density Based Spatial Clustering Application with Noise

Density-Based Spatial Clustering of Applications with Noise [35] is a popular data grouping algorithm that uses automated analysis techniques to similar group information together. It can be used to find clusters of any shape in a data collection with noise and outliers. Clusters are dense sections of data space separated by regions with a lower density of points.

The goal is to define dense areas that can be determined by the number of things in close proximity to a place. It is crucial to understand that DBSCAN has two important parameters that are required for work.

1. Epsilon (ϵ): Determines how close points in a cluster can be seen to each other.
2. Minimum points (MinPts): The minimal amount of points required to produce a concentrated area.

The underlying premise is that there must be at least a certain number of points in the vicinity of a particular radius for each given group. The ϵ parameter determines the surrounding radius around a given point; every point x in the data set is tagged as a central point with a neighbor above or equal to MinPts; x is a limit point if the number of neighbors is less than MinPts. Finally, if a point is neither a center nor a border, it is referred to as a point of noise or an outlying point.

When users do not know much about the data being modified, this technique proves helpful in metaheuristics. See pseudo-code in Algorithms 1 and 2 to understand how this strategy works.

The computed groups are shown on the right side of Figure 1. Each iteration of the process adjusts the positions until the algorithm converges. It is important to mention that the noise points are those that do not remain in any cluster at the time of iterating the algorithm and we can visualize them as the white points around the clusters identified in the graph (Red, Yellow, Blue and Green), all these noise points are grouped into a single cluster, and in this way all the points will belong to a specific cluster.

Algorithm 1: Cuckoo Search pseudo-code.

```
Input: $\alpha, \beta, n, Pa, T$ ;
Output: a set of solutions ;
//Produce the first generation of $n$ nests ;
foreach nest $n_i$, ($\forall i = 1, \ldots, n$) do
    foreach dimension $d$, ($\forall d = 1, \ldots, m$) do
        $x_i^d \leftarrow Random\{0,1\}$ ;
    end
    $f_i \leftarrow z(x_i)$ ;
end
$globalfit \leftarrow +\infty$ ;
//Produce $T$-generations of $n$ nests ;
while $t < T$ do
    //Evaluate the fitness and update the best solution ;
    $\{minfit, minindex\} \leftarrow min(f)$;
    if $minfit < globalfit$ then
        $globalfit \leftarrow minfit$ ;
        $\hat{x}^d(t) \leftarrow x_{minindex}^d(t)$ ;
    end
    //the bird abandon the nest with probability $Pa$;
    foreach nest $n_i$, ($\forall i = 1, \ldots, n$) do
        foreach dimension $d$, ($\forall d = 1, \ldots, m$) do
            if $rand > Pa$ then
                { Select the worst nest according to $Pa[0,1]$ and replace them for
                  new random solutions };
            end
        end
    end
    //the bird fly to another new nest with Lévy flight;
    foreach nest $n_i$, ($\forall i = 1, \ldots, n$) do
        foreach dimension $d$, ($\forall d = 1, \ldots, m$) do
            if $rand > Pa$ then
                { Generate new solutions through Equations (1) and 2};
                $x_i^d(t+1) \leftarrow x_i^d(t) + \alpha \otimes$ Lévy($\beta$) ;
                //In previous step, the value generated belongs to the real domain
                  an it must be brought to a binary domain ;
                $x_i^d(t+1) \leftarrow T(x_i^d(t+1))$ ;
            end
        end
    end
end
Postprocess results and visualizations;
```

Algorithm 2: DBSCAN pseudo-code.

```
Input: dataset of P = {p1....pn};
Input: eps;
Input: MinPts;
Output: a set of k clusters.;
foreach unvisited point in P do
    determine the state as visited;
    neighborPts = getNeighbors(point, eps);
    C = 0
    if count(neighborPts) < MinPts then
        set the state as noise;
    else
        C = next cluster;
        expandCluster(point, neighborPts, C, eps, MinPts)
    end
end
expandCluster(point, neighborPts, C, eps, MinPts);
add point to cluster C;
foreach pnt in neighborPts do
    if !visitedPoints(pnt) then
        add pnt to visited ones;
        newNeighborPts = getNeighbors(pnt, eps);
        if sizeof(neighborPts) >= MinPts then
            neighborPts = neighborPts ⋃ newNeighborPts
        end
    end
    if pnt is not yet member of any cluster then
        add pnt to cluster C
    end
end
getNeighbors(point, eps);
return points within pnt eps-neighborhood
```

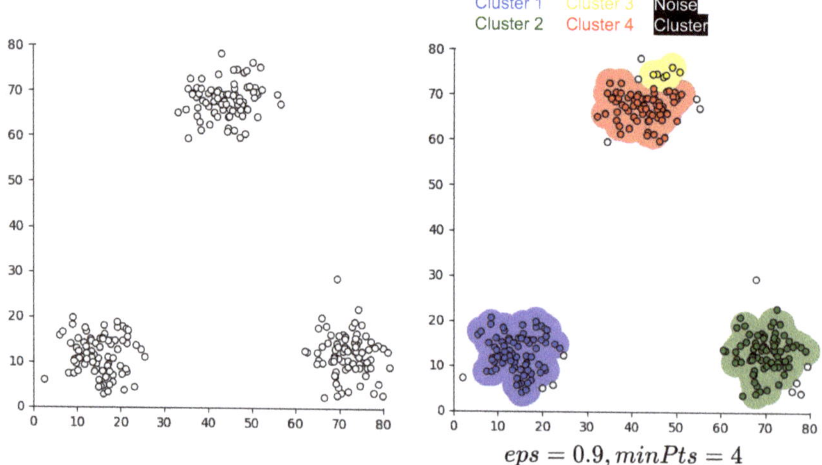

Figure 1. Application of the DBSCAN algorithm to the solution space as an example of clustering.

4. Proposed Approach: Integrating DBSCAN in CSA

In this section, we describe how DBSCAN was integrated on CSA, in addition to all the keys elements for this technique to work. We add a shortcode at the end of each iteration to decide whether intervention is appropriate to implement parameter control intervention and use the DBSCAN algorithm to calculate the values of those parameters (see Algorithm 3).

Algorithm 3: Integration CSA with DBSCAN.

```
Input: α, β, n, Pa, T ;
Output: a set of solutions ;
//Produce the first generation of n nests ;
foreach nest n_i, (∀i = 1, ..., n) do
    foreach dimension d, (∀d = 1, ..., m) do
        x_i^d ← Random{0,1} ;
    end
    f_i ← z(x_i) ;
end
globalfit ← +∞ ;
//Produce T-generations of n nests ;
while t < T do
    //the bird abandon the nest with probability Pa;
    foreach nest n_i, (∀i = 1, ..., n) do
        foreach dimension d, (∀d = 1, ..., m) do
            if rand > Pa then
                { Select the worst nest according to Pa[0,1] and replace them for new
                  random solutions };
            end
        end
    end
    //the bird fly to another new nest with Lévy flight;
    foreach nest n_i, (∀i = 1, ..., n) do
        foreach dimension d, (∀d = 1, ..., m) do
            if rand > Pa then
                { Generate new solutions through Equations (1) and 2};
                x_i^d(t + 1) ← x_i^d(t) + α⊗ Lévy(β) ;
                //In previous step, the value generated belongs to the real domain an it
                  must be brought to a binary domain ;
                x_i^d(t + 1) ← T(x_i^d(t + 1)) ;
            end
        end
    end
    //Evaluate the fitness and update the best solution ;
    {minfit, minindex} ← min(f);
    if minfit < globalfit then
        globalfit ← minfit ;
        x̂^d(t) ← x_minindex^d(t) ;
    end
    //Integration of DBSCAN parameter tuning ;
    if LibertyParameter % t == 0 and t > 1 then
        { Run DBSCAN with nest solutions and evaluate results to set parameter settings};
    end
end
Post-process results and visualizations;
```

In the following sections, we explain some relevant topics that are important to mention to understand our approach. Section 4.1 explains under which criteria the DBSCAN intervening in the metaheuristic to make an analysis on the search space and make the clustering on the solutions. Then, Section 4.2 indicates how the noise cluster is fundamental to determinate the abandon probabilities.

4.1. Free Execution Parameter

We decide to include a variable to control the moment to intervene the parameter values of the CSA to make the metaheuristic maintain its independence of executions and its specific behavior, in this case, we consider it prudent to set it on one hundred iterations of a free run. If these execution values are reached, then the algorithm performs a procedure to update the CSA parameter values, as is described in line 39 of the Algorithm 3.

4.2. Online Parameter Setting

As we mention in the previous section, the update of the values of the parameters in the CSA occurs when the value of the free execution parameter has reached its limit, then the DBSCAN algorithm can be run and used the generated clusters to infer in the parameter values of the metaheuristic. How to associate the parameters of the Pa and $Nest$ is detailed in the following sections.

4.2.1. Probability Abandon Nest

To set the value of the probability of nest abandon, we use the value of the noise point obtained from the DCSCAN execution. That value indicates the points that are excluded from any cluster, so we can associate the number of noise points to make the metaheuristic consider those probabilities to explore new points of the search space. To make this possible, we associate the percentage number of noise points to set the abandon probabilities value. To let the metaheuristic keep the normal execution we use bounds between 10 to 40 percentage. That indicates, that if the noise points are more than 40% then, the Pa values are set to 0.4, in the same way, if the noise point is less than 10%, the Pa values are set to 0.1. In other cases, the values are set to the percentage values of noise points.

4.2.2. Number of Nests

To determine when varying the number of nests on the CSA, our approach considers keeping in memory the last best fitness value to compare with the new candidate solutions, if the best value of fitness does not vary on the fourth intervention of the DBSCAN, we consider that it is necessary to increase the number of nests, to amplify the search space. So, in this case, the number of nests increases in five on five every time this scenario occurs. On the other hand, if the global best value improves four times consecutively, then the number of nest decrease, eliminating the worst five.

4.3. Exploration Influence

During the tests, we realize that the CSA solution space converges to a single large cluster that considers all possible solutions while evaluating the executions, in some instances, certain clusters appear to have values compressed very close to each other. We deem it entirely appropriate in this scenario to have CSA explore the search space. For this, half of the cluster points are renovated to new random ones, allowing the metaheuristic to diversify the search space. The replacement criteria are according to the following Formula (3)

$$ExpInfl = \frac{\sum_{ClusterSol=1}^{n} Fitness_{ClusterSol}}{n} \quad (3)$$

where $ClusterSol$ corresponding to all solutions in the current cluster evaluates. $ExplInfl$ is evaluated with the best possible solution of the global population, if the absolute value of the difference between both is larger than one, then we renew a half-point on those clusters.

4.4. Set Covering Problem

Many studies use the Set Covering Problem (SCP) to represent real scenarios into the studies, as we can see in the area of management of crew on airlines [36], in the optimization of the location of the emergency buildings [37], manufacturing [38] and production [39]. As we can see, SCP is a classical combinatorial optimization problem. It is belong to the NP-hard class [40] and is formally defined as well: let $A = (a_{ij})$ be a binary matrix with M-rows ($\forall\ i \in I = \{1, \ldots, M\}$) and N-columns ($\forall\ j \in J = \{1, \ldots, N\}$), and let $C = (c_j)$ be a vector representing the cost of each column j, assuming that $c_j > 0$, $\forall\ j = \{1, ..., N\}$. Then, it observes that a column j covers a row i if $a_{ij} = 1$. Therefore, it has:

$$a_{ij} = \begin{cases} 1, & \text{if row } i \text{ can be covered by column } j \\ 0, & \text{otherwise} \end{cases}$$

The SCP entails identifying a group of materials that can be used to address a lot of purposes for the least amount of money. A feasible solution corresponds to a subset of columns in its matrix form, and the demands are associated with rows and regarded as constraints. The goal of the challenge is to find the columns that best cover all of the rows.

The Set Covering Problem identifies a low-cost subset S of columns that covers each row with at least one column from S. The SCP can be expressed using integer programming as follows:

$$\begin{aligned} &\texttt{minimize} \sum_{j=1}^{n} c_j x_j \\ &\text{subject to:} \\ &\sum_{j=1}^{n} a_{ij} x_j \geq 1 \quad \forall\ i \in I \\ &x_j \in \{0, 1\} \quad \forall\ j \in J \end{aligned} \tag{4}$$

Instances: We use 65 instances from Beasley's OR-library, which are arranged into 11 sets, to evaluate the algorithm's performance when solving the SCP. To represent the instances, we present on Table 1 the following details: instance group, the number of rows M, number of columns N, the cost range, density (percentage of non-zeroes in the matrix).

Table 1. Instances taken from the Beasley's OR-Library.

Instance Group	M	N	Cost Range	Density (%)	Best Known
4	200	1000	[1, 100]	2	Known
5	200	2000	[1, 100]	2	Known
6	200	1000	[1, 100]	5	Known
A	300	3000	[1, 100]	2	Known
B	300	3000	[1, 100]	5	Known
C	400	4000	[1, 100]	2	Known
D	400	4000	[1, 100]	5	Known
NRE	500	5000	[1, 100]	10	Unknown (except NRE.1)
NRF	500	5000	[1, 100]	20	Unknown (except NRF.1)
NRG	1000	10,000	[1, 100]	2	Unknown (except NRG.1)
NRH	1000	10,000	[1, 100]	5	Unknown

Reducing the instance size of SCP: In [41] different pre-processing approaches have been proposed in particular to reduce the size of the SCP, with Column Domination and Column Inclusion being the most effective. These methods are used to accelerate the processing of the algorithm.

Column Domination is the process of removing unnecessary columns from a problem in such a way that the final solution is unaffected.

Steps:

- All the columns are ordered according to their cost in ascending order.
- If there are equal cost columns, these are sorted in descending order by the number of rows that the column j covers.
- Verify if the column j whose rows can be covered by a set of other columns with a cost less than c_j (cost of the column j).
- It is said that column j is dominated and can be eliminated from the problem.

Column Inclusion: when the domination process has terminated, the process of inclusion is performed, which means if a row is covered only by one column, means that there is no best column to cover those rows, which implies, that column must be included in the optimal solution. All of this process will be included to data of instances and let the new solutions satisfying the constraints.

5. Experimental Results

To evaluate the performance of our proposal, we test the instances of the SCP [42], comparing the original CS with different configurations, versus the SACSDBCAN.

5.1. Methodology

To adequately evaluate the performance of metaheuristics, a performance analysis is required [43]. For this work, we compare the supplied best solution of the CSA to the best-known result of the benchmark retrieved from the OR-Library [44]. Figure 2 depicts the procedures involved in doing a thorough examination of the enhanced metaheuristic. We create objectives and recommendations for the experimental design to show that the proposed approach is a viable alternative for determining metaheuristic parameters. Then, as a vital indicator for assessing future results, we evaluate the best value. We use ordinal analysis and statistical testing to evaluate whether a strategy is significantly better in this circumstance. Lastly, we detail the hardware and software aspects that were used to replicate computational experiments, and we present all of the results in tables and graphs.

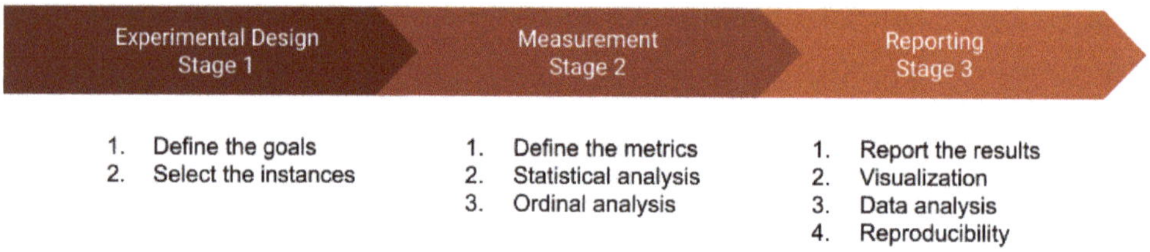

Figure 2. Evaluation stages to determine the performance of an metaheuristic.

As a result, we conduct a contrast statistical test for each case, using the Kolmogorov–Smirnov–Lilliefors process [45] to measure sample autonomy and the Mann–Whitney–Wilcoxon [46] test to statistically evaluate the data, in Figure 3 we describe and determinate the organization.

The Kolmogorov-Smirnov-Lilliefors test allows us to assess sample independence by calculating the Z_{MIN} or Z_{MAX} (depending on whether the task is minimization or maximization) obtained from each instance's 31 executions.

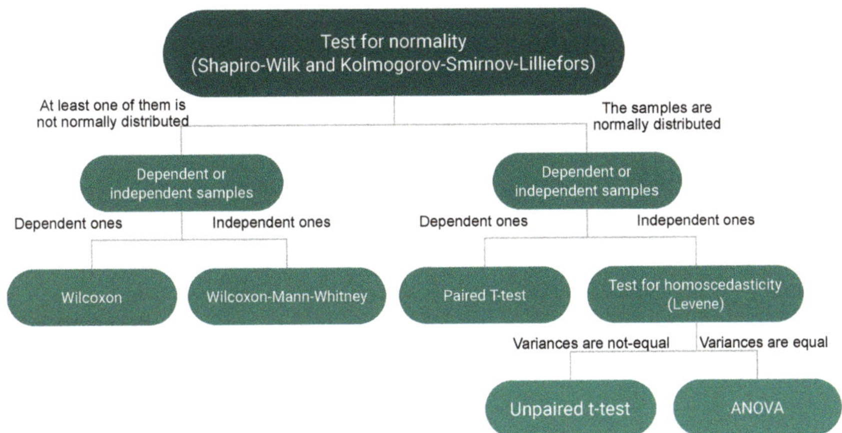

Figure 3. Statistical significance test.

The relative percentage deviation is used to assess the results (RPD). The RPD value computes the difference between the objective value Z_min and the minimal best-known value Z_opt for each instance in our experiment, and it is determined as follows:

$$RPD = \left(\frac{Z_{min} - Z_{opt}}{Z_{opt}} \right)$$

5.2. Set Covering Problem Results

Infrastructure: Java 1.8 was used to implement SACSDBSCAN. The Personal Computer (PC) has the common attributes: MacOS with a 2.7 GHz Intel Core i7 CPU and 16 GB of RAM.

Setup variables: The configuration for our suggested approach is shown in Table 2 below.

Table 2. SACSDBSCAN Parameters for SCP.

Population Initial	Abandon Probability Pa	α	Max Iterations	L_b and U_b
10	0.1–0.4	0.01	5000	0 and 1

Sixty-five SCP instances were considered, each of which was run 31 times; the results are presented in Tables 3 and 4.

Overview: The algorithms are ranked in order of Z_{min} achieved. The instances that obtained Z_{min} are also displayed.

- Score
  1. CS got $38/65$ Z_{min}.
  2. SACSDBSCAN got $10/65$ Z_{min}.
- The two algorithm got Z_{min} in the same instances : 5.4–6.4–A.4–B.2–B.3–B.4–B.5–D.3–D.5–NRE.1–NRE.2–NRE.3–NRF.1–NRF.3–NRF.4–NRF.5–NRH.5

As can be seen in the results of the algorithms that solved SCP, we compare the distribution of the samples of each instance using a violin plot, which allows us to observe the entire distribution of the data. We provide and discuss the most difficult instances of each group to create a resume of all the instances below (4.10, 5.10, 6.5, A.5, B.5, C.5, D.5, NRE.5, NRF.5, NRG.5 and NRH.5):

In [47], the authors display the results obtained, the information is detailed, and the configuration that they use. The information is organized as follows: MIN: the minimum value reached, MAX: the maximum value reached, AVG: the average value, BKS: the best-known solution, RPD is determined by Equation (43), and lastly the average of the fitness obtained.

The first method was the standard cuckoo search algorithm with various settings, and the second was SACSDBSCAN, as previously indicated. Tables 5–8 show the behavior of our proposed algorithm versus the original algorithm. The best results are highlighted with underline and maroon color. For example, in the instance 4.1, the best reached solution by our proposal overcomes than the classical CS algorithm. The same strategy is used in all comparisons.

Table 3. Results of SACSDBSCAN for instance 4, 5, 6 and A.

Instance	SACSDBSCAN						
	Nest = 10 to 50						
	Pa = 0.1 to 0.45						
	BKS	**Fit**	**RDP**	**AVG**	**MIN**	**MAX**	**STD**
4.1	429	429	0.00	429	429	430	0.352
4.2	512	512	0.00	513	512	518	1.552
4.3	514	516	0.00	517	516	520	1.407
4.4	494	494	0.00	495	494	497	0.775
4.5	512	512	0.00	512	512	514	0.828
4.6	560	560	0.00	560	560	564	1.121
4.7	430	430	0.00	430	430	433	0.799
4.8	492	492	0.00	495	492	499	2.640
4.9	641	641	0.00	646	641	653	3.680
4.10	514	514	0.00	514	514	516	0.743
5.1	253	253	0.00	253	253	254	0.516
5.2	302	303	0.00	306	303	312	3.355
5.3	226	226	0.00	227	226	229	1.345
5.4	242	242	0.00	243	242	245	1.234
5.5	211	211	0.00	211	211	212	0.458
5.6	213	213	0.00	213	213	213	0.000
5.7	293	293	0.00	293	293	296	0.816
5.8	288	288	0.00	288	288	289	0.258
5.9	279	279	0.00	280	279	281	0.632
5.10	265	265	0.00	266	265	268	1.047
6.1	138	138	0.00	141	138	144	1.807
6.2	146	146	0.00	148	146	150	1.598
6.3	145	145	0.00	148	145	148	1.060
6.4	131	131	0.00	131	131	133	0.737
6.5	161	161	0.00	163	161	167	1.751
A.1	253	254	0.00	256	254	259	1.234
A.2	252	254	0.01	256	254	259	1.502
A.3	232	232	0.00	234	232	236	1.060
A.4	234	235	0.00	238	235	242	2.586
A.5	236	236	0.00	237	236	238	0.640

Table 4. Results of SACSDBSCAN for instance B, C, D, NRE, NRF, NRG and NRH.

Instance	SACSDBSCAN						
	Nest = 10 to 50						
	Pa = 0.1 to 0.45						
	BKS	**Fit**	**RDP**	**AVG**	**MIN**	**MAX**	**STD**
B.1	69	73	0.05	76	73	79	2.610
B.2	76	76	0.00	80	76	87	3.218
B.3	80	80	0.00	84	80	88	2.673
B.4	79	79	0.00	82	79	88	2.560
B.5	72	72	0.00	73	72	78	1.710
C.1	227	227	0.00	229	227	234	1.993
C.2	219	219	0.00	221	219	225	1.685
C.3	243	243	0.00	246	243	250	2.131
C.4	219	219	0.00	221	219	224	1.309
C.5	215	215	0.00	216	215	217	0.799
D.1	60	61	0.02	64	61	66	1.414
D.2	66	70	0.06	70	70	70	0.000
D.3	72	76	0.05	79	76	82	1.859
D.4	62	63	0.02	66	63	67	1.486
D.5	61	63	0.03	64	63	66	0.910
NRE.1	29	29	0.00	30	29	30	0.516
NRE.2	30	31	0.03	32	31	34	0.862
NRE.3	27	28	0.04	29	28	31	0.976
NRE.4	28	30	0.07	31	30	33	0.834
NRE.5	28	29	0.03	29	29	30	0.516
NRF.1	14	15	0.07	15	15	16	0.458
NRF.2	15	15	0.00	16	15	17	0.458
NRF.3	14	16	0.13	17	16	17	0.507
NRF.4	14	15	0.07	15	15	16	0.389
NRF.5	15	15	0.00	15	15	16	0.258
NRG.1	176	188	0.06	194	188	197	2.789
NRG.2	154	160	0.04	165	160	167	2.282
NRG.3	166	180	0.08	182	180	183	0.862
NRG.4	168	177	0.05	183	177	186	2.728
NRG.5	168	181	0.07	183	181	184	0.799
NRH.1	63	69	0.09	71	69	72	0.976
NRH.2	63	66	0.05	67	66	67	0.414
NRH.3	59	66	0.11	67	66	68	0.738
NRH.4	63	63	0.00	64	63	66	1.528
NRH.5	55	59	0.07	60	59	61	0.707

Table 5. Comparison results for set of groups 4, 5, 6, A. SACSDBSCAN v/s CS Nest = 20, 30.

Instance	BKS	SACSDBSCAN			CS															CS														
		Nest = 10 to 50			Nest = 20															Nest = 30														
		Pa = 0.1 to 0.45			Pa = 0.1			Pa = 0.15			Pa = 0.25			Pa = 0.35			Pa = 0.45			Pa = 0.1			Pa = 0.15			Pa = 0.25			Pa = 0.35			Pa = 0.45		
		Fit	RPD	AVG	Fit	AVG	RPD	Fit	AVG	RPD	Fit	AVG	RPD	Fit	AVG	RPD	Fit	AVG	RPD	Fit	AVG	RPD	Fit	AVG	RPD	Fit	AVG	RPD	Fit	AVG	RPD	Fit	AVG	RPD
4.1	429	429	0	429	468	476	9.09	471	476	9.79	471	476	9.79	468	476	9.09	468	475	9.09	470	476	9.56	470	476	9.56	470	476	9.56	474	477	10.49	471	476	9.79
4.2	512	512	0	513	604	612	17.97	608	613	18.75	597	614	16.60	597	613	16.60	597	613	16.60	597	613	16.60	606	612	18.36	597	613	16.60	597	612	16.60	597	613	16.60
4.3	516	516	0	517	578	586	12.45	575	587	11.87	578	585	12.45	575	586	11.87	575	585	11.87	578	585	12.45	575	585	11.87	568	584	10.51	575	586	11.87	575	584	11.87
4.4	494	494	0	495	568	581	14.98	560	581	13.36	571	581	15.59	571	581	15.59	560	580	13.36	560	579	13.36	560	579	13.36	568	580	14.98	560	581	13.36	568	579	14.98
4.5	512	512	0	512	575	615	12.30	592	618	15.63	601	619	17.38	592	617	15.63	594	616	16.02	591	613	15.43	594	614	16.02	596	617	16.41	592	615	15.63	598	620	16.80
4.6	560	560	0	560	643	653	14.82	637	651	13.75	637	652	13.75	637	652	13.75	643	652	14.82	637	651	13.75	637	650	13.75	643	652	14.82	637	651	13.75	637	651	13.75
4.7	430	430	0	430	518	527	20.47	518	527	20.47	518	527	20.47	518	526	20.47	518	526	20.47	518	524	20.47	518	526	20.47	514	526	19.53	515	525	19.77	518	526	20.47
4.8	492	492	0	495	552	558	12.20	546	557	10.98	547	557	11.18	552	558	12.20	552	558	12.20	546	555	10.98	546	555	10.98	544	557	10.57	546	556	10.98	550	556	11.79
4.9	641	641	0	646	763	788	19.03	763	787	19.03	763	788	19.03	763	785	19.03	763	788	19.03	762	785	18.88	752	782	17.32	763	786	19.03	763	787	19.03	763	787	19.03
4.10	514	514	0	514	588	608	14.40	588	609	14.40	588	608	14.40	588	607	14.40	588	608	14.40	588	609	14.40	588	609	14.40	588	607	14.40	588	607	14.40	595	608	15.76
5.1	253	253	0	253	303	311	19.76	305	310	20.55	303	310	19.76	303	311	19.76	305	311	20.55	300	309	18.58	302	310	19.37	303	310	19.76	303	309	19.76	303	310	19.76
5.2	302	303	0.003	306	374	382	23.84	374	383	23.84	374	382	23.84	370	383	22.52	374	381	23.84	369	383	22.19	370	382	22.52	374	382	23.84	368	383	21.85	374	382	23.84
5.3	226	226	0	227	257	262	13.72	259	263	14.60	257	262	13.72	257	262	13.72	257	263	13.72	257	262	13.72	259	263	14.60	257	262	13.72	257	263	13.72	257	262	13.72
5.4	242	242	0	243	271	275	11.98	269	275	11.16	268	274	10.74	271	274	11.98	271	275	11.98	270	274	11.57	268	274	10.74	271	275	11.98	271	274	11.98	268	274	10.74
5.5	211	211	0	211	254	258	20.38	257	260	21.80	254	258	20.38	256	259	21.33	252	259	19.43	252	258	19.43	252	258	19.43	254	258	20.38	251	259	18.96	252	258	19.43
5.6	213	213	0	213	256	268	20.19	260	269	22.07	259	268	21.60	260	268	22.07	256	268	20.19	256	267	20.19	256	267	20.19	265	269	24.41	256	267	20.19	260	269	22.07
5.7	293	293	0	293	344	352	17.41	345	352	17.75	341	352	16.38	344	352	17.41	344	352	17.41	341	350	16.38	342	351	16.72	343	351	17.06	343	351	17.06	341	351	16.38
5.8	288	288	0	288	355	365	23.26	359	365	24.65	355	364	23.26	357	365	23.96	359	365	24.65	355	364	23.26	359	365	24.65	355	363	23.26	355	363	23.26	355	364	23.26
5.9	279	279	0	280	338	356	21.15	346	356	24.01	340	356	21.86	338	357	21.15	343	356	22.94	338	355	21.15	338	355	21.15	338	355	21.15	343	356	22.94	338	356	21.15
5.10	265	265	0	266	304	310	14.72	306	310	15.47	307	310	15.85	306	310	15.47	308	310	16.23	305	310	15.09	305	310	15.09	306	310	15.47	305	310	15.09	306	310	15.47
6.1	138	138	0	141	169	176	22.46	173	177	25.36	169	176	22.46	169	176	22.46	169	176	22.46	169	176	22.46	169	176	22.46	169	177	22.46	169	175	22.46	169	176	22.46
6.2	146	146	0	148	168	175	15.07	168	174	15.07	168	176	15.07	168	175	15.07	168	175	15.07	168	174	15.07	168	173	15.07	168	174	15.07	168	174	15.07	168	174	15.07
6.3	145	145	0	148	166	175	14.48	169	176	16.55	169	176	16.55	171	176	17.93	169	176	16.55	166	175	14.48	169	174	16.55	169	175	16.55	166	174	14.48	166	174	14.48
6.4	131	131	0	131	142	149	8.40	142	149	8.40	147	149	12.21	147	149	12.21	142	149	8.40	142	149	8.40	145	149	10.69	146	149	11.45	142	149	8.40	142	149	8.40
6.5	161	161	0	163	196	205	21.74	195	205	21.12	202	206	25.47	196	205	21.74	196	205	21.74	200	205	24.22	199	204	23.60	196	204	21.74	196	204	21.74	198	204	22.98
A.1	253	254	0.004	256	277	279	9.49	274	279	8.30	274	279	8.30	274	279	8.30	277	279	9.49	274	278	8.30	277	279	9.49	274	278	8.30	274	278	8.30	274	279	8.30
A.2	252	254	0.01	256	315	318	25.00	310	317	23.02	310	318	23.02	310	317	23.02	313	317	24.21	314	317	24.60	310	316	23.02	314	317	24.60	312	318	23.81	311	317	23.41
A.3	232	232	0	234	263	267	13.36	256	266	10.34	263	266	13.36	258	266	11.21	263	267	13.36	258	266	11.21	261	266	12.50	258	266	11.21	263	266	13.36	262	266	12.93
A.4	234	235	0.004	238	274	277	17.09	273	278	16.67	272	277	16.24	274	277	17.09	271	277	15.81	271	276	15.81	273	277	16.67	275	277	17.52	271	277	15.81	271	276	15.81
A.5	236	236	0	237	271	276	14.83	271	276	14.83	266	276	12.71	269	276	13.98	271	276	14.83	267	275	13.14	266	274	12.71	271	276	14.83	267	275	13.14	271	275	14.83

Table 6. Comparison results for set of groups B, C, D, NRE, NRF, NRG, NRH. SACSDBSCAN v/s CS Nest = 20, 30.

Instance	BKS	SACSDBSCAN			CS															CS														
		Nest = 10 to 50			Nest = 20															Nest = 30														
		Pa = 0.1 to 0.45			*Pa* = 0.1			*Pa* = 0.15			*Pa* = 0.25			*Pa* = 0.35			*Pa* = 0.45			*Pa* = 0.1			*Pa* = 0.15			*Pa* = 0.25			*Pa* = 0.35			*Pa* = 0.45		
		Fit	RPD	AVG	Fit	AVG	RPD	Fit	AVG	RPD	Fit	AVG	RPD	Fit	AVG	RPD	Fit	AVG	RPD	Fit	AVG	RPD	Fit	AVG	RPD	Fit	AVG	RPD	Fit	AVG	RPD	Fit	AVG	RPD
B.1	69	73	0.04	76	88	91	27.54	88	91	27.54	89	91	28.99	87	91	26.09	88	92	27.54	88	91	27.54	89	91	28.99	88	91	27.54	89	92	28.99	87	92	26.09
B.2	76	76	0	80	96	98	26.32	95	99	25.00	96	99	26.32	94	99	23.68	96	99	26.32	94	98	23.68	95	98	25.00	94	99	23.68	95	99	25.00	95	99	25.00
B.3	80	80	0	84	91	93	13.75	91	93	13.75	91	93	13.75	91	93	13.75	91	93	13.75	91	92	13.75	90	92	12.50	91	93	13.75	91	93	13.75	91	93	13.75
B.4	79	79	0	82	102	107	29.11	102	107	29.11	101	107	27.85	104	108	31.65	104	108	31.65	100	106	26.58	101	107	27.85	100	107	26.58	102	107	29.11	102	106	29.11
B.5	72	72	0	73	86	89	19.44	87	90	20.83	86	90	19.44	87	89	20.83	85	90	18.06	86	89	19.44	86	89	19.44	88	89	22.22	85	89	18.06	83	89	15.28
C.1	227	227	0	229	267	270	17.62	264	269	16.30	267	270	17.62	267	270	17.62	265	269	16.74	265	269	16.74	264	268	16.30	267	270	17.62	264	269	16.30	265	269	16.74
C.2	219	219	0	221	266	273	21.46	268	273	22.37	267	273	21.92	267	274	21.92	266	272	21.46	268	272	22.37	266	272	21.46	267	272	21.92	266	271	21.46	269	272	22.83
C.3	243	243	0	246	308	313	26.75	307	313	26.34	308	314	26.75	306	313	25.93	309	314	27.16	306	312	25.93	309	313	27.16	306	313	25.93	306	313	25.93	305	313	25.51
C.4	219	219	0	221	263	267	20.09	264	266	20.55	262	267	19.63	262	266	19.63	263	267	20.09	262	266	19.63	262	266	19.63	263	266	20.09	263	266	20.09	263	267	20.09
C.5	215	215	0	216	248	255	15.35	248	256	15.35	250	256	16.28	248	257	15.35	251	256	16.74	250	256	16.28	251	255	16.74	248	255	15.35	248	255	15.35	251	256	16.74
D.1	60	61	0.02	64	70	72	16.67	70	72	16.67	70	73	16.67	70	73	16.67	71	73	18.33	71	72	18.33	70	72	16.67	70	72	16.67	70	72	16.67	70	72	16.67
D.2	66	70	0.06	70	79	80	19.70	79	80	19.70	78	80	18.18	79	81	19.70	79	81	19.70	78	80	18.18	77	80	16.67	79	80	19.70	79	80	19.70	79	80	19.70
D.3	72	76	0.05	79	86	88	19.44	87	89	20.83	87	89	20.83	87	89	20.83	85	88	18.06	86	88	19.44	86	88	19.44	86	88	19.44	84	88	16.67	87	89	20.83
D.4	62	63	0.02	66	72	73	16.13	71	72	14.52	70	72	12.90	69	73	11.29	71	73	14.52	69	72	11.29	70	72	12.90	71	72	14.52	70	72	12.90	70	73	12.90
D.5	61	63	0.03	64	70	71	14.75	70	71	14.75	71	72	16.39	71	72	16.39	70	72	14.75	70	71	14.75	70	71	14.75	70	71	14.75	70	71	14.75	70	72	14.75
NRE.1	29	29	0	30	31	33	6.90	32	33	10.34	31	33	6.90	32	33	10.34	32	33	10.34	32	33	10.34	32	33	10.34	31	32	6.90	32	33	10.34	32	33	10.34
NRE.2	30	31	0.03	32	37	38	23.33	35	38	16.67	36	38	20	37	38	23.33	36	38	20	36	37	20	36	38	20	37	38	23.33	36	38	20	36	38	20
NRE.3	27	28	0.04	29	35	36	29.63	35	36	29.63	34	36	25.93	34	36	25.93	34	36	25.93	35	36	29.63	34	36	25.93	35	36	29.63	35	36	29.63	35	36	29.63
NRE.4	28	30	0.07	31	34	35	21.43	34	35	21.43	35	35	25.00	34	35	21.43	35	36	25.00	34	35	21.43	34	35	21.43	34	35	21.43	34	35	21.43	34	35	21.43
NRE.5	28	29	0.03	29	33	35	17.86	33	35	17.86	34	35	21.43	33	35	17.86	33	35	17.86	32	35	14.29	32	35	14.29	33	35	17.86	34	35	21.43	32	35	14.29
NRF.1	14	15	0.07	15	18	18	28.57	17	18	21.43	18	18	28.57	18	18	28.57	18	18	28.57	18	18	28.57	18	18	28.57	18	18	28.57	18	18	28.57	17	18	21.43
NRF.2	15	15	0	16	18	19	20	18	19	20	18	19	20	18	19	20	18	19	20	18	19	20	18	19	20	18	19	20	18	19	20	18	19	20
NRF.3	14	16	0.13	17	20	21	42.86	19	21	35.71	20	21	42.86	20	21	42.86	20	21	42.86	19	21	35.71	20	21	42.86	19	21	35.71	19	21	35.71	19	21	35.71
NRF.4	14	15	0.07	15	18	19	28.57	19	20	35.71	19	20	35.71	19	19	35.71	19	20	35.71	18	19	28.57	18	19	28.57	18	19	28.57	18	19	28.57	18	19	28.57
NRF.5	15	15	0	15	16	17	6.67	16	17	6.67	16	17	6.67	16	17	6.67	16	17	6.67	16	16	6.67	16	17	6.67	16	16	6.67	16	17	6.67	16	17	6.67
NRG.1	176	188	0.06	194	230	235	30.68	228	235	29.55	230	235	30.68	232	235	31.82	230	236	30.68	229	234	30.11	228	236	29.55	229	235	30.11	228	235	29.55	229	235	30.11
NRG.2	154	160	0.04	165	187	192	21.43	189	192	22.73	187	192	21.43	187	192	21.43	188	193	22.08	187	191	21.43	185	191	20.13	188	191	22.08	187	192	21.43	188	192	22.08
NRG.3	166	180	0.08	182	198	200	19.28	198	200	19.28	198	200	19.28	197	200	18.67	196	199	18.07	198	200	19.28	198	199	19.28	196	199	18.07	198	200	19.28	197	199	18.67
NRG.4	168	177	0.05	183	216	220	28.57	213	221	26.79	217	221	29.17	216	221	28.57	215	222	27.98	216	220	28.57	214	219	27.38	216	220	28.57	214	219	27.38	217	221	29.17
NRG.5	168	181	0.07	183	219	223	30.36	218	224	29.76	218	223	29.76	220	224	30.95	212	224	26.19	219	223	30.36	218	223	29.76	221	224	31.55	220	224	30.95	216	224	28.57
NRH.1	63	69	0.09	71	82	84	30.16	82	84	30.16	83	85	31.75	82	85	30.16	83	85	31.75	82	85	30.16	82	85	30.16	83	84	31.75	82	85	30.16	82	85	30.16
NRH.2	63	66	0.05	67	79	82	25.40	79	82	25.40	77	82	22.22	80	82	26.98	79	82	25.40	78	82	23.81	79	81	25.40	78	82	23.81	79	82	25.40	80	82	26.98
NRH.3	59	66	0.11	67	74	77	25.42	76	77	28.81	76	77	28.81	76	77	28.81	75	77	27.12	75	77	27.12	75	77	27.12	75	77	27.12	75	77	27.12	75	77	27.12
NRH.4	63	63	0	64	74	76	17.46	74	76	17.46	75	76	19.05	72	76	14.29	74	76	17.46	74	76	17.46	72	75	14.29	74	76	17.46	74	76	17.46	74	76	17.46
NRH.5	55	59	0.07	60	67	69	21.82	67	69	21.82	67	69	21.82	67	69	21.82	67	69	21.82	67	69	21.82	67	69	21.82	67	69	21.82	66	69	20	67	69	21.82

Table 7. Comparison results for set of groups 4, 5, 6, A. SACSDBSCAN v/s CS Nest = 40, 50.

Instance	BKS	SACSDBSCAN			CS															CS														
		Nest = 10 to 50			Nest = 40															Nest = 50														
		Pa = 0.1 to 0.45			*Pa* = 0.1			*Pa* = 0.15			*Pa* = 0.25			*Pa* = 0.35			*Pa* = 0.45			*Pa* = 0.1			*Pa* = 0.15			*Pa* = 0.25			*Pa* = 0.35			*Pa* = 0.45		
		Fit	RPD	AVG	Fit	AVG	RPD	Fit	AVG	RPD	Fit	AVG	RPD	Fit	AVG	RPD	Fit	AVG	RPD	Fit	AVG	RPD	Fit	AVG	RPD	Fit	AVG	RPD	Fit	AVG	RPD	Fit	AVG	RPD
4.1	429	429	0	429	470	475	9.56	471	476	9.79	467	476	8.86	471	476	9.79	469	475	9.32	470	475	9.56	470	475	9.56	470	475	9.56	471	476	9.79	470	476	9.56
4.2	512	512	0	513	597	611	16.60	597	611	16.60	597	611	16.60	597	611	16.60	597	612	16.60	602	611	17.58	597	611	16.60	597	612	16.60	606	613	18.36	597	611	16.60
4.3	516	516	0	517	578	585	12.45	575	585	11.87	575	585	11.87	575	584	11.87	568	585	10.51	575	584	11.87	574	584	11.67	575	584	11.87	575	583	11.87	565	584	9.92
4.4	494	494	0	495	567	579	14.78	568	579	14.98	568	580	14.98	560	581	13.36	569	579	15.18	560	577	13.36	571	582	15.59	560	578	13.36	568	579	14.98	560	578	13.36
4.5	512	512	0	512	575	616	12.30	575	612	12.30	589	616	15.04	589	615	15.04	594	616	16.02	592	614	15.63	598	617	16.80	592	615	15.63	592	616	15.63	592	619	15.63
4.6	560	560	0	560	639	651	14.11	639	651	14.11	629	649	12.32	639	652	14.11	637	650	13.75	637	650	13.75	637	649	13.75	639	651	14.11	637	650	13.75	637	650	13.75
4.7	430	430	0	430	517	524	20.23	515	525	19.77	515	524	19.77	509	525	18.37	518	525	20.47	518	524	20.47	516	525	20	517	523	20.23	518	525	20.47	511	524	18.84
4.8	492	492	0	495	543	555	10.37	552	556	12.20	547	556	11.18	546	555	10.98	539	556	9.55	546	556	10.98	546	556	10.98	546	555	10.98	546	555	10.98	546	555	10.98
4.9	641	641	0	646	763	784	19.03	763	785	19.03	763	786	19.03	763	784	19.03	763	780	19.03	760	781	18.56	763	787	19.03	753	785	17.47	760	785	18.56	761	784	18.72
4.10	514	514	0	514	588	608	14.40	588	608	14.40	596	606	15.95	588	608	14.40	588	605	14.40	588	607	14.40	595	607	15.76	588	605	14.40	588	607	14.40	588	605	14.40
5.1	253	253	0	253	303	310	19.76	305	310	20.55	303	309	19.76	303	310	19.76	303	309	19.76	301	309	18.97	299	309	18.18	303	310	19.76	303	309	19.76	299	309	18.18
5.2	302	303	0.003	306	374	381	23.84	374	381	23.84	373	383	23.51	371	381	22.85	371	381	22.85	369	381	22.19	369	381	22.19	370	382	22.52	373	380	23.51	371	381	22.85
5.3	226	226	0	227	257	262	13.72	257	262	13.72	257	262	13.72	257	262	13.72	257	262	13.72	257	262	13.72	257	262	13.72	257	261	13.72	256	262	13.27	259	262	14.60
5.4	242	242	0	243	270	274	11.57	268	274	10.74	268	274	10.74	268	274	10.74	268	273	10.74	271	274	11.98	268	273	10.74	270	275	11.57	270	274	11.57	270	274	11.57
5.5	211	211	0	211	254	259	20.38	252	258	19.43	252	259	19.43	255	259	20.85	255	259	20.85	252	258	19.43	255	259	20.85	252	258	19.43	253	257	19.91	252	258	19.43
5.6	213	213	0	213	265	269	24.41	260	268	22.07	256	268	20.19	265	268	24.41	256	268	20.19	256	267	20.19	260	268	22.07	260	269	22.07	259	268	21.60	256	268	20.19
5.7	293	293	0	293	341	351	16.38	341	350	16.38	341	350	16.38	341	351	16.38	341	351	16.38	341	349	16.38	346	352	18.09	341	351	16.38	343	350	17.06	341	350	16.38
5.8	288	288	0	288	355	363	23.26	355	361	23.26	356	364	23.61	355	364	23.26	357	364	23.96	355	363	23.26	355	362	23.26	354	363	22.92	352	362	22.22	359	363	24.65
5.9	279	279	0	280	338	354	21.15	341	355	22.22	338	354	21.15	343	354	22.94	326	356	16.85	346	354	24.01	337	356	20.79	338	355	21.15	343	356	22.94	338	355	21.15
5.10	265	265	0	266	305	309	15.09	306	310	15.47	306	309	15.47	305	310	15.09	306	310	15.47	303	309	14.34	302	309	13.96	305	309	15.09	302	310	13.96	305	309	15.09
6.1	138	138	0	141	165	176	19.57	168	174	21.74	165	174	19.57	167	175	21.01	169	175	22.46	169	176	22.46	162	175	17.39	169	175	22.46	169	175	22.46	167	174	21.01
6.2	146	146	0	148	168	175	15.07	168	174	15.07	168	174	15.07	168	173	15.07	168	175	15.07	168	172	15.07	168	172	15.07	168	174	15.07	168	172	15.07	168	172	15.07
6.3	145	145	0	148	166	174	14.48	165	174	13.79	169	175	16.55	169	175	16.55	171	175	17.93	168	173	15.86	166	173	14.48	169	174	16.55	168	173	15.86	169	174	16.55
6.4	131	131	0	131	142	148	8.40	142	149	8.40	145	149	10.69	142	149	8.40	146	149	11.45	142	148	8.40	142	149	8.40	148	149	12.98	140	148	6.87	147	149	12.21
6.5	161	161	0	163	196	203	21.74	196	204	21.74	199	204	23.60	196	204	21.74	196	205	21.74	198	204	22.98	196	204	21.74	196	203	21.74	199	204	23.60	196	204	21.74
A.1	253	254	0.004	256	273	278	7.91	274	278	8.30	278	279	9.88	273	278	7.91	273	278	7.91	277	279	9.49	274	278	8.30	273	278	7.91	274	278	8.30	274	278	8.30
A.2	252	254	0.01	256	308	317	22.22	307	316	21.83	313	317	24.21	310	316	23.02	312	317	23.81	310	316	23.02	311	316	23.41	310	316	23.02	310	316	23.02	310	316	23.02
A.3	232	232	0	234	263	266	13.36	264	266	13.79	258	266	11.21	258	266	11.21	258	265	11.21	258	266	11.21	261	266	12.50	264	266	13.79	262	266	12.93	258	266	11.21
A.4	234	235	0.004	238	272	276	16.24	271	276	15.81	274	277	17.09	273	276	16.67	273	277	16.67	271	276	15.81	271	276	15.81	274	277	17.09	272	276	16.24	271	276	15.81
A.5	236	236	0	237	267	275	13.14	269	275	13.98	270	274	14.41	271	274	14.83	270	275	14.41	271	274	14.83	269	274	13.98	271	275	14.83	267	274	13.14	266	274	12.71

Table 8. Comparison results for set of groups B, C, D, NRE, NRF, NRG, NRH. SACSDBSCAN v/s CS Nest = 40, 50.

Instance	BKS	SACSDBSCAN			CS														
		Nest = 10 to 50			Nest = 40														
		Pa = 0.1 to 0.45			Pa = 0.1			Pa = 0.15			Pa = 0.25			Pa = 0.35			Pa = 0.45		
		Fit	RPD	AVG	Fit	AVG	RPD	Fit	AVG	RPD	Fit	AVG	RPD	Fit	AVG	RPD	Fit	AVG	RPD
B.1	69	73	0.04	76	88	91	27.54	88	91	27.54	87	91	26.09	88	91	27.54	87	90	26.09
B.2	76	76	0	80	94	98	23.68	94	98	23.68	94	98	23.68	94	98	23.68	95	98	25.00
B.3	80	80	05	84	90	92	12.50	91	92	13.75	89	92	11.25	91	92	13.75	91	93	13.75
B.4	79	79	0	82	102	106	29.11	102	107	29.11	103	107	30.38	100	106	26.58	101	106	27.85
B.5	72	72	0	73	87	89	20.83	85	89	18.06	85	89	18.06	85	89	18.06	86	89	19.44
C.1	227	227	0	229	267	269	17.62	263	269	15.86	264	269	16.30	265	269	16.74	264	269	16.30
C.2	219	219	0	221	268	272	22.37	264	272	20.55	265	271	21.00	266	271	21.46	260	272	18.72
C.3	243	243	0	246	308	312	26.75	307	312	26.34	306	312	25.93	299	312	23.05	309	313	27.16
C.4	219	219	0	221	263	266	20.09	263	266	20.09	263	266	20.09	263	266	20.09	264	267	20.55
C.5	215	215	0	216	247	254	14.88	248	255	15.35	247	253	14.88	249	255	15.81	251	256	16.74
D.1	60	61	0.02	64	70	72	16.67	70	72	16.67	71	72	18.33	70	72	16.67	69	72	15.00
D.2	66	70	0.06	70	78	80	18.18	78	80	18.18	78	80	18.18	78	80	18.18	78	80	18.18
D.3	72	76	0.05	79	86	88	19.44	86	88	19.44	86	88	19.44	85	88	18.06	86	88	19.44
D.4	62	63	0.02	66	71	72	14.52	71	72	14.52	71	72	14.52	70	72	12.90	71	72	14.52
D.5	61	63	0.03	64	70	71	14.75	70	72	14.75	70	71	14.75	70	71	14.75	70	71	14.75
NRE.1	29	29	0	30	32	32	10.34	32	33	10.34	32	33	10.34	31	33	6.90	31	33	6.90
NRE.2	30	31	0.03	32	36	38	20	36	38	20	36	38	20	36	38	20	36	38	20
NRE.3	27	28	0.04	29	34	36	25.93	35	36	29.63	34	36	25.93	35	36	29.63	34	36	25.93
NRE.4	28	30	0.07	31	34	35	21.43	34	35	21.43	34	35	21.43	34	35	21.43	34	35	21.43
NRE.5	28	29	0.03	29	32	34	14.29	32	35	14.29	33	35	17.86	32	35	14.29	33	35	17.86
NRF.1	14	15	0.07	15	17	18	21.43	18	18	28.57	18	18	28.57	18	18	28.57	17	18	21.43
NRF.2	15	15	0	16	18	19	20	18	19	20	18	19	20	18	19	20	18	19	20
NRF.3	14	16	0.13	17	19	21	35.71	19	21	35.71	19	21	35.71	20	21	42.86	20	21	42.86
NRF.4	14	15	0.07	15	18	19	28.57	18	19	28.57	18	19	28.57	18	19	28.57	18	19	28.57
NRF.5	15	15	0	15	16	16	6.67	16	16	6.67	16	16	6.67	16	16	6.67	16	17	6.67
NRG.1	176	188	0.06	194	228	234	29.55	229	234	30.11	227	234	28.98	229	234	30.11	230	234	30.68
NRG.2	154	160	0.04	165	186	191	20.78	181	190	17.53	188	191	22.08	187	191	21.43	187	191	21.43
NRG.3	166	180	0.08	182	198	199	19.28	197	199	18.67	195	199	17.47	197	199	18.67	195	199	17.47
NRG.4	168	177	0.05	183	215	219	27.98	213	219	26.79	214	220	27.38	214	221	27.38	216	221	28.57
NRG.5	168	181	0.07	183	220	222	30.95	215	222	27.98	220	223	30.95	218	223	29.76	218	223	29.76
NRH.1	63	69	0.09	71	82	84	30.16	81	84	28.57	82	84	30.16	83	84	31.75	83	84	31.75
NRH.2	63	66	0.05	67	79	82	25.40	78	81	23.81	79	81	25.40	80	82	26.98	80	82	26.98
NRH.3	59	66	0.11	67	75	77	27.12	75	77	27.12	75	77	27.12	75	77	27.12	75	77	27.12
NRH.4	63	63	0	64	73	75	15.87	73	75	15.87	74	76	17.46	74	75	17.46	73	75	15.87
NRH.5	55	59	0.07	60	67	68	21.82	67	68	21.82	67	68	21.82	67	68	21.82	67	69	21.82

Instance	CS														
	Nest = 50														
	Pa = 0.1			Pa = 0.15			Pa = 0.25			Pa = 0.35			Pa = 0.45		
	Fit	AVG	RPD	Fit	AVG	RPD	Fit	AVG	RPD	Fit	AVG	RPD	Fit	AVG	RPD
B.1	87	90	26.09	87	90	26.09	88	91	27.54	87	91	26.09	87	91	26.09
B.2	96	98	26.32	94	97	23.68	94	98	23.68	94	98	23.68	94	98	23.68
B.3	91	92	13.75	90	92	12.50	91	92	13.75	90	92	12.50	91	93	13.75
B.4	102	106	29.11	100	106	26.58	101	106	27.85	101	106	27.85	101	106	27.85
B.5	85	89	18.06	86	89	19.44	86	89	19.44	86	89	19.44	86	89	19.44
C.1	265	269	16.74	265	268	16.74	264	269	16.30	267	269	17.62	263	268	15.86
C.2	268	272	22.37	262	271	19.63	266	271	21.46	267	271	21.92	267	272	21.92
C.3	308	313	26.75	302	312	24.28	306	312	25.93	302	311	24.28	304	312	25.10
C.4	262	266	19.63	262	266	19.63	262	265	19.63	263	265	20.09	263	266	20.09
C.5	250	254	16.28	248	254	15.35	248	255	15.35	248	254	15.35	248	255	15.35
D.1	70	72	16.67	70	72	16.67	70	72	16.67	70	72	16.67	70	72	16.67
D.2	78	80	18.18	78	80	18.18	78	80	18.18	78	80	18.18	78	80	18.18
D.3	85	88	18.06	86	88	19.44	87	88	20.83	86	88	19.44	84	88	16.67
D.4	70	72	12.90	70	72	12.90	71	72	14.52	71	72	14.52	71	72	14.52
D.5	70	71	14.75	69	71	13.11	70	71	14.75	70	71	14.75	70	71	14.75
NRE.1	32	32	10.34	32	33	10.34	32	32	10.34	31	33	6.90	32	32	10.34
NRE.2	37	38	23.33	36	38	20	36	38	20	36	38	20	37	38	23.33
NRE.3	34	36	25.93	34	36	25.93	34	36	25.93	34	36	25.93	34	36	25.93
NRE.4	34	35	21.43	34	35	21.43	34	35	21.43	34	35	21.43	34	35	21.43
NRE.5	33	35	17.86	33	34	17.86	33	34	17.86	32	34	14.29	33	34	17.86
NRF.1	17	18	21.43	18	18	28.57	18	18	28.57	18	18	28.57	17	18	21.43
NRF.2	18	19	20	18	19	20	18	19	20	18	19	20	18	19	20
NRF.3	20	21	42.86	20	21	42.86	20	21	42.86	19	21	35.71	20	21	42.86
NRF.4	18	19	28.57	18	19	28.57	19	19	35.71	18	19	28.57	18	19	28.57
NRF.5	16	16	6.67	16	16	6.67	16	16	6.67	16	16	6.67	16	16	6.67
NRG.1	232	234	31.82	229	234	30.11	230	234	30.68	227	234	28.98	231	235	31.25
NRG.2	187	191	21.43	187	191	21.43	186	191	20.78	187	191	21.43	185	190	20.13
NRG.3	197	199	18.67	197	199	18.67	197	199	18.67	196	199	18.07	197	200	18.67
NRG.4	215	219	27.98	213	220	26.79	212	219	26.19	215	219	27.98	217	220	29.17
NRG.5	218	222	29.76	218	222	29.76	220	223	30.95	219	222	30.36	219	223	30.36
NRH.1	80	84	26.98	81	84	28.57	83	84	31.75	82	84	30.16	82	84	30.16
NRH.2	78	81	23.81	77	81	22.22	79	81	25.40	79	81	25.40	79	81	25.40
NRH.3	75	77	27.12	75	77	27.12	75	77	27.12	75	77	27.12	75	77	27.12
NRH.4	73	75	15.87	73	75	15.87	74	75	15.87	73	75	17.46	74	75	17.46
NRH.5	67	68	21.82	67	68	21.82	67	68	21.82	67	68	21.82	65	68	18.18

The distribution of the data in all instances (Figures 4–14), shows that the performance of our proposal is better than the traditional the cuckoo search optimizer. Concentrating the largest distribution of results in the optimal values, while in original CS they are visibly distant. For example, in the instance B.5, we can see that the distribution is better in our proposal, but at the same time, reflex the behavior to move the result on the best values in his executions, getting the center of the distribution in the best value quartile. Other instance that shows this behaviour can be observed in C.5. Here, our proposal generates again a large number of optimum results.

The scenario in D.5 and NRE.5, is similar in all of them, the behavior of the ASCSDB-SCAN show that reach best results compare to CS. In instance NRF.5 we can see that the behavior of both algorithms obtains a similar figure, reflexing the results obtained from the nature of the sample data. In this scenario, we can rescue results in our proposal, obtains betters solutions. The instance NRG.5 is the only scenario where the results of this instance are at six point of distance to obtain an Z_{opt} in SACSDBSCAN. Finally, for the instance NRH.5, the behavior of the ASCSDBSCAN is again superior than CS.

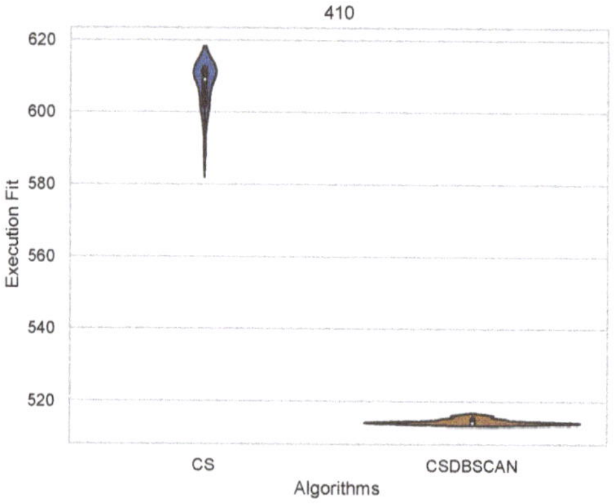

Figure 4. Instance 4.10 distribution.

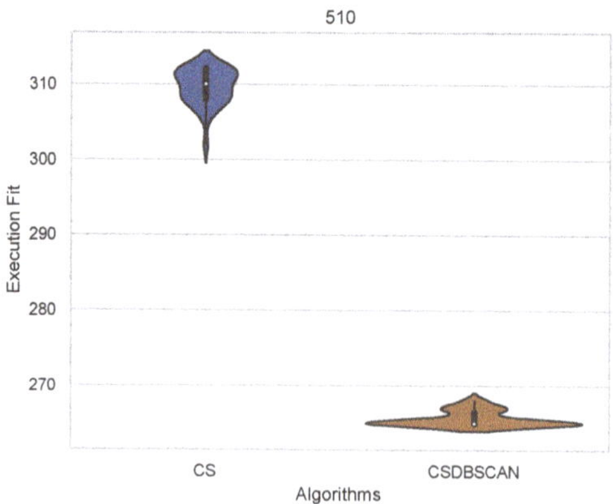

Figure 5. Instance 5.10 distribution.

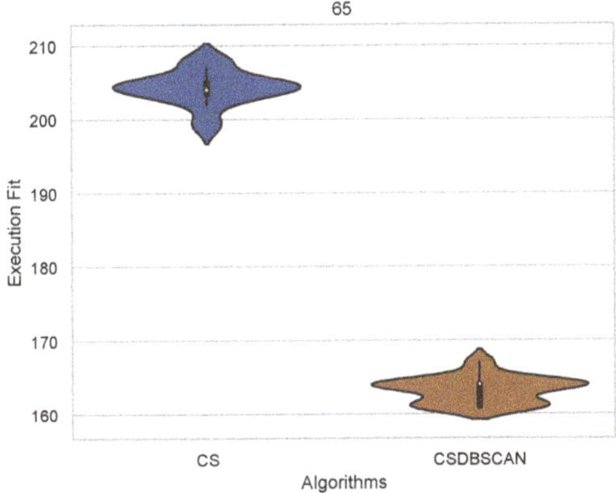

Figure 6. Instance 6.5 distribution.

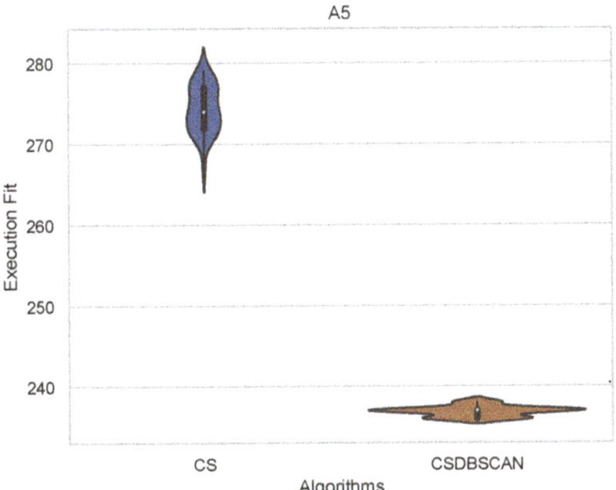

Figure 7. Instance A.5 distribution.

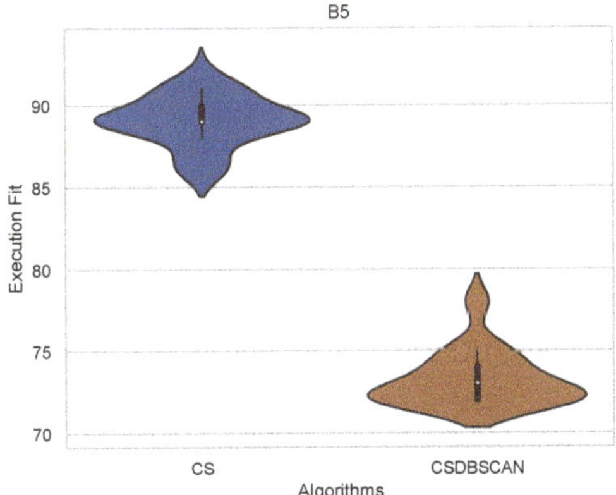

Figure 8. Instance B.5 distribution.

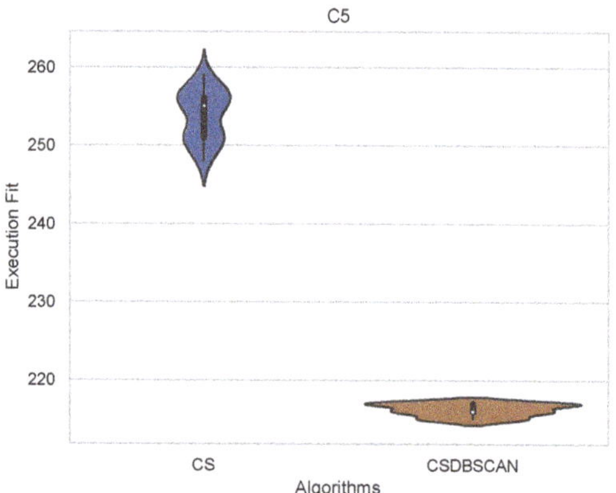

Figure 9. Instance C.5 distribution.

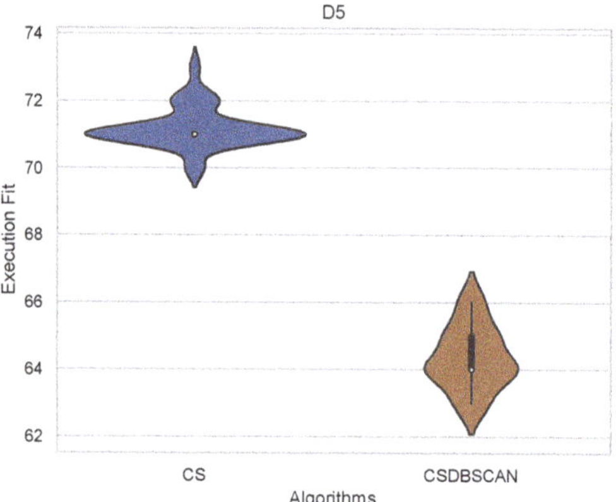

Figure 10. Instance D.5 distribution.

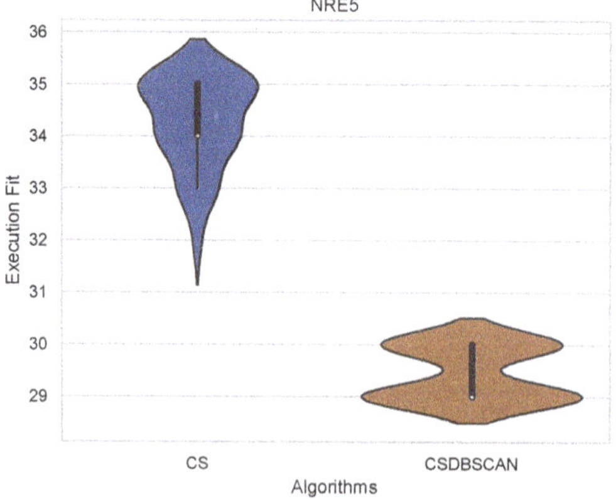

Figure 11. Instance NRE.5 distribution.

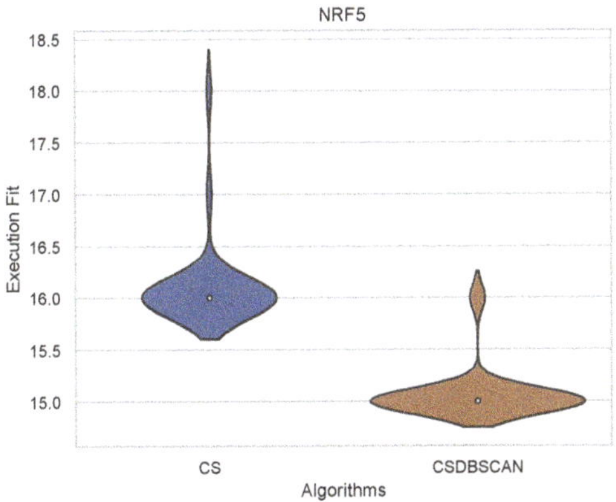

Figure 12. Instance NRF.5 distribution.

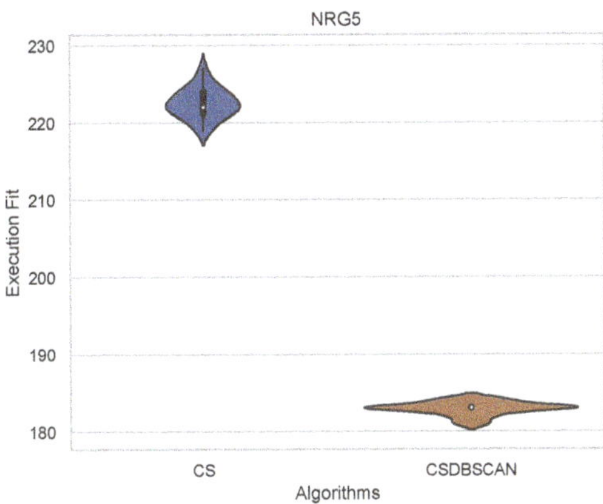

Figure 13. Instance NRG.5 distribution.

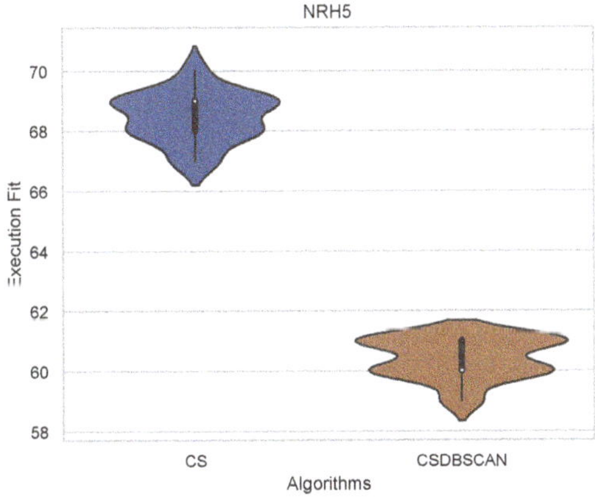

Figure 14. Instance NRH.5 distribution.

5.3. Statistical Test

As previously stated, we offer the following hypotheses in order to determine independence:

- H_0: states that Z_{min}/Z_{max} follows a normal distribution.
- H_1: states the opposite.

The test performed has yielded *p_value* lower than 0.05; therefore, H_0 cannot be assumed. Now that we know that the samples are independent, and it cannot be assumed that they follow a normal distribution, it is not feasible to use the central limit theorem. Therefore, for evaluating the heterogeneity of samples we use a non-parametric evaluation called Mann–Whitney–Wilcoxon test to compare all the results of the hardest instances we propose the following hypotheses:

- H_0: CS is better than SACSDBSCAN.
- H_1: states the opposite.

Finally, the statistical contrast test reveals which technique is considerably superior.

The Wilcoxon signed rank test was used to compare SCP on the algorithms techniques for the hardest instances (Tables 9–19). Smaller p-values than 0.05 define that H_0 cannot be assumed because the significance level is also set to 0.05.

To conduct the test run that supports the study, we use a method from the PISA system. We specify all data distributions (each in a file and each data in a line) in this procedure, and the algorithm returns a p-value for the hypotheses.

The following tables show the result of the Mann–Whitney–Wilcoxon test. To understand them, it is necessary to know the following acronyms:

- SWS = Statistically without significance.

Table 9. p-values for instance 4.10.

	CS	**SACSDBSCAN**
CS	Not applicable	SWS
SACSDBSCAN	1.711×10^{-12}	Not applicable

Table 10. p-values for instance 5.10.

	CS	**SACSDBSCAN**
CS	Not applicable	SWS
SACSDBSCAN	2.228×10^{-12}	Not applicable

Table 11. p-values for instance 6.5.

	CS	**SACSDBSCAN**
CS	Not applicable	SWS
SACSDBSCAN	1.711×10^{-12}	Not applicable

Table 12. p-values for instance A.5.

	CS	**SACSDBSCAN**
CS	Not applicable	SWS
SACSDBSCAN	3.055×10^{-12}	Not applicable

Table 13. p-values for instance b.5.

	CS	**SACSDBSCAN**
CS	Not applicable	SWS
SACSDBSCAN	3.417×10^{-12}	Not applicable

Table 14. *p*-values for instance c.5.

	CS	**SACSDBSCAN**
CS	Not applicable	SWS
SACSDBSCAN	4.100×10^{-12}	Not applicable

Table 15. *p*-values for instance d.5.

	CS	**SACSDBSCAN**
CS	Not applicable	SWS
SACSDBSCAN	1.288×10^{-12}	Not applicable

Table 16. *p*-values for instance NRE.5.

	CS	**SACSDBSCAN**
CS	Not applicable	SWS
SACSDBSCAN	1.915×10^{-12}	Not applicable

Table 17. *p*-values for instance NRF.5.

	CS	**SACSDBSCAN**
CS	Not applicable	SWS
SACSDBSCAN	2.406×10^{-12}	Not applicable

Table 18. *p*-values for instance NRG.5.

	CS	**SACSDBSCAN**
CS	Not applicable	SWS
SACSDBSCAN	3.296×10^{-12}	Not applicable

Table 19. *p*-values for instance NRH.5.

	CS	**SACSDBSCAN**
CS	Not applicable	SWS
SACSDBSCAN	2.482×10^{-12}	Not applicable

In all the cases, as mentioned above, the *p*-values reported are less than 0.05, and SWS suggests that they have no statistical significance. So, with this knowledge, in each instance mentioned, we can see the SACSDBSCAN algorithm was better than the original CS.

If we focus on the instances where our proposal improves the result obtained in comparison to the original CS algorithm, we can infer that the solutions achieved are distributed in a centered way on their optimal value, which reflects that the behavior of this algorithm is very positive. This is reflected in the violin Figure 8 or Figures 10 and 14.

5.4. *Comparison Results in Similar Hybrid Algorithms*

Within the literature, recent studies can be found that use hybrid algorithms that solve the coverage problem [5,48–50]. However, to compare a hybrid algorithm that resembles our proposal, we have considered making a comparison of results with hybrid algorithms that work with bio-inspired metaheuristics, improved by ML, and they solve the set covering problem. In this scheme, we have three algorithms. The first one is the crow search algorithm boosted by the DBSCAN method (called CSADBSCAN). The second studied approach was the integration between the crow search algorithm and the Kmean method (CSAKmean). Both hybridizations were proposed by Valdivia et al. in [51]. Finally, we employ an improved version of the cuckoo search algorithm with the Kmean transition algorithm (KMTA), recently proposed by García et al. in [52].

Tables 20 and 21 present best values reached in CSADBSCAN, CSAKmean and KMTA. Those algorithms implement different strategies to improve metaheuristics with ML. To resume and centering the results of the best values obtained, we add the AVG measure in the final row of each table. Unfortunately, KMTA only reports results of the first of each family instance, so N/R means Not Reported.

Table 20. Best values reached by an improved bio-inspired algorithm with ML that solve the set covering problem Instances 4, 5, 6 and A.

Instance	BKS	SACSDBSCAN		CSADBSCAN		CSAKmean		KMTA	
		Best	RDP	Best	RDP	Best	RDP	Best	RDP
4.1	429	429	0.000	429	0.000	429	0.000	430	0.002
4.2	512	512	0.000	512	0.000	513	0.002	N/R	N/R
4.3	516	516	0.000	516	0.000	516	0.000	N/R	N/R
4.4	494	494	0.000	494	0.000	495	0.002	N/R	N/R
4.5	512	512	0.000	512	0.000	514	0.004	N/R	N/R
4.6	560	560	0.000	560	0.000	560	0.000	N/R	N/R
4.7	430	430	0.000	430	0.000	430	0.000	N/R	N/R
4.8	492	492	0.000	492	0.000	493	0.002	N/R	N/R
4.9	641	641	0.000	641	0.000	645	0.006	N/R	N/R
4.10	514	514	0.000	514	0.000	513	0.002	N/R	N/R
5.1	253	253	0.000	253	0.000	253	0.000	253	0.000
5.2	302	303	0.003	302	0.000	308	0.020	N/R	N/R
5.3	226	226	0.000	226	0.000	228	0.009	N/R	N/R
5.4	242	242	0.000	242	0.000	242	0.000	N/R	N/R
5.5	211	211	0.000	211	0.000	211	0.000	N/R	N/R
5.6	213	213	0.000	213	0.000	213	0.000	N/R	N/R
5.7	293	293	0.000	293	0.000	293	0.000	N/R	N/R
5.8	288	288	0.000	288	0.000	288	0.000	N/R	N/R
5.9	279	279	0.000	279	0.000	279	0.000	N/R	N/R
5.10	265	265	0.000	265	0.000	267	0.008	N/R	N/R
6.1	138	138	0.000	140	0.014	140	0.014	138	0.000
6.2	146	146	0.000	146	0.000	146	0.000	N/R	N/R
6.3	145	145	0.000	145	0.000	147	0.014	N/R	N/R
6.4	131	131	0.000	131	0.000	131	0.000	N/R	N/R
6.5	161	161	0.000	162	0.006	163	0.012	N/R	N/R
A.1	253	254	0.004	254	0.004	254	0.004	254	0.004
A.2	252	254	0.008	256	0.016	257	0.020	N/R	N/R
A.3	232	232	0.000	233	0.004	235	0.013	N/R	N/R
A.4	234	235	0.004	236	0.009	235	0.004	N/R	N/R
A.5	236	236	0.000	236	0.000	236	0.000	N/R	N/R
AVG	–	–	0.001	–	0.002	–	0.005	–	0.002

Table 20 shows how SACSDBSCAN obtain better results average in comparison with the other algorithms for the instances 4, 5, 6 y A. Table 21 shows how SACSDBSCAN obtain better results average in comparison with CSADBSCAN and CSAKmean in a very wide value to CSADBSCAN, a difference to 1.141 and 0.038 distance with CSAKmean. Only KMTA obtain the best AVG value with the reported best values with 0.018 difference.

Table 21. Best values reached by an improved bio-inspired algorithm with ML that solve the set covering problem Instances B, C, D, NRE, NRF, NRG, and NRH.

Instance	BKS	SACSDBSCAN		CSADBSCAN		CSAKmean		KMTA	
		Best	RDP	Best	RDP	Best	RDP	Best	RDP
B.1	69	73	0.058	69	0.000	74	0.072	69	0.000
B.2	76	76	0.000	81	0.066	83	0.092	N/R	N/R
B.3	80	80	0.000	82	0.025	84	0.050	N/R	N/R
B.4	79	79	0.000	83	0.051	84	0.063	N/R	N/R
B.5	72	72	0.000	78	0.083	72	0.000	N/R	N/R
C.1	227	227	0.000	233	0.026	228	0.004	229	0.009
C.2	219	219	0.000	226	0.032	226	0.032	N/R	N/R
C.3	243	243	0.000	253	0.041	254	0.045	N/R	N/R
C.4	219	219	0.000	224	0.023	225	0.027	N/R	N/R
C.5	215	215	0.000	222	0.033	215	0.000	N/R	N/R
D.1	60	61	0.017	68	0.133	66	0.100	60	0.000
D.2	66	70	0.061	73	0.106	71	0.076	N/R	N/R
D.3	72	76	0.056	82	0.139	82	0.139	N/R	N/R
D.4	62	63	0.016	70	0.129	67	0.081	N/R	N/R
D.5	61	63	0.033	72	0.180	66	0.082	N/R	N/R
NRE.1	29	29	0.000	70	1.414	30	0.034	29	0.000
NRE.2	30	31	0.033	83	1.767	34	0.133	N/R	N/R
NRE.3	27	28	0.037	74	1.741	34	0.259	N/R	N/R
NRE.4	28	30	0.071	83	1.964	33	0.179	N/R	N/R
NRE.5	28	29	0.036	92	2.286	30	0.071	N/R	N/R
NRF.1	14	15	0.071	372	25.571	17	0.214	14	0.000
NRF.2	15	15	0.000	74	3.933	18	0.200	N/R	N/R
NRF.3	14	16	0.143	335	22.929	19	0.357	N/R	N/R
NRF.4	14	15	0.071	52	2.714	18	0.286	N/R	N/R
NRF.5	15	15	0.000	65	3.333	16	0.067	N/R	N/R
NRG.1	176	188	0.068	287	0.631	197	0.119	176	0.000
NRG.2	154	160	0.039	208	0.351	168	0.091	N/R	N/R
NRG.3	166	180	0.084	211	0.271	183	0.102	N/R	N/R
NRG.4	168	177	0.054	250	0.488	186	0.107	N/R	N/R
NRG.5	168	181	0.077	230	0.369	183	0.089	N/R	N/R
NRH.1	63	69	0.095	N/R	N/R	71	0.127	64	0.016
NRH.2	63	66	0.048	N/R	N/R	71	0.127	N/R	N/R
NRH.3	59	66	0.119	N/R	N/R	69	0.169	N/R	N/R
NRH.4	63	63	0.000	N/R	N/R	68	0.079	N/R	N/R
NRH.5	55	59	0.073	N/R	N/R	61	0.109	N/R	N/R
AVG	–	–	0.021	–	1.162	–	0.059	–	0.003

6. Conclusions

In this paper, we can conclude that the use of the machine learning technique to make a metaheuristic autonomous parameters setting has the 38/65 min values fitness on the test that we make, in one single configuration over the 20 different configurations, that demonstrates that with our proposed algorithm, it is not necessary make the complex task to find the best parameter setting of the metaheuristic CS, that is, in most of the time, in a try-and-error way. The result of the experiment demonstrates that it is positive to use the DBSCAN algorithm to infer his result and use that information to let us make changes to the values of the parameters. The comparison results with other bio-inspired hybrid algorithms applied on the set covering problem demonstrate that the use of DBSCAN obtains better results on average fitness values in comparison with studies that report all the best results values. The exploration criteria that we use can let the algorithm vary the search space to find another best candidate, as we saw in the box graphs and the distribution of the instances. In addition, the use of the noise points associate with the *Pa*, lets the SACSDBSCAN keep the variety on his behavior and not forget the stochastic factor that characterizes a metaheuristic.

The free execution parameter allows the metaheuristic to maintain its natural behavior across executions. At the time of reaching the freedom parameter, we can analyze the results of the metaheuristics and classify its results space to be able to make its classification and corresponding intervention of the possible candidate solutions, eliminating the worst for possible new better solutions.

As future work, we consider implementing an improvement to the criterion of population increase and decrease using clusterization strategies. In another line of work, we want to implement different machine learning techniques to be able to perform algorithms that allow effective use of the population increase/decrease in metaheuristics, and thus be able to deliver tools that make these algorithms more efficient. In addition, we are considering contributing new works that provide comparisons of this algorithm with other hybrid variants of cuckoo search, and other types of metaheuristics that use ML to improve their performance, different from those presented in this work, used to solve continuous problems.

Author Contributions: Formal analysis, N.C., R.S. and B.C.; Investigation, N.C., R.S., B.C., S.V. and R.O.; Methodology, R.S., R.O. and S.V.; Resources, R.S. and B.C.; Software, N.C., S.V. and R.O.; Validation, B.C. and S.V.; writing—original draft, N.C., S.V. and R.O.; writing—review and editing, N.C., R.S., B.C., S.V. and R.O. All authors have read and agreed to the published version of the manuscript.

Funding: Ricardo Soto is supported by grant CONICYT/FONDECYT/REGULAR/1190129. Broderick Crawford is supported by grant ANID/FONDECYT/REGULAR/1210810. Nicolás Caselli is supported by grant INF-PUCV 2019–2021.

Institutional Review Board Statement: Not applicable.

Informed Consent Statement: Not applicable.

Data Availability Statement: No new data were created or analyzed in this study. Data sharing is not applicable to this article.

Conflicts of Interest: The authors declare no conflict of interest. The founding sponsors had no role in the design of the study; in the collection, analyses, or interpretation of data; in the writing of the manuscript, and in the decision to publish the results.

Abbrevations

Symbol	
α	Step size to generate the next solution
Pa	Probability abandon of CSA
β	Scalar number between 0 and 3
ϵ	Max Distance with each other point
T	Max number of Iterations
t	Current Iteration
L_b	Lower Bound
U_b	Upper Bound
$X_i^d(t)$	Solution *i*th in dimension *d* at iteration *t*
Acronyms	
ML	Machine Learning
CSA	Cuckoo Search Algorithm
SACSADBSCAN	Self-Adaptive Cuckoo Search DBSCAN

DBSCAN	Spatial Clustering Based on Noise Application Density
SCP	Set Covering Problem
RPD	Relative Percentage Deviation
CSADBSCAN	Crow Search Algorithm boosted by the DBSCAN method
CSAKmean	crow search algorithm and the Kmean method
KMTA	Kmean transition algorithm

References

1. Olivares, R.; Muñoz, F.; Riquelme, F. A multi-objective linear threshold influence spread model solved by swarm intelligence-based methods. *Knowl.-Based Syst.* **2021**, *212*, 106623. [CrossRef]
2. Soto, R.; Crawford, B.; Olivares, R.; Carrasco, C.; Rodriguez-Tello, E.; Castro, C.; Paredes, F.; de la Fuente-Mella, H. A reactive population approach on the dolphin echolocation algorithm for solving cell manufacturing systems. *Mathematics* **2020**, *8*, 1389. [CrossRef]
3. Taramasco, C.; Olivares, R.; Munoz, R.; Soto, R.; Villar, M.; de Albuquerque, V.H.C. The patient bed assignment problem solved by autonomous bat algorithm. *Appl. Soft Comput.* **2019**, *81*, 105484. [CrossRef]
4. Munoz, R.; Olivares, R.; Taramasco, C.; Villarroel, R.; Soto, R.; Alonso-Sánchez, M.F.; Merino, E.; de Albuquerque, V.H.C. A new EEG software that supports emotion recognition by using an autonomous approach. *Neural Comput. Appl.* **2018**, *32*, 11111–11127. [CrossRef]
5. Crawford, B.; Soto, R.; Olivares, R.; Embry, G.; Flores, D.; Palma, W.; Castro, C.; Paredes, F.; Rubio, J.M. A binary monkey search algorithm variation for solving the set covering problem. *Nat. Comput.* **2019**, *19*, 825–841. [CrossRef]
6. Huang, C.; Li, Y.; Yao, X. A survey of automatic parameter tuning methods for metaheuristics. *IEEE Trans. Evol. Comput.* **2019**, *24*, 201–216. [CrossRef]
7. Song, H.; Triguero, I.; Özcan, E. A review on the self and dual interactions between machine learning and optimisation. *Prog. Artif. Intell.* **2019**, *8*, 143–165. [CrossRef]
8. Ghorban, F.; Milani, N.; Schugk, D.; Roese-Koerner, L.; Su, Y.; Müller, D.; Kummert, A. Conditional multichannel generative adversarial networks with an application to traffic signs representation learning. *Prog. Artif. Intell.* **2018**, *8*, 73–82. [CrossRef]
9. Caruso, G.; Gattone, S.; Fortuna, F.; Di Battista, T. Cluster Analysis for mixed data: An application to credit risk evaluation. *Socio-Econ. Plan. Sci.* **2021**, *73*, 100850. [CrossRef]
10. Michalewicz, Z.; Fogel, D.B. *How to Solve It: Modern Heuristics*; Springer Science & Business Media: Berlin, Germany, 2013.
11. D'Adamo, I.; González-Sánchez, R.; Medina-Salgado, M.S.; Settembre-Blundo, D. E-Commerce Calls for Cyber-Security and Sustainability: How European Citizens Look for a Trusted Online Environment. *Sustainability* **2021**, *13*, 6752. [CrossRef]
12. García, J.; Yepes, V.; Martí, J.V. A Hybrid k-Means Cuckoo Search Algorithm Applied to the Counterfort Retaining Walls Problem. *Mathematics* **2020**, *8*, 555. [CrossRef]
13. Rosli, S.J.; Rahim, H.A.; Abdul Rani, K.N.; Ngadiran, R.; Ahmad, R.B.; Yahaya, N.Z.; Abdulmalek, M.; Jusoh, M.; Yasin, M.N.M.; Sabapathy, T.; et al. A Hybrid Modified Method of the Sine Cosine Algorithm Using Latin Hypercube Sampling with the Cuckoo Search Algorithm for Optimization Problems. *Electronics* **2020**, *9*, 1786. [CrossRef]
14. Lin, H.; Siu, S.W.I. A Hybrid Cuckoo Search and Differential Evolution Approach to Protein—Ligand Docking. *Int. J. Mol. Sci.* **2018**, *19*, 3181. [CrossRef] [PubMed]
15. Saeed, S.; Ong, H.C.; Sathasivam, S. Self-Adaptive Single Objective Hybrid Algorithm for Unconstrained and Constrained Test functions: An Application of Optimization Algorithm. *Arab. J. Sci. Eng.* **2019**, *44*, 3497–3513. [CrossRef]
16. Thirugnanasambandam, K.; Prakash, S.; Subramanian, V.; Pothula, S.; Thirumal, V. Reinforced cuckoo search algorithm-based multimodal optimization. *Appl. Intell.* **2019**, *49*, 2059–2083. [CrossRef]
17. Dhabal, S.; Venkateswaran, P. An Improved Global-Best-Guided Cuckoo Search Algorithm for Multiplierless Design of Two-Dimensional IIR Filters. *Circuits Syst. Signal Process.* **2019**, *38*, 805–826. [CrossRef]
18. Li, J.; Xiao, D.D.; Lei, H.; Zhang, T.; Tian, T. Using Cuckoo Search Algorithm with Q-Learning and Genetic Operation to Solve the Problem of Logistics Distribution Center Location. *Mathematics* **2020**, *8*, 149. [CrossRef]
19. Jaballah, A.; Meddeb, A. A new variant of cuckoo search algorithm with self adaptive parameters to solve complex RFID network planning problem. *Wirel. Netw.* **2019**, *25*, 1585–1601. [CrossRef]
20. Ma, H.S.; Li, S.X.; Li, S.F.; LV, Z.N.; Wang, J.S. An Improved dynamic self-adaption cuckoo search algorithm based on collaboration between subpopulations. *Neural Comput. Appl.* **2019**, *31*, 1375–1389. [CrossRef]
21. Mlakar, U.; Fister, I.; Fister, I. Hybrid self-adaptive cuckoo search for global optimization. *Swarm Evol. Comput.* **2016**, *29*, 47–72. [CrossRef]
22. Senjyu, T.; Saber, A.; Miyagi, T.; Shimabukuro, K.; Urasaki, N.; Funabashi, T. Fast technique for unit commitment by genetic algorithm based on unit clustering. *IEE Proc.-Gener. Transm. Distrib.* **2005**, *152*, 705–713. [CrossRef]
23. Lee, C.; Gen, M.; Kuo, W. Reliability optimization design using a hybridized genetic algorithm with a neural-network technique. *IEICE Trans. Fundam. Electron. Commun. Comput. Sci.* **2001**, *84*, 627–637.
24. Bouktif, S.; Fiaz, A.; Ouni, A.; Serhani, M.A. Multi-Sequence LSTM-RNN Deep Learning and Metaheuristics for Electric Load Forecasting. *Energies* **2020**, *13*, 391. [CrossRef]

25. Cicceri, G.; Inserra, G.; Limosani, M. A Machine Learning Approach to Forecast Economic Recessions—An Italian Case Study. *Mathematics* **2020**, *8*, 241. [CrossRef]
26. Luan, F.; Cai, Z.; Wu, S.; Liu, S.Q.; He, Y. Optimizing the Low-Carbon Flexible Job Shop Scheduling Problem with Discrete Whale Optimization Algorithm. *Mathematics* **2019**, *7*, 688. [CrossRef]
27. Ly, H.B.; Le, T.T.; Le, L.M.; Tran, V.Q.; Le, V.M.; Vu, H.L.T.; Nguyen, Q.H.; Pham, B.T. Development of Hybrid Machine Learning Models for Predicting the Critical Buckling Load of I-Shaped Cellular Beams. *Appl. Sci.* **2019**, *9*, 5458. [CrossRef]
28. Korytkowski, M.; Senkerik, R.; Scherer, M.M.; Angryk, R.A.; Kordos, M.; Siwocha, A. Efficient Image Retrieval by Fuzzy Rules from Boosting and Metaheuristic. *J. Artif. Intell. Soft Comput. Res.* **2020**, *10*, 57–69. [CrossRef]
29. Hoang, N.D. Image processing based automatic recognition of asphalt pavement patch using a metaheuristic optimized machine learning approach. *Adv. Eng. Inform.* **2019**, *40*, 110–120. [CrossRef]
30. Bui, Q.T.; Van, M.P.; Hang, N.T.T.; Nguyen, Q.H.; Linh, N.X.; Ha, P.M.; Tuan, T.A.; Cu, P.V. Hybrid model to optimize object-based land cover classification by meta-heuristic algorithm: An example for supporting urban management in Ha Noi, Viet Nam. *Int. J. Digit. Earth* **2019**, *12*, 1118–1132. [CrossRef]
31. Valdivia, S.; Soto, R.; Crawford, B.; Caselli, N.; Paredes, F.; Castro, C.; Olivares, R. Clustering-Based Binarization Methods Applied to the Crow Search Algorithm for 0/1 Combinatorial Problems. *Mathematics* **2020**, *8*, 70. [CrossRef]
32. Lewis, R. A survey of metaheuristic-based techniques for University Timetabling problems. *OR Spectr.* **2008**, *30*, 167–190. [CrossRef]
33. Tzanetos, A.; Fister, I.; Dounias, G. A comprehensive database of Nature-Inspired Algorithms. *Data Brief* **2020**, *31*, 105792. [CrossRef]
34. Yang, X.S.; Deb, S. Cuckoo Search via Levy Flights. *Res. Gate* **2009**, 210–214. [CrossRef]
35. Ester, M.; Kriegel, H.P.; Sander, J.; Xu, X. *A Density-Based Algorithm for Discovering Clusters in Large Spatial Databases with Noise*; AAAI Press: Portland, OR, USA, 1996; pp. 226–231.
36. Smith, B.M. IMPACS—A Bus Crew Scheduling System Using Integer Programming. *Math. Program.* **1988**, *42*, 181–187. [CrossRef]
37. Toregas, C.; Swain, R.; ReVelle, C.; Bergman, L. The Location of Emergency Service Facilities. *Oper. Res.* **1971**, *19*, 1363–1373. [CrossRef]
38. Foster, B.A.; Ryan, D.M. An Integer Programming Approach to the Vehicle Scheduling Problem. *J. Oper. Res. Soc.* **1976**, *27*, 367–384. [CrossRef]
39. Vasko, F.J.; Wolf, F.E.; Stott, K.L. A set covering approach to metallurgical grade assignment. *Eur. J. Oper. Res.* **1989**, *38*, 27–34. [CrossRef]
40. Garey, M.R.; Johnson, D.S. *Computers and Intractability: A Guide to the Theory of NP-Completeness*; W. H. Freeman & Co.: New York, NY, USA, 1979.
41. Caprara, A.; Toth, P.; Fischetti, M. Algorithms for the set covering problem. *Ann. Oper. Res.* **2000**, *98*, 353–371. [CrossRef]
42. Gass, S.; Fu, M. Set-covering Problem. In *Encyclopedia of Operations Research and Management Science*; Springer: Cham, Switzerland, 2013; p 1393.
43. Bartz-Beielstein, T.; Preuss, M. Experimental research in evolutionary computation. In Proceedings of the 9th Annual Conference Companion on Genetic and Evolutionary Computation, London, UK, 7–11 July 2007; pp. 3001–3020.
44. Beasley, J. OR-Library. 1990. Available online: https://goo.gl/lO1UQ6 (accessed on 3 August 2021).
45. Lilliefors, H. On the Kolmogorov-Smirnov Test for Normality with Mean and Variance Unknown. *J. Am. Stat. Assoc.* **1967**, *62*, 399–402. [CrossRef]
46. Mann, H.; Donald, W. On a Test of Whether one of Two Random Variables is Stochastically Larger than the Other. *Ann. Math. Stat.* **1947**, *18*, 50–60. [CrossRef]
47. Soto, R.; Crawford, B.; Olivares, R.; Niklander, S.; Johnson, F.; Paredes, F.; Olguín, E. Online control of enumeration strategies via bat algorithm and black hole optimization. *Nat. Comput.* **2016**, *16*, 241–257. [CrossRef]
48. Castillo, M.; Soto, R.; Crawford, B.; Castro, C.; Olivares, R. A Knowledge-Based Hybrid Approach on Particle Swarm Optimization Using Hidden Markov Models. *Mathematics* **2021**, *9*, 1417. [CrossRef]
49. Soto, R.; Crawford, B.; Olivares, R.; Taramasco, C.; Figueroa, I.; Gómez, Á.; Castro, C.; Paredes, F. Adaptive Black Hole Algorithm for Solving the Set Covering Problem. *Math. Probl. Eng.* **2018**, *2018*, 1–23. [CrossRef]
50. Crawford, B.; Soto, R.; Olivares, R.; Riquelme, L.; Astorga, G.; Johnson, F.; Cortes, E.; Castro, C.; Paredes, F. A self-adaptive biogeography-based algorithm to solve the set covering problem. *RAIRO-Oper. Res.* **2019**, *53*, 1033–1059. [CrossRef]
51. Valdivia, S.; Crawford, B.; Soto, R.; Lemus-Romani, J.; Astorga, G.; Misra, S.; Salas-Fernández, A.; Rubio, J.M. Bridges Reinforcement Through Conversion of Tied-Arch Using Crow Search Algorithm. In *Computational Science and Its Applications—ICCSA 2019*; Springer International Publishing: Cham, Switzerland, 2019; pp. 525–535. [CrossRef]
52. García, J.; Moraga, P.; Valenzuela, M.; Crawford, B.; Soto, R.; Pinto, H.; Peña, A.; Altimiras, F.; Astorga, G. A Db-Scan Binarization Algorithm Applied to Matrix Covering Problems. *Comput. Intell. Neurosci.* **2019**, *2019*, 1–16. [CrossRef] [PubMed]

MDPI

Article

Carbon Trading Mechanism, Low-Carbon E-Commerce Supply Chain and Sustainable Development

Liang Shen [1], Xiaodi Wang [1], Qinqin Liu [2], Yuyan Wang [3,*], Lingxue Lv [3] and Rongyun Tang [4]

1 School of Public Finance and Taxation, Shandong University of Finance and Economics, Jinan 250014, China; 20067433@sdufe.edu.cn (L.S.); 20049447@sdufe.edu.cn (X.W.)
2 Physical Education College, Shandong University of Finance and Economics, Jinan 250014, China; 20065959@sdufe.edu.cn
3 School of Management Science and Engineering, Shandong University of Finance and Economics, Jinan 250014, China; 192106011@mail.sdufe.edu.cn
4 Department of Industrial and System Engineering, University of Tennessee, Knoxville, TN 37996, USA; rtang7@vols.utk.edu
* Correspondence: wangyuyan1224@126.com or 20088164@sdufe.edu.cn

Abstract: Considering the carbon trading mechanism and consumers' preference for low-carbon products, a game decision-making model for the low-carbon e-commerce supply chain (LCE-SC) is constructed. The influences of commission and carbon trading on the optimal decisions of LCE-SC are discussed and then verified through numerical analysis. On this basis, the influence of carbon trading on regional sustainable development is empirically analyzed. The results show that the establishment of carbon trading pilots alleviates the negative impact of unfair profit distribution. Increasing the commission rate in a reasonable range improves the profitability of LCE-SC. Nevertheless, with the enhancement of consumers' low-carbon preference, a lower commission rate is more beneficial to carbon emission reduction. The total carbon emission is positively related to the commission rate. However, the unit carbon emission decreases first and then increases with the commission rate. The influence of the carbon price sensitivity coefficient on the service level is first positive and then negative, while the influence on the manufacturer's profit goes the opposite. The empirical analysis confirms that the implementation of carbon trading is conducive to regional sustainable development and controlling environmental governance intensity promotes carbon productivity.

Keywords: carbon emission; carbon trading; e-commerce supply chain; sustainable development

Citation: Shen, L.; Wang, X.; Liu, Q.; Wang, Y.; Lv, L.; Tang, R. Carbon Trading Mechanism, Low-Carbon E-Commerce Supply Chain and Sustainable Development. *Mathematics* **2021**, 9, 1717. https://doi.org/10.3390/math9151717

Academic Editor: Frank Werner

Received: 28 May 2021
Accepted: 18 July 2021
Published: 21 July 2021

1. Introduction

At present, global climate change caused by greenhouse gas has become a serious threat to sustainable development [1], and it has become a global consensus to take reasonable and effective measures to control carbon emission [2]. As early as 1960, Coase [3] proposed that the problem of externalities can be solved by defining property rights and trading voluntarily in the market. Stern [4] and Yang et al. [5] also mentioned that establishing a carbon trading market pricing by the market is an effective emission-reduction measure since the external cost for carbon emission can be internalized. Since the establishment of carbon trading pilots in Beijing, Shanghai, Tianjin, Fujian Province, Guangdong Province, Hubei Province, and Chongqing in 2013, the Chinese carbon trading market has been active. At the end of 2017, carbon trading was officially launched in China [6]. According to data from taipaifang.com (http://www.tanjiaoyi.com, accessed date 20 July 2021), the total transaction volume of the seven pilots amounted to RMB 94.9 million in 2019.

Implementing the carbon trading mechanism exerts manifold impacts. For low-emission enterprises, production costs are directly suppressed and even economic benefits can be obtained through carbon trading. For high-emission enterprises, excessive carbon emission brings greater production costs and social pressure, which impels them to invest more manual labor and materials in carbon emission reduction (CER) and promote the transformation of development mode [7]. For instance, as the first wave of enterprises included in the emission management in Hubei Province, Huaxin Cement emitted carbon exceeding the quota by 1.153 million tons and spent more than RMB 30 million to purchase the carbon quota in 2014. In 2015, the enterprise achieved a surplus carbon quota of 424,000 tons through investment in CER, and its net income from carbon trading exceeded RMB 9 million [8].

As the burgeoning commercial form, the e-commerce platform has changed the operation mode of the traditional supply chain. Due to convenience and efficiency, the e-commerce platform has won the favor of lots of consumers. Data from the National Bureau of Statistics of China show that Chinese online retail sales amounted to RMB 10.63 trillion in 2019, an increase of 16.5% over 2018. In practice, there are two working forms of the e-commerce platform, reselling and agency selling. Since the enterprise can directly decide key factors such as retail price and thus control market demand through pricing power, the majority of e-commerce platform's suppliers prefer agency selling. As a result, the e-commerce platform and the supplier form an e-commerce supply chain (E-SC) that is different from the offline one [9]. As the revolution of the supply chain in the Internet era, the E-SC has become the main supply chain operation mode and the most important network economic carrier.

Considering the impact of implementing carbon trading on corporate profitability and the change of enterprise operation mode in the Internet era, the decision-making of emission-dependent enterprises has changed. However, existing research on the low-carbon supply chain centers on enterprises in traditional supply chains [10–12], so it is innovative to discuss the decision-making of the E-SC under the carbon trading mechanism. Thus, our research focuses on the following issues. Firstly, in the context of carbon trading, how should the low-carbon e-commerce supply chain (LCE-SC) make decisions? Secondly, what impacts does fluctuation of the carbon market and platform fee exert on the LCE-SC's optimal decisions, including the decisions on production and environmental protection? Thirdly, as an important measure to achieve sustainable development, how does the carbon trading mechanism influence corporate economic and environmental performance, and further influence regional sustainable development? The goals of our work are to identify the operation mode of LCE-SC under carbon trading and examine the aforementioned issues. Then, we expect to verify the significance of carbon trading implementation and put forward relevant suggestions for enterprise operation and government policymaking.

Game theory is a typical method to study the decision-making of e-commerce supply chains and low-carbon supply chains. For example, the literature [13,14] uses game theory to solve the equilibrium decision of the e-commerce supply chain model, proving the feasibility of using game theory to explore the decision-making of LCE-SC. Besides, game theory can well reflect the confrontation between the e-commerce platform and suppliers. PSM-DID is recognized as an excellent empirical method to study the implementation effect of a policy. For example, using PSM-DID, Jia et al. [15] demonstrated the positive impact of high-speed rail construction on China's regional economic development, and Liu et al. [16] analyzed the impact of environmental regulation on enterprise green innovation. However, the two methods have not been combined to study carbon trading. This is exactly our innovation, that is, to combine micro modeling research with macro empirical research to explore the implementation effect of carbon trading.

Specifically, this paper firstly constructs an LCE-SC decision-making model which consists of a manufacturer with a certain carbon cap and an agency-selling e-commerce platform that provides the manufacturer with sales service. Then, the influence of carbon trading on the decisions and performance of LCE-SC is discussed. On this basis, empirical analysis is conducted to further study the influence on the sustainable development of the enterprise group, that is, regional sustainable development. The results are as follows.

Different from existing studies [17–19], an LCE-SC considering the carbon trading mechanism and consumers' low-carbon preference is constructed in this paper. It is found that the increased carbon price sensitivity coefficient leads to an increase and then a decline in the e-commerce platform's service level. The influence of the commission rate on the total carbon emission is positive, but the influence on unit carbon emission is first negative and then positive. Compared with the low-cap manufacturer, the high-cap one's sales price is higher but profit is lower. The e-commerce platform cooperating with the high-cap manufacturer can make more profits.

The moderating effect of carbon parameters is discussed. Different from the existing research conclusion [20], it is found that within a certain range, with the increase in the commission rate, the e-commerce platform's and supply chain system's profits both increase. However, when the threshold is exceeded, the overall profit of LCE-SC decreases; and although the e-commerce platform gains a high percentage of profit, its actual profit declines. Besides, as consumers' low-carbon preference is enhanced, a lower commission rate is more beneficial to reducing emissions and improving supply chain profit. This study also indicates that the carbon price sensitivity coefficient exerts a non-linear effect on the manufacturer's profit: with the increase in the coefficient, the manufacturer's profit decreases first and then increases. Moreover, the higher the coefficient, the more significant the marginal impacts of optimizing emission-reduction cost on the manufacturer's and the e-commerce platform's profits.

This paper combines micro research with macro research and adopts theoretical modeling and empirical analysis to study the implementation effect of carbon trading. Based on numerical simulation, this paper proposes a preliminary assumption that the carbon trading mechanism positively affects regional carbon productivity. Using panel data of 30 provincial administrative regions in China from 2009 to 2017, an empirical analysis is conducted. It is found that implementing the carbon trading mechanism is conducive to regional sustainable development in China's provinces. Resource endowment, industrial structure, environmental governance, and demographic factors also have a certain impact on regional sustainable development. It is also noteworthy that excessive environmental regulation is not beneficial to regional sustainable development, which shows that environmental governance programs should be optimized.

The rest of this paper consists of the following parts. Firstly, a literature review is provided in Section 2. Section 3 is the description and assumptions of the LCE-SC decision-making model. The optimal decisions of LCE-SC are deduced and the influence mechanism of commission rate and carbon trading is discussed in Section 4. Section 5 is the empirical research on the influence of carbon trading on regional sustainable development. The conclusions and managerial insights are proposed in Section 6.

2. Literature Review

The literature closely related to this study is organized into the following three streams: decision-making of LCE-SC, the influence of the carbon trading mechanism on decisions of low-carbon supply chains, and the influence of carbon trading on regional sustainable development.

2.1. Decision-Making of LCE-SC

The decision-making problem of LCE-SC is a hot spot in current research. Ji and Sun [21] constructed four decision-making models of e-commerce delivery strategies with diverse emission restriction intensities and analyzed the influence of the restriction intensity on e-commerce enterprises' decisions. Considering customers' low-carbon awareness, Han and Wang [20] discussed the pricing strategy of LCE-SC and designed a coordination mechanism of the system. Wang and Huang [22] studied the return strategy, pricing, and CER decisions under online sales and carbon tax. Wang et al. [23] discussed the impact of government low-carbon subsidy on the recycling strategy of the closed-loop E-SC. These studies focused on the influence of CER on the supply chain operation, without consideration of carbon trading. Unlike existing research, this paper explores the influence mechanism of carbon trading on LCE-SC decision-making.

2.2. Influence of Carbon Trading on Decisions of Low-Carbon Supply Chains

The carbon policy is internalized in the operation cost and influences the supply chain decision-making together with economic factors [24]. More and more scholars tend to consider the impacts of both the carbon trading market and product market on supply chain members' performances. The aim is to establish a supply chain model that follows the Triple Bottom Line Principle of economic-social-environmental [25,26]. Focusing on the management of the two-echelon supply chain, Dong et al. [27] discussed the impact of carbon trading on the output and sales price. Du et al. [28] found that the carbon trading policy is easier to implement and more effective to save public resources than other government punitive measures. Xu et al. [29] studied the decision-making of CER and coordination in the supply chain considering carbon trading. Xu et al. [30] discussed the influence of carbon trading on the production and pricing decisions of the make-to-order supply chain. Wang et al. [31] focused on the fresh supply chain and discussed its optimal decisions under cap-and-trade. The above literature studies the influence of emission policies on the decision-making of offline supply chains. Unlike existing studies, this paper extends carbon trading to the rapidly developing LCE-SC and discusses its influence mechanism.

2.3. Influence of Carbon Trading on Regional Sustainable Development

The influence of carbon trading on regional sustainable development has been studied. The carbon trading market was first launched in the United States, the United Kingdom, and the European Union, and has contributed to CER [32]. Wang et al. [33] built a CGE model for Guangdong Province to analyze the influence of carbon trading on the province's economy of China and found that the mechanism can effectively reduce GDP losses and achieve the strict emission-reduction targets. Zhao et al. [34] constructed a dynamic simulation model and found that the negative effect of carbon trading on the GDP of the Beijing-Tianjin-Hebei region is far less than the positive effect on energy saving and CER. Zhou et al. [35] proved empirically that the implementation of carbon trading has caused a decline in China's carbon intensity. However, from the provincial perspective, the establishment of pilots only has an obvious negative effect on the carbon intensities of Beijing and Guangdong Province [36]. Using semi-structured interviewing, Hamzah et al. [37] confirmed that the implementation of carbon trading is consistent with Malaysia's sustainable development goals. With carbon emission intensity as one of the control variables, the carbon trading mechanism affects both CER and economic growth, and the effect extent is diverse in different provinces [38]. Zheng et al. [39] adopted a multi-agents technique and found by model simulation that the carbon trading mechanism harms the growth of GDP while reducing emission. It is recommended that in order to maintain economic stability, different regions need to set different emission restrictions.

Most of the existing literature uses carbon intensity as the dependent variable to empirically study the implementation effect of carbon trading. However, carbon intensity emphasizes CER rather than economic growth, which is the opposite of what developing countries seek. Therefore, this paper adopts carbon productivity as the dependent variable to explore the impact of the carbon trading mechanism on enterprise group behavior, that is, the impact on regional sustainable development.

The differences between our research and the related literature are shown in Table 1.

Table 1. Papers that are most related to our research.

Author(s)	Supply Chain System	Policy	Research Method	Customers' Environmental Awareness	Emission Reduction Investment	Variable Carbon Price
Xu et al. [29]	A manufacturer and a retailer	Carbon trading	Modeling and numerical simulation	Yes	Yes	No
Fan et al. [11]	A manufacturer and a retailer	Carbon trading	Modeling and numerical simulation	No	Yes	No
Xu et al. [40]	A manufacturer and a retailer	Governmental subsidy	Modeling and numerical simulation	Yes	Yes	No
Ma et al. [41]	A supplier, a third-party logistics service provider, and a retailer	Carbon trading and carbon tax	Modeling and numerical simulation	No	No	No
Wang et al. [31]	A supplier and multiple retailers	Carbon trading	Modeling and numerical simulation	No	No	No
Xia et al. [42]	A manufacturer; an ordinary manufacturer and a low-carbon manufacturer	Carbon trading	Modeling and numerical simulation	Yes	No	No
Liu et al. [43]	A manufacturer and a retailer	Power control structure	Modeling and numerical simulation	Yes	Yes	-
This paper	A manufacturer and an e-commerce platform	Carbon trading	Modeling, numerical simulation, and empirical analysis	Yes	Yes	Yes

3. Problem Description and Assumptions

An LCE-SC model composed of an emission-dependent manufacturer and an agency-selling e-commerce platform is constructed in this paper and the model structure is shown in Figure 1. The manufacturer reaches cooperation with the e-commerce platform before production and sells products to online consumers through the platform. In return, the e-commerce platform charges a constant proportion of commission per unit of product [44]. Exogenous commission rate has been widely used in the research of the e-commerce supply chain [45]. In this working form, the manufacturer has the pricing power and can control the market demand through sales price. Settling in the platform, the manufacturer produces products and decides sales price. Carbon emissions are generated during production. If the carbon emissions are excessive, the manufacturer needs to purchase the carbon quota from other enterprises; otherwise, the surplus carbon quota is sold. The e-commerce platform provides the manufacturer with sales promotion services, such as online-store display, advertising, online customer service, and credit maintenance. Online consumers purchase products through the platform. Subsequently, the platform transmits the order to the manufacturer in charge of delivery. After consumers receive the product, the platform returns the payment to the manufacturer and charges a commission [46].

Assume that the carbon trading market has been established in the country or region, where the manufacturer can sell or purchase carbon emission. Similar to the product market, carbon trading price is affected by carbon emission.

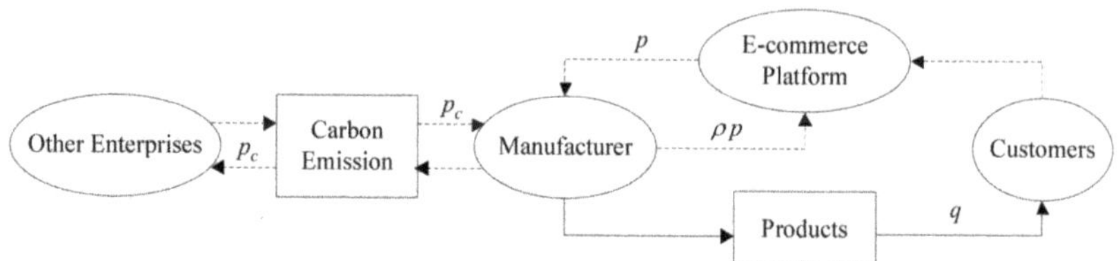

Figure 1. Model structure of LCE-SC.

The model symbols are described as follows:

c—unit production cost without consideration of the cost of CER.

p—unit sales price of products, the manufacturer's decision variable.

ρ—commission rate, which represents the platform fee for the unit sales. The parameter is assumed to satisfy $0 < \rho < 1 - \frac{c}{p}$ to ensure that manufacturing is more lucrative.

s—service level of sales promotion for low-carbon products, the e-commerce platform's decision variable. According to Nair and Narasimhan [47], the cost function is assumed to be $C(s) = ls^2/2$, among which $l(l > 0)$ is the coefficient of service cost, specifically referring to the cost of improving the unit service level.

Y—total carbon emission, the manufacturer's decision variable. Assuming that the emission cap of the manufacturer is Y^U, which is the amount of carbon emission produced by the conventional production, and there is $Y \leq Y^U$. As the manufacturer invests in emission-reduction equipment, the cost for reducing carbon emission is assumed as $I(Y) = h(Y^U - Y)^2$ $I(Y) = h(Y^U - Y)^2$ $I(Y) = h\left(Y^U - Y\right)^2$, where $h(h > 0)$ is the coefficient of emission-reduction cost. There are $I(Y) \geq 0$, $I'(Y) \leq 0$, $I^{''(Y)} > 0$, which mean that as CER increases (i.e., carbon emission decreases), the cost increases, and the marginal cost increases.

q—market demand for low-carbon products. Since the e-commerce platform's sales service directly affects consumers' online shopping experience, the market demand is sensitive to service level. Drawing on the demand function form of Xia et al. [48], it is supposed that the demand function is

$$q = \alpha - \beta p + \gamma s + k\left(Y^U - Y\right) \tag{1}$$

where α is the size of the product market, β, γ and k, respectively, represent sensitivity coefficients of price, service level, and CER for product demand.

p_c—carbon price. The pricing mechanism of the carbon market is similar to the product market, which means that p_c is affected by supply and demand. The inverse demand function is

$$p_c = \alpha_c - rY \tag{2}$$

where α_c is the scale of the carbon market, and r is the carbon price sensitivity coefficient which measures the punishment intensity for carbon emissions.

To ensure that the manufacturer who emits much carbon is punished, p_c in the inverse demand function (i.e., Equation (2)) is allowed to be negative. $p_c > 0$ means that the manufacturer invests in CER and sells the redundant carbon emission to obtain profits. The higher the carbon emission, the lower the profit. When $p_c < 0$, namely the emission is higher than α_c/r, the manufacturer is obliged to purchase carbon emission or get punished due to high emission, thus paying a cost.

To ensure the practical meaning of the research problem, it is assumed that $a_c - rY^U < 0$, which means the high-emission manufacturer suffers punishment. Moreover, $4(h + r)\beta - k^2(1 - \rho) > 0$, $2rY^U - \alpha_c > ck$, $\alpha(1 - \rho) > c\beta$, $ck + 4hY^U + 2\alpha_c > \frac{Akl(1-\rho)\left(\alpha+kY^U\right)}{l\beta A-2B}$, which can ensure that the optimal solutions of the model exist and are positive.

On the basis of the model assumptions, the manufacturer's profit function in the product market is:

$$\pi_{M_1} = (p - c - \rho p)\left[\alpha - \beta p + \gamma s + k\left(Y^U - Y\right)\right] \tag{3}$$

The manufacturer's profit function in the carbon market is:

$$\pi_{M_2} = (a_c - rY)Y - h(Y^U - Y)^2 \tag{4}$$

The manufacturer's total profit function is:

$$\pi_M = \pi_{M_1} + \pi_{M2} = (p - c - \rho p)\left[\alpha - \beta p + \gamma s + k\left(Y^U - Y\right)\right] + (a_c - rY)Y - h(Y^U - Y)^2 \tag{5}$$

The e-commerce platform's profit function is:

$$\pi_E = \rho p\left[\alpha - \beta p + \gamma s + k\left(Y^U - Y\right)\right] - ls^2/2 \tag{6}$$

4. Optimal Decisions of LCE-SC and the Influence Mechanism of Carbon Trading

4.1. Optimal Decisions

In practice, the e-commerce platform formulates and publishes the conditions of entry for the manufacturer to settle in the platform, and most large-scale e-commerce companies, such as Tmall (https://www.tmall.com, accessed date 20 July 2021) and Youpin (https://www.xiaomiyoupin.com, accessed date 20 July 2021), set higher entry thresholds to maintain brand benefits. Only when the manufacturer satisfies the conditions can it enter and cooperate with the platform. Thus, with consideration of the actual operation of e-business, the leading enterprise in LCE-SC is assumed to be the e-commerce platform. The manufacturer, as the follower, follows the sale rules to sell low-carbon products. Thus, the platform and the manufacturer constitute a Stackelberg game model. In decision-making, the platform first decides its service level s. The manufacturer subsequently makes decisions on the carbon emission Y and the sales price p. The solutions of backward induction are shown as follows.

It can be derived from Equation (5) that the Hessian matrix of π_M is $H = \begin{bmatrix} \frac{\partial^2\pi_M}{\partial p^2} & \frac{\partial^2\pi_M}{\partial p\partial Y} \\ \frac{\partial^2\pi_M}{\partial Y\partial p} & \frac{\partial^2\pi_M}{\partial Y^2} \end{bmatrix}$ $= \begin{bmatrix} -2(1-\rho)\beta & -k(1-\rho) \\ -k(1-\rho) & -2(h+r) \end{bmatrix}$ and $det(H) = 4(h+r)\beta(1-\rho) - k^2(1-\rho)^2 > 0$. Besides, since $\frac{\partial^2\pi_M}{\partial p^2} < 0$, there is a maximum of π_M. The reaction functions of p and Y are the simultaneous solution of $\partial\pi_M/\partial p = 0$ and $\partial\pi_M/\partial Y = 0$.

$$p = \frac{2(h+r)c\beta + \left[2(h+r)(\alpha + s\gamma) + k\left(2rY^U - \alpha_c - ck\right)\right](1-\rho)}{4(h+r)\beta(1-\rho) - k^2(1-\rho)^2} \tag{7}$$

$$Y = \frac{\left(ck + 4hY^U + 2\alpha_c\right)\beta - k(1-\rho)\left(kY^U + \alpha + s\gamma\right)}{4(h+r)\beta - k^2(1-\rho)}. \tag{8}$$

Substituting Equations (7) and (8) into Equation (6), $\partial^2\pi_E/\partial s^2 = -l < 0$ can be derived, so the maximum of π_E exists. According to $\partial\pi_E/\partial s = 0$, the e-commerce platform's optimal service level is

$$s^* = \frac{2B\left[4\alpha(h+r) - ck^2 + 2k\left(2rY^U - \alpha_c\right)\right]}{\gamma\left[A^2 l - 8B(h+r)\right]} \tag{9}$$

The optimal sales price is derived by substituting Equation (9) into Equation (7).

$$p^* = \frac{F_2 + cF_3 + F_6}{(1-\rho)F_3}$$

Similarly, according to Equations (8) and (9), the optimal carbon emission is

$$Y^* = \frac{(Al\beta - 2B)(ck + 4hY^U + 2\alpha_c) - Akl(1-\rho)(\alpha + kY^U)}{A^2 l - 8B(h+r)}$$

According to $y = Y/q$, the optimal unit carbon emission of the product can be calculated.

$$y^* = \frac{(1-\rho)F_4}{\beta[8Bc(h+r) + F_2 + F_6]}$$

Correspondingly, the manufacturer's profit in the product market is

$$\pi^*_{M_1} = \frac{2(h+r)\beta(F_1+F_2)[2(h+r)F_1 + F_2]}{F_3^2(1-\rho)}$$

The manufacturer's profit in the carbon market is

$$\pi^*_{M_2} = \frac{\alpha_c F_3 F_4 - rF_4^2 - h(Y^U F_3 - F_4)^2}{F_3^2}$$

The manufacturer's total profit is

$$\pi^*_M = \frac{2(h+r)\beta(F_1+F_2)[2(h+r)F_1 + F_2]}{F_3^2(1-\rho)} + \frac{\alpha_c F_3 F_4 - rF_4^2 - h(Y^U F_3 - F_4)^2}{F_3^2}$$

The e-commerce platform's profit is

$$\pi^*_E = \frac{1}{F_3^2}\left\{\frac{\beta\rho[2(h+r)F_1 + F_2][2(h+r)F_1 + F_2 + cF_3]}{(1-\rho)^2} - 2l(h+r)\beta\rho F_5^2\right\}$$

The common factors are as follows: $A = 4(h+r)\beta - k^2(1-\rho)$, $B = (h+r)\beta\gamma^2\rho$, $F_1 = 2Bc + Al(\alpha - c\beta - \alpha\rho)$, $F_2 = Akl(1-\rho)(2rY^U - \alpha_c)$, $F_3 = A^2 l - 8B(h+r)$, $F_4 = (Al\beta - 2B)(ck + 4hY^U + 2\alpha_c) - Akl(1-\rho)(\alpha + kY^U)$, $F_5 = 4\alpha(h+r) - ck^2 + 2k(2rY^U - \alpha_c)$, $F_6 = 2Al(h+r)(\alpha - c\beta - \alpha\rho)$.

4.2. Analysis of LCE-SC Model

Proposition 1. *Manufacturer's optimal carbon emission Y^* and the unit carbon emission y^* are positively related to Y^U, α_c, and h, while Y^* and y^* are negatively related to k and r. Y^* is positively related to ρ, while there are two cases of the relationship between y^* and ρ: when ρ satisfies $\rho < 1 - \frac{[F_2 + 8(h+r)Bc + F_6]\{kl(kY^U+\alpha)(1-\rho)[2A - k^2(1-\rho)] + (ck + 4hY^U + 2\alpha_c)\beta F_7\}}{F_4\{2c\beta(h+r)[4(h+r)\gamma^2 - k^2 l] - l[2(h+r)\beta - k^2(1-\rho)](F_5 + ck^2)\}}$, y^* is negatively related to ρ; otherwise, y^* is positively related to ρ.*

Note that $F_7 = 2k^2 l(1-\rho) + 2(h+r)\gamma^2(2\rho - 1) - 4(h+r)l\beta$.

Proof of Proposition 1. See Appendix A. □

According to Proposition 1, except ρ, the correlations of other parameters with Y^* and y^* are similar. Thus, these two decision variables are represented by carbon emission here.

For the manufacturer, the carbon emission increases with the emission cap Y^U. Therefore, the carbon emission of the high-cap manufacturer is still high under carbon trading and CER. With the increase in the coefficient of emission-reduction cost h, the manufacturer pays more for CER, and the revenues from the products market and carbon market are insufficient to cover the cost. Therefore, the manufacturer prefers high carbon emissions to ensure profit.

With the increase in the commission rate ρ, in order to obtain a high profit after paying commission, appropriate depression of CER is an alternative method for the manufacturer to control costs. However, the unit carbon emission decreases first and then increases with the commission rate. The reason lies in the significant increase in product demand caused by the increase in the commission. Hence, the total carbon emission increases, but the unit emission decreases. Once the threshold of the commission rate is exceeded, a serious distribution inequity erodes the manufacturer's enthusiasm for production, which means that the market demand drops off and the unit carbon emission goes up. The practical significance of this conclusion is that increasing the commission rate leads to environmental deterioration, and once the commission rate is too high, the market share of the product will be seriously damaged. It is more advantageous to control the commission rate in a lower range to realize the benign operation of the supply chain. In addition, the higher the sensitivity coefficient of CER k, the higher the market demand for low-carbon products, which further motivates the manufacturer to control emissions.

Changes in the scale of the carbon market α_c and the carbon price sensitivity coefficient r reflect the influence mechanism of the carbon market. A larger scale of carbon market means a higher threshold for the manufacturer to bear high-emission punishment, which lowers the enthusiasm for CER and results in higher carbon emissions. Moreover, a larger carbon price sensitivity coefficient means that high emissions can lead to an excessively low or negative carbon price, causing substantial economic loss. In this situation, the manufacturer tends to reduce emissions to obtain profit. This shows that the carbon market directly affects the carbon emission of the manufacturer, and government departments can regulate the carbon emissions of manufacturing enterprises by adjusting the supply and demand relationship in the carbon market.

Proposition 2. *The e-commerce platform's optimal service level s^* is positively related to Y^U, ρ, and k, while s^* is negatively related to α_c and h. There are two cases of the relationship between s^* and r: when r satisfies $ck + 4hY^U + 2\alpha_c > 2klA(1-\rho)F_5/F_3$, s^* is positively related to r; otherwise, s^* is negatively related to r.*

Proof of Proposition 2. See Appendix B. □

As can be seen from Proposition 2, with the increase in the emission cap Y^U, carbon emission and emission reduction $(Y^U - Y^*)$ both increase. The e-commerce platform is willing to provide a better sales promotion service for low-carbon products. It is indicated that in Proposition 1, a higher coefficient of emission-reduction cost h leads to more carbon emission. As a result, the e-commerce platform puts less emphasis on the products, and the service level decreases accordingly. With the increase in the sensitivity coefficient of CER k, the market demand increases. The increasing profit impels the platform to improve sales service to promote its brand value. When referring to the carbon market, the influence mechanism of the scale of the carbon market α_c is found to be similar to h. Besides, the correlation between the optimal service level and the carbon price sensitivity coefficient r is first positive and then negative. When r is small, increasing the coefficient impels the manufacturer to reduce emission, and its low-carbon products can gain more favor from e-commerce platform; when r is large, the manufacturer's profit gained for emission reduction is far less than the cost, so the increment of emission reduction decreases, and a lower service level is provided by the e-commerce platform. The carbon price sensitivity coefficient indirectly affects the sales promotion service of the e-commerce platform, that is, blindly strengthening the punishment intensity for carbon emissions is not conducive to improving the service level, which will affect consumers' online shopping experience. The practical significance of this conclusion is that the government should control environmental regulation within a certain intensity range.

Proposition 3. *The optimal sale price p^* is positively related to Y^{U}, ρ, k, and r, while p^* is negatively related to α_c and h.*

Proof of Proposition 3. Similar to that of Proposition 1. □

According to Proposition 1, as the emission cap Y^{U} increases, the manufacturer suffers more punishment in the carbon market. Therefore, Proposition 3 shows that the sales price increases to make up for this loss. This means that the high-cap manufacturer's sales price is higher than the low-cap one's. However, an increased coefficient of emission-reduction cost h erodes the enthusiasm of the manufacturer for reducing emission, and the decline in variable costs leads to a lower sales price. Similarly, increasing the sensitivity coefficient of CER k can help lower carbon emissions. For the sake of maximizing its profit, the manufacturer increases sales price to compensate for the emission-reduction cost. As for the commission rate ρ, the increase in this parameter means that the e-commerce platform divides more profit. As a result, the manufacturer tends to guarantee its own profit by increasing sales price. Since the increase in the carbon-market scale α_c implies a decrease in the punishment for carbon emissions, the impact of α_c on the sales price is the same as the impact of h. On the contrary, increasing the carbon price sensitivity coefficient r causes the high-emission manufacturer to purchase carbon emissions and the low-emission one to pay more for CER according to the law of increasing marginal cost. As a result, the sales price increases. This conclusion is consistent with the research of Xing et al. [49] which shows that increasing sales price is the optimal strategy for the manufacturer under carbon trading.

Proposition 4. *(1) The optimal manufacturer's profit π_M^* is positively related to α_c and k, while π_M^* is negatively related to Y^{U}, h, and ρ. (2) The optimal e-commerce platform's profit π_E^* is positively related to Y^{U}, r, and k, while π_E^* is negatively related to α_c and h.*

Proof of Proposition 4. Similar to that of Proposition 1. □

As can be seen from Proposition 4, with the increase in the emission cap Y^{U}, the manufacturer's profit decreases, and the e-commerce platform's profit increases. In the context of carbon trading and CER, the profit of the high-cap manufacturer is lower, and the e-commerce platform cooperating with the high-cap manufacturer gains a higher profit. Moreover, as the coefficient of emission-reduction cost h increases, the sales price and market demand of the products decrease but the carbon emission increases. The manufacturer's revenues in the product market and the carbon market are insufficient to cover the increasing emission-reduction costs, which results in a decline in the profits of both members in LCE-SC. On the contrary, increasing the sensitivity coefficient of CER k causes an increase in the product demand, which boosts the manufacturer's revenue in the product market as well as the e-commerce platform's profit. Similarly, the carbon price sensitivity coefficient r increases the platform's profit. Understandably, a higher commission rate ρ brings less profit to the manufacturer. Since the expansion of the carbon market α_c causes a decrease in sales price and demand, the sales revenue and the profit shared by e-platform decline. However, the increase in the manufacturer's carbon revenue and the decline in the cost of CER boost its own profit.

Different from the previous conclusion that reducing carbon emission harms the profits of supply chain members [41,50], this paper points out that emission reduction does not erode the profits of supply chain members by comparing Proposition 1 and Proposition 4. On the contrary, by enhancing consumers' low-carbon preference or reducing the manufacturer's emissions-reduction cost, profits of the members in LCE-SC all increase, and carbon emission decreases, thus improving supply chain operation.

4.3. Numerical Analysis

In order to verify the above propositions and further discuss the influence of the parameters on the decision-making of LCE-SC, numerical examples are given below. Drawing on the research of Shen and Wang [51] and Wang et al. [23], the base parameters are supposed to be $\alpha = 10000$, $\beta = 5$, $\gamma = 2$, $c = 500$, $l = 1$.

Analyze the impact of commission rate ρ and the sensitivity coefficient of CER k on the LCE-SC's performances. Based on the base parameters, assume that $Y^U = 6000$, $\alpha_c = 5000$, $h = 1$, and $r = 1$, and take ρ and k as independent variables. The changing surfaces of economic and environmental performances are shown in Figure 2.

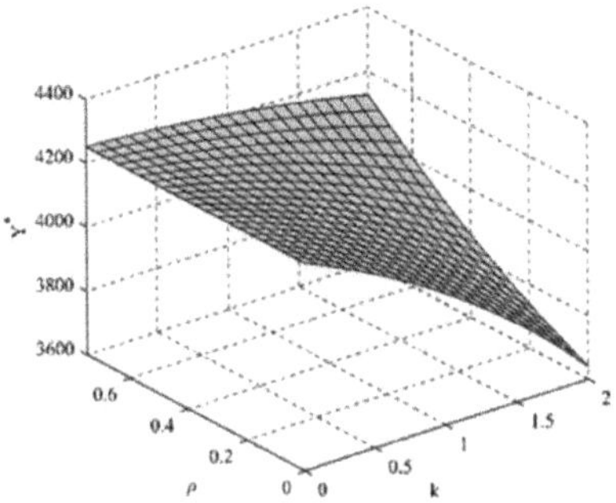

(**a**) Changes in carbon emission

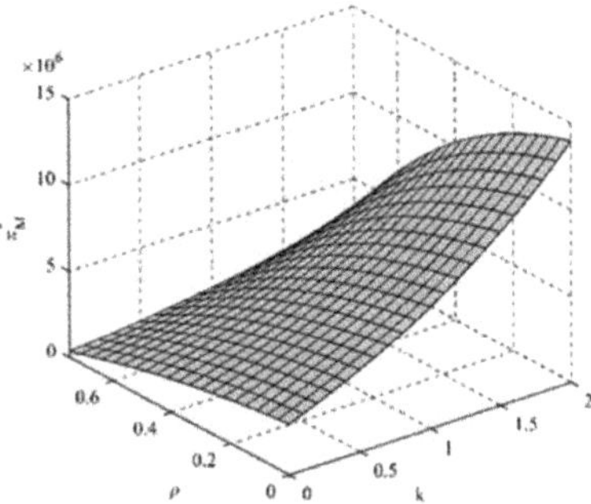

(**b**) Changes in the profit of manufacturer

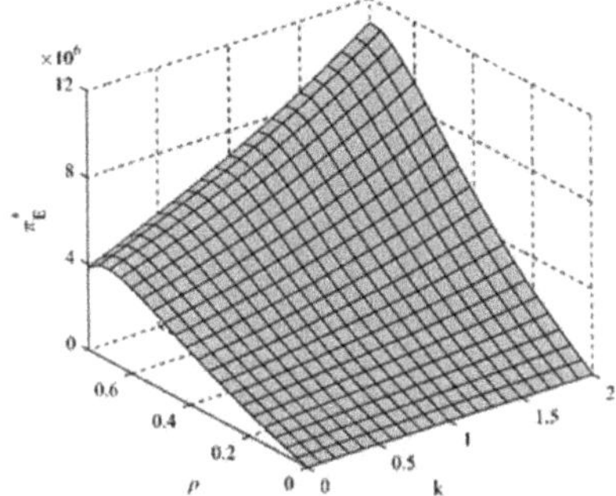

(**c**) Changes in the profit of e-commerce platform

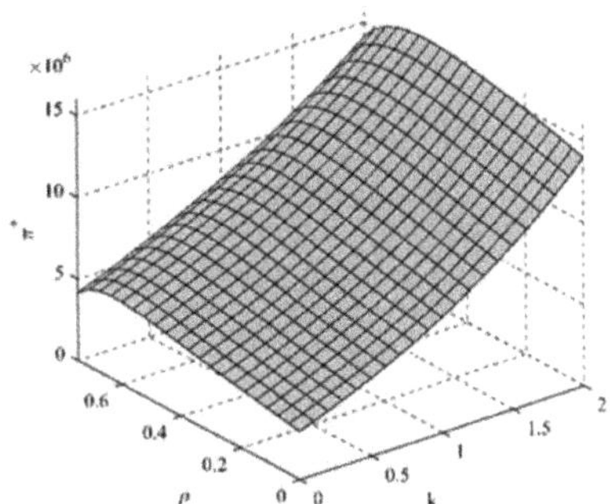

(**d**) Changes in the profit of LCE-SC

Figure 2. Changes in the LCE-SC's performances with ρ and k.

As can be seen from Figure 2d, with the appropriate increase in the commission rate ρ, the optimal profit of the LCE-SC increases. This is because the increase within a reasonable range in the commission rate causes an increase in market demand, which makes the supply chain maintain high profitability. It can be concluded that for LCE-SC, the implementation of carbon trading alleviates the negative impact of the unfair profit distribution. However, a high commission rate loses its coordinating role, and market demand drops significantly, which results in a decline in the overall profit of LCE-SC. Although the e-commerce platform obtains a high percentage of profit, its actual profit declines, which can explain why the commission rates set by major e-commerce platforms often do not exceed 30%. Additionally, no matter how the commission rate changes, increasing consumers' preference for low-carbon products is conducive to CER and also increases the profits of both parties. Moreover, the lower the commission rate, the more significant the marginal influence of the sensitivity coefficient of CER k on reducing emission and manufacturer's profit, which shows that as consumers' low-carbon preference is enhanced, a lower commission rate is beneficial to environmental protection.

Analyze the impact of the coefficient of emission-reduction cost h and carbon price sensitivity coefficient r on the LCE-SC's performances. Based on the base parameters, assume that $\rho = 0.05$, $k = 1$, $\alpha_c = 5000$, and $Y^U = 6000$, and take h and r as independent variables. The changing surfaces of economic and environmental performances are shown in Figure 3.

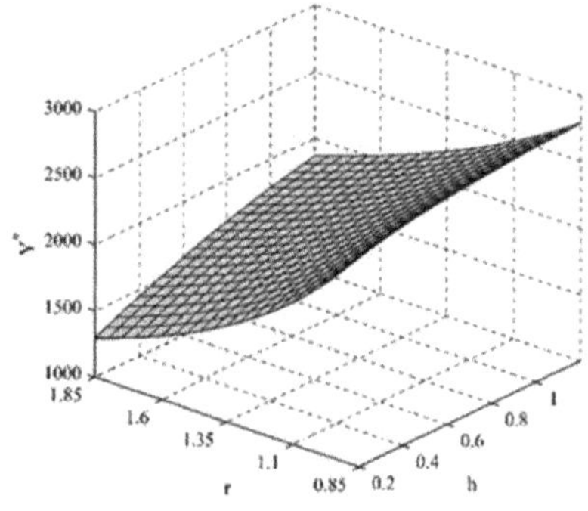

(**a**) Changes in carbon emission

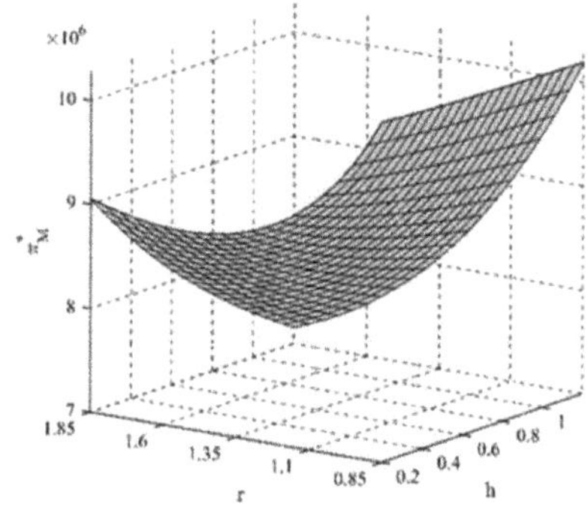

(**b**) Changes in the profit of manufacturer

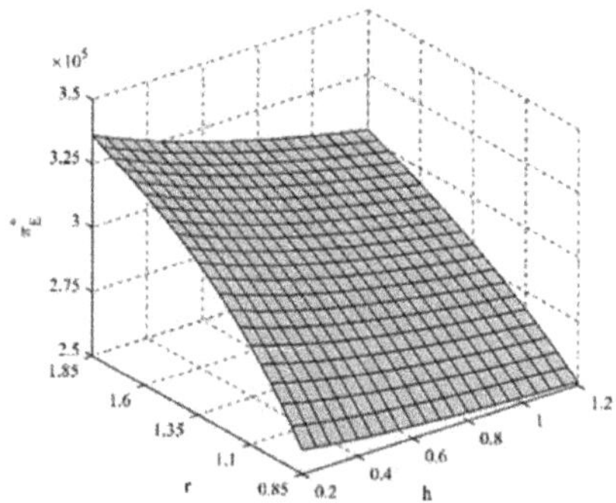

(**c**) Changes in the profit of e-commerce platform

Figure 3. Changes in optimal decisions with h and r.

It is graphically shown in Figure 3b that the influence of the carbon price sensitivity coefficient r on the manufacturer's profit is first negative and then positive. With the increase in r, carbon emission decreases according to Figure 3a, and the manufacturer's revenue is not enough to cover the ever-growing emission-reduction cost, resulting in a decrease in its profit. When r is too high, the manufacturer still chooses to reduce emission, but the extent of reduction becomes smaller. At this time, boosting sales revenue can make up for the emission-reduction cost, and the profit increases. It is illustrated in Figure 3b,c that, as r increases, the marginal impacts of emission-reduction cost coefficient h on profits of both members in LCE-SC increase. This shows that when r is in the low-value range, the method to optimize the emission-reduction program for reducing cost has a certain limitation; when r is larger, the marginal profit by optimizing the emission-reduction program is higher.

Analyze the impact of carbon emission cap Y^U and the scale of the carbon market α_c on the LCE-SC's performances. Based on the base parameters, assume that $\rho = 0.05$, $k = 1$, $h = 1$, and $r = 1$, and take Y^U and α_c as independent variables. The changing surfaces of economic and environmental performances are shown in Figure 4.

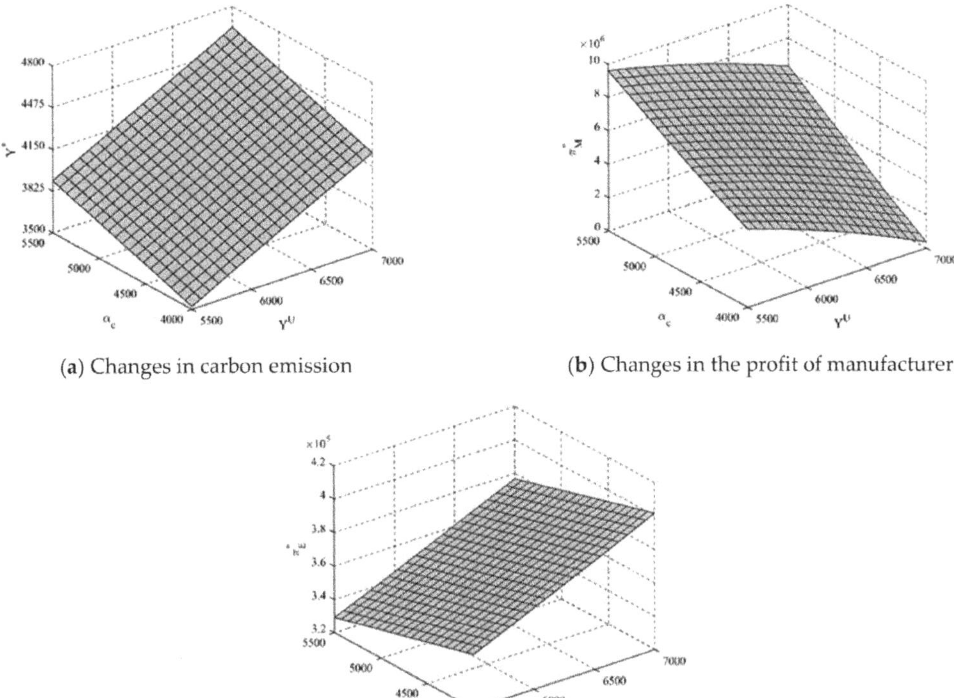

(**a**) Changes in carbon emission

(**b**) Changes in the profit of manufacturer

(**c**) Changes in the profit of e-commerce platform

Figure 4. Changes in optimal decisions with Y^U and α_c.

As illustrated in Figure 4, the changing surface of each variable with the carbon emission cap Y^U and the scale of the carbon market α_c is almost flat. It can be concluded that the changes in the carbon-market scale have the same impacts on the carbon emissions of both high-cap and low-cap enterprises and the profits of both members in LCE-SC. According to Proposition 1, the government can regulate carbon emissions through the scale of the carbon market, and the above conclusion indicates that this measure cannot produce differentiated CER effects for high-cap and low-cap enterprises, which causes a certain limitation.

With the rapid growth of the national economy, environmental issues related to the over-consumption of resources and energy have become more serious in China. Excess carbon emission has caused the greenhouse effect, which attracts widespread attention [52]. In the context of low carbon, enterprises seek a balance between maintaining profit growth and controlling carbon emission. However, the existing research centers on the impact of carbon trading on the industry from a macro perspective or the impact on corporate decision-making from a micro perspective [11], and there is no research combining the two. On the one hand, as important micro-units in the regional economy [53], enterprises are the main force to drive regional economic development. On the other hand, industrial enterprises are the focus to control carbon emissions, and their total amount of CER reflects the whole industry's effort to conserve energy and reduce emissions. Therefore, it is of practical significance to expand the micro research to the macro level.

The key to balancing economic development and emission reduction is to increase carbon productivity which is also the unique way for developing countries to achieve sustainable development [54]. To ascertain the influence of carbon trading on carbon productivity, assuming that $\rho = 0.05$, $k = 1$, $h = 1$, $r = 1$, and $Y^U = 6000$, α_c is taken as the explanatory variable. Changes in LCE-SC's carbon productivity ($\frac{\pi_M^* + \pi_E^*}{Y^*}$) and emission-dependent manufacturer's carbon productivity ($\frac{\pi_M^*}{Y^*}$) are shown in Figure 5. As shown in this figure, increasing the scale of the carbon market promotes the carbon productivities of both LCE-SC and the manufacturer. It can be preliminarily inferred from the micro perspective that enterprises' carbon productivity is improved due to the implementation of carbon trading. Then, what is the implementation effect of the carbon trading mechanism on the macro level? In the next section, the issue is addressed by empirical analysis. Considering the authority and availability of statistical data, regional carbon productivity is selected as the indicator to measure the sustainable development level of the enterprise group. The research relationships from the micro perspective and the macro perspective are shown in Figure 6.

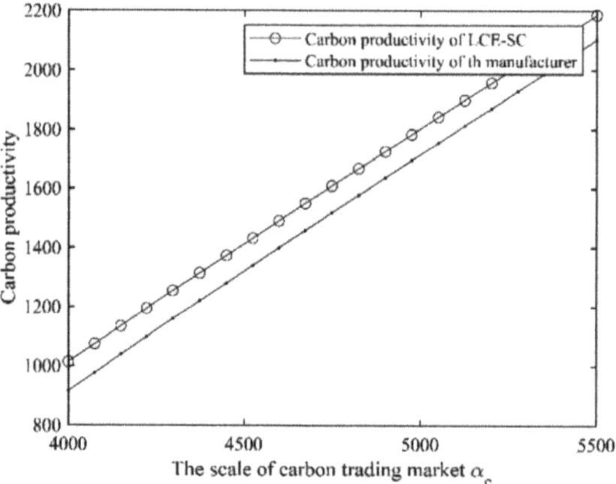

Figure 5. Changes in carbon productivity with the scale of carbon market.

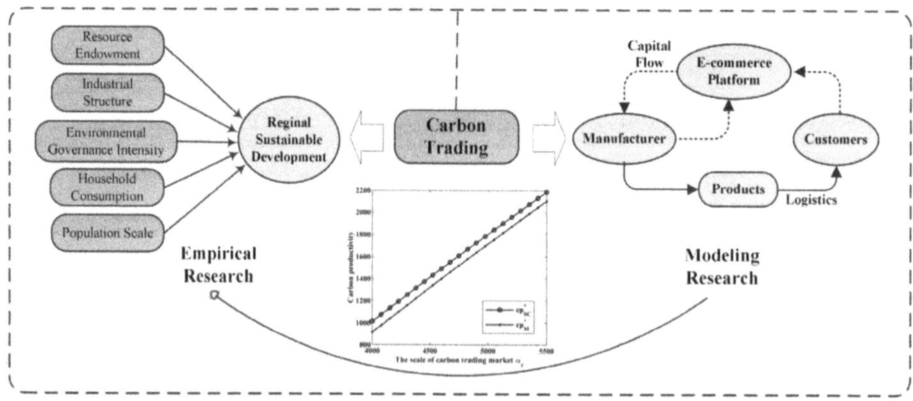

Figure 6. Relationship between the micro research and the macro research.

5. An Empirical Analysis of the Influence of Carbon Trading on Regional Enterprises' Sustainable Development

5.1. The Method of Empirical Analysis

Based on natural experiments and pooled cross-sections, the Difference-in-Difference method (DID) is widely adopted in evaluating the implementation effect of specific policies [55]. The advantage of DID lies in controlling the discrepancy between the experimental group and the control group before and after implementing the policy to eliminate some uncontrollable and unpredictable factors [56]. The basic form of DID is: $Y_{gt} = \beta_0 + \beta_1 T_g + \beta_2 P_t + \beta_3 (T_g \times P_t) + \varepsilon_{gt}$, among which T_g and P_t are dummy variables with $g = 1, \ldots, G$ indexing cross-sectional units and $t = 1, \ldots, T$ indexing periods. The interaction item is the estimation of the treatment effect under the parallel trend assumption [57]. In recent years, the DID gradually became the mainstream method for measuring the effect of carbon trading. For example, Zhang et al. [36] and Dong et al. [58] empirically analyzed the implementation effect of carbon trading policy on the basis of provincial panel data and the DID method. Zhu et al. [59] explored whether the carbon trading policy promotes green development efficiency in China by DID.

However, the selection of carbon trading pilots is not arbitrary. Instead, it is dependent on the regional economic level, historical data of carbon emission, environmental regulation, and other important indicators, which causes heterogeneity. Among such observational studies, scholars tend to choose the Propensity Score Matching (PSM) to overcome the selective bias in causality assessment [60]. Rosenbaum and Rubin [61] proved that among the observation subjects that match the propensity scores, the treatment group and the control group have similar baseline characteristic distribution. Therefore, scholars tend to combine the PSM and DID methods to verify the policy effect. For example, based on provincial panel data, Zhou et al. [35] conducted an empirical study to assess the influence of carbon trading policy using PSM-DID. With reference to previous research, this paper adopts PSM-DID to empirically analyze the influence of the carbon trading mechanism on regional sustainable development.

5.2. Variable Selection and Data Sources

5.2.1. Carbon Productivity

The research object of this section is the influence of the carbon trading mechanism on regional sustainable development. With reference to Zhang et al. [62], it is found that scholars prefer to choose carbon intensity as the dependent variable for empirical research. However, for developing countries, carbon intensity is more suited to measuring carbon emission reduction rather than stressing economic development. Thus, referring to Wang et al. [63], carbon productivity, which reflects GDP per unit carbon emission, is selected as the dependent variable. The higher carbon productivity means greater economic output and lower carbon emission, so carbon productivity can measure a country's or a region's effort to deal with the global warming problem, and place more emphasis on economic growth.

5.2.2. Control Variables

With reference to Yan et al. [64] and Hu et al. [65], resource endowment and industrial structure are selected as control variables in this paper. According to Proposition 2, environmental governance intensity is added. On this basis, referring to Xu et al. [66], household consumption and population scale are supplemented to reduce the endogenous deviation between the treatment group and the control group due to demographic factors. Table 2 shows the specific measurement methods of these variables.

Table 2. Measurement methods of control variables.

Variable	Measurement Method	Symbol
Resource Endowment	Proportion of investment in fixed assets of the mining industry (excluding rural households) to the total investment in fixed assets	*RE*
Industrial Structure	Proportion of secondary industry to regional GDP	*IS*
Environmental Governance Intensity	Proportion of investment completed in the treatment of industrial pollution to regional GDP	*EGI*
Household Consumption	Household consumption in the total consumption of energy	*HC*
Population Scale	Population at year-end	*POP*

5.2.3. Data Sources and Descriptive Statistics

Taking into account the availability and timeliness of the data, provincial panel data from 2009 to 2017 are used. The data of Tibet, Hong Kong, Macao, and Taiwan are excluded from the research since there are relatively more defaults. The original data sources for calculating the variables are as follows: GDP, investment in fixed assets of the mining industry (excluding rural households), investment of fixed assets in the whole society, GDP of the secondary industry, investment completed in the treatment of industrial pollution, and population at year-end come from the China Statistical Yearbook of 2010 to 2018 and the provincial annual database of the National Bureau of Statistics. The household consumption is from the China Energy Statistical Yearbook of 2010 to 2018. Table 3 shows the descriptive statistics of variables in the model.

Table 3. Descriptive statistics of variables.

Variables		*ln CP*	*RE*	*IS*	*EGI*	*ln HC*	*ln POP*
2009	Mean	7.76602	0.04620	0.47461	0.00159	6.86159	8.16080
	Std. Dev.	0.52569	0.04531	0.07627	0.00116	0.69372	0.76387
2010	Mean	7.82837	0.04374	0.49071	0.00113	6.95988	8.17059
	Std. Dev.	0.52532	0.04516	0.07586	0.00089	0.70520	0.75837
2011	Mean	7.91161	0.04554	0.49565	0.00107	7.04084	8.17772
	Std. Dev.	0.54293	0.04507	0.08064	0.00067	0.70480	0.75428
2012	Mean	7.96974	0.04119	0.48648	0.00110	7.11616	8.18518
	Std. Dev.	0.55357	0.03921	0.07903	0.00085	0.69778	0.75045
2013	Mean	8.01021	0.03832	0.46781	0.00172	7.12947	8.19240
	Std. Dev.	0.60977	0.03522	0.07952	0.00131	0.67814	0.74697
2014	Mean	8.06601	0.03313	0.45984	0.00189	7.15885	8.19891
	Std. Dev.	0.61777	0.03036	0.07815	0.00180	0.66334	0.74442
2015	Mean	8.11762	0.02778	0.43255	0.00122	7.23768	8.20539
	Std. Dev.	0.62495	0.02502	0.07790	0.00073	0.66524	0.74327
2016	Mean	8.18463	0.01971	0.41553	0.00133	7.29991	8.21142
	Std. Dev.	0.67653	0.01824	0.07769	0.00142	0.66808	0.74330
2017	Mean	8.21442	0.01861	0.40707	0.00090	7.35082	8.21709
	Std. Dev.	0.67278	0.01852	0.07578	0.00074	0.67553	0.74345
Mean		8.00763	0.03491	0.45892	0.00133	7.12836	8.19106

According to Table 3, from 2009 to 2017, carbon productivity increased year by year, while resource endowment generally showed a downward trend, but rebounded slightly in 2011. Industrial structure showed an upward trend from 2009 to 2011 and began to decline in 2012. Environmental governance intensity was in an unstable fluctuation, peaking in 2014, and it reached the lowest in 2017. Household consumption and population scale increased from 2009 to 2017. It can be seen that the changing trends of each control variable and the explained variable are not exactly the same or the opposite. Therefore, whether these control variables have a significant impact on carbon productivity needs to be further verified.

5.3. Model Construction

Calculate the ring growth of China's carbon productivity, and draw a line chart in time series. As shown in Figure 7, before 2013, the changes in China's carbon productivity were in an unstable fluctuation, but since the carbon trading mechanism was implemented, carbon productivity has been steadily increasing. It is preliminarily inferred that the carbon trading mechanism improves carbon productivity. Therefore, the implementation of the carbon trading mechanism is regarded as a natural trial, where pilot provinces and cities constitute the treatment group and the control group includes other provinces. Specifically, Beijing, Shanghai, Guangdong Province (Shenzhen is included in Guangdong Province), Tianjin, Hubei Province, Chongqing, and Fujian Province compose the treatment group, while the control group consists of other provinces excluding Tibet, Hong Kong, Macao, and Taiwan. Since the pilots were initiated in 2013, 2009–2012 is regarded as the pre-implementation period with 2013–2017 as the implementation period. Carbon trading's implementation effect is evaluated by contrasting the changes in the two periods between the treatment group and the control group. The absolute variables such as carbon productivity, household consumption, and population scale are logarithmically processed. The regression model based on the DID method with the control variables added is as follows:

$$ln\ CP_{it} = \beta_0 + \beta_1 period_t + \beta_2 treated_i + \beta_3 did_{it} + \beta_4 CV_{it} + \mu_{it}$$

Among the model, i indexes region, hile t indexes year. $did_{it} = period_t \times treated_i$ is the interactive item that reports the net implementation effect of carbon trading and is the core explanatory variable of the regression model. $period_t$ is the time dummy variable. $period_t = 1$ indicates that carbon trading has been implemented that year, while $period_t = 0$ indicates the opposite. $treated_i$ is the region dummy variable. $treated_i = 1$ indicates the province or city is the pilot area, while $treated_i = 0$ indicates the opposite. $ln\ CP_{it}$ represents the logarithm of carbon productivity, and CV_{it} represents the set of control variables.

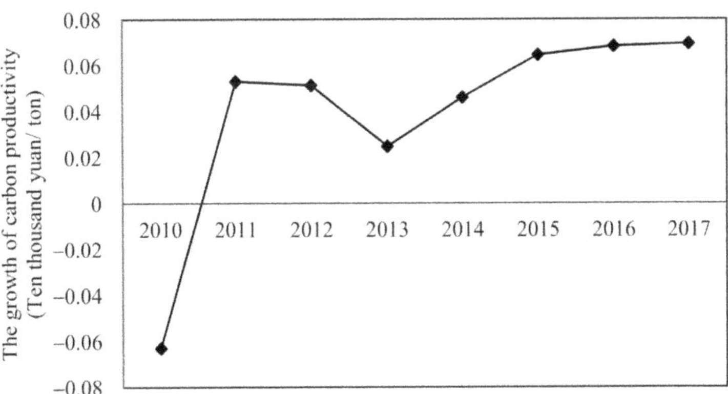

Figure 7. Annual carbon productivity growth.

5.4. Analysis of Regressive Results

5.4.1. Preliminary DID Analysis

The influence of carbon trading on regional carbon productivity is analyzed by DID. Table 3 shows the regressive results of Stata15.1. The results point out that the effect of the interactive item on the carbon productivity of pilot regions is positive with the 1% significant level, which implies that the carbon trading mechanism has a positive impact on regional sustainable development.

Although the regressive results preliminarily verify the implementation effect of carbon trading, the parallel trend assumption that the pilot regions and other regions possess the same changing trend of carbon productivity should be satisfied. Therefore, data in the respective four years before and after implementing the carbon trading mechanism are selected to test the parallel trend. A regression model is constructed with the interaction item and $ln\ CP$ as the independent and dependent variables, respectively. As shown in Table 4, $Before$ variables are all significant and the parallel trend assumption is not satisfied, which implies the self-selection bias in the pilot region cannot be ruled out.

Table 4. The results of preliminary DID regression.

	ln CP
period	0.122 * (2.53)
treated	0.396 *** (5.55)
did	0.233 ** (2.67)
RE	−6.122 *** (−8.93)
IS	−0.881 ** (−3.36)
EGI	−143.706 *** (−7.18)
ln HC	−0.229 *** (−4.18)
ln POP	0.283 *** (5.12)
_cons	7.934 *** (33.35)
N	270
R^2	0.7598

* $p < 0.05$, ** $p < 0.01$, *** $p < 0.001$.

5.4.2. Analysis of PSM Results

In order to eliminate the self-selection bias and make the pilot regions and other regions meet the parallel trend assumption, the PSM method is selected to improve the matching degree between the two groups. After the nearest neighbor matching, the estimated value of the average treatment effect on the treated (ATT) is 0.405 and the t value is 3.48 at the 1% significant level. Therefore, the ATT is significantly positive, which indicates that the establishment of carbon trading pilots significantly promotes regional carbon productivity.

To test whether the distribution of each control variable in the treatment group and the control group is balanced after matching, a t-test is adopted. As shown in Table 5, after matching, the biases of resource endowment, environmental governance intensity, and household consumption are reduced by more than 90%, and deviations of industrial structure and population scale have also been improved to a certain extent. Moreover, the p values of each control variable do not pass the test of significance at the level of 10%. It can be deduced from the results that the null hypothesis that there is no systematic difference between the two groups is accepted. The result of PSM is valid, and PSM-DID can be used to estimate the implementation effect of carbon trading.

Table 5. The results of the parallel trend test in the DID model.

	ln CP
Before4	0.457 * (2.09)
Before3	0.514 * (2.35)
Before2	0.608 ** (2.78)
Before1	0.692 ** (3.17)
Current	0.796 *** (3.64)
After1	0.863 *** (3.95)
After2	0.929 *** (4.25)
After3	1.097 *** (5.02)
After4	1.093 *** (5.00)
_cons	7.825 *** (76.63)
N	270
R^2	0.3551

* $p < 0.05$, ** $p < 0.01$, *** $p < 0.001$.

5.4.3. PSM-DID Regression Analysis

After PSM, observations that did not satisfy the common support assumption are deleted. Then the DID method is used for regression, the results are shown in Table 6 where control variables are not added in Model 1 but added in Model 2. Comparatively analyzing the two models, it is found that there is an obvious improvement of R^2 in Model 2 compared with Model 1, which shows that adding control variables increases the goodness of fit. Thus, analysis of the influence of each control variable on carbon productivity has practical significance.

Table 6. Validity test of PSM.

Variable	Mean Control	Mean Treated	Reduct |bias|	*t* Value	*p* Value
RE	0.01424	0.01408	99.5	−0.05	0.958
IS	0.47500	0.45169	41.7	−1.52	0.132
EGI	0.00094	0.00090	94.4	−0.36	0.723
ln HC	7.28740	7.28770	99.8	0.00	0.998
ln POP	8.27600	8.23570	55.6	−0.26	0.797

The results in Table 7 are as follows:

Table 7. The results of PSM-DID regression.

	Model 1	Model 2
period	0.174 *	0.160 **
	(2.55)	(3.27)
_treated	0.420 ***	0.363 ***
	(4.00)	(5.24)
did	0.287 *	0.202 *
	(2.11)	(2.35)
RE		−8.478 ***
		(−6.22)
IS		−0.933 **
		(−3.16)
EGI		−163.073 ***
		(−5.81)
ln HC		−0.255 ***
		(−4.31)
ln POP		0.297 ***
		(4.42)
_cons	7.890 ***	8.101 ***
	(146.95)	(31.84)
N	206	206
$adjR^2$	0.3179	0.7350

* $p < 0.05$, ** $p < 0.01$, *** $p < 0.001$.

Whether or not the regression model is added control variables, the coefficient of the core explanatory variable, *did*, is significantly positive at the 5% significant level. This shows that after eliminating the self-selection bias as much as possible, the net implementation effect of carbon trading on carbon productivity is significantly positive. This suggests that carbon trading is conducive to regional sustainable development, and the model is robust.

The higher the fixed assets investment of the mining industry, the lower the regional carbon productivity; the higher the gross annual value of the secondary industry, the more disadvantageous to regional sustainable development. The impact of *RE* on carbon productivity is significantly negative at the level of 0.1%, which indicates that resource endowment is an important factor that affects carbon productivity. It is also worth noting that dependence on resources is not conducive to the development of technologically innovative industries. Since the impact of *IS* is significantly negative at the 1% significant level, it is imperative to promote the industrial transformation to achieve sustainable development.

The greater the intensity of environmental governance, the lower the regional carbon productivity. The impact of *EGI* on carbon productivity is significantly negative at the 0.1% significant level, which indicates that excessive environmental regulation is unbeneficial to regional development. It is of significance to optimizing environmental governance programs.

The enlargement of the population scale is conducive to regional sustainable development, while the increase in household consumption drops regional carbon productivity. The impacts of *ln HC* and *ln POP* on *ln CP* are significant at the level of 0.1%, but the former's coefficient is negative and the latter's is positive. It is shown that the low-carbon lifestyle plays a positive role in promoting carbon productivity, and appropriate population growth fills the labor shortage of high-tech industries and promotes regional sustainable development.

5.4.4. Placebo Test

In order to verify the validation of the DID model after PSM, a placebo test was conducted. We randomly assigned the treated group and the control group, randomly assigned the time node of policy implementation, and re-estimated the model. This process is repeated 1000 times. If the estimated coefficient of the interaction item is not significant, the regression model of PSM-DID in Section 5.4.3 is proved valid. The result of the placebo test is shown in Figure 8. It can be seen that the estimated coefficient of the interaction term is distributed around 0 and obeys the normal distribution, which shows that the DID model passes the placebo test.

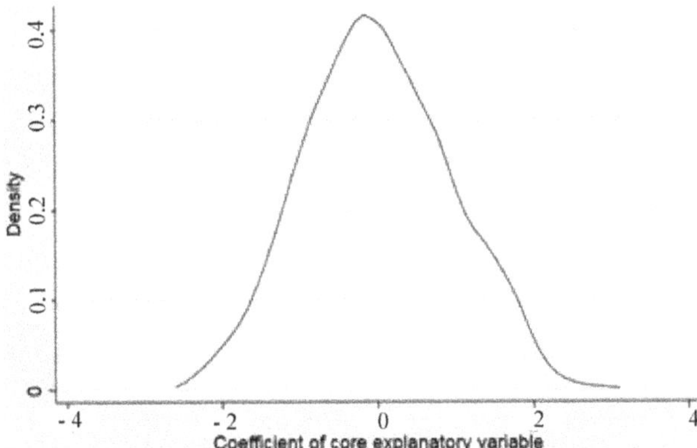

Figure 8. Placebo test.

6. Results and Discussion

By establishing the LCE-SC model and conducting empirical analysis, we achieve the goal of finding the influences of carbon trading on optimal decisions and sustainable development. The important results are summarized below.

The existing literature shows that increasing the commission rate will harm the revenue of the E-SC [20]. However, we found that in the LCE-SC, increasing the commission rate within a reasonable range improves the profitability of LCE-SC, which indicates that implementing carbon trading can effectively alleviate the negative impact of unfair profit distribution. Moreover, as consumers' low-carbon preference is gradually enhanced, a lower commission rate is more conducive to CER. However, the unit carbon emission decreases first and then increases with the commission rate. Compared with the low-cap manufacturer, the high-cap one's sales price is higher but profit is lower. The e-commerce platform cooperating with the high-cap manufacturer can make more profits.

When discussing carbon trading, the extant literature assumes that the carbon price is a fixed exogenous variable [40–43]. However, we assumed that the carbon price is affected by Demand and Supply and found that with the increase in the carbon price sensitivity coefficient, the e-commerce platform's sales promotion service first increases and then declines, and the manufacturer's profit first declines and then increases. The higher the carbon price sensitivity coefficient, the higher the marginal profits that members in LCE-SC can obtain by optimizing the coefficient of emission-reduction cost. Besides, changes in the carbon market scale have the same impacts on the decisions of both high-cap and low-cap enterprises.

We find from both the modeling analysis and empirical research that the implementation of carbon trading significantly improves regional carbon productivity, and it is noteworthy that controlling the intensity of environmental regulation is conducive to regional sustainable development. In addition, promoting industrial transformation, advocating a low-carbon lifestyle, and appropriately relaxing population restriction are empirically proven to be effective in increasing carbon productivity.

As the impact of carbon trading expands, and the cultivation of sinking markets promotes the continuous and rapid development of the e-commerce industry, it is imperative to explore the impact of carbon trading on LCE-SC and regional sustainable development. Compared with existing studies, the innovation of our research includes two aspects: on the one hand, the existing research only considers the production and inventory decisions of the traditional offline supply chain affected by carbon trading, such as in [67–69]. This paper introduces the carbon trading mechanism into the e-commerce supply chain for the first time, complementing the research on the interaction of carbon trading and supply chain management.

On the other hand, the existing research discussing the policy effect adopts micro research methods only, such as in [50,70] showing the influences of carbon tax and carbon quota by modeling analysis, respectively, or macro research methods only, such as in [71,72] studying the effects of subsidy and carbon trading by empirical analysis respectively, lacking the transition between the two. This paper combined micro modeling analysis with macro empirical research and found it a feasible way to better study the implementation effect of a policy.

7. Conclusions

In the low-carbon context, the carbon trading mechanism and consumers' low-carbon preference are introduced into the decision-making model of LCE-SC, which differs from the models in existing studies. The optimal decisions of LCE-SC have been gained by the Stackelberg game. On this basis, this paper analyzes how the commission rate and carbon trading influence the decision-making and performance of LCE-SC. The moderating effects of these parameters are discussed by numerical simulation. Then, based on the initial hypothesis that the carbon trading mechanism promotes carbon productivity, further empirical research on the implementation effect of carbon trading is conducted. As expected, we got the following conclusions that play a directive role in enterprise operation and policymaking.

Firstly, for emission-dependent manufacturers, producing low-carbon products not only generates higher variable costs but also brings environmental benefits, that is, better brand image and higher profits for enterprises. Therefore, considering consumers' low-carbon preference and the e-commerce platform's sales rules, manufacturers should optimize emission-reduction programs to control emission-reduction costs for the sake of higher profits and fewer carbon emissions.

Secondly, CER relies on the investment of low-carbon manufacturers, as well as the cooperation of e-commerce enterprises. As the core factor for coordinating e-commerce platforms and settled-in manufacturers, the commissions need to be kept in a low range to reduce emissions. Thus, e-commerce platforms should consider the impact of the commission rate on CER and set an appropriate commission rate for the win-win result of economic benefits and environmental performance in LCE-SC.

Thirdly, as implementing carbon trading contributes to both the economy and the environment, the government should actively promote the carbon trading mechanism with a way to standardize the operation of pilot carbon trading markets. By doing this, the role of carbon trading in mitigating unfair profit distribution and promoting regional sustainability can be fully utilized. Besides, the government should appropriately relax population restrictions and foster more high-tech professionals. Meanwhile, the intensity of environmental governance should also be controlled. Since the cost of CER brings pressure for emission-dependent enterprises, the government can consider providing emission-reduction subsidies.

Finally, increased consumers' low-carbon preference can improve low-carbon enterprises' economic and environmental benefits. Therefore, the government should help promote the publicity of low-carbon products and advocate the low-carbon lifestyle, which is advantageous to regional sustainable development. In the current big data era, e-commerce platforms have the most direct contact with consumers. They can track consumers' consumption behavior and analyze their information by technical means for personally recommending low-carbon products.

Although our research fills the gap of the literature on carbon trading to some extent, there are certain research limitations. Since the carbon price is influenced by the government instead of determined entirely by the market, more factors can be considered in LCE-SC, such as subsidy for the carbon price and differential carbon pricing for enterprises in different regions. The consideration that can make the model more realistic is the direction of future work.

Author Contributions: Conceptualization, Y.W.; methodology, Y.W. and L.S.; software, L.L. and X.W.; formal analysis, L.L. and Q.L.; data curation, Q.L.; Visualization, X.W.; writing—original draft preparation, X.W., Y.W. and L.S.; writing—review and editing, L.S. and R.T.; funding acquisition Y.W. All authors have read and agreed to the published version of the manuscript.

Funding: This research was funded by National Natural Science Foundation of China, grant number 7197129 and Science and Technology Support Program for Youth Innovation of Colleges and Universities in Shandong Province, grant number 2019RWG017.

Institutional Review Board Statement: Not applicable.

Informed Consent Statement: Not applicable.

Data Availability Statement: The data presented in this study are available on request from the corresponding author.

Conflicts of Interest: The authors declare no conflict of interest.

Appendix A

Proof of Proposition 1.
$\frac{\partial Y^*}{\partial Y^U} = \frac{4h(l\beta A-2B)-k^2 lA(1-\rho)}{F_3} > 0$; $\frac{\partial Y^*}{\partial \alpha_c} = \frac{2(l\beta A-2B)}{F_3} > 0$; $\frac{\partial Y^*}{\partial \rho} = \frac{kl\beta\left[4(h+r)\alpha+2k\left(2rY^U-\alpha_c\right)-ck^2\right]\left\{k^4 l(1-\rho)^2+8(h+r)^2\beta\gamma^2-2k^2(h+r)(1-\rho)\left[4l\beta-\gamma^2(1+\rho)\right]\right\}}{F_3^2} > 0$. The same procedure is adapted to prove that $\frac{\partial Y^*}{\partial h} > 0$, $\frac{\partial Y^*}{\partial r} < 0$, $\frac{\partial Y^*}{\partial k} < 0$, $\frac{\partial y^*}{\partial Y^U} > 0$, $\frac{\partial y^*}{\partial \alpha_c} > 0$, $\frac{\partial y^*}{\partial h} > 0$, $\frac{\partial y^*}{\partial r} < 0$, $\frac{\partial y^*}{\partial k} < 0$.

According to $\frac{\partial y^*}{\partial \rho} = \frac{\beta[8Bc(h+r)+F_2+F_6]\left\{kl\left(kY^U+\alpha\right)(1-\rho)\left[2A-k^2(1-\rho)\right]+\left(ck+4hY^U+2\alpha_c\right)\beta F_7\right\}}{\beta^2[8Bc(h+r)+F_2+F_6]^2} - \frac{\beta(1-\rho)F_4\left\{2c\beta(h+r)\left[4(h+r)\gamma^2-k^2 l\right]-l[2(h+r)\beta-k^2(1-\rho)]\left(F_5+ck^2\right)\right\}}{\beta^2[8Bc(h+r)+F_2+F_6]^2}$, if $1-\rho > \frac{[F_2+8(h+r)Bc+F_6]\left\{kl\left(kY^U+\alpha\right)(1-\rho)\left[2A-k^2(1-\rho)\right]+\left(ck+4hY^U+2\alpha_c\right)\beta F_7\right\}}{F_4\left\{2c\beta(h+r)[4(h+r)\gamma^2-k^2 l]-l[2(h+r)\beta-k^2(1-\rho)](F_5+ck^2)\right\}}$, $\frac{\partial y^*}{\partial \rho} < 0$; if $1-\rho < \frac{[F_2+8(h+r)Bc+F_6]\left\{kl\left(kY^U+\alpha\right)(1-\rho)\left[2A-k^2(1-\rho)\right]+\left(ck+4hY^U+2\alpha_c\right)\beta F_7\right\}}{F_4\left\{2c\beta(h+r)[4(h+r)\gamma^2-k^2 l]-l[2(h+r)\beta-k^2(1-\rho)](F_5+ck^2)\right\}}$, $\frac{\partial y^*}{\partial \rho} > 0$. □

Appendix B

Proof of Proposition 2.
$\frac{\partial s^*}{\partial Y^U} = \frac{8Bkr}{F_3\gamma} > 0$, $\frac{\partial s^*}{\partial \rho} = \frac{2l(h+r)\beta\gamma AF_5\left[4(h+r)\beta-k^2(1+\rho)\right]}{F_3^2} > 0$, $\frac{\partial s^*}{\partial \alpha_c} = -\frac{4Bk}{F_3\gamma} < 0$. The same procedure is adapted to prove that $\frac{\partial s^*}{\partial k} > 0$, $\frac{\partial s^*}{\partial h} < 0$.

According to $\frac{\partial s^*}{\partial r} = \frac{2\beta\gamma\rho k\left\{F_3\left(4hY^U+2\alpha_c+ck\right)-2klA(1-\rho)F_5\right\}}{F_3^2}$, if $4hY^U+2\alpha_c+ck > \frac{2klA(1-\rho)F_5}{F_3}$, $\frac{\partial s^*}{\partial r} > 0$; if $4hY^U+2\alpha_c+ck < \frac{2klA(1-\rho)F_5}{F_3}$, $\frac{\partial s^*}{\partial r} < 0$. □

References

1. Zhang, L.; Zhou, H.; Liu, Y.; Lu, R. The optimal carbon emission reduction and prices with cap and trade mechanism and competition. *Int. J. Environ. Res. Public Health* **2018**, *15*, 2570. [CrossRef] [PubMed]
2. Ding, H.; Huang, H.; Tang, O. Sustainable supply chain collaboration with outsourcing pollutant-reduction service in power industry. *J. Clean. Prod.* **2018**, *186*, 215–228. [CrossRef]
3. Coase, R.H. The Problem of Social Cost. *Econ. Anal. Law Sel. Read.* **2007**, 1–13. [CrossRef]
4. Stern, N. The Economics of Climate Change. *Am. Econ. Rev. Pap. Proc.* **2008**, *98*, 1–37. [CrossRef]

5. Yang, L.; Zheng, C.; Xu, M. Comparisons of low carbon policies in supply chain coordination. *J. Syst. Sci. Syst. Eng.* **2014**, *23*, 342–361. [CrossRef]
6. Zhou, K.; Li, Y. Influencing factors and fluctuation characteristics of China's carbon emission trading price. *Phys. A Stat. Mech. Its Appl.* **2019**, *524*, 459–474. [CrossRef]
7. Qiyan, Z.; Weibin, C.; Xiang, W. The Effect and Influence of Carbon-Emission Trading on the Concept and Action of Energy Conservation and Emission Reduction of Enterprises. Available online: http://www.cncete.com/magazine/show-525.html (accessed on 15 October 2016).
8. Zhang, K.; Yao, Y.; Liang, Q.; Saren, G. How should China prioritize the deregulation of electricity prices in the context of carbon pricing? A computable general equilibrium analysis. *Energy Econ.* **2021**, *96*, 105187. [CrossRef]
9. Pagès-Bernaus, A.; Ramalhinho, H.; Juan, A.A.; Calvet, L. Designing e-commerce supply chains: A stochastic facility–location approach. *Int. Trans. Oper. Res.* **2019**, *26*, 507–528. [CrossRef]
10. Cao, K.; Xu, X.; Wu, Q.; Zhang, Q. Optimal production and carbon emission reduction level under cap-and-trade and low carbon subsidy policies. *J. Clean. Prod.* **2017**, *167*, 505–513. [CrossRef]
11. Fan, Y.; Wang, M.; Zhao, L. Production-inventory and emission reduction investment decision under carbon cap-and-trade policy. *RAIRO Oper. Res.* **2018**, *52*, 1043–1067. [CrossRef]
12. Jiang, W.; Yuan, L.; Wu, L.; Guo, S. Carbon emission reduction and profit distribution mechanism of construction supply chain with fairness concern and capand-trade. *PLoS ONE* **2019**, *14*, e0224153. [CrossRef]
13. Dumrongsiri, A.; Fan, M.; Jain, A.; Moinzadeh, K. A supply chain model with direct and retail channels. *Eur. J. Oper. Res.* **2008**, *187*, 691–718. [CrossRef]
14. Yan, N.; Liu, Y.; Xu, X.; He, X. Strategic dual-channel pricing games with e-retailer finance. *Eur. J. Oper. Res.* **2020**, *283*, 138–151. [CrossRef]
15. Jia, S.; Zhou, C.; Qin, C. No difference in effect of high-speed rail on regional economic growth based on match effect perspective? *Transp. Res. Part A Policy Pract.* **2017**, *106*, 144–157. [CrossRef]
16. Liu, Y.; Wang, A.; Wu, Y. Environmental regulation and green innovation: Evidence from China's new environmental protection law. *J. Clean. Prod.* **2021**, *297*, 126698. [CrossRef]
17. Yang, L.; Wang, G.; Ke, C. Remanufacturing and promotion in dual-channel supply chains under cap-and-trade regulation. *J. Clean. Prod.* **2018**, *204*, 939–957. [CrossRef]
18. Kang, K.; Zhao, Y.; Zhang, J.; Qiang, C. Evolutionary game theoretic analysis on low-carbon strategy for supply chain enterprises. *J. Clean. Prod.* **2019**, *230*, 981–994. [CrossRef]
19. Mishra, M.; Hota, S.K.; Ghosh, S.K.; Sarkar, B. Controlling waste and carbon emission for a sustainable closed-loop supply chain management under a cap-and-trade strategy. *Mathematics* **2020**, *8*, 466. [CrossRef]
20. Han, Q.; Wang, Y. Decision and coordination in a low-carbon E-supply chain considering the manufacturer's carbon emission reduction behavior. *Sustainability* **2018**, *10*, 1686. [CrossRef]
21. Ji, S.; Sun, Q. Low-carbon planning and design in B & R logistics service: A case study of an E-commerce big data platform in China. *Sustainability* **2017**, *9*, 2052. [CrossRef]
22. Wang, J.; Huang, X. The optimal carbon reduction and return strategies under carbon tax policy. *Sustainability* **2018**, *10*, 2471. [CrossRef]
23. Wang, Y.; Fan, R.; Shen, L.; Miller, W. Recycling decisions of low-carbon e-commerce closed-loop supply chain under government subsidy mechanism and altruistic preference. *J. Clean. Prod.* **2020**, *259*, 120883. [CrossRef]
24. Benjaafar, S.; Li, Y.; Daskin, M. Carbon footprint and the management of supply chains: Insights from simple models. *IEEE Trans. Autom. Sci. Eng.* **2013**, *10*, 99–116. [CrossRef]
25. Barbosa-Póvoa, A.P.; da Silva, C.; Carvalho, A. Opportunities and challenges in sustainable supply chain: An operations research perspective. *Eur. J. Oper. Res.* **2018**, *268*, 399–431. [CrossRef]
26. Chai, Q.; Xiao, Z.; Lai, K.H.; Zhou, G. Can carbon cap and trade mechanism be beneficial for remanufacturing? *Int. J. Prod. Econ.* **2018**, *203*, 311–321. [CrossRef]
27. Dong, C.; Shen, B.; Chow, P.S.; Yang, L.; Ng, C.T. Sustainability investment under cap-and-trade regulation. *Ann. Oper. Res.* **2016**, *240*, 509–531. [CrossRef]
28. Du, S.; Ma, F.; Fu, Z.; Zhu, L.; Zhang, J. Game-theoretic analysis for an emission-dependent supply chain in a 'cap-and-trade' system. *Ann. Oper. Res.* **2015**, *228*, 135–149. [CrossRef]
29. Xu, X.; He, P.; Xu, H.; Zhang, Q. Supply Chain Coordination with Green Technology under Cap-and-Trade Regulation. *Int. J. Prod. Econ.* **2017**, *183 Pt B*, 433–442. [CrossRef]
30. Xu, X.; Zhang, W.; He, P.; Xu, X. Production and pricing problems in make-to-order supply chain with cap-and-trade regulation. *Omega (UK)* **2017**, *66*, 248–257. [CrossRef]
31. Wang, M.; Zhao, L.; Herty, M. Joint replenishment and carbon trading in fresh food supply chains. *Eur. J. Oper. Res.* **2019**, *277*, 561–573. [CrossRef]
32. Smith, S.; Swierzbinski, J. Assessing the performance of the UK Emissions Trading Scheme. *Environ. Resour. Econ.* **2007**, *37*, 131–158. [CrossRef]
33. Wang, P.; Dai, H.C.; Ren, S.-Y.; Zhao, D.-Q.; Masui, T. Achieving Copenhagen target through carbon emission trading: Economic impacts assessment in Guangdong Province of China. *Energy* **2015**, *79*, 212–227. [CrossRef]

34. Zhao, X.; Zhang, Y.; Liang, J.; Li, Y.; Jia, R.; Wang, L. The sustainable development of the economic-energy-environment (3E) system under the carbon trading (CT) mechanism: A Chinese case. *Sustainability* **2018**, *10*, 98. [CrossRef]
35. Zhou, B.; Zhang, C.; Song, H.; Wang, Q. How does emission trading reduce China's carbon intensity? An exploration using a decomposition and difference-in-differences approach. *Sci. Total Environ.* **2019**, *676*, 514–523. [CrossRef] [PubMed]
36. Zhang, W.; Zhang, N.; Yu, Y. Carbon mitigation effects and potential cost savings from carbon emissions trading in China's regional industry. *Technol. Forecast. Soc. Chang.* **2019**, *141*, 1–11. [CrossRef]
37. Hamzah, T.A.A.T.; Zainuddin, Z.; Yusoff, M.M.; Osman, S.; Abdullah, A.; Saini, K.M.; Sisun, A. The conundrum of carbon trading projects towards sustainable development: A review from the palm oil industry in Malaysia. *Energies* **2019**, *12*, 3530. [CrossRef]
38. Zhang, H.; Zhang, R.; Li, G.; Li, W.; Choi, Y. Sustainable feasibility of carbon trading policy on heterogenetic economic and industrial development. *Sustainability* **2019**, *11*, 6869. [CrossRef]
39. Zheng, J.; Yang, M.; Ma, G.; Xu, Q.; He, Y. Multi-agents-based modeling and simulation for carbon permits trading in China: A regional development perspective. *Int. J. Environ. Res. Public Health* **2020**, *17*, 301. [CrossRef] [PubMed]
40. Xu, L.; Wang, C.; Miao, Z.; Chen, J. Governmental subsidy policies and supply chain decisions with carbon emission limit and consumer's environmental awareness. *RAIRO Oper. Res.* **2019**, *53*, 1675–1689. [CrossRef]
41. Ma, X.; Wang, J.; Bai, Q.; Wang, S. Optimization of a three-echelon cold chain considering freshness-keeping efforts under cap-and-trade regulation in Industry 4.0. *Int. J. Prod. Econ.* **2020**, *220*. [CrossRef]
42. Xia, X.; Li, C.; Zhu, Q. Game analysis for the impact of carbon trading on low-carbon supply chain. *J. Clean. Prod.* **2020**, *276*, 123220. [CrossRef]
43. Liu, X.; Du, W.; Sun, Y. Green supply chain decisions under different power structures: Wholesale price vs. revenue sharing contract. *Int. J. Environ. Res. Public Health* **2020**, *17*, 7737. [CrossRef]
44. Zhang, J.; Cao, Q.; He, X. Contract and product quality in platform selling. *Eur. J. Oper. Res.* **2019**, *272*, 928–944. [CrossRef]
45. Tan, Y.R.; Carrillo, J.E. Strategic Analysis of the Agency Model for Digital Goods. *Prod. Oper. Manag.* **2017**, *26*, 724–741. [CrossRef]
46. Adida, E.; Ratisoontorn, N. Consignment contracts with retail competition. *Eur. J. Oper. Res.* **2011**, *215*, 136–148. [CrossRef]
47. Nair, A.; Narasimhan, R. Dynamics of competing with quality- and advertising-based goodwill. *Eur. J. Oper. Res.* **2006**, *175*, 462–474. [CrossRef]
48. Xia, L.; Hao, W.; Qin, J.; Ji, F.; Yue, X. Carbon emission reduction and promotion policies considering social preferences and consumers' low-carbon awareness in the cap-and-trade system. *J. Clean. Prod.* **2018**, *195*, 1105–1124. [CrossRef]
49. Xing, E.; Shi, C.; Zhang, J.; Cheng, S.; Lin, J.; Ni, S. Double third-party recycling closed-loop supply chain decision under the perspective of carbon trading. *J. Clean. Prod.* **2020**, *259*. [CrossRef]
50. Wang, W.; Zhou, C.; Li, X. Carbon reduction in a supply chain via dynamic carbon emission quotas. *J. Clean. Prod.* **2019**, *240*, 118244. [CrossRef]
51. Shen, L.; Wang, Y. Supervision mechanism for pollution behavior of Chinese enterprises based on haze governance. *J. Clean. Prod.* **2018**, *197*, 571–582. [CrossRef]
52. Zhang, J.; Jiang, H.; Liu, G.; Zeng, W. A study on the contribution of industrial restructuring to reduction of carbon emissions in China during the five Five-Year Plan periods. *J. Clean. Prod.* **2018**, *176*, 629–635. [CrossRef]
53. Li, X.; Luo, Q.; Hai, B.; Li, L. Progress in the micro-perspective studies of economic geography in China since 1980s. *J. Geogr. Sci.* **2016**, *26*, 1041–1056. [CrossRef]
54. Zhang, K.; Xu, D.; Li, S.; Zhou, N.; Xiong, J. Has China's pilot emissions trading scheme influenced the carbon intensity of output? *Int. J. Environ. Res. Public Health* **2019**, *16*, 1854. [CrossRef] [PubMed]
55. Wang, L.; Watanabe, T. Effects of environmental policy on public risk perceptions of haze in Tianjin City: A difference-in-differences analysis. *Renew. Sustain. Energy Rev.* **2019**, *109*, 199–212. [CrossRef]
56. Li, J.; Lin, B. Environmental impact of electricity relocation: A quasi-natural experiment from interregional electricity transmission. *Environ. Impact Assess. Rev.* **2017**, *66*, 151–161. [CrossRef]
57. Wing, C.; Simon, K.; Bello-Gomez, R.A. Designing Difference in Difference Studies: Best Practices for Public Health Policy Research. *Annu. Rev. Public Health* **2018**, *39*, 453–469. [CrossRef] [PubMed]
58. Dong, F.; Dai, Y.; Zhang, S.; Zhang, X.; Long, R. Can a carbon emission trading scheme generate the Porter effect? Evidence from pilot areas in China. *Sci. Total Environ.* **2019**, *653*, 565–577. [CrossRef]
59. Zhu, B.; Zhang, M.; Huang, L.; Wang, P.; Su, B.; Wei, Y.M. Exploring the effect of carbon trading mechanism on China's green development efficiency: A novel integrated approach. *Energy Econ.* **2020**, *85*, 104601. [CrossRef]
60. Austin, P.C. Some methods of propensity-score matching had superior performance to others: Results of an empirical investigation and monte carlo simulations. *Biom. J.* **2009**, *51*, 171–184. [CrossRef]
61. Rosenbaum, P.R.; Rubin, D.B. The central role of the propensity score in observational studies for causal effects. *Matched Sampl. Causal Eff.* **2006**, 170–184. [CrossRef]
62. Zhang, C.; Wang, Q.; Shi, D.; Li, P.; Cai, W. Scenario-based potential effects of carbon trading in China: An integrated approach. *Appl. Energy* **2016**, *182*, 177–190. [CrossRef]
63. Wang, H.; Chen, Z.; Wu, X.; Nie, X. Can a carbon trading system promote the transformation of a low-carbon economy under the framework of the porter hypothesis?—Empirical analysis based on the PSM-DID method. *Energy Policy* **2019**, *129*, 930–938. [CrossRef]

64. Yan, D.; Kong, Y.; Ren, X.; Shi, Y.; Chiang, S.W. The determinants of urban sustainability in Chinese resource-based cities: A panel quantile regression approach. *Sci. Total Environ.* **2019**, *686*, 1210–1219. [CrossRef] [PubMed]
65. Hu, Y.; Ren, S.; Wang, Y.; Chen, X. Can carbon emission trading scheme achieve energy conservation and emission reduction? Evidence from the industrial sector in China. *Energy Econ.* **2020**, *85*, 104590. [CrossRef]
66. Xu, S.C.; He, Z.X.; Long, R.Y. Factors that influence carbon emissions due to energy consumption in China: Decomposition analysis using LMDI. *Appl. Energy* **2014**, *127*, 182–193. [CrossRef]
67. Chang, X.; Xia, H.; Zhu, H.; Fan, T.; Zhao, H. Production decisions in a hybrid manufacturing-remanufacturing system with carbon cap and trade mechanism. *Int. J. Prod. Econ.* **2015**, *162*, 160–173. [CrossRef]
68. Halat, K.; Hafezalkotob, A. Modeling carbon regulation policies in inventory decisions of a multi-stage green supply chain: A game theory approach. *Comput. Ind. Eng.* **2019**, *128*, 807–830. [CrossRef]
69. Xiong, S.; Feng, Y.; Huang, K. Optimal MTS and MTO hybrid production system for a single product under the cap-and-trade environment. *Sustainability* **2020**, *12*, 2426. [CrossRef]
70. Konstantaras, I.; Skouri, K.; Benkherouf, L. Optimizing inventory decisions for a closed–loop supply chain model under a carbon tax regulatory mechanism. *Int. J. Prod. Econ.* **2021**, *239*, 108185. [CrossRef]
71. Kong, D.; Xia, Q.; Xue, Y.; Zhao, X. Effects of multi policies on electric vehicle diffusion under subsidy policy abolishment in China: A multi-actor perspective. *Appl. Energy* **2020**, *266*, 114887. [CrossRef]
72. Wang, L.; Cui, L.; Weng, S.; Liu, C. Promoting industrial structure advancement through an emission trading scheme: Lessons from China's pilot practice. *Comput. Ind. Eng.* **2021**, *157*, 107339. [CrossRef]

MDPI

Article

Cryptocurrency Portfolio Selection—A Multicriteria Approach

Zdravka Aljinović, Branka Marasović * and Tea Šestanović

Faculty of Economics, Business and Tourism, University of Split, 21000 Split, Croatia; zdravka.aljinovic@efst.hr (Z.A.); tea.sestanovic@efst.hr (T.Š.)
* Correspondence: branka.marasovic@efst.hr; Tel.: +385-2143-0697

Abstract: This paper proposes the PROMETHEE II based multicriteria approach for cryptocurrency portfolio selection. Such an approach allows considering a number of variables important for cryptocurrencies rather than limiting them to the commonly employed return and risk. The proposed multiobjective decision making model gives the best cryptocurrency portfolio considering the daily return, standard deviation, value-at-risk, conditional value-at-risk, volume, market capitalization and attractiveness of nine cryptocurrencies from January 2017 to February 2020. The optimal portfolios are calculated at the first of each month by taking the previous 6 months of daily data for the calculations yielding with 32 optimal portfolios in 32 successive months. The out-of-sample performances of the proposed model are compared with five commonly used optimal portfolio models, i.e., naïve portfolio, two mean-variance models (in the middle and at the end of the efficient frontier), maximum Sharpe ratio and the middle of the mean-CVaR (conditional value-at-risk) efficient frontier, based on the average return, standard deviation and VaR (value-at-risk) of the returns in the next 30 days and the return in the next trading day for all portfolios on 32 dates. The proposed model wins against all other models according to all observed indicators, with the winnings spanning from 50% up to 94%, proving the benefits of employing more criteria and the appropriate multicriteria approach in the cryptocurrency portfolio selection process.

Keywords: cryptocurrency; portfolio selection; return and risk measures; market capitalization; volume; attractiveness; PROMETHEE II; multicriteria model

Citation: Aljinović, Z.; Marasović, B.; Šestanović, T. Cryptocurrency Portfolio Selection—A Multicriteria Approach. *Mathematics* **2021**, *9*, 1677. https://doi.org/10.3390/math9141677

Academic Editor: Frank Werner

Received: 26 June 2021
Accepted: 14 July 2021
Published: 16 July 2021

1. Introduction

As a response to the everlasting changes in the surroundings, investors adjust the structure of their portfolios in order to maximize the targeted return and risk ratio. In periods of persistently low interest rates, as exhibited in the last decade in the world, traditional investments become less interesting and investors seek alternative forms of investment in the pursuit of higher returns and possibly a lower risk obtained by diversification of the portfolio. In this regard, cryptocurrencies as an alternative form of investment, obtained increasing attention of many investors and this paper. The basic requirement that each new-alternative form of investment should meet is the contribution in terms of Markowitz diversification, i.e., contribution to a more favourable relationship between return and risk of the portfolio, which is exactly what this paper is trying to examine for cryptocurrency portfolio only.

Over the last few years, a number of papers have been published on this topic. Some consider the contribution of particular cryptocurrencies, mostly Bitcoin, to portfolios including other assets being either traditional ones or combinations of traditional and alternative investments [1–5]. Some consider cryptocurrencies' contribution through a set of cryptocurrencies, mostly represented by the cryptocurrency index (CRIX) or its subsets, regarding certain criterion like market capitalization [6–10]. All of them confirm that cryptocurrencies contribute to a better return/risk ratio of portfolios. Recognizing good features of cryptocurrencies as an asset class, some studies have evaluated pure cryptocurrency portfolios, using different approaches to portfolio selection and comparing their performances. Most common strategies for portfolio selection are: the $1/N$ equal weighted rule, so called naïve diversification, Markowitz's mean-variance optimization strategy, risk parity principle, maximum Sharpe ratio or just taking the CRIX portfolio [11–14]. There are different winning strategies depending on the observed period and the sample. The question arises about the "quality" and characteristics the cryptocurrencies should have to be included in the portfolio. It is to be expected that not all available cryptocurrencies are equally favourable for investment, due to their specificities in a number of features. In the previously mentioned papers, the prevailing strategy for selecting a cryptocurrency sample was focusing on a number of top cryptocurrencies in terms of market capitalization. Sometimes the selection was limited to, e.g., the portfolio represented by CRIX or a portfolio containing a number of "most popular" currencies. Only within a few studies, the criterion of liquidity was added in the process of cryptocurrencies selection. Trimborn et al. [8] included cryptocurrencies in their portfolios, combined with stocks from the US, German and Portuguese capital markets. Given the high volatility and a relatively low liquidity, instead of the standard mean-variance model, they propose the LIBRO (liquidity bounded risk–return optimization) method. Garcia et al. [15] extend the stochastic mean-semivariance model to a fuzzy multiobjective model. In addition to return and risk, liquidity is also considered as a portfolio performance measure. The proposed methodology is tested on a data set of assets from the Latin American integrated market, showing the effectiveness and efficiency of the model. Variations in volatility and return but also in other asset specific indicators like liquidity and attractiveness should be included in the appropriate, comprehensive manner in portfolio optimization models that include cryptocurrencies. That is why in this paper a set of different and cryptocurrency-specific criteria are considered. This is a rather novel approach in cryptocurrency portfolio selection since some papers used liquidity only as a precondition for selecting assets into the portfolio, while attractiveness has only been proved in the time-series analysis to influence the cryptocurrency prices and returns.

The traditionally applied portfolio optimization approach based on the Markowitz mean-variance model is not appropriate for cryptocurrency portfolio selection, since standard model assumptions like normal return distribution or a quadratic utility function are not met for this investment class. Thus, in the process of portfolio selection it is more suitable to consider alternative risk measures, which take into account also higher moments of distribution. This issue was partially recognized and accepted in the previously analysed papers, mostly by taking the conditional value-at-risk (CVaR) as the risk measure, i.e., by applying mean-CVaR strategy for portfolio selection. Moreover, for this class of risky assets, it is useful to include more different risk measures in the portfolio selection model.

There are numerous studies and resulting findings on including suitable alternative risk measures in models. Methods of mathematical programming and multicriteria decision making (MCDM) make it possible to incorporate all recognized specificities and constraints into the model, in order to find out which assets with the assigned weights should be selected for the optimal cryptocurrencies portfolio. Generally, MCDM methods can be classified into two categories, discrete multiattribute decision making (MADM) and continuous multiobjective decision making (MODM) methods. MODM methods are used where alternatives are non-predetermined. The aim is to design the optimal alternative by considering a set of quantifiable objectives, i.e., well-defined design constraints. Thus, MODM methods deal with the design process and the number of alternatives is infinite (continuous) [16]. Since variables other than return and risk are considered as important, the selection of the optimal portfolio becomes a multiobjective problem in which we have to design the best portfolio out of an infinite number of feasible portfolios. Sometimes before selecting the optimal portfolio, it is necessary to reduce the sample of possible constituents of the portfolio by taking exclusively those with the best properties. In that case, when security analysis is required, MADM methods are very useful.

In the last fifty years a large number of multiple criteria methods have been applied in the field of stock portfolio selection. One of the newer bibliographic reviews of papers that apply MCDM methods and procedures for stock portfolio selection was carried out by

Aouni et al. [17] who, as a result, offered a classification of a range of MCDA techniques used in the security analysis/evaluation part and in portfolio construction/optimization parts. The review indicates analytic hierarchy process (AHP)-based techniques, ELECTRE-based approaches and Technique for Order of Preference by Similarity to Ideal Solution (TOPSIS) approaches as the most popular in the security analysis phase and goal programming as the most popular in the portfolio construction phase. The main advantage of multicriteria methods over mean-risk models for portfolio selection is that they take into consideration a number of conflicting criteria, not only risk and return. It is important to emphasize that without a decision-maker there is no solution to multicriteria models. Weights of criteria are determined by the decision-makers' opinion and then the chosen multicriteria method formulates the best compromise solution for that decision-maker. Therefore, the subjectivity of the decision-maker represents the often-mentioned disadvantage of these models. An effective way of reducing subjectivity in these models is by increasing the number of participating experts in the process of decision-making.

This paper focuses on the selection of the optimal cryptocurrency portfolio, which should be observed as a multicriteria programming problem and which should be solved using the appropriate techniques. In this paper a modified and adjusted multiobjective programming model based on the PROMETHEE II approach is proposed. It will be applied and tested using a sample of cryptocurrencies for which all required data are available in the period from 2017 to 2020. Although already used for other assets, the proposed model has never been applied to cryptocurrency portfolio selection, which due to their specifics require special attention in criteria selection, and preference function types and criteria weights. The weights of the seven chosen criteria are estimated using the AHP method or, more precisely, its eigenvalue procedure, incorporating the opinion of different experts' for more objective decision-making.

Based on the problem defined though literature inspection, a research hypothesis can be defined: A multicriteria approach for cryptocurrency portfolio selection based on the PROMETHEE II model yields better out of sample performances compared to the five most commonly used portfolio optimization models in different performances aspects.

Therefore, this paper contributes to the existing literature in several ways. Firstly, by defining the appropriate model for cryptocurrency portfolio selection due to their specificities, i.e., a multicriteria approach based on the PROMETHEE II model. Secondly, by incorporating criteria that, to the best of our knowledge, have never been used before in portfolio optimization. Thirdly, by engaging different experts to obtain results that are more objective. Finally, by examining whether the proposed multicriteria model yields better out-of-sample performances compared to the five most commonly used portfolio optimization models while using different important out-of-sample performances measures.

The remainder of the paper is organized as follows. Section 2 provides a short overview of cryptocurrencies. Section 3 describes the data and offers a detailed description of the criteria used in the paper, together with descriptive statistics for the selected cryptocurrencies in the observed period. Section 4 presents the proposed multicriteria model, its implementation is presented in Section 5 and the results of the research in Section 6. Section 7 provides a discussion and interpretation of the obtained results. The last section concludes the paper. The four tables with results of out-of-sample comparisons of different models are given in the Appendix A.

2. Cryptocurrencies—A Short Overview

Cryptocurrency emerged as a byproduct of another invention. Satoshi Nakamoto, the unknown founder of Bitcoin, the most famous cryptocurrency, presented his invention as a "Peer-to-Peer Electronic Cash System" [18]. Numerous attempts made during the 1990s to establish a decentralized digital money system finally succeeded and resulted in the introduction of a new currency—cryptocurrency. While only 6 or 7 years ago most professionals in the world of finance still considered cryptocurrency as something untrustworthy and the scientific approach to studying it was scarce, we are currently witnessing a real surge of interest in cryptocurrencies. The Internet abounds in platforms that provide information about basic concepts, data, features, ways of trading, very interesting thoughts and analyses offered by professionals on many aspects of cryptocurrencies. Nonetheless, there is a clear need for scientific research and scientifically based analyses to provide answers to a number of issues that have recently been raised and that have drawn our attention.

Unlike money, which is tangible, the currency you can take along, cryptocurrency is digital money, a digital asset that can be exchanged. Using cryptocurrency instead of paper currency means avoiding bank intermediation and verification and transaction costs [19]. The prefix "crypto" indicates the use of cryptography for security and verification purposes in the course of creating and transferring money. The cryptocurrency transactions are processed and completed using the blockchain technology—the technology that underpins many of the innovations that are currently revolutionising the financial services sector around the world [20]. The most notable application of blockchain is in the development and operation of cryptocurrencies, but there is a space and opportunity for its application in other sectors such as international trade, taxation, supply chain management, business operations and governance. Authors demonstrate how organizations and regulators can leverage blockchain to improve business operations and efficiency while reducing operational costs.

The supply of cryptocurrency is limited, i.e., cryptocurrencies are mined and are created by decryption—complex mathematical tasks are solved by the power of computers. After finding the solution, the miner builds a block and adds it to the chain for which they are rewarded with a certain amount of cryptocurrency. For example, Bitcoin's algorithm determines the rate at which new bitcoins are created over time until they reach the maximum of 21 million bitcoins, which should be reached by the year 2140 [19]. The author discusses a possibility of introducing a Bitcoin standard and, despite some benefits it would have over the current fiat money standards, it concludes that it is unlikely that the Bitcoin standard will stem out, since the authorities will take actions to prevent it. The reason for this is twofold. The first is to protect the seigniorage revenues gained from the costless money creation, while the second is to preserve the ability to affect their domestic economies by implementing the interest rate policies.

Theoretical roots of the decentralization of money offered by today's digital currencies we can find even back in the seventies in the Friedrich von Hayek's theory of private money. Economic implications of the theory for digital currencies are investigated in the paper [21]. In the digital economy, cash can actually disappear and payments can centre on social and economic platforms, weakening traditional monetary policy channels. The article confirms that stable digital money is preferable for foresight, calculation and accounting.

However, there is no consensus among professionals dealing with cryptocurrencies about their classification and evaluation [22]. Authors propose refining the existing standards and introducing rules for classification and evaluation of cryptocurrencies. They also indicate that the best solution is to develop new international financial reporting standards for the accounting of cryptocurrencies.

Today, for Bitcoin and the rest of ever more numerous cryptocurrencies, we could say that they can fulfil their "fundamental task" of (crypto) currency, digital money as a means of payment. A continuously growing number of companies accept Bitcoins as payment for their goods and services [23]. There are Crypto ATMs—according to https://coinatmradar.com/ (accessed on 19 April 2021) there were 18,541 ATMs in 72 countries, 275,795 establishments offering other services (exchange offices, shops and various other services accepting cryptocurrency). Currently there are 270 web-based digital currency exchanges according to https://coin.market/exchanges, (accessed on 19 April 2021).

There is an opinion that cryptographic assets do not fully satisfy the conditions to be a currency and that they are more similar to an asset class [24]. White et al. [25] claim "that Bitcoin's behaviour more closely resembles..., an emerging asset class, rather than a currency...". Furthermore, [26] have shown that cryptocurrencies behave more like an investment instrument than a currency and [27] that Bitcoins are mostly used as a speculative investment and not as an alternative currency and medium of exchange. Bouri et al. [28] conclude that Bitcoin is a poor hedge but contributes to a well-diversified portfolio. In the focus of our interest and in that of many potential investors are cryptocurrencies as an alternative form of investment. The "infrastructure" to support cryptocurrencies as a potential special investment class exists: we can discuss the cryptocurrency market in terms of the total supply of and demand for cryptocurrencies. Furthermore, we can also talk about the primary and secondary market of cryptocurrencies: the primary market refers to newly issued cryptocurrencies that raise capital for the issuer's needs, most often start-ups based on blockchain technology, while the secondary market relates to trading in already established cryptocurrencies. Based on stock, bond and other indices, the cryptocurrency index (CRIX) has been established as a benchmark for the cryptocurrency market. The index was created as a result of the joint project of Humboldt University Berlin from Germany, SKBI School of Business, Singapore Management University and CoinGecko. A special, new methodology was required for creating the new index, given the specificity of the cryptocurrency market. The process of generating the index, the specificity of the approach and methodology is described in [29]. Besides that, the indexing methodology and information on current index composition along with other relevant data can be found at https://www.coingecko.com/en/crypto_index/crix (accessed on 5 November 2020) and/or https://thecrix.de/ (accessed on 5 November 2020).

Some papers study the characteristics of cryptocurrencies as a special class of investment by examining the relationship between return and risk, the correlation of return with that of other classes and the politicoeconomic determinants. Burniske and White [30], inspired by Greer's [31] classification of investments in superior asset classes and criteria for their identification, highlight four distinct features distinguishing between different types of investment: investability, politicoeconomic features, correlation of returns, i.e., price independence and risk–reward profile. First, for a particular type of investment there should be so called investability, which also implies a certain level of liquidity. Second, a special type of investment has a special political and economic profile that follows from the source value of investment, investment management and its primary purpose. Third, the market value of the investment should be independent of other types of investment, indicating the absence of or low correlation of their returns. The previous three characteristics should lead to differentiated risk–reward profiles, which can then be further "broken down" into the specificity of returns and volatility of each particular class. Thus, for example, ordinary shares and bonds make different types of investments: after meeting the first criterion of investability, they differ according to the other three criteria. The authors consider Bitcoin as the main representative of cryptocurrencies in the course of 5 years and conclude that Bitcoin represents a special type of investment due to the observed indicators. It should be borne in mind that cryptocurrencies are not a reliable store of value and they do not have a stable purchasing power over a long period of time, unlike fiat money. Moreover, cryptocurrencies are not able to ensure the stream of payments to the owner, unlike other assets such as real estate, stocks or bonds [32].

The risk and the return of cryptocurrencies Bitcoin, Ripple and Ethereum are considered in an extensive paper [33]. The ratio between the return and risk of cryptocurrencies differs from those found in shares, ordinary currencies and precious metals. Cryptocurrencies are exposed neither to the factors most frequently affecting the stock market nor to macroeconomic factors, their market is influenced by other specific impacts. Sajter [34] compares the returns of the same three cryptocurrencies with the returns of six major world equity indices and concludes that the observed cryptocurrencies can be considered a new, specific form of investment, since the trend of their values or returns is not related to the trend of equity index returns. Similarly, Ankenbrand and Bieri [35] concluded that cryptocurrencies can be seen as an individual asset class. Additional studies [11,36,37] offer valuable insides and facts about cryptocurrencies as a new financial asset, which makes them an effective diversification tool.

Many studies named in the Introduction of the paper have already shown the benefits of including cryptocurrencies into portfolio optimisation processes. The task of this study is to propose appropriate methodology for selecting optimal cryptocurrency portfolio, taking into consideration a broader set of features of this new investible instrument.

3. Data and Criteria

The starting sample consisted of the top 20 cryptocurrencies ranked by market capitalization, calculated on the basis of circulating supply, according to https://coinmarketcap.com/all/views/all/ (accessed on 12 February 2020). As the first criterion we took the most common one—market capitalization. Market capitalization is a metric that indicates the market value and size of a cryptocurrency [38]. Since the cryptocurrency price can give an inaccurate measure of its total value, the market capitalization can identify the value of a cryptocurrency and accurately compare it to other cryptocurrencies. This concept is the same as the one from the stock market, where stock market capitalization is the current stock price multiplied by the total number of existing stocks. Accordingly, cryptocurrency market capitalization equals the total number of circulating coins multiplied by its current price. The most common metrics for the total number of circulating coins is the circulating supply, defined as the number of coins currently circulating in the market available to the general public. Although market capitalization is often taken as a starting criterion for the cryptocurrency sample selection, due to its importance for practitioners, it should be also taken as a separate criterion in the process of portfolio optimization.

The portfolio construction can be significantly obstructed by the problem of liquidity of assets and in this sense a special attention has to be put on cryptocurrencies, since they have far lower daily trading amounts than traditional financial assets [8]. The possibility of trading the assets on the reallocation date and selling or buying between two reallocation dates is of crucial importance in portfolio management. In a recent study [39] analyse four cryptocurrencies in order to identify the determinants of their liquidity. They have found the number of transactions to be one of the most important liquidity drivers, while the most commonly used financial market variables have not proved to have explanatory power. Since spread data for cryptocurrencies are not easily available, Trimborn et al. [8] used the turnover value as a proxy for liquidity. It is calculated for a set period, for example 24 h, as the sum of products of the number of assets and its price. That data is actually registered as the volume (24 h) on the https://coinmarketcap.com/all/views/all/ (accessed on 25 August 2020) the amount of cryptocurrency that has been traded during a certain period of time, i.e., 24 h in this case. As the second criterion we selected the trading volume, following the opinion and practice that the trading volume can also be used as a liquidity measure. It shows how easily the stock can be bought and sold. A low trading volume indicates the infrequent trading with a cryptocurrency and consequently the difficulty to purchase or to sell shares. A high trading volume means that a cryptocurrency is highly liquid and may be bought or sold easily. Besides that, among practitioners it can be seen that the volume presents one of the most valuable pieces of data, which can show the direction and movement of the cryptocurrency and predict the future price and its demand.

However, as shown by the CRIX values (according to https://thecrix.de/ (accessed on 12 February 2020)), the Bitcoin prices, market capitalization and trading volumes (https://coinmarketcap.com/currencies/bitcoin/ (accessed on 12 February 2020)), in the period from January 2015 to February 2020, the market was characterised by a sluggish movement of prices, market capitalization and volumes from 2015 to 2017. The surge in their values started in 2017, they peaked in 2018, followed by the intensive trading period characterized by high volatility with continuous fluctuations in their values until February 2020. Therefore, the period of monotonous price movements and trading from 2015 to 2017 is excluded from further calculations since it is more interesting and challenging to test the proposed model in more volatile periods.

Cryptocurrencies display high expected returns with large volatilities [11]. The mean-variance analysis is limited because of the highly non-normal return distribution of cryptocurrencies [6]. Kajtazi and Moro [2] found, in line with previous research, that bitcoin exhibits large kurtosis and is positively skewed (albeit to a much lesser extent than previously reported). Moreover, among several unique properties, [40] found that cryptocurrencies have leverage effects and Student's t error distributions. The study [41] uses the symmetric (GARCH 1,1) and asymmetric (EGARCH, TGARCH and PGARCH) models to measure the volatility of cryptocurrencies. The results prove again the high volatility of cryptocurrencies and in most cases, the asymmetric PGARCH with Student's t distribution provides a better fit. In accordance with these findings, we continued with the combination of alternative risk measures, which should be used for this class of risky assets. Besides volumes and market capitalization, daily closing prices are collected from https://coinmarketcap.com/ (accessed on 12 February 2020). Based on daily closing prices, the values of four more criteria: expected daily return, standard deviation, value-at-risk (VaR) and conditional value-at-risk (CVaR) were calculated.

From the day of its appearance within JP Morgan Bank, the value-at-risk (VaR) as a risk measure has been attracting immense attention. It has become one of the most controversial financial instruments. Despite being criticised, it became a very popular and widely used risk measure because of its simplicity, applicability and universality [42]. In addition, controlling authorities have imposed regulatory constraints on the asset allocations of financial institutions based on the estimation of VaR [43]. VaR is a statistical measure that assesses the risk of an asset or the whole portfolio, expressed with one number. It shows the worst estimated loss for a certain time horizon and a certain confidence level. VaR represents the difference between the invested amount of money and the value that is not going to be failed in α% cases—the value that corresponds to the 1-α percentile of the distribution. However, VaR does not provide any information about the values from the tail of the distribution, i.e., the values that exceed the value of VaR with small probability, but high losses. The risk measure that provides such information is conditional value-at-risk (CVaR). For a given time horizon and confidence level α, CVaR is defined as the conditional expectation of losses greater than VaR. VaR has a drawback of not being a coherent risk measure, i.e., it does not fulfil the subadditivity condition: $\rho(X+Y) \leq \rho(X) + \rho(Y)$. This means that VaR of a portfolio is greater than the sum of VaRs of its constituents [44], which can discourage portfolio diversification and lead to the dangerous risk concentration. Ref. [45] compared VaR, variance and CVaR and concluded that only CVaR is a coherent risk measure. Moreover, CVaR has superior mathematical characteristics over VaR; it keeps good properties of VaR and overcomes its shortages. In agreement with experts, it is decided to proceed with both measures as criteria, VaR for estimating loss for a specific time horizon and a certain confidence level, CVaR as a coherent risk measure, which gives valuable information about losses from the tail of distribution, which exceed VaR.

Guided by opinions and findings indicating that the phenomenon of cryptocurrency created considerable interest and became an appealing investment due to its unique qualities [46], as the last, seventh criterion we took the popularity or attractiveness of cryptocurrencies. Ref. [23] studied Bitcoin attractiveness for investors and users finding its significant impact on Bitcoin price with variation over time. Sovbetov [47] concluded that attractiveness of cryptocurrencies matters, also finding that its recognition is subjected to the time factor. Positive correlation of cryptocurrency attractiveness and its price has also been confirmed by other research [46,48]. The attractiveness of cryptocurrencies is usually measured by the amount of the cryptocurrency-related posts in social media like Twitter, Google Trends, Yahoo, Wikipedia and others. Although, the fact that somebody is interested in gaining information from social media does not necessarily mean active participation in the market, many studies prove significant influences and connections between data offered by social media and trading and prices of assets. Matta et al.'s [49] study proved that the investment professionals in Bitcoin use social media activity and information extracted by a web search and found it helpful. Investors search social media when making decisions since it is proved that sentiment analysis captures information not embedded in prices [50]. News and information extracted from online social media (blogs, Twitter feeds, etc.) can be used to predict changes in various economic and business indicators, which is supposed to have an impact also for the Bitcoin price [51]. Stolarski et al. [52] studied cryptocurrency perception using Wikipedia and Google Trends although cryptocurrencies seem to have embraced Twitter as a major channel of communication. Park and Lee [53] investigate the Twitter-mediated communication behaviours among cryptocurrencies, finding that cryptocurrencies' active networking strategies affected their credit scores. Kaminski [54] had already shown that the microblogging platform Twitter may be interpreted as a virtual trading floor that emotionally reflects Bitcoin's market movement. Therefore, quickly recognizing and incorporating the impact of tweets on the price direction in the trading strategy can provide both a purchasing and selling advantage [55]. Kraaijeveld and De Smedt [56] study the predictive power of the Twitter sentiment for various cryptocurrencies, finding that it has predictive power for the returns of Bitcoin, Bitcoin Cash and Litecoin and for EOS and TRON. Due to the importance of Twitter in studying cryptocurrencies, its transparency and simplicity regarding data collection, we proceed with the number of Tweets, obtained from https://bitinfocharts.com/comparison/tweets-btc-eth-ltc-xrp.html (accessed on 25 August 2020), as the attractiveness measure.

Finally, the dataset contains nine cryptocurrencies: Bitcoin—BTC, Dash, Ethereum Classic—ETC, Ethereum—ETH, Litecoin—LTC, Monero—XMR, Neo, Stellar—XLM and Ripple—XRP, for which all required data are available in the period from January 2017 to February 2020. Namely, some of the top 20 cryptocurrencies by market capitalization on 12 February 2020 are not traded in the proposed period from 2017 to 2020 and for some there are no comparable and available data for the number of tweets.

Table 1 gives the overview of descriptive statistics for nine cryptocurrencies along with the Jarque–Bera (JB) test for normality. The null hypothesis of the JB test is a joint hypothesis of both the skewness and the excess kurtosis being zero, i.e., matching a normal distribution. From Table 1 it can be concluded that for all cryptocurrencies, the null hypothesis can be rejected at 1, 5 and 10% significance levels, i.e., the returns are not normally distributed. This can be corroborated by mostly all cryptocurrencies having positively skewed distribution (except for Bitcoin, which has a somewhat negatively skewed distribution, although the coefficient of skewness is roughly around zero). Moreover, all cryptocurrencies show positive excess kurtosis indicating leptokurtic distribution, meaning that the tails on this distribution is heavier than that of a normal distribution, indicating a higher degree of risk and higher probability of extreme values. For that reason, in the process of portfolio selection it is more appropriate to take into consideration alternative risk measures and other criteria, as we anticipated.

Table 1. Descriptive statistics for the selected cryptocurrencies from 1 January 2017 to 11 February 2020.

	BTC	DASH	ETC	ETH	LTC	XMR	NEO	XLM	XRP
Min	−0.2075	−0.2432	−0.4353	−0.3155	−0.3952	−0.2932	−0.4610	−0.3664	−0.6163
q1	−0.0159	−0.0275	−0.0240	−0.0213	−0.0263	−0.0262	−0.0364	−0.0325	−0.0241
Me	0.0023	−0.0012	0.0000	0.0001	−0.0008	−0.0004	−0.0017	−0.0021	−0.0026
q3	0.0215	0.0290	0.0272	0.0257	0.0268	0.0287	0.0328	0.0308	0.0205
Max	0.2251	0.4377	0.4577	0.2901	0.5103	0.4303	0.8012	0.7231	1.0274
μ	0.0020	0.0021	0.0019	0.0030	0.0025	0.0016	0.0040	0.0030	0.0033
σ	0.0426	0.0629	0.0664	0.0571	0.0625	0.0616	0.0850	0.0826	0.0778
α_3	−0.05	0.96	0.15	0.25	1.14	0.39	1.62	1.99	2.90
α_4	3.56	6.72	6.94	4.29	9.52	4.72	14.88	16.19	37.41
JB	592.47 ***	2289.7 ***	2263.1 ***	871.8 ***	4495.8 ***	1072.9 ***	10874 ***	13042 ***	67245 ***

Source: The authors' calculations in R Studio (*** indicate significance at the 0.01 level).

Values of other criteria for the whole sample are given in Table 2. VaR, measuring the level of financial risk for each cryptocurrency over the whole sample, indicates that the highest possible loss can be obtained with Monero, Neo and Stellar and the lowest possible loss with Bitcoin. The values of CVaR, measuring the mean of tail risk, is the lowest for Bitcoin and the highest for Neo, Stellar and Ripple. The biggest values of the mean volume (MVlm), mean market capitalization (MMC) and mean number of tweets (MoT) can be observed for Bitcoin and the lowest values for Monero, Ethereum Classic and Stellar respectively.

Table 2. Values of other criteria for the whole sample.

	BTC	DASH	ETC	ETH	LTC	XMR	NEO	XLM	XRP
VaR	0.0676	0.0921	0.0957	0.0848	0.0852	0.1026	0.1100	0.1078	0.0900
CVaR	0.1013	0.1634	0.1683	0.1475	0.1664	0.1496	0.2304	0.2212	0.2167
MVlm	9.16×10^9	1.86×10^8	3.73×10^8	3.64×10^9	1.30×10^9	6.76×10^7	2.00×10^8	1.33×10^8	8.42×10^8
MMC	1.11×10^{11}	1.78×10^9	1.14×10^9	2.91×10^{10}	4.30×10^9	1.64×10^9	1.59×10^9	2.45×10^9	1.55×10^{10}
MoT	37,033.30	3299.03	121.85	11,778.24	2397.39	508.05	1181.00	37.97	4296.33

Source: The authors' calculations in MATLAB and R Studio.

4. The Multicriteria Model

The multicriteria (MC) model based on the PROMETHEE II approach [57] is applied. According to the PROMETHEE II model, each alternative P, in this case the cryptocurrencies portfolio, is evaluated with two flows. The positive flow $\Phi^+(P)$ indicates how much one cryptocurrency portfolio is better than other cryptocurrency portfolios in all criteria. The higher the $\Phi^+(P)$ is, the better is the cryptocurrency portfolio. The negative flow $\Phi^-(P)$ indicates how much better the cryptocurrency portfolio is over other cryptocurrency portfolios. The lesser the $\Phi^-(P)$ is, the better the cryptocurrency portfolio is. Finally, the net flow Φ is the difference between these positive and negative flows: i.e.,

$$\Phi(P) = \Phi^+(P) - \Phi^-(P) \tag{1}$$

The higher the net flow $\Phi(P)$ is, the better the cryptocurrency portfolio is. Positive and negative flows are calculated by pairwise comparisons of all the cryptocurrency portfolios and for every criterion simultaneously.

Since the number of possible portfolios that can be made up from a sample of cryptocurrencies is infinite, it is impossible to compare all pairs of portfolios. Therefore, this study employs the procedure introduced by Khoury and Martel [58] and Zmitri et al. [59], following the applications of the procedure in [60,61].

Each cryptocurrency portfolio (its positive and negative flow) is compared to two imaginary portfolios: ideal ($\overline{P}$) and anti-ideal ($\underline{P}$). Compared to the anti-ideal, the positive flow $\Phi^+(P)$ is obtained. The higher the $\Phi^+(P)$ is, the better is the cryptocurrency portfolio since it is more distant from the anti-ideal. The lower the $\Phi^-(P)$ is, the better the cryptocurrency portfolio is since it is closer to the ideal, Moreover, the higher the net flow Φ is, the better the cryptocurrency portfolio is.

For each criterion C_j, $(j = 1, 2, \ldots, n)$, which has to be maximized, the ideal is:

$$C_j(\overline{P}) = \max_i C_j(A_i), \tag{2}$$

where $A = \{A_1, A_2, \ldots, A_N\}$ is the set of N alternatives, in this case nine cryptocurrencies. For the same criterion, which has to be maximized, the anti-ideal is:

$$C_j(\underline{P}) = \min_i C_j(A_i) \tag{3}$$

Without the loss of generality, we can suppose that all criteria are to be maximized.

The set of feasible solutions is the set of cryptocurrency portfolios, which can be formed from the observed cryptocurrencies. The evaluation of the cryptocurrency portfolio P according to criterion j is obtained by multiplying the share invested in each cryptocurrency A_i in the portfolio P, i.e., a_i, with the evaluation of cryptocurrency i according to criterion

j. Obviously, the sum of all shares invested in each cryptocurrency A_i in the portfolio P equals 1.

$$C_j(P) = \sum_{i=1}^{N} a_i C_j(A_i) \quad (4)$$

For each criterion C_j the preference functions are defined as in the PROMETHEE method where indifference q and preference p thresholds are predefined numbers from the interval $\left[0, C_j(\overline{P}) - C_j(\underline{P})\right]$, where $0 \leq q, p \leq C_j(\overline{P}) - C_j(\underline{P})$ and $q \leq p$ is always true. Moreover, q and p always have economic significance.

In this particular application we assumed that the highest value of preference threshold p cannot exceed half the span between the anti-ideal and ideal according to that criterion, i.e., $p \in \left[q, \frac{C_j(\overline{P}) - C_j(\underline{P})}{2}\right]$.

Like in [62], we used the same preference function for all criteria and in this study that is the linear preference function with the indifference threshold (type V), as it most generally displays the relations between pairs of cryptocurrency portfolios. It has the following general form:

$$\Psi(d) = \begin{cases} 0, \ d \leq q \\ \frac{d-q}{p-q}, \ q < d \leq p \\ 1, \ d > p \end{cases} \quad (5)$$

where d is the difference in the evaluation of the two alternatives by the same criterion.

In this application, when the cryptocurrency portfolio P is compared to the anti-ideal ($\underline{P}$), i.e., when we calculate $\Phi_j^+(P)$, the difference d presents the "distance" from the anti-ideal (by j-criterion). Taking that $d_j(P) = C_j(P) - C_j(\underline{P})$ for $\Phi_j^+(P)$ we have [61]:

$$\Phi_j^+(P) = \begin{cases} 0, & C_j(P) \leq C_j(\underline{P}) + q_j^- \\ \frac{C_j(P) - C_j(\underline{P}) - q_j^-}{p_j^- - q_j^-}, & C_j(\underline{P}) + q_j^- < C_j(P) \leq C_j(\underline{P}) + p_j^- \\ 1, & C_j(P) > C_j(\underline{P}) + p_j^- \end{cases} \quad (6)$$

Analogously, taking that $d_j(P) = C_j(\overline{P}) - C_j(P)$ for $\Phi_j^-(P)$ we have:

$$\Phi_j^-(P) = \begin{cases} 0, & C_j(P) \geq C_j(\overline{P}) - q_j^+ \\ \frac{C_j(\overline{P}) - C_j(P) - q_j^+}{p_j^+ - q_j^+}, & C_j(\overline{P}) - p_j^+ \leq C_j(P) < C_j(\overline{P}) - q_j^+ \\ 1, & C_j(P) < C_j(\overline{P}) - p_j^+ \end{cases} \quad (7)$$

Finally, for $\Phi_j(P)$ we have:

$$\Phi_j(P) = \begin{cases} -1, & C_j(P) \leq C_j(\underline{P}) + q_j^- \\ \frac{C_j(P) - C_j(\underline{P}) - p_j^-}{p_j^- - q_j^-}, & C_j(\underline{P}) + q_j^- < C_j(P) \leq C_j(\underline{P}) + p_j^- \\ 0, & C_j(\underline{P}) + p_j^- < C_j(P) \leq C_j(\overline{P}) - p_j^+ \\ \frac{p_j^+ - C_j(\overline{P}) + C_j(P)}{p_j^+ - q_j^+}, & C_j(\overline{P}) - p_j^+ \leq C_j(P) < C_j(\overline{P}) - q_j^+ \\ 1, & C_j(P) \geq C_j(\overline{P}) - q_j^+ \end{cases} \quad (8)$$

or graphically as in paper [61] p. 62.

Positive and negative flows, $\Phi_j^+(P)$ and $\Phi_j^-(P)$, have to be calculated separately for each criterion $C_j, j = 1, 2, \ldots, n$. Then, the net flow is obtained as a weighted sum of the difference between positive and negative flows, i.e.,

$$\Phi(P) = \sum_{j=1}^{n} w_j\left(\Phi_j^+(P) - \Phi_j^-(P)\right) \tag{9}$$

where relation (10) follows for any possible cryptocurrency portfolio P,

$$\Phi(\underline{P}) \leq \Phi(P) \leq \Phi(\overline{P}) \tag{10}$$

The weights w_j of the criteria are obtained in agreement with the decision maker and/or by some of the methods for determining the weights of criteria. For obtaining the weights in this case, the AHP method and its eigenvalue procedure with pairwise comparisons obtained by a group of experts is applied.

Finally, the optimal cryptocurrency portfolio $(a_1, a_2, \ldots, a_N)$ is one that finds the maximum net flow, i.e.,

$$Max\ \Phi(P) \tag{11}$$

subject to:

$$\sum_{i=1}^{N} a_i = 1 \tag{12}$$

$$0 \leq a_i \leq a_{M_i}, \tag{13}$$

where a_i is the share invested in A_i in the cryptocurrency portfolio, a_{M_i} is the maximum proportion to invest in cryptocurrency A_i in cryptocurrency portfolio P and N is the number of cryptocurrencies, which can be included in cryptocurrency portfolio P.

5. Implementation of the Model

The presented model was used for the selection of optimal portfolios of cryptocurrencies based on the sample of nine cryptocurrencies: Bitcoin—BTC, Dash, Ethereum Classic—ETC, Ethereum—ETH, Litecoin—LTC, Monero—XMR, Neo, Stellar—XLM and Ripple—XRP and seven criteria: daily return, standard deviation, value-at-risk (VaR), conditional value-at-risk (CVaR), volume, market capitalization and attractiveness.

For criteria selection and determination of their weights, twelve experts were engaged, some of them professionals dealing with cryptocurrencies and others scientists, including the authors of the paper. By engaging twelve experts, we reduced subjectivity in the process of weights calculation. Weights of the chosen criteria were estimated using the Saaty's AHP method, its eigenvalue procedure [63,64]. After the complete Saaty matrix was obtained by the experts, the weight of criteria was calculated using Expert Choice and are given in the last row of Table 3. The reported inconsistency is 0.02.

Table 3. Heading of decision matrices.

Criterion	μ	σ	**VaR**	**CVaR**	**MVlm**	**MMC**	**MoT**
Min/Max	max	min	min	min	max	max	max
Type	V	V	V	V	V	V	V
Weights	0.208	0.141	0.321	0.183	0.057	0.055	0.035

We can see that the biggest accent in the evaluators' judgements for the importance of criteria is given to the possible losses by investing in cryptocurrency portfolios. Due to the generally accepted and confirmed opinion of high riskiness of cryptocurrencies, the VaR has taken the highest percentage, as much as 32%. The other risk measures, together with the expected return are also highly esteemed: expected return—$E(R)$ 21%, CVaR 18% and standard deviation—St.dev. 14%. Volume—Vol and market capitalization—MMC have lower and almost equally estimated importance, 5.7% and 5.5% respectively. The lowest weight is given to the criterion of attractiveness, measured by the number of tweets (MoT), 3.5%.

For choosing the preference functions for the observed criteria, the same group of experts was consulted and it was decided to proceed with the linear preference function with the indifference threshold (type V) for all criteria, as it most generally displays the relations between the pairs of alternatives. Thresholds q^-, q^+, p^- and p^+ were calculated in the following manner for all criteria, which is in accordance with that previously said about the intervals of the threshold's values:

$$q_j^- = (s_j(2) - s_j(1)), q_j^+ = (s_j(9) - s_j(8)),$$

$$p_j^- = \frac{\left(s_j(9) - q_j^+ - s_j(1) - q_j^-\right)}{2} + q_j^-,$$

$$p_j^+ = \frac{\left(s_j(9) - q_j^+ - s_j(1) - q_j^-\right)}{2} + q_j^+, \forall j = 1, 2, \ldots 7,$$

where $s_j(1), s_j(2), \ldots, s_j(9)$ are sorted values of the evaluation of alternatives according to criterion j. Finally, the heading of all decision matrices is given in Table 3.

For the period from January 2017 to February 2020, using the rolling window of 6 months of daily returns, volumes and market capitalization, the mean daily returns, standard deviations, VaR, CVaR and average volumes and average market capitalization were calculated. The optimal portfolios using the described MC model were calculated at the first of each month by taking the previous 6 months of daily data for the calculations. Moreover, the attractiveness, measured by the number of tweets, was taken for each cryptocurrency on the day before the calculation of the optimal portfolio. Therefore, 32 optimal portfolios in 32 successive months were obtained. The maximum proportion to invest in cryptocurrency A_i in the optimal cryptocurrency portfolio P was limited to $a_{Mi} = 0.5 \; \forall i = 1, 2, \ldots, 9$.

6. Results

The resulting portfolios with the weights of each cryptocurrency are given in Table 4. We can see that the most favourable cryptocurrency is Bitcoin, the fact that could be discerned from the data and results given in Tables 1 and 2. It is followed by Ethereum, Litecoin, Ripple, Dash and Ethereum Classic, depending on the date and period. In general, it can be said that there was a high level of diversification, which was also supported with the constraint (13).

Table 4. Optimal portfolios with the weights of each cryptocurrency.

Date	BTC	DASH	ETC	ETH	LTC	XMR	NEO	XLM	XRP
1 July 2017	0.2322	0.0169	0.2200	0.4971	0.0068	0.0022	0.0067	0.0023	0.0158
1 August 2017	0.2890	0.0191	0.0596	0.4944	0.1186	0.0026	0.0077	0.0017	0.0073
1 September 2017	0.4893	0.0025	0.0266	0.1093	0.2581	0.0017	0.1099	0.0006	0.0020
1 October 2017	0.4842	0.0291	0.0188	0.2111	0.0338	0.0153	0.1947	0.0040	0.0089
1 November 2017	0.4656	0.0886	0.0317	0.1163	0.0553	0.0396	0.1539	0.0111	0.0379
1 December 2017	0.4990	0.4086	0.0007	0.0014	0.0594	0.0024	0.0274	0.0003	0.0007
1 January 2018	0.4783	0.1963	0.0069	0.0130	0.0936	0.1759	0.0120	0.0083	0.0156
1 February 2018	0.4545	0.0044	0.0015	0.4811	0.0037	0.0069	0.0204	0.0238	0.0038
1 March 2018	0.4709	0.0003	0.0002	0.4962	0.0082	0.0003	0.0004	0.0228	0.0007
1 April 2018	0.1742	0.0003	0.0002	0.3038	0.4956	0.0004	0.0004	0.0246	0.0005
1 May 2018	0.3200	0.0005	0.0003	0.4991	0.1567	0.0008	0.0006	0.0211	0.0008
1 June 2018	0.3893	0.0001	0.0001	0.4998	0.1017	0.0001	0.0002	0.0085	0.0004
1 July 2018	0.4686	0.0003	0.0004	0.4987	0.0009	0.0004	0.0008	0.0293	0.0005
1 August 2018	0.4721	0.0361	0.0074	0.0210	0.4140	0.0089	0.0133	0.0077	0.0195
1 September 2018	0.4938	0.0222	0.0191	0.0662	0.1051	0.0125	0.0062	0.2239	0.0510
1 October 2018	0.4935	0.0378	0.0304	0.0130	0.0367	0.0119	0.0052	0.2561	0.1155

Table 4. *Cont.*

Date	BTC	DASH	ETC	ETH	LTC	XMR	NEO	XLM	XRP
1 November 2018	0.4658	0.0098	0.0259	0.0100	0.0120	0.0168	0.0064	0.3960	0.0573
1 December 2018	0.4921	0.0008	0.0009	0.0005	0.0007	0.0011	0.0004	0.3042	0.1995
1 January 2019	0.4817	0.0113	0.0115	0.0094	0.0233	0.0169	0.0066	0.0983	0.3409
1 February 2019	0.3563	0.0415	0.0350	0.0341	0.0861	0.0461	0.0318	0.0541	0.3150
1 March 2019	0.3634	0.0008	0.0004	0.0009	0.1370	0.0006	0.0006	0.0004	0.4958
1 April 2019	0.3890	0.0541	0.0372	0.0545	0.1975	0.0505	0.0412	0.0607	0.1152
1 May 2019	0.4853	0.0050	0.0032	0.0086	0.1316	0.0048	0.0036	0.0041	0.3539
1 June 2019	0.4997	0.0003	0.0002	0.0007	0.3894	0.0002	0.0003	0.0002	0.1091
1 July 2019	0.4534	0.0936	0.0310	0.0677	0.1890	0.0560	0.0294	0.0342	0.0458
1 August 2019	0.4994	0.4289	0.0005	0.0013	0.0485	0.0066	0.0004	0.0009	0.0134
1 September 2019	0.4995	0.4596	0.0005	0.0006	0.0005	0.0030	0.0002	0.0019	0.0341
1 October 2019	0.4418	0.2611	0.0011	0.0021	0.0015	0.0015	0.0006	0.0030	0.2873
1 November 2019	0.4995	0.2797	0.0003	0.0005	0.0005	0.0004	0.0002	0.0003	0.2186
1 December 2019	0.4992	0.0006	0.0012	0.0125	0.0014	0.0025	0.0069	0.0007	0.4750
1 January 2020	0.4994	0.0004	0.0010	0.0014	0.0005	0.0029	0.0021	0.0015	0.4908
1 February 2020	0.5000	0.0000	0.4997	0.0000	0.0000	0.0000	0.0000	0.0000	0.0003

Source: the authors' calculations in MATLAB 2017a, The MathWorks, Inc.

For the evaluation of the results, out-of-sample testing was performed. The optimal portfolios obtained by the multicriteria (MC) model are compared with the naïve portfolio (NAIVE), two mean-variance (MV) models (first one in the middle of the efficient frontier having the average variance—MV middle and the second at the end of the efficient frontier having the maximum variance and containing only one cryptocurrency—MV max). It is also compared to a portfolio obtained using the maximum Sharpe ratio (Max Sharp) and the one with mean-CVaR optimization (MCVaR), from the middle of the mean-CVaR efficient frontier. The nominated principles of portfolio selection are commonly found in related research and in portfolio optimization in general.

Brauneis and Mestel [13] compared risk and return of different MV portfolios to single cryptocurrency investments and two benchmarks, the naively diversified portfolio and the CRIX. They found that in terms of the Sharpe ratio and certainty equivalent returns, the naïvely diversified portfolio outperforms single cryptocurrencies and more than 75% of MV portfolios. The similar study by Platanakis et al. [12] concluded that naïve diversification is as good, if not better, than MV diversification. Weiyi [14] concludes that none of the observed models (minimum variance, risk parity, MV, maximum Sharpe and maximum utility) is consistently better than the $1/N$ rule in the Sharpe ratio. We took into consideration confirmed good features of the naïve portfolio and included the $1/N$ principle of portfolio selection as one of the models to be compared with the proposed MC model.

Within the group of papers considering the contribution of cryptocurrencies to portfolios with the rest of assets being either traditional ones or combinations of traditional and alternative investments, there are studies that considered the conditional value-at-risk (CVaR) as the appropriate risk measure: [2,6,8,9]. Namely, it is necessary to take into consideration the alternative risk measures. In accordance with the approach of the mentioned studies, we proceed with comparison of performances of the mean–CVaR model (Table 5) and proposed MC model.

In almost all mentioned studies, the Sharpe ratio is a standard metric for measuring risk-adjusted model performance. It is also used as a principle in the process of selecting the optimal (cryptocurrency) portfolio [9,14]. Together with NAÏVE and mean–CVaR model, in the out-of-sample testing and comparison, we used the maximum Sharpe ratio model, given in Table 5.

Table 5. Five portfolio optimization models.

	Model	Abbreviation	Objective Function	Constraints
1	$1/N$ rule	NAIVE		
2	Markowitz model	MV middle	$Max\ E(R_P)$	$\sigma_P \leq s$ $\sum_{i=1}^{N} a_i = 1$ $a_i \geq 0$
3	Markowitz model	MV max	$Max\ E(R_P)$	$\sum_{i=1}^{N} a_i = 1$ $a_i \geq 0$
4	Maximum Sharpe ratio model	Max Sharp	$Max\ \frac{E(R_P)}{\sigma_P}$	$\sum_{i=1}^{N} a_i = 1$ $a_i \geq 0$
5	Mean–CVaR model	MCVaR middle	$Max\ E(R_P)$	$CVaR_P \leq c$ $\sum_{i=1}^{N} a_i = 1$ $a_i \geq 0$

s is the standard deviation of middle portfolio on the mean-variance efficient frontier; c is the CVaR of the middle portfolio on the mean-CVaR efficient frontier.

An inevitable model in the analysis is the Markowitz mean-variance (MV) model. In this study we employed two MV models, the first one in the middle of the efficient frontier having the average variance—MV middle and the second at the end of the efficient frontier having the maximum variance and containing only one cryptocurrency—MV max.

Applied portfolio optimization models for the purpose of out-of-sample testing and comparison are presented in Table 5.

The average return in the next 30 days, the standard deviation of the returns in the next 30 days, VaR of the returns in the next 30 days and return in the next trading day for 32 portfolios (on 32 dates) obtained as results of the six different optimization models are given in Appendix A, in Tables A1–A4 respectively. In Table 6 of the MC model winnings, we summarized in how many cases the MC model was better than other models, considering different indicators.

Table 6. MC model winnings.

Indicator	Number of Cases When MC Model Is Better Than				
	NAIVE	MV Middle	MV Max	Max Sharp	MCVaR
Average return in the next 30 days	17	18	18	18	21
Standard deviation of the returns in the next 30 days	26	22	24	22	30
VaR of the returns in the next 30 days	25	21	23	22	25
Return in the next trading day	20	18	19	16	17

7. Discussion and Interpretation of Results

Regarding the average return in the next 30 days NAÏVE portfolio performed rather close to the MC model with 15 (47%) winnings, while in all other comparisons it loses more convincingly (Table 6). This is a completely opposite finding to those of [4,12,13] that found undeniable superiority of the naïve portfolio.

When comparing performances of the mean–CVaR model and proposed MC model, only regarding the return in the next trading day the mean–CVaR model came closer to the MC model with 15 (47%) winnings, while in all other comparisons it loses more convincingly (Table 6). This proves the necessity of including more risk measures in the portfolio selection.

Looking at MC model winnings against the three models, according to all performance indicators, we could find that Max Sharp performed better than the previous two models, however the MC model won in all categories, except in the return in the next trading day. There the MC model and the maximum Sharpe ratio model had the same number of winnings (Table 6).

Finally, both MV middle and MV max models were outperformed by the MC model (Table 6) as in [60].

The obtained results show that the proposed MC model functioned really well: it outperformed all other models in all four out-of-sample comparisons in a very convincing manner. It won at least in 50% of cases (MC model against Max Sharp in the comparison of returns of the portfolios in the next trading day, Table 6) up to even 94% of cases (MC model against MCVaR in the comparison of standard deviations of the returns in the next 30 days, Table 6). The best performance MC model shows in the case of comparisons of standard deviations of the returns in the next 30 days, from 22 to 30 winnings from 32 possible cases, i.e., in 69–94% of cases. It is not only the percentage of winnings that should be considered, but also the differences in the values of risk measured by standard deviation, which are mostly in favour of the MC model—they can be rather big in cases when the MC model wins and rather small in opposite cases, Table A2. This excellent performance of the MC model is followed by another comparison of risk, measured by the value-at-risk, as shown in Table A3 where the MC model had from 21 to 25 winnings, i.e., in 66–78% of cases, also with notable differences of values of VaR in favour of the MC model. A slightly lesser "success" of the MC model, but still very notable, is according to the comparisons of returns: from 17 to 21 winnings, i.e., in 53–66% of cases, when comparisons of the average return in the next 30 days were considered (Table A1), and from 16 to 20 winnings, i.e., 50–67% of cases, when comparisons of the returns of portfolios in the next trading day were observed (Table A4).

The excellent performance of the presented multicriteria model for the selection of cryptocurrencies portfolio was even stronger if we took into consideration not only the partial results from the four comparisons, but also the return–risk ratios.

Undoubtedly, the inclusion of more criteria—features of cryptocurrencies, among them more risk estimators, since simple descriptive statistics pointed out the specificities of cryptocurrencies and unfulfilled assumptions of other models—and the adoption of appropriate methodology, the appropriate multicriteria decision-making model, helped us to design the best method for cryptocurrency portfolio selection.

8. Conclusions

In the last few years numerous studies have confirmed cryptocurrencies as valuable constituents of optimal portfolios, parallel with equally remarkable amounts of research finding that cryptocurrencies have differentiated risk–reward profiles, with the absence of or very low correlation of returns with other types of investment. Depending on the sample and period, different portfolio optimization models were winners and, so, different models were recommended for cryptocurrency portfolio selection. Some of those studies recognized the need to introduce other criteria, besides return and risk. The first one is the constraint of liquidity, while market capitalization constraint is only partially adopted in the selection of the cryptocurrency sample where it is usually those with the highest market capitalization that are selected. Sometimes the observed cryptocurrency benchmark portfolio is CRIX or some number of its most weighted constituents. In that way, the criterion of market capitalization is partially introduced. Additionally, the need of applying alternative risk measures in the cryptocurrency portfolio selection process is only partially adopted. Due to the non-normal distribution of cryptocurrencies' returns, there is a clear need to overcome the mean-variance framework and to employ appropriate risk measures.

Since cryptocurrencies undoubtedly contribute to the better risk–return performances of portfolios, the issue is to recognize criteria, beside the return and risk, important for the selection of cryptocurrencies, and to propose the model for cryptocurrency portfolio selection, which will adopt the recommended criteria. These were the issues of this research.

The seven criteria have been recognized and adopted for the cryptocurrency portfolio selection. Beside the expected return and variance as a risk measure, additional risk measures, value-at-risk and conditional value-at-risk are recognized as important for this job and employed in the process. Moreover, market capitalization and volumes of cryptocurrencies are considered as very important market indicators for many investors. The seventh criterion is attractiveness, in this case measured by number of tweets.

The proposed model is a multiobjective programming model based on the PROMETHEE II method, while weights of the seven chosen criteria are estimated using the AHP method, its eigenvalue procedure.

The model is demonstrated and tested using a sample of cryptocurrencies for which all required data were available in the period from January 2017 to February 2020.

Out-of-sample testing results and comparisons with performances of commonly used models, according to different indicators, gave a significant advantage to the proposed method for cryptocurrency portfolio selection, confirming the advantages of including a range of criteria, besides return and variance, and the use of appropriate multicriteria decision methodology.

Working on these issues, a number of questions and ideas arise. In the process of criteria evaluation, the group of consulted experts put the attention primarily to VaR and other risk measures together with the expected return. While exploring other criteria, using mostly Internet and social media as sources of information, we met practitioners' opinions where they were very much oriented towards market volume, market capitalization and/or attractiveness, together with returns, and less to the risk and appropriate risk measures. This provides an incentive to explore the importance of criteria, consulting a wider set of practitioners, not necessarily formally accepted experts, to test the model under such conditions. In the process of exploring the criteria other indicators that might be interesting as additional criteria in the model, like a hash rate and transaction costs were encountered. The importance and influence of such indicators and ways for their evaluation and inclusion in the model should be investigated. Moreover, it is worth considering other MCDM models for comparison purposes. Finally, the proposed model together with all other models studied in this paper were considering the cryptocurrency market under relatively normal conditions, before the big COVID-19 crisis. Future research should consider the behaviour and previously confirmed "independency" features of cryptocurrencies, which made them desirable portfolio constituents and the model performances under new conditions.

Author Contributions: Conceptualization, Z.A., B.M. and T.Š.; methodology, B.M.; software, B.M.; validation, Z.A., B.M. and T.Š.; formal analysis, Z.A., B.M. and T.Š.; investigation, Z.A., B.M. and T.Š.; resources, Z.A., B.M. and T.Š.; data curation, T.Š.; writing—original draft preparation, Z.A.; writing—review and editing, Z.A., B.M. and T.Š.; visualization, Z.A., B.M. and T.Š.; supervision, Z.A.; project administration, Z.A.; funding acquisition, Z.A. All authors have read and agreed to the published version of the manuscript.

Funding: This work is supported by the Croatian Science Foundation (CSF) under Grant [IP-2019-04-7816].

Institutional Review Board Statement: Not applicable.

Informed Consent Statement: Not applicable.

Data Availability Statement: The data presented in this study are openly available in FigShare at https://doi.org/10.6084/m9.figshare.14546247.v1 (accessed on 19 April 2021), reference number 14546247.

Conflicts of Interest: The authors declare no conflict of interest. The funders had no role in the design of the study; in the collection, analyses or interpretation of data; in the writing of the manuscript, or in the decision to publish the results.

Appendix A

Table A1. Mean return in the next 30 days for 6 different models.

Date	MC Model	NAIVE	MV Middle	MV Max	Max Sharp	MCVaR
1 July 2017	−0.00846	−0.00720	−0.01004	−0.00666	−0.00951	−0.00981
1 August 2017	0.01831	0.02118	0.03384	0.05164	0.02166	0.03836
1 September 2017	−0.00430	−0.00613	−0.00458	−0.00410	−0.00492	−0.00444
1 October 2017	0.00431	0.00208	−0.00140	−0.00281	0.00337	−0.00144
1 November 2017	0.01793	0.01934	0.01497	0.01278	0.01590	0.01024
1 December 2017	0.01934	0.02716	0.01697	0.01703	0.01704	0.03068
1 January 2018	−0.01059	0.00131	−0.00742	−0.00883	−0.01040	0.01044
1 February 2018	−0.00639	−0.00487	−0.01148	−0.01911	−0.00893	−0.00612
1 March 2018	−0.01426	−0.01784	−0.01557	−0.01677	−0.01630	−0.02305
1 April 2018	0.00745	0.01100	0.01384	0.01948	0.01507	0.01174
1 May 2018	−0.00612	−0.01076	−0.00860	−0.01028	−0.01018	−0.01046
1 June 2018	−0.00753	−0.00885	−0.01051	−0.01215	−0.01215	−0.01319
1 July 2018	0.00515	0.00399	0.00985	0.01724	0.01724	0.00109
1 August 2018	−0.01040	−0.01383	−0.00656	−0.00656	−0.00656	−0.01950
1 September 2018	0.00298	0.00494	−0.00005	−0.00005	−0.00005	0.00652
1 October 2018	−0.00356	−0.00543	−0.00348	−0.00501	−0.00501	−0.00562
1 November 2018	−0.01107	−0.01438	−0.01314	−0.01314	−0.01314	−0.01625
1 December 2018	−0.00679	−0.00543	−0.00706	−0.00540	−0.00540	−0.00498
1 January 2019	−0.00401	−0.00306	−0.00399	−0.00406	−0.00406	−0.00137
1 February 2019	0.00410	0.00537	0.00215	0.00123	0.00123	0.00403
1 March 2019	−0.00036	0.00105	0.00023	−0.00221	−0.00221	−0.00098
1 April 2019	0.00723	0.00619	0.00887	0.00856	0.00856	0.00314
1 May 2019	0.01003	0.00856	0.00907	0.00561	0.00561	0.00633
1 June 2019	0.00837	0.00365	0.01015	0.01308	0.01126	0.00297
1 July 2019	−0.00280	−0.00630	−0.00425	−0.00992	−0.00130	−0.00385
1 August 2019	−0.00283	−0.00490	−0.00533	−0.00880	−0.00058	−0.00745
1 September 2019	−0.00102	0.00100	−0.00119	−0.00227	−0.00227	0.00129
1 October 2019	−0.00748	−0.00963	−0.00803	−0.00820	−0.00820	−0.00898
1 November 2019	0.00047	0.00348	0.00146	0.00244	0.00244	0.00966
1 December 2019	−0.01043	−0.01316	−0.00938	−0.00853	−0.00853	−0.01360
1 January 2020	0.00825	0.01545	0.00872	0.00910	0.00910	0.01045
1 February 2020	0.00984	0.00939	0.01109	0.01365	0.01365	0.00962
MC model better		17	18	18	18	21

Source: the authors' calculations in MATLAB.

Table A2. Standard deviations of the returns in the next 30 days.

Date	MC Model	NAIVE	MV Middle	MV Max	Max Sharp	MCVaR
1 July 2017	0.07496	0.07034	0.08456	0.09640	0.07659	0.08502
1 August 2017	0.02971	0.03042	0.07466	0.13974	0.02929	0.08411
1 September 2017	0.08286	0.08559	0.10009	0.14038	0.08286	0.10368
1 October 2017	0.02294	0.02917	0.04006	0.06587	0.02291	0.04376
1 November 2017	0.02931	0.03416	0.05672	0.08712	0.03975	0.03611
1 December 2017	0.06198	0.07373	0.07155	0.10676	0.06230	0.10038
1 January 2018	0.07012	0.08032	0.07903	0.08970	0.07036	0.09869
1 February 2018	0.07227	0.08305	0.08522	0.08559	0.07905	0.08818
1 March 2018	0.04150	0.04706	0.04925	0.05854	0.05060	0.05228
1 April 2018	0.05874	0.06295	0.06776	0.07955	0.07089	0.06727
1 May 2018	0.03771	0.03840	0.04498	0.05410	0.05242	0.04872
1 June 2018	0.04323	0.04423	0.05136	0.05276	0.05276	0.04925
1 July 2018	0.03391	0.03815	0.04317	0.05796	0.05796	0.04673
1 August 2018	0.03502	0.04523	0.02895	0.02895	0.02895	0.06235
1 September 2018	0.03518	0.04872	0.02362	0.02362	0.02362	0.06286
1 October 2018	0.02748	0.03353	0.03006	0.04139	0.04139	0.03769

Table A2. *Cont.*

Date	MC Model	NAIVE	MV Middle	MV Max	Max Sharp	MCVaR
1 November 2018	0.04095	0.04527	0.03705	0.03705	0.03705	0.04776
1 December 2018	0.05492	0.06052	0.05309	0.04829	0.04829	0.06143
1 January 2019	0.03969	0.05116	0.04060	0.04527	0.04527	0.05446
1 February 2019	0.02653	0.02979	0.02501	0.02969	0.02969	0.02797
1 March 2019	0.02655	0.03019	0.02918	0.02886	0.02886	0.03839
1 April 2019	0.03636	0.03649	0.04521	0.05675	0.05675	0.03620
1 May 2019	0.04675	0.04806	0.04614	0.05193	0.05193	0.05490
1 June 2019	0.03360	0.03284	0.03676	0.04620	0.03992	0.03908
1 July 2019	0.06093	0.05895	0.06195	0.06508	0.06314	0.06872
1 August 2019	0.02973	0.02882	0.03370	0.03898	0.03230	0.03474
1 September 2019	0.02418	0.02615	0.02208	0.02248	0.02248	0.02790
1 October 2019	0.03216	0.03724	0.03032	0.02840	0.02840	0.04025
1 November 2019	0.03067	0.02952	0.03456	0.03739	0.03739	0.04333
1 December 2019	0.02573	0.03019	0.02428	0.02389	0.02389	0.03456
1 January 2020	0.03087	0.03795	0.02997	0.03032	0.03032	0.03732
1 February 2020	0.03878	0.02656	0.04639	0.06295	0.06295	0.03108
MC model better		26	22	24	22	30

Source: the authors' calculations in MATLAB.

Table A3. VaR of the returns in the next 30 days.

Date	MC Model	NAIVE	MV Middle	MV Max	Max Sharp	MCVaR
1 July 2017	0.14720	0.11604	0.13227	0.14674	0.13408	0.13272
1 August 2017	0.03533	0.02392	0.06561	0.13305	0.02524	0.06911
1 September 2017	0.27211	0.28985	0.24010	0.32852	0.26675	0.24837
1 October 2017	0.03294	0.05378	0.08710	0.11948	0.03510	0.09091
1 November 2017	0.07482	0.07651	0.10852	0.12399	0.08613	0.07650
1 December 2017	0.16536	0.18522	0.20809	0.25023	0.16609	0.17318
1 January 2018	0.21151	0.25978	0.23845	0.25880	0.21708	0.29431
1 February 2018	0.17755	0.19189	0.19366	0.18522	0.18750	0.23017
1 March 2018	0.11269	0.12871	0.13160	0.15434	0.13431	0.13820
1 April 2018	0.13467	0.14447	0.14563	0.16561	0.15232	0.14870
1 May 2018	0.08546	0.09116	0.10130	0.11899	0.11694	0.09309
1 June 2018	0.11563	0.12065	0.12794	0.12948	0.12948	0.11978
1 July 2018	0.07847	0.08296	0.08595	0.10819	0.10819	0.08638
1 August 2018	0.08055	0.10988	0.06853	0.06853	0.06853	0.13807
1 September 2018	0.11266	0.16303	0.08042	0.08042	0.08042	0.19286
1 October2018	0.10553	0.14092	0.11317	0.15436	0.15436	0.16628
1 November 2018	0.12795	0.15882	0.14356	0.14356	0.14356	0.16667
1 December 2018	0.12172	0.11357	0.11108	0.08417	0.08418	0.12669
1 January 2019	0.10461	0.13097	0.10311	0.10881	0.10881	0.12652
1 February 2019	0.04699	0.05833	0.04236	0.04825	0.04825	0.05481
1 March 2019	0.09956	0.11617	0.10688	0.09842	0.09842	0.12041
1 April 2019	0.07099	0.07830	0.08055	0.10363	0.10363	0.08677
1 May 2019	0.07546	0.07203	0.07128	0.07143	0.07143	0.07445
1 June 2019	0.06173	0.07472	0.06425	0.06511	0.06457	0.09343
1 July 2019	0.14050	0.13559	0.14333	0.13756	0.14634	0.13558
1 August 2019	0.07357	0.08857	0.09535	0.10416	0.08327	0.11402
1 September 2019	0.05368	0.06360	0.05207	0.05984	0.05984	0.04936
1 October 2019	0.14063	0.16482	0.13385	0.12099	0.12099	0.16620
1 November 2019	0.06858	0.06168	0.07058	0.07231	0.07231	0.05574
1 December 2019	0.08079	0.08534	0.06071	0.04909	0.04909	0.09815
1 January 2020	0.02731	0.03345	0.02903	0.03027	0.03027	0.03496
1 February 2020	0.06595	0.04800	0.08808	0.13353	0.13353	0.05235
MC model better		25	21	23	22	25

Source: the authors' calculations in MATLAB.

Table A4. Returns of the portfolios in the next trading day.

Date	MC Model	NAIVE	MV Middle	MV Max	Max Sharp	MCVaR
1 July 2017	−0.05645	−0.06292	−0.07465	−0.09359	−0.06811	−0.07560
1 August 2017	0.03577	0.02565	0.02260	0.02486	0.03253	0.02559
1 September 2017	0.00297	0.01380	0.00572	−0.01678	0.01417	0.00383
1 October 2017	−0.01723	−0.02591	−0.03030	−0.03836	−0.01826	−0.02971
1 November 2017	0.04395	0.02462	0.03310	0.01488	0.05000	0.01074
1 December 2017	0.01040	0.10020	−0.02426	−0.05243	0.00404	0.02641
1 January 2018	−0.07168	−0.04928	−0.09462	−0.10738	−0.06419	−0.00381
1 February 2018	0.03349	0.00930	−0.00893	−0.03345	0.00466	0.02299
1 March 2018	0.04778	0.03813	0.02425	0.02016	0.01826	0.08550
1 April 2018	−0.01446	0.00911	0.00616	0.01479	0.00598	0.01621
1 May 2018	−0.03046	−0.03979	−0.00839	0.00906	0.00489	−0.04919
1 June 2018	−0.01244	−0.00694	−0.02743	−0.02966	−0.02966	0.00158
1 July 2018	−0.04448	−0.05134	−0.06004	−0.05867	−0.05867	−0.05018
1 August 2018	−0.02712	−0.02684	−0.02849	−0.02849	−0.02849	−0.02431
1 September 2018	0.00013	−0.00348	0.00641	0.00641	0.00641	−0.00126
1 October 2018	−0.05305	−0.06265	−0.05678	−0.08304	−0.08304	−0.06307
1 November 2018	−0.00680	−0.00333	0.00310	0.00310	0.00310	0.00029
1 December 2018	−0.02196	−0.01661	−0.03203	−0.04476	−0.04476	−0.00428
1 January 2019	0.01353	0.04387	0.01459	0.02443	0.02443	0.05353
1 February 2019	0.00689	0.01316	0.00035	−0.00406	−0.00406	0.00622
1 March 2019	−0.02904	−0.03510	−0.03372	−0.03543	−0.03543	−0.04990
1 April 2019	0.00806	0.00734	0.01564	0.02526	0.02526	0.01167
1 May 2019	0.00893	0.00780	0.00419	−0.00493	−0.00493	0.01672
1 June 2019	−0.03842	−0.05909	−0.03652	−0.03673	−0.03660	−0.06384
1 July 2019	0.04833	0.03980	0.04017	0.02479	0.04818	0.03693
1 August 2019	−0.00234	−0.01228	−0.01201	−0.00979	−0.01507	−0.02272
1 September 2019	−0.02812	−0.01824	−0.00708	−0.01410	−0.01410	−0.02744
1 October 2019	−0.00146	−0.01488	−0.00119	0.00666	0.00666	−0.01956
1 November 2019	−0.00238	−0.00461	0.00050	0.00192	0.00192	−0.00002
1 December 2019	−0.03441	−0.04339	−0.03297	−0.03182	−0.03182	−0.04782
1 January 2020	0.07870	0.07880	0.08613	0.09150	0.09150	0.09386
1 February 2020	0.07537	0.04447	0.09245	0.12755	0.12755	0.04793
MC model better		20	18	19	16	17

Source: the authors' calculations in MATLAB.

References

1. Briere, M.; Oosterlinck, K.; Szafarz, A. Virtual currency, tangible return: Portfolio diversification with bitcoin. *J. Asset Manag.* **2015**, *16*, 365–373. [CrossRef]
2. Kajtazi, A.; Moro, A. The role of bitcoin in well diversified portfolios: A comparative global study. *Int. Rev. Financ. Anal.* **2019**, *61*, 143–157. [CrossRef]
3. Symitsi, E.; Chalvatzis, K.J. The economic value of bitcoin: A portfolio analysis of currencies, gold, oil and stocks. *Res. Int. Bus. Financ.* **2019**, *48*, 97–110. [CrossRef]
4. Platanakis, E.; Urquhart, A. Should investors include Bitcoin in their portfolios? A portfolio theory approach. *Br. Account. Rev.* **2020**, *52*. [CrossRef]
5. Li, J.P.; Naqvi, B.; Rizvi, S.K.A.; Chang, H.L. Bitcoin: The biggest financial innovation of fourth industrial revolution and a portfolio's efficiency booster. *Technol. Forecast. Soc. Chang.* **2021**, *162*, 120383. [CrossRef]
6. Lee, D.K.C.; Guo, L.; Wang, Y. Cryptocurrency: A new investment opportunity? *J. Altern. Invest.* **2018**, *20*, 16–40. [CrossRef]
7. Krückeberg, S.; Scholz, P. Cryptocurrencies as an asset class. In *Cryptofinance and Mechanisms of Exchange. Contributions to Management Science*; Goutte, S., Guesmi, K., Saadi, S., Eds.; Springer: Cham, Switzerland, 2019; pp. 1–28. [CrossRef]
8. Trimborn, S.; Li, M.; Härdle, W.K. Investing with Cryptocurrencies—A Liquidity Constrained Investment Approach. *J. Financ. Econom.* **2019**, *18*, 280–306. Available online: https://ssrn.com/abstract=2999782 (accessed on 1 November 2020). [CrossRef]
9. Petukhina, A.; Trimborn, S.; Härdle, W.K.; Elendner, H. Investing with Cryptocurrencies—Evaluating their potential for portfolio allocation strategies. *SSRN 3274193* **2020**, 1–61. Available online: https://ssrn.com/abstract=3274193 (accessed on 1 November 2020). [CrossRef]
10. Ma, Y.; Ahmad, F.; Liu, M.; Wang, Z. Portfolio optimization in the era of digital financialization using cryptocurrencies. *Technol. Forecast. Soc. Chang.* **2020**, *161*, 120265. [CrossRef]

11. Elendner, H.; Trimborn, S.; Ong, B.; Lee, T.M. Chapter 7—The cross-section of crypto-currencies as financial assets: Investing in crypto-currencies beyond bitcoin. In *Handbook of Blockchain, Digital Finance, and Inclusion. Cryptocurrency, FinTech, InsurTech, and Regulation*; Lee, D.K.C., Deng, R., Eds.; Academic Press: Cambridge, MA, USA, 2018; Volume 1, pp. 145–173. [CrossRef]
12. Platanakis, E.; Sutcliffe, C.; Urquhart, A. Optimal vs. naïve diversification in cryptocurrencies. *Econ. Lett.* **2018**, *171*, 93–96. [CrossRef]
13. Brauneis, A.; Mestel, R. Cryptocurrency-portfolios in a mean-variance framework. *Financ. Res. Lett.* **2019**, *28*, 259–264. [CrossRef]
14. Weiyi, L. Portfolio diversification across cryptocurrencies. *Financ. Res. Lett.* **2019**, *29*, 200–205. [CrossRef]
15. García, F.; González-Bueno, J.; Guijarro, F.; Oliver, J. A multiobjective credibilistic portfolio selection model. Empirical study in the Latin American integrated market. *Entrep. Sustain. Issues* **2020**, *8*, 1027–1046. [CrossRef]
16. Zavadskas, E.K.; Turskis, Z.; Kildiene, S. State of art surveys of overviews on MCDM/MADM methods. *Technol. Econ. Dev. Econ.* **2014**, *20*, 165–179. [CrossRef]
17. Aouni, B.; Doumpos, M.; Pérez Gladish, B.M.; Steuer, R.E. On the increasing importance of multiple criteria decision aid methods for portfolio selection. *J. Oper. Res. Soc.* **2018**, *69*, 1525–1542. [CrossRef]
18. Nakamoto, S. Bitcoin: A Peer-to-Peer Electronic Cash System. 2008. Available online: https://bitcoin.org/bitcoin.pdf (accessed on 4 November 2020).
19. Weber, W.E. A bitcoin standard: Lessons from the gold standard. *Bank Can. Staff. Work. Pap.* **2016**, *14*, 1–37. Available online: https://www.bankofcanada.ca/wp-content/uploads/2016/03/swp2016-14.pdf (accessed on 23 November 2020).
20. Kimani, D.; Adams, K.; Attah-Boakye, R.; Ullah, S.; Frecknall-Hughesd, J.; Kim, J. Blockchain, business and the fourth industrial revolution: Whence, whither, wherefore and how? *Technol. Forecast. Soc. Chang.* **2020**, *161*, 120254. [CrossRef]
21. Mikhaylov, A. Development of Friedrich von Hayek's theory of private money and economic implications for digital currencies. *Terra Econ.* **2021**, *19*, 53–62. [CrossRef]
22. Morozova, T.; Akhmadeev, R.; Lehoux, L.; Yumashev, A.; Meshkova, G.; Lukiyanova, M. Crypto asset assessment models in financial reporting content typologies. *Entrep. Sustain. Issues* **2020**, *7*, 2196–2212. [CrossRef]
23. Ciaian, P.; Rajcaniova, M.; Kancs, D. The economics of Bitcoin price formation. *Appl. Econ.* **2016**, *48*, 1799–1815. [CrossRef]
24. Holovatiuk, O. Cryptocurrencies as an asset class in portfolio optimisation. *Cent. Eur. Econ. J.* **2020**, *7*, 33–55. [CrossRef]
25. White, R.; Marinakis, Y.; Islam, N.; Walsh, S. Is Bitcoin a currency, a technology-based product, or something else? *Technol. Forecast. Soc. Chang.* **2020**, *151*, 119877. [CrossRef]
26. İçellioğlu, C.S.; Öner, S. An investigation on the volatility of cryptocurrencies by means of heterogeneous panel data analysis. *Procedia Comput. Sci.* **2019**, *158*, 913–920. [CrossRef]
27. Baur, D.G.; Hong, K.; Lee, A.D. Bitcoin: Medium of exchange or speculative assets? *J. Int. Financ. Mark. Inst. Money* **2018**, *54*, 177–189. [CrossRef]
28. Bouri, E.; Molnár, P.; Azzi, G.; Roubaud, D.; Hagfors, L.I. On the hedge and safe haven properties of Bitcoin: Is it really more than a diversifier? *Financ. Res. Lett.* **2017**, *20*, 192–198. [CrossRef]
29. Trimborn, S.; Hardle, W.K. CRIX an Indeks for cryptocurrencies. *J. Empir. Financ.* **2018**, *49*, 107–122. [CrossRef]
30. Burniske, C.; White, A. Bitcoin: Ringing the Bell for a New Asset Class. *ARK Invest* **2017**, 1–24. Available online: https://research.ark-invest.com/hubfs/1_Download_Files_ARK-Invest/White_Papers/Bitcoin-Ringing-The-Bell-For-A-New-Asset-Class.pdf (accessed on 13 October 2020).
31. Greer, R. What is an asset class anyway? *J. Portf. Manag.* **1997**, *23*, 83–91. [CrossRef]
32. Mikhaylov, A. Cryptocurrency market analysis from the open innovation perspective. *J. Open Innov. Technol. Mark. Complex.* **2020**, *6*, 197. [CrossRef]
33. Liu, Y.; Tsyvinski, A. Risks and returns of cryptocurrency. *NBER Work. Pap.* **2018**, w24877. Available online: https://www.nber.org/papers/w24877.pdf (accessed on 5 October 2020).
34. Sajter, D. Time-series analysis of the most common cryptocurrencies. *Econ. Thought Pract.* **2019**, *1*, 267–282. Available online: https://hrcak.srce.hr/221035 (accessed on 8 November 2020).
35. Ankenbrand, T.; Bieri, D. Assessment of cryptocurrencies as an asset class by their characteristics. *Invest. Manag. Financ. Innov.* **2018**, *15*, 169–181. [CrossRef]
36. Corbet, S.; Lucey, B.; Urquhart, A.; Yarovayad, L. Cryptocurrencies as a financial asset: A systematic analysis. *Int. Rev. Financ. Anal.* **2019**, *62*, 182–199. [CrossRef]
37. Hu, A.S.; Parlour, C.A.; Rajan, U. Cryptocurrencies: Stylized facts on a new investible instrument. *Financ. Manag.* **2019**, *48*, 1049–1068. [CrossRef]
38. O'Connor, M. What Is "Market Cap" in Crypto, and Why Is It Important? 2019. Available online: https://dailyhodl.com/2019/03/06/what-is-market-cap-in-crypto-and-why-is-it-important/ (accessed on 20 November 2020).
39. Brauneis, A.; Mestel, R.; Theissen, E. What drives the liquidity of cryptocurrencies? A long-term analysis. *Financ. Res. Lett.* **2020**, *39*, 101537. [CrossRef]
40. Phillip, A.; Chan, J.S.K.; Peiris, S. A new look at cryptocurrencies. *Econ. Lett.* **2018**, *163*, 6–9. [CrossRef]
41. Salamat, S.; Lixia, N.; Naseem, S.; Mohsin, M.; Zia-ur-Rehman, M.; Baig, S.A. Modeling cryptocurrencies volatility using Garch models: A comparison based on normal and student's t-error distribution. *Entrep. Sustain. Issues* **2020**, *7*, 1580–1596. [CrossRef]
42. Aljinović, Z.; Trgo, A. CVaR in measuring sector's risk on the croatian stock exchange. *Bus. Syst. Res.* **2018**, *9*, 8–17. [CrossRef]
43. Hafsa, H. CVaR in portfolio optimization: An essay on the French market. *Int. J. Financ. Res.* **2015**, *6*, 101–111. [CrossRef]

44. Dowd, K. *Measuring Market Risk*; John Wiley and Sons: Chichester, NY, USA, 2002.
45. Artzner, P.; Delbaen, F.; Eber, J.M.; Heath, D. Coherent measures of risk. *Math. Financ.* **1999**, *9*, 203–228. [CrossRef]
46. Goczek, L.; Skliarov, I. What drives the Bitcoin price? A factor augmented error correction mechanism investigation. *Appl. Econ.* **2019**, *51*, 6393–6410. [CrossRef]
47. Sovbetov, Y. Factors influencing cryptocurrency prices: Evidence from Bitcoin, Ethereum, Dash, Litcoin, and Monero. *J. Econ. Financ. Anal.* **2018**, *2*, 1–27. Available online: https://mpra.ub.uni-muenchen.de/85036/1/MPRA_paper_85036.pdf (accessed on 1 February 2020).
48. Poyser, O. Exploring the Dynamics of Bitcoin's Price: A Bayesian Structural Time Series Approach. Ph.D. Thesis, Universitat Autònoma de Barcelona, 2018. Available online: https://www.researchgate.net/publication/317356728_Exploring_the_determinants_of_Bitcoin\T1\textquoterights_price_an_application_of_Bayesian_Structural_Time_Series (accessed on 11 November 2020).
49. Matta, M.; Lunesu, I.; Marchesi, M. Bitcoin spread prediction using social and web search media. In *User Modeling, Adaptation, and Personalization, Proceedings of the 23rd UMAP 2015 Conference, Dublin, Ireland, 29 June–3 July 2015*; Springer: Berlin/Heidelberg, Germany, 2015; Available online: https://www.semanticscholar.org/paper/Bitcoin-Spread-Prediction-Using-Social-and-Web-Matta-Lunesu/1345a50edee28418900e2c1a4292ccc51138e1eb (accessed on 12 November 2020).
50. Teti, E.; Dallocchio, M.; Aniasi, A. The relationship between Twitter and stock prices. Evidence from the US technology industry. *Technol. Forecast. Soc. Chang.* **2019**, *149*, 119747. [CrossRef]
51. Letra, I. What Drives Cryptocurrency Value? A Volatility and Predictability Analysis. Master's Thesis, Lisbon School of Economics & Management, 2016. Available online: https://www.repository.utl.pt/bitstream/10400.5/12556/1/DM-IJSL-2016.pdf (accessed on 11 November 2020).
52. Stolarski, P.; Lewoniewski, W.; Abramowicz, W. Cryptocurrencies perception using Wikipedia and Google trends. *Information* **2020**, *11*, 234. [CrossRef]
53. Park, H.W.; Lee, Y. How are twitter activities related to top cryptocurrencies' performance? Evidence from social media network and sentiment analysis. *Društvena Istraživanja* **2019**, *28*, 435–460. [CrossRef]
54. Kaminski, J.C. Nowcasting the Bitcoin market with Twitter signals. *arXiv* **2014**, arXiv:abs/1406.7577.
55. Abraham, J.; Higdon, D.; Nelson, J.; Ibarra, J. Cryptocurrency price prediction using tweet volumes and sentiment analysis. *SMU Data Sci. Rev.* **2018**, *1*, 1–21. Available online: https://scholar.smu.edu/datasciencereview/vol1/iss3/1 (accessed on 6 September 2020).
56. Kraaijeveld, O.; de Smedt, J. The predictive power of public Twitter sentiment for forecasting cryptocurrency prices. *J. Int. Financ. Mark. Inst. Money* **2020**, *65*, 1–22. [CrossRef]
57. Brans, J.P.; Mareschal, B.; Vincke, P. PROMETHEE: A new family of outranking methods in multicriteria analysis. In *Operational research '84, Proceedings of the Tenth IFORS International Conference on Operational Research, Washington, DC, USA, 6–10 August 1984*; Elsevier: Amsterdam, The Netherlands, 1984; pp. 477–490.
58. Khoury, N.; Martel, J.M. The relationship between risk-return characteristics of mutual funds and their size. *Financ. Rev. L'association Fr. Financ.* **1990**, *11*, 67–82.
59. Zmitri, R.; Martel, J.M.; Dumas, Y. Un indice multicritère de santé financière pour les succursales bancaires. *FINÉCO* **1998**, *8*, 107–121.
60. Aljinović, Z.; Marasović, B.; Tomić-Plazibat, N. Multi-criterion approach versus Markowitz in selection of the optimal portfolio. In Proceedings of the SOR '05: 8th Symposium on Operational Research in Slovenia, Nova Gorica, Slovenia, 2005; Studio LUMINA: Lubljana, Slovenia, 2005; pp. 261–266.
61. Marasović, B.; Babić, Z. Two-step multi-criteria model for selecting optimal portfolio. *Int. J. Prod. Econ.* **2011**, *134*, 58–66. [CrossRef]
62. Bouri, A.; Martel, J.M.; Chabchoub, H. A multi-criterion approach for selecting attractive portfolio. *J. Multi Criteria Decis. Anal.* **2002**, *11*, 269–277. [CrossRef]
63. Babić, Z. *Modeli i Metode Poslovnog Odlučivanja*; Sveučilište u Splitu, Ekonomski Fakultet: Split, Croatia, 2011.
64. Saaty, T.L. *Decision Making for Leaders: The Analytic Hierarchy Process for Decision in a Complex World*, 3rd ed.; RWS Publications: Pittsburgh, PA, USA, 2012.

MDPI

Article

Perishable Inventory System with N-Policy, MAP Arrivals, and Impatient Customers

R. Suganya [1,†], Lewis Nkenyereye [2,*,†], N. Anbazhagan [1,*], S. Amutha [3], M. Kameswari [4], Srijana Acharya [5] and Gyanendra Prasad Joshi [6]

1 Department of Mathematics, Alagappa University, Karaikudi 630003, Tamil Nadu, India; saisugan92@gmail.com
2 Department of Computer and Information Security, Sejong University, Seoul 05006, Korea
3 Ramanujan Centre for Higher Mathematics, Alagappa University, Karaikudi 630003, Tamil Nadu, India; amuthas@alagappauniversity.ac.in
4 Department of Mathematics, School of Advanced Sciences, Kalasalingam Academy of Research and Education, Krishnankoil, Srivilliputhur 626128, Tamil Nadu, India; m.kameshwari@klu.ac.in
5 Department of Convergence Science, Kongju National University, Gongju 32588, Korea; sriz@ynu.ac.kr
6 Department of Computer Science and Engineering, Sejong University, Seoul 05006, Korea; joshi@sejong.ac.kr
* Correspondence: nkenyele@sejong.ac.kr (L.N.); anbazhagann@alagappauniversity.ac.in (N.A.)
† Authors have equal contributions.

Abstract: In this study, we consider a perishable inventory system that has an (s, Q) ordering policy, along with a finite waiting hall. The single server, which provides an item to the customer after completing the required service performance for that item, only begins serving after N customers have arrived. Impatient demand is assumed in that the customers waiting to be served lose patience and leave the system if the server's idle time overextends or if the arriving customers find the system to be full and will not enter the system. This article analyzes the impatient demands caused by the N-policy server to an inventory system. In the steadystate, we obtain the joint probability distribution of the level of inventory and the number of customers in the system. We analyze some measures of system performance and get the total expected cost rate in the steadystate. We present a beneficial cost function and confer the numerical illustration that describes the impact of impatient customers caused by N-policy on the inventory system's total expected cost rate.

Keywords: (s, Q)-policy; Markovian Arrival Process; N-policy; impatient customers

Citation: Suganya, R.; Nkenyereye, L.; Anbazhagan, N.; Amutha, S.; Kameswari, M.; Acharya, S.; Joshi, G.P. Perishable Inventory System with N-Policy, MAP Arrivals, and Impatient Customers. *Mathematics* **2021**, *9*, 1514. https://doi.org/10.3390/math9131514

Academic Editor: Frank Werner

Received: 21 May 2021
Accepted: 20 June 2021
Published: 28 June 2021

1. Introduction

Perishable inventory system research draws inspiration from Nahmias' [1] seminal piece on ordering policies for perishable inventory. Nahmias studied the ordering policies for fixed and random shelf lifetime perishable inventory. Earlier inventory systems research usually assumed that the stock items are non-perishable. However, this is not realistic, thus creating the need to study perishable inventory systems. For more details on perishable inventory, we refer interested readers to Aijun Liu et al. [2], Darestani [3], Ioannidis [4], Kalpakam and Arivarignan [5], Liu and Lian [6], Sung-Seok Ko [7], Weiss [8], and Zhang et al. [9].

Generally, in the literature on inventory models, customers receive the stock demanded instantaneously only when the stock is available; otherwise, waiting is the norm. In the case of the inventory maintained at a service facility, customers usually wait for the item demanded because some service is performed on it, for instance, a fast food outlet or hospital dispensary. Further, due to the complexity and uniqueness of a customer's order, the service time may stretch and be variable, such as special medicinal preparation for liver-impaired patients or gluten-free dietary requests. This then builds a queue in the service system, often leading to impatient customers, with those customers sometimes reneging or balking from the service. Recognizing that queues can form during stock-out situations, Berman et al. [10] examined an inventory model with a service facility where both the demand and service rates are known and constant. They determined the optimal order quantity for the minimal expected total cost. Since then, there has been keen interest in the perishable queueing-inventory system and impatient customers (see, for example, Amirthakodi and Sivakumar [11], Arivarignan et al. [12], Hamadi et al. [13], Manuel et al. [14], and Lawrence et al. [15]).

For many inventory systems with service activities, the setup can require several minutes, and these setup activities incur costs to the inventory system. One way to reduce the setup cost is to employ an N-policy, i.e., if the system is empty, the server is on vacation. When there are at least N customers in the system, the server begins service. Yadin and Naor [16] suggested the N-policy concept. Heyman [17] first analyzed the N-policy system with an $M/G/1$ queue. The N-policy has been extended by others, such as Ke [18], Kella [19], and Wang and Ke [20], to a queueing network. Krishnamoorthy and Anbazhagan [21] have considered a finite waiting hall perishable inventory system under an N-policy. Similarly, Jeganathan et al. [22] considered a perishable inventory system with a finite waiting hall and customer service under an N-policy, but they allowed the server to take multiple vacations, assuming that the customers reach the service station in a Poisson manner and inventory replenishment is instantaneous.

All previous references about N-policy in the inventory system focused on the setup cost reduced in the system. Herein, we examine another fact that the cost of customers lost. It is a significant component of the total expected cost rate.

Despite the fact that the N-policy successfully lessens the inventory system's general arrangement cost, it can nevertheless bring about waiting time vulnerability for the primary *N-1* customers. For instance, the first customer arrives at a vacant waiting hall, and the service channel withholds the service until the other *N-1* customers arrive into the system. Assuming the customer appearance rate is moderate, there is a probability of developing customer impatience. Our work is motivated by this perception. Specifically, we investigated the effect of N-policy on the arriving customers to the inventory system and focused on showing the possible results of increasingly impatient customers' impact on the total expected cost rate of the system.

In real life, you can see some rides in theme and amusement parks, theaters in malls, as well as adventure activities like skydiving, scuba diving, rafting, and parasailing starting to sell tickets to customers after some customers come to their systems. In these systems, the first customers have to wait for other arrivals. They become easily impatient, so they go for other systems.

We examine a perishable inventory system with a finite waiting capacity, and the customers arrive as a Markovian Arrival Process. We assume that the server provides service only when there are N customers in the system; otherwise, the server remains idle. If the customers arrive and find the system to be full, they will not enter the system. At the same time, the customer who is waiting for service and finds the server to be idle becomes impatient and may exit the system.

The remainder of this paper is structured as follows. Section 2 presents the notation used in the paper and the corresponding model development. In Section 3, the steady state analysis of the model is presented. In Section 4, we derive the measures of system performance under steady-state analysis. In Section 5, the total expected system cost rate is obtained. A cost analysis is provided in Section 6. Section 7 presents the numerical illustration. Section 8 concludes the paper.

2. Model

The following notation will be used in this paper:

$\mathbf{0}$: Zero matrix.
I : Identity matrix.
I_x : Identity matrix of order x.

$[\mathbb{P}]_{ij}$: Entry at $(i,j)^{th}$ position of a matrix $\mathbb{P}$.
$F_{i_{(x \times y)}}$: Size of matrix F_i is x row and y column.
e : Unit column vector of appropriate dimension.
$I(t)$: Inventory level at time t.
$T(t)$: Server status at time t.
$C(t)$: Number of customers waiting and being served at time t.
$J(t)$: Phase of the arrival process at time t.

$$T(t) = \begin{cases} 0, & \text{if server is idle} \\ 1, & \text{if server is busy.} \end{cases}$$

Consider that a perishable inventory system contains a limited waiting hall size $H(< \infty)$ (including the service receiver) with at most S items as inventory and a single server. When the customer demand reaches a predetermined level N $(0 < N < H)$, the server begins service. The customers request for one item each. The customer only receives the requested item after certain service activities are performed on that item. Service time is a negative exponential distribution with parameter $\mu(> 0)$. For replenishment, an order quantity $Q(= S - s > s + 1)$ is placed when the inventory level drops to the reorder level s and the items are received only after a random time, which has a negative exponential distribution with parameter $\beta(> 0)$. The customers who are waiting for service may exit the system while the server is idle, these impatient(reneging) customers are assumed to leave the system after a random time, which is distributed as a negative exponential with parameter $\alpha(> 0)$. If the waiting hall is full, then all new arriving customers are considered to be lost. The lifetime of each item has a negative exponential distribution with parameter $\gamma(> 0)$. We assume that the item does not perish when it is in service.

The MAP is a rich class of point processes that include many well-known processes such as the Poisson process. As is notable, the Poisson measure is the least complex and most manageable one, which is utilized widely in stochastic modeling.The possibility of the MAP is to fundamentally sum up the Poisson process and still save the manageability for modeling purposes. Hence, the MAP is a convenient tool for modeling both renewal and non-renewal arrivals. While MAP is defined for both discrete and continuous times, here we use only the continuous time case. For the description of the arrival process, we use the MAP's description as given in Lucantoni et al. [23]. Consider a continuous-time Markov chain on the state space 1, 2,. . ., x. When the chain is in state i, $1 \leq i \leq$ x, it remains for an exponential time with parameter v_i. When the sojourn time ends, the chain may transition in two ways. First, if the transition is with a customer arrival, then the chain enters state j with probability $c_{ij}, 1 \leq j \leq$ x. Second, if the transition is without a customer arrival, then the chain enters state j with probability $d_{ij}, 1 \leq j \leq$ x, $i \neq j$. Note that the chain can remain in the same state (i.e., from state i to state i) when an arrival occurs. Consider the matrices $F_f, f = 0, 1$ of size x as $[F_0]_{ii} = -v_i$ and $[F_0]_{ij} = v_i d_{ij}, i \neq j$, $[F_1]_{ij} = v_i c_{ij}, 1 \leq i, j \leq$ x. Clearly, $F = F_0 + F_1$ is an infinitesimal generator of a continuous-time Markov chain. We assume that F is irreducible and $F_0 e \neq 0$.

Let φ be the stationary probability vector of a continuous-time Markov chain with generator F. Then, φ is the unique probability vector satisfying $\varphi F = 0$, $\varphi e = 1$.

Suppose ω is the primary probability vector of the hidden Markov chain dependent on the MAP. Then we can obtain the time epochs by picking an appropriate ω, such as an independent arrival point, the end of the interval of at least k arrivals, and where the system is in a particular state such as the beginning or end of a busy period.

Setting $\omega = \varphi$, we obtain the stationary distribution of the MAP. The constant $\lambda = \varphi F_1 e$ is the fundamental rate, which provides the mean of the customer arrivals in unit time.

For more details on the MAP, we refer the interested reader to Latouche and Ramaswami [24], Lee and Jeon [25], and Chakravarthy and Dudin [26].

3. Analysis

Let *L(t)*,*T(t)*,*C(t)* and *J(t)*, respectively, denote the inventory level, server status, number of customer waiting and being served and phase of the arrival process at time t. From the assumptions made on the input and output processes, it can be shown that the quadruple $\{(I(t), T(t), C(t), J(t)), t \geq 0\}$ is a Markov process whose state space is

$\mathbb{E} = E_1 \cup E_2 \cup E_3 \cup E_4$, with
$E_1 = \{(i, 0, 0, r) : 1 \leq i \leq S; 1 \leq r \leq \text{x}\}$
$E_2 = \{(i, k, m, r) : 1 \leq i \leq S; k = 0, 1; 1 \leq m \leq N-1; 1 \leq r \leq \text{x}\}$
$E_3 = \{(i, k, m, r) : 1 \leq i \leq S; k = 1; N \leq m \leq H; 1 \leq r \leq \text{x}\}$
$E_4 = \{(0, 0, m, r) : 0 \leq m \leq H; 1 \leq r \leq \text{x}\}$

We order the elements of $\mathbb{E}$lexicographically. Then the infinitesimal generator $\mathbb{P}$ of the Markov process $\{(I(t), T(t), C(t), J(t)), t \geq 0\}$ has the following block partitioned form:

$$[\mathbb{P}]_{ij} = \begin{cases} \mathbb{Y}_i, j = i-1, i = 1, 2, \ldots S \\ \mathbb{X}_i, j = i, \quad i = 0, 1, \ldots S \\ \mathbb{Z}, j = i+Q, i = 1, 2, \ldots s \\ \mathbb{Z}' \; j = i+Q, \quad i = 0 \\ 0, \text{otherwise} \; . \end{cases}$$

where

$$\mathbb{Z}' = \begin{matrix} & 0 & 1 \\ 0 & \Big(F_{2((H+1)x \times Nx)} & \mathbf{0}_{((H+1)x \times Hx)} \Big) \end{matrix}$$

Submatrix F_2 is

$$F_2 = \begin{matrix} & \begin{matrix} 0 & 1 & \cdots & N-1 \end{matrix} \\ \begin{matrix} 0 \\ 1 \\ \vdots \\ N-1 \\ N \\ \vdots \\ H \end{matrix} & \begin{pmatrix} \beta \mathbf{I_x} & \mathbf{0} & \cdots & \mathbf{0} \\ \mathbf{0} & \beta \mathbf{I_x} & \cdots & \mathbf{0} \\ \vdots & \vdots & \vdots & \vdots \\ \mathbf{0} & \mathbf{0} & \cdots & \beta \mathbf{I_x} \\ \mathbf{0} & \mathbf{0} & \cdots & \mathbf{0} \\ \vdots & \vdots & \vdots & \vdots \\ \mathbf{0} & \mathbf{0} & \cdots & \mathbf{0} \end{pmatrix} \end{matrix}$$

$$Z = \begin{matrix} & \begin{matrix} 0 & 1 \end{matrix} \\ \begin{matrix} 0 \\ 1 \end{matrix} & \begin{pmatrix} F_{3(Nx \times Nx)} & \mathbf{0}_{(Nx \times Hx)} \\ \mathbf{0}_{(Hx \times Nx)} & F_{4(Hx \times Hx)} \end{pmatrix} \end{matrix}$$

Submatrices F_3 and F_4 are

$$F_3 = \begin{matrix} & \begin{matrix} 0 & 1 & \cdots & N-1 \end{matrix} \\ \begin{matrix} 0 \\ 1 \\ \vdots \\ N-1 \end{matrix} & \begin{pmatrix} \beta \mathbf{I_x} & \mathbf{0} & \cdots & \mathbf{0} \\ \mathbf{0} & \beta \mathbf{I_x} & \cdots & \mathbf{0} \\ \vdots & \vdots & \vdots & \vdots \\ \mathbf{0} & \mathbf{0} & \cdots & \beta \mathbf{I_x} \end{pmatrix} \end{matrix}$$

$$F_4 = \begin{matrix} & \begin{matrix} 0 & 1 & \cdots & H \end{matrix} \\ \begin{matrix} 0 \\ 1 \\ \vdots \\ H \end{matrix} & \begin{pmatrix} \beta \mathbf{I_x} & \mathbf{0} & \cdots & \mathbf{0} \\ \mathbf{0} & \beta \mathbf{I_x} & \cdots & \mathbf{0} \\ \vdots & \vdots & \vdots & \vdots \\ \mathbf{0} & \mathbf{0} & \cdots & \beta \mathbf{I_x} \end{pmatrix} \end{matrix}$$

For $i = 0$

$$\mathbb{X}_i = \begin{matrix} & 0 \\ 0 & \Big(F_{5((H+1)\text{x} \times (H+1)\text{x})} \Big) \end{matrix}$$

Submatrix F_5 is

$$F_5 = \begin{array}{c} \\ 0 \\ 1 \\ \vdots \\ N-1 \\ N \\ \vdots \\ H-1 \\ H \end{array} \begin{array}{c} \begin{array}{cccccccccccc} 0 & 1 & 2 & \cdots & N-2 & N-1 & N & \cdots & H-2 & H-1 & H \end{array} \\ \left(\begin{array}{ccccccccccc} F_0-\beta \boldsymbol{I_x} & F_1 & \mathbf{0} & \cdots & \mathbf{0} & \mathbf{0} & \mathbf{0} & \cdots & \mathbf{0} & \mathbf{0} & \mathbf{0} \\ \alpha \boldsymbol{I_x} & F_0-(\alpha+\beta)\boldsymbol{I_x} & F_1 & \cdots & \mathbf{0} & \mathbf{0} & \mathbf{0} & \cdots & \mathbf{0} & \mathbf{0} & \mathbf{0} \\ \vdots & \vdots & \vdots & \vdots & \vdots & \vdots & \vdots & \vdots & \vdots & \vdots & \vdots \\ \mathbf{0} & \mathbf{0} & \mathbf{0} & \cdots & \alpha \boldsymbol{I_x} & F_0-(\alpha+\beta)\boldsymbol{I_x} & F_1 & \cdots & \mathbf{0} & \mathbf{0} & \mathbf{0} \\ \mathbf{0} & \mathbf{0} & \mathbf{0} & \cdots & \mathbf{0} & \alpha \boldsymbol{I_x} & F_0-\alpha \boldsymbol{I_x} & \cdots & \mathbf{0} & \mathbf{0} & \mathbf{0} \\ \vdots & \vdots & \vdots & \vdots & \vdots & \vdots & \vdots & \vdots & \vdots & \vdots & \vdots \\ \mathbf{0} & \mathbf{0} & \mathbf{0} & \vdots & \mathbf{0} & \mathbf{0} & \mathbf{0} & \vdots & \alpha \boldsymbol{I_x} & F_0-\alpha \boldsymbol{I_x} & F_1 \\ \mathbf{0} & \mathbf{0} & \mathbf{0} & \vdots & \mathbf{0} & \mathbf{0} & \mathbf{0} & \vdots & \mathbf{0} & \alpha \boldsymbol{I_x} & F_0-\alpha \boldsymbol{I_x} \end{array} \right) \end{array}$$

For $i = 1,2, \ldots$ s

$$\mathbb{X}_i = \begin{array}{c} \\ 0 \\ 1 \end{array} \begin{array}{c} \begin{array}{cc} 0 & 1 \end{array} \\ \left(\begin{array}{cc} F_{6_{(Nx \times Nx)}} & \mathbf{0}_{(Nx \times Hx)} \\ \mathbf{0}_{(Hx \times Nx)} & F_{7_{(Hx \times Hx)}} \end{array} \right) \end{array}$$

Submatrices F_6 and F_7 are

$$F_6 = \begin{array}{c} \\ 0 \\ 1 \\ \vdots \\ N-2 \\ N-1 \end{array} \begin{array}{c} \begin{array}{cccccc} 0 & 1 & 2 & \cdots & N-2 & N-1 \end{array} \\ \left(\begin{array}{cccccc} F_0-(i\gamma+\beta)\boldsymbol{I_x} & F_1 & \mathbf{0} & \cdots & \mathbf{0} & \mathbf{0} \\ \alpha \boldsymbol{I_x} & F_0-(\alpha+i\gamma+\beta)\boldsymbol{I_x} & F_1 & \cdots & \mathbf{0} & \mathbf{0} \\ \vdots & \vdots & \vdots & \vdots & \vdots & \vdots \\ \mathbf{0} & \mathbf{0} & \mathbf{0} & \cdots & F_0-(\alpha+i\gamma+\beta)\boldsymbol{I_x} & F_1 \\ \mathbf{0} & \mathbf{0} & \mathbf{0} & \cdots & \alpha \boldsymbol{I_x} & F-(\alpha+i\gamma+\beta)\boldsymbol{I_x} \end{array} \right) \end{array}$$

$$F_7 = \begin{array}{c} \\ 1 \\ 2 \\ \vdots \\ H-1 \\ H \end{array} \begin{array}{c} \begin{array}{ccccc} 1 & 2 & \cdots & H-1 & H \end{array} \\ \left(\begin{array}{ccccc} F_0-(i\gamma+\mu+\beta)\boldsymbol{I_x} & F_1 & \cdots & \mathbf{0} & \mathbf{0} \\ \mathbf{0} & F_0-(i\gamma+\mu+\beta)\boldsymbol{I_x} & \cdots & \mathbf{0} & \mathbf{0} \\ \vdots & \vdots & \vdots & \vdots & \vdots \\ \mathbf{0} & \mathbf{0} & \cdots & F_0-(i\gamma+\mu+\beta)\boldsymbol{I_x} & F_1 \\ \mathbf{0} & \mathbf{0} & \cdots & \mathbf{0} & F-(i\gamma+\mu+\beta)\boldsymbol{I_x} \end{array} \right) \end{array}$$

For $i = s + 1, \ldots S$

$$\mathbb{X}_i = \begin{array}{c} \\ 0 \\ 1 \end{array} \begin{array}{c} \begin{array}{cc} 0 & 1 \end{array} \\ \left(\begin{array}{cc} F_{8_{(Nx \times Nx)}} & \mathbf{0}_{(Nx \times Hx)} \\ \mathbf{0}_{(Hx \times Nx)} & F_{9_{(Hx \times Hx)}} \end{array} \right) \end{array}$$

Submatrices F_8 and F_9 are

$$F_8 = \begin{array}{c} \\ 0 \\ 1 \\ \vdots \\ N-2 \\ N-1 \end{array} \begin{array}{c} \begin{array}{cccccc} 0 & 1 & 2 & \cdots & N-2 & N-1 \end{array} \\ \left(\begin{array}{cccccc} F_0-(i\gamma)\boldsymbol{I_x} & F_1 & \mathbf{0} & \cdots & \mathbf{0} & \mathbf{0} \\ \alpha \boldsymbol{I_x} & F_0-(\alpha+i\gamma)\boldsymbol{I_x} & F_1 & \cdots & \mathbf{0} & \mathbf{0} \\ \vdots & \vdots & \vdots & \vdots & \vdots & \vdots \\ \mathbf{0} & \mathbf{0} & \mathbf{0} & \cdots & F_0-(\alpha+i\gamma)\boldsymbol{I_x} & F_1 \\ \mathbf{0} & \mathbf{0} & \mathbf{0} & \cdots & \alpha \boldsymbol{I_x} & F-(\alpha+i\gamma)\boldsymbol{I_x} \end{array} \right) \end{array}$$

$$F_9 = \begin{array}{c} \\ 1 \\ 2 \\ \vdots \\ H-1 \\ H \end{array} \begin{array}{c} \begin{array}{ccccc} 1 & 2 & \cdots & H-1 & H \end{array} \\ \left(\begin{array}{ccccc} F_0-(i\gamma+\mu)\boldsymbol{I_x} & F_1 & \cdots & \mathbf{0} & \mathbf{0} \\ \mathbf{0} & F_0-(i\gamma+\mu)\boldsymbol{I_x} & \cdots & \mathbf{0} & \mathbf{0} \\ \vdots & \vdots & \vdots & \vdots & \vdots \\ \mathbf{0} & \mathbf{0} & \cdots & F_0-(i\gamma+\mu)\boldsymbol{I_x} & F_1 \\ \mathbf{0} & \mathbf{0} & \cdots & \mathbf{0} & F-(i\gamma+\mu)\boldsymbol{I_x} \end{array} \right) \end{array}$$

For $i = 1$

$$\mathbb{Y}_i = \begin{array}{c} \\ 0 \\ 1 \end{array} \begin{array}{c} 0 \\ \left(\begin{array}{c} F_{10_{(Nx \times (H+1)x)}} \\ F_{11_{(Hx \times (H+1)x)}} \end{array} \right) \end{array}$$

Submatrices F_{10} and F_{11} are

$$F_{10} = \begin{array}{c} \\ 0 \\ 1 \\ \vdots \\ N-1 \end{array} \begin{array}{c} \begin{array}{cccccccc} 0 & 1 & \cdots & N-1 & N & \cdots & H \end{array} \\ \left(\begin{array}{ccccccc} i\gamma \mathbf{I}_x & \mathbf{0} & \cdots & \mathbf{0} & \mathbf{0} & \cdots & \mathbf{0} \\ 0 & i\gamma \mathbf{I}_x & \cdots & \mathbf{0} & \mathbf{0} & \cdots & \mathbf{0} \\ \vdots & \vdots & \vdots & \vdots & \vdots & \vdots & \vdots \\ \mathbf{0} & \mathbf{0} & \cdots & i\gamma \mathbf{I}_x & \mathbf{0} & \cdots & \mathbf{0} \end{array} \right) \end{array}$$

$$F_{11} = \begin{array}{c} \\ 0 \\ 1 \\ \vdots \\ H-1 \\ H \end{array} \begin{array}{c} \begin{array}{cccccc} 0 & 1 & 2 & \cdots & H-1 & H \end{array} \\ \left(\begin{array}{cccccc} \mu \mathbf{I}_x & i\gamma \mathbf{I}_x & \mathbf{0} & \cdots & \mathbf{0} & \mathbf{0} \\ \mathbf{0} & \mu \mathbf{I}_x & i\gamma \mathbf{I}_x & \cdots & \mathbf{0} & \mathbf{0} \\ \vdots & \vdots & \vdots & \vdots & \vdots & \vdots \\ \mathbf{0} & \mathbf{0} & \mathbf{0} & \cdots & i\gamma \mathbf{I}_x & \mathbf{0} \\ \mathbf{0} & \mathbf{0} & \mathbf{0} & \cdots & \mu \mathbf{I}_x & i\gamma \mathbf{I}_x \end{array} \right) \end{array}$$

For $i = 2, \ldots S$

$$\mathbb{Y}_i = \begin{array}{c} \\ 0 \\ 1 \end{array} \begin{array}{c} \begin{array}{cc} 0 & 1 \end{array} \\ \left(\begin{array}{cc} F_{12_{(Nx \times Nx)}} & \mathbf{0}_{(Nx \times Hx)} \\ F_{13(Hx \times Nx)} & F_{14_{(Hx \times Hx)}} \end{array} \right) \end{array}$$

Submatrices F_{12}, F_{13}, and F_{14} are

$$F_{12} = \begin{array}{c} \\ 0 \\ 1 \\ \vdots \\ N-1 \end{array} \begin{array}{c} \begin{array}{cccc} 0 & 1 & \cdots & N-1 \end{array} \\ \left(\begin{array}{cccc} i\gamma \mathbf{I}_x & \mathbf{0} & \cdots & \mathbf{0} \\ \mathbf{0} & i\gamma \mathbf{I}_x & \cdots & \mathbf{0} \\ \vdots & \vdots & \vdots & \vdots \\ \mathbf{0} & \mathbf{0} & \cdots & i\gamma \mathbf{I}_x \end{array} \right) \end{array}$$

$$F_{13} = \begin{array}{c} \\ 1 \\ 2 \\ \vdots \\ H \end{array} \begin{array}{c} \begin{array}{cccc} 0 & 1 & \cdots & H \end{array} \\ \left(\begin{array}{cccc} \mu \mathbf{I}_x & \mathbf{0} & \cdots & \mathbf{0} \\ \mathbf{0} & \mathbf{0} & \cdots & \mathbf{0} \\ \vdots & \vdots & \vdots & \vdots \\ \mathbf{0} & \mathbf{0} & \cdots & \mathbf{0} \end{array} \right) \end{array}$$

$$F_{14} = \begin{array}{c} \\ 1 \\ 2 \\ \vdots \\ H \end{array} \begin{array}{c} \begin{array}{ccccc} 1 & 2 & \cdots & \mathrm{H}-1 & H \end{array} \\ \left(\begin{array}{ccccc} i\gamma \mathbf{I}_x & \mathbf{0} & \cdots & \mathbf{0} & \mathbf{0} \\ \mu \mathbf{I}_x & i\gamma \mathbf{I}_x & \cdots & \mathbf{0} & \mathbf{0} \\ \vdots & \vdots & \vdots & \vdots & \vdots \\ \mathbf{0} & \mathbf{0} & \cdots & \mu \mathbf{I}_x & i\gamma \mathbf{I}_x \end{array} \right) \end{array}$$

It is noted that matrix $\mathbb{Z}'$ is of the order $((H + 1)\mathrm{x}) \times (N\mathrm{x} + H\mathrm{x})$, matrix $\mathbb{Z}$ is of the order $(N\mathrm{x} + H\mathrm{x}) \times (N\mathrm{x} + H\mathrm{x})$, matrices $\mathbb{X}_i, i = 1,2,\ldots S$ are of order $(N\mathrm{x} + H\mathrm{x}) \times (N\mathrm{x} + H\mathrm{x})$, matrix $\mathbb{X}_0$ is of the order $((H + 1)\mathrm{x}) \times ((H + 1)\mathrm{x})$, matrices $\mathbb{Y}_i, i = 2,3,\ldots S$ are of order $(N\mathrm{x} + H\mathrm{x}) \times (N\mathrm{x} + H\mathrm{x})$, and matrix $\mathbb{Y}_1$ is of the order $(N\mathrm{x} + H\mathrm{x}) \times ((H + 1)\mathrm{x})$, respectively.

4. Study of Steady-State Vector

The process $\{I(t), T(t), C(t), J(t); t \geq 0\}$ is a continuous-time Markov chain (CTMC) having the state space E. Hence, the steady-state vector

$$\Xi(i,k,m,r) = \lim_{t \to \infty} Pr[I(t) = i, T(t) = k, C(t) = m, J(t) = r / I(0), T(0), C(0), J(0)]$$

exists, and is independent of the initial state.

Let $\Xi = (\Xi(0), \Xi(1), \ldots, \Xi(S))$,
$where\ \Xi(i) = (\Xi(i,0), \Xi(i,1)), i = 0,1,\ldots S$
$with\ \Xi(i,k) = (\Xi(i,k,0), \Xi(i,k,1), \ldots \Xi(i,k,H)), k = 0,1$
$with\ \Xi(i,k,m) = (\Xi(i,k,m,1), \Xi(i,k,m,2), \ldots \Xi(i,k,m,x)), m = 0,1,\ldots H$

Then, the steady state vector Ξ satisfies $\Xi\mathbb{P} = 0$, $\Xi e = 1$.

Lemma 1. *For the Markov process, the steady-state vector Ξ whose rate matrix is $\mathbb{P}$ is defined by*

$$\Xi(i) = \Xi(Q)\nabla_i,\ i = 0,1,\ldots S$$

where

$$\nabla_i = \begin{cases} (-1)^{Q-i}\mathbb{Y}_Q\mathbb{X}_{Q-1}^{-1}\mathbb{Y}_{Q-1}\ldots\mathbb{Y}_{i+1}\mathbb{X}_i^{-1}, & i = 0,1,\ldots Q-1; \\ \boldsymbol{I}, & i = Q; \\ (-1)^{2Q-i+1}\sum\limits_{j=1}^{S-i}\left\{ \begin{array}{c} \left(\mathbb{Y}_Q\mathbb{X}_{Q-1}^{-1}\mathbb{Y}_{Q-1}\ldots\mathbb{Y}_{s+1-j}\mathbb{X}_{s-j}^{-1}\right)\mathbb{Z}\mathbb{X}_{S-j}^{-1} \\ \left(\mathbb{Y}_{S-j}\mathbb{X}_{S-j-1}^{-1}\mathbb{Y}_{S-j-1}\ldots\mathbb{Y}_{i+1}\mathbb{X}_i^{-1}\right) \end{array} \right\}, & i = Q+1,\ldots S; \end{cases}$$

and $\Xi(Q)$ can be attained by workout the following two equations:

$$\Xi(Q)\left(\left\{(-1)^Q\textstyle\sum_{j=0}^{S-1}\left\{ \begin{array}{c} \left(\mathbb{Y}_Q\mathbb{X}_{Q-1}^{-1}\mathbb{Y}_{Q-1}\ldots\mathbb{Y}_{s+1-j}\mathbb{X}_{s-j}^{-1}\right) \\ \mathbb{Z}\mathbb{X}_{S-j}^{-1}\left(\mathbb{Y}_{S-j}\mathbb{X}_{S-j-1}^{-1}\mathbb{Y}_{S-j-1}\ldots\mathbb{Y}_{Q+2}\mathbb{X}_{Q+1}^{-1}\right) \end{array}\right\}\right\}\mathbb{Y}_{Q+1} + \mathbb{X}_Q + \left\{(-1)^Q\mathbb{Y}_Q\mathbb{X}_{Q-1}^{-1}\mathbb{Y}_{Q-1}\ldots\mathbb{Y}_1\mathbb{X}_0^{-1}\right\}\mathbb{Z}'\right) = 0$$

and

$$\Xi(Q)\left(\sum_{i=0}^{Q-1}\left\{(-1)^{Q-i}\mathbb{Y}_Q\mathbb{X}_{Q-1}^{-1}\mathbb{Y}_{Q-1}\ldots\mathbb{Y}_{i+1}\mathbb{X}_i^{-1}\right\} + \boldsymbol{I} + \sum_{i=Q+1}^{S}\{(-1)^{2Q-i+1}\sum_{j=0}^{S-i}\{\left(\mathbb{Y}_Q\mathbb{X}_{Q-1}^{-1}\mathbb{Y}_{Q-1}\ldots\mathbb{Y}_{s+1-j}\mathbb{X}_{s-j}^{-1}\right)\mathbb{Z}\mathbb{X}_{S-j}^{-1}\right.$$
$$\left(\mathbb{Y}_{S-j}\mathbb{X}_{S-j-1}^{-1}\mathbb{Y}_{S-j-1}\ldots\mathbb{Y}_{i+1}\mathbb{X}_i^{-1}\right)\}\})\boldsymbol{e} = 1$$

Proof. The well-known equations are,

$$\Xi\mathbb{P} = 0\ and\ \Xi\boldsymbol{e} = 1.$$

The equation $\Xi\mathbb{P} = 0$ can be written as

$$\begin{array}{c} \Xi(i+1)\mathbb{Y}_{i+1} + \Xi(i)\mathbb{X}_i = 0,\ i = 0,1,\ldots Q-1 \\ \Xi(i+1)\mathbb{Y}_{i+1} + \Xi(i)\mathbb{X}_i + \Xi(i-Q)\mathbb{Z}' = 0,\ i = Q \\ \Xi(i+1)\mathbb{Y}_{i+1} + \Xi(i)\mathbb{X}_i + \Xi(i-Q)\mathbb{Z} = 0,\ i = Q+1, Q+2,\ldots S-1 \\ \Xi(i)\mathbb{X}_i + \Xi(i-Q)\mathbb{Z} = 0,\ i = S \end{array} \quad (1)$$

Except (1), the above equations can be solved recursively, yielding

$$\Xi(i) = \Xi(Q)\nabla_i,\ i = 0,1,\ldots S.$$

where

$$\nabla_i = \begin{cases} (-1)^{Q-i}\mathbb{Y}_Q\mathbb{X}_{Q-1}^{-1}\mathbb{Y}_{Q-1}\ldots\mathbb{Y}_{i+1}\mathbb{X}_i^{-1}, & i = 0,1,\ldots Q-1; \\ \boldsymbol{I}, & i = Q; \\ (-1)^{2Q-i+1}\sum\limits_{j=0}^{S-i}\left\{ \begin{array}{c} \left(\mathbb{Y}_Q\mathbb{X}_{Q-1}^{-1}\mathbb{Y}_{Q-1}\ldots\mathbb{Y}_{s+1-j}\mathbb{X}_{s-j}^{-1}\right)\mathbb{Z}\mathbb{X}_{S-j}^{-1} \\ \left(\mathbb{Y}_{S-j}\mathbb{X}_{S-j-1}^{-1}\mathbb{Y}_{S-j-1}\ldots\mathbb{Y}_{i+1}\mathbb{X}_i^{-1}\right) \end{array} \right\}, & i = Q+1,\ldots S; \end{cases}$$

Solving Equation (1) and normalizing the condition after putting the value of ∇_i in that equation, we obtain $\Xi(Q)$, i.e.,

$$\Xi(Q)\left(\left\{(-1)^Q\sum_{j=0}^{S-1}\left\{\begin{array}{c}\left(\mathbb{Y}_Q\mathbb{X}_{Q-1}^{-1}\mathbb{Y}_{Q-1}\cdots\mathbb{Y}_{s+1-j}\mathbb{X}_{s-j}^{-1}\right)\\ \mathbb{Z}\mathbb{X}_{S-j}^{-1}\left(\mathbb{Y}_{S-j}\mathbb{X}_{S-j-1}^{-1}\mathbb{Y}_{S-j-1}\cdots\mathbb{Y}_{Q+2}\mathbb{X}_{Q+1}^{-1}\right)\end{array}\right\}\right\}\mathbb{Y}_{Q+1}+\mathbb{X}_Q+\left\{(-1)^Q\mathbb{Y}_Q\mathbb{X}_{Q-1}^{-1}\mathbb{Y}_{Q-1}\cdots\mathbb{Y}_1\mathbb{X}_0^{-1}\right\}\mathbb{Z}'\right)=0$$

and

$$\Xi(Q)(\sum_{i=0}^{Q-1}\left\{(-1)^{Q-i}\mathbb{Y}_Q\mathbb{X}_{Q-1}^{-1}\mathbb{Y}_{Q-1}\cdots\mathbb{Y}_{i+1}\mathbb{X}_i^{-1}\right\}+\boldsymbol{I}+\sum_{i=Q+1}^{S}\left\{(-1)^{2Q-i+1}\sum_{j=0}^{S-i}\left\{\left(\mathbb{Y}_Q\mathbb{X}_{Q-1}^{-1}\mathbb{Y}_{Q-1}\cdots\mathbb{Y}_{s+1-j}\mathbb{X}_{s-j}^{-1}\right)\mathbb{Z}\mathbb{X}_{S-j}^{-1}\right.\right.$$

$$\left.\left.\left(\mathbb{Y}_{S-j}\mathbb{X}_{S-j-1}^{-1}\mathbb{Y}_{S-j-1}\cdots\mathbb{Y}_{i+1}\mathbb{X}_i^{-1}\right)\right\}\right\})e=1$$

□

5. Derivation of System Performance Measures

We infer some performance measures of this system during a steady state. It is seen that $\Xi(i)$ is the steady-state probability vector for the inventory level being i with every constituent mentioned: server status in the system, the number of customers, waiting and being served, and the phase of the arrival process. Hence, $\Xi(i)e$ provides the probability that the inventory level in a steadystate is i. Similarly, $\Xi(i,k,m)e$ is the probability that the inventory level i, server status j, and customers waiting (including being served) k are in a steadystate.

5.1. Mean Inventory Level

Let M_L be the mean inventory level in a steadystate, which can be expressed as

$$M_L=\sum_{i=1}^{S}i\left(\sum_{k=0}^{1}\sum_{m=1}^{N-1}\Xi(i,k,m)\right)e+\sum_{i=1}^{S}i\left(\sum_{m=N}^{H}\Xi(i,1,m)\right)e+\sum_{i=1}^{S}i(\Xi(i,0,0))e.$$

5.2. Mean Reorder Rate

Let M_{RO} be the mean reorder rate in a steady state. If a demand service is completed or any of the (s + 1) items fails, then the inventory level drops to s from level (s + 1), a stock reorder is triggered. This then leads to

$$\begin{aligned}M_{RO}=\mu\sum_{m=1}^{H}\Xi(s+1,1,m)e+(s+1)\gamma\sum_{k=0}^{1}\sum_{m=1}^{N-1}\Xi(s+1,k,m)e\\+(s+1)\gamma\sum_{m=N}^{H}\Xi(s+1,1,m)e+(s+1)\gamma\Xi(s+1,0,0)e.\end{aligned}$$

5.3. Mean Perishable Rate

Let M_P be the mean perishable rate in a steadystate, which is given by

$$M_P=\sum_{i=1}^{S}\sum_{k=0}^{1}\sum_{m=1}^{N-1}i\gamma\Xi(i,k,m)e+\sum_{i=1}^{S}\sum_{m=N}^{H}i\gamma\Xi(i,1,m)e+\sum_{i=1}^{S}i\gamma\Xi(i,0,0)e.$$

5.4. Mean Balking Rate

Let M_B be the mean balking rate in a steadystate, which can be stated as

$$M_B=\frac{1}{\lambda}\sum_{i=1}^{S}\Xi(i,1,H)F_1e+\frac{1}{\lambda}\Xi(0,0,H)F_1e.$$

5.5. Mean Reneging Rate

Let M_R be the mean reneging rate in a steadystate, which is given by

$$M_R=\sum_{i=0}^{S}\sum_{m=1}^{N-1}m\alpha\Xi(i,0,m)e+\sum_{m=1}^{H}m\alpha\Xi(0,0,m)e.$$

5.6. Mean Waiting Time

Let M_W be the mean waiting time of the customers in the waiting hall in a steady state. Then, by Little's formula,

$$M_W = \frac{L}{\lambda_a}$$

where $L = \sum_{m=1}^{N-1} m\left(\sum_{i=1}^{S} \sum_{k=0}^{1} \Xi(i,k,m)\right)e + \sum_{m=N}^{H} m\left(\sum_{i=1}^{S} \Xi(i,1,m)\right)e + \sum_{m=1}^{H} m\Xi(0,0,m)e$

and the effective arrival rate (Ross [27]) λ_a is given by

$$\lambda_a = \frac{1}{\lambda}\sum_{i=1}^{S}\sum_{k=0}^{1}\sum_{m=1}^{N-1}\Xi(i,k,m)F_1 e + \frac{1}{\lambda}\sum_{i=1}^{S}\sum_{m=N}^{H-1}\Xi(i,1,m)F_1 e + \frac{1}{\lambda}\sum_{i=1}^{S}\Xi(i,0,0)F_1 e + \frac{1}{\lambda}\sum_{m=0}^{H=1}\Xi(0,0,m)F_1 e.$$

6. Cost Analysis

In order to calculate the total expected cost per unit time, we consider the following cost components.

C_C: Unit inventory carrying cost per unit time
C_S: Setup cost per order
C_B: Balking cost per customer per unit time
C_P: Perishable cost per item per unit time
C_R: Reneging cost per customer per unit time

Using the system performance measures from Section 5, the long-run expected system cost rate is given by

$$TC(S,s,H) = C_C M_L + C_S M_{RO} + C_B M_B + C_P M_P + C_R M_R + C_W M_W$$

where M_L, M_{RO}, M_P, M_R, and M_W are given in Section 5.

7. Numerical Illustration

This section presents some numerical experimentations that feature the convexity of the total expected system cost rate. In particular, we show the calculability of the outcomes inferred in our work and uncover the presence of local optima when the total cost function is a bivariate function. It is difficult to show convexity as the computations of Ξ/s are recursive. The arrival process is Erlang, and as an MAP, its parameters are given by (F_0, F_1), with

$$F_0 = \begin{pmatrix} -1 & 1 & 0 \\ 0 & -1 & 1 \\ 0 & 0 & -1 \end{pmatrix} \text{ and } F_1 = \begin{pmatrix} 0 & 0 & 0 \\ 0 & 0 & 0 \\ 1 & 0 & 0 \end{pmatrix}$$

In Tables 1–3, each row has a value in bold, and each column has a value that is underlined to represent the minima of the row and column, respectively. The value that is bold and underlined is then the least cost rate of the inventory system. Therefore, we have a (local) optimum for the related cost function of the table.

Table 1. Total expected cost rate interms of S and s.

S/s	8	9	10	11	12	13	14
30	3.2593	2.5439	2.1758	**2.0335**	2.0755	2.3231	2.8720
31	2.0682	1.5162	1.2608	**1.1835**	1.2480	1.4639	1.8954
32	<u>1.4396</u>	0.9747	<u>0.8044</u>	<u>**0.7869**</u>	<u>0.8831</u>	<u>1.1048</u>	<u>1.5060</u>
33	1.4966	<u>0.9729</u>	**0.8102**	0.8291	0.9707	1.2480	1.7235
34	2.6559	1.7741	**1.4900**	1.5060	1.7365	2.1750	2.8915

Table 2. Total expected cost rate interms of s and *H*.

H/s	3	4	5	6	7
3	0.9623	**0.9560**	0.9772	1.1089	1.3626
4	0.8967	**0.8933**	0.9065	1.0349	1.2932
5	3.8915	3.7048	**3.5771**	3.5779	3.7260
6	6.6794	6.4398	6.2803	**6.2629**	6.3866
7	9.7410	9.5087	9.3401	**9.2999**	9.3926

Table 3. Total expected cost rate in terms of *S* and *H*.

S/H	6	7	8	9	10
9	32.0697	**30.3781**	32.8875	38.2844	44.1008
10	31.8089	**29.1228**	32.8407	37.5364	42.5232
11	29.5793	**28.5009**	33.1264	37.1302	41.1433
12	30.5683	**28.1685**	33.3080	36.8679	39.9816
13	31.3798	**28.3862**	33.4184	36.7971	39.1530
14	32.0893	**28.6924**	33.7491	36.6116	38.6069
15	32.6321	**29.2989**	34.2411	36.8371	38.0025
16	33.1071	**29.8624**	34.7161	37.2047	37.9498
17	33.5074	**30.6399**	35.2495	37.5077	37.8875
18	33.9234	**31.3228**	35.7665	37.9689	38.0269

Let $H = 8$, $N = 5$, $\beta = 0.95$, $\mu = 1.04$, $\gamma = 0.6$, $\alpha = 0.35$, $\lambda = 0.8$ and $C_C = 0.1$, $C_S = 0.8$, $C_B = 0.07$, $C_P = 0.05$, $C_R = 0.1$, $C_W = 0.1$.

In Table 1, the values of *TC(S,s,8)* are shown.

The numerical example suggests that $TC(S,s,8)$ in (S,s) is convex and that the local optimum occurs at $(S,s) = (32,11)$, as displayed in Table 1 and Figure 1.

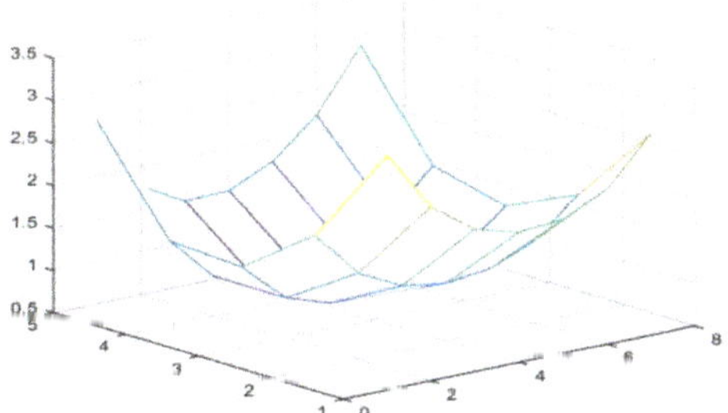

Figure 1. Total expected cost rate of S and s.

Let $S = 40$, $N = 2$, $\beta = 0.11$, $\mu = 1$, $\gamma = 0.235$, $\alpha = 0.59$, $\lambda = 0.93$ and $C_C = 0.011$, $C_S = 0.001$, $C_B = 0.03$, $C_P = 0.01$, $C_R = 0.4$, $C_W = 0.05$.

From Table 2, the numerical example suggests that *TC(40,s,H)* in *(s,H)* is convex and that the local optimum occurs at *(s,H)* = (4,4).

Let $s = 2$, $N = 2$, $\beta = 0.46$, $\mu = 1.25$, $\gamma = 0.14$, $\alpha = 0.1$, $\lambda = 0.24$ and $C_C = 0.17$, $C_S = 0.005$, $C_B = 0.97$, $C_P = 0.03$, $C_R = 0.08$, $C_W = 0.06$.

TC(S,2,H) values are displayed in Table 3.

The numerical example suggests that $TC(S,2,H)$ in (S,H) is convex and that the local optimum occurs at $(S,H) = (12,7)$.

Figure 2 grants the impact of the impatient customer rates(α), on the total expected cost rate TC via five curves thatrelate to N = 2,3,4,5,6. The acquired values for the remaining parameters and costs are displayed in the actual figure. Because ofFigure 2, we perceive that the total cost value decreases when the customer requirements for service begin (*i.e.*, N) increases and the impatient customers' rate(α) increases.

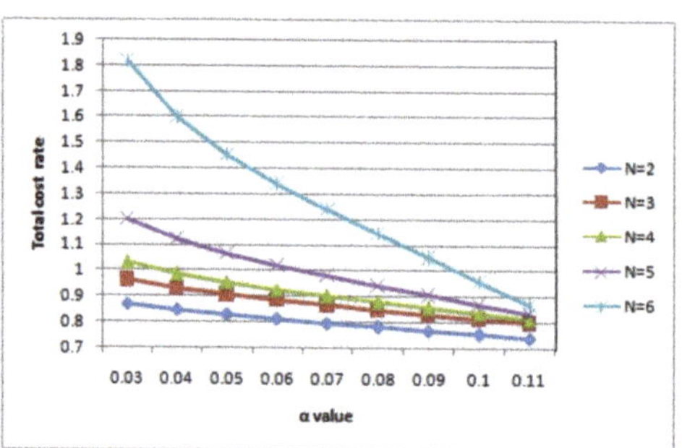

Figure 2. $TC(32,11)$ vs. α, $H = 8$, $\beta = 0.95$, $\mu = 1.04$, $\gamma = 0.6$, $C_C = 0.1$, $C_S = 0.8$, $C_B = 0.07$, $C_P = 0.05$, $C_R = 0.1$, $C_W = 0.1$.

Figure 3 grants the impact of the impatience customer rates (α), on the total expected cost rate TC via three curves thatrelate to μ = 2,3,4. Because ofFigure 3, we perceive that the total cost value decreases when the service rate (μ) decreases, and the impatient customer rate(α) increases.

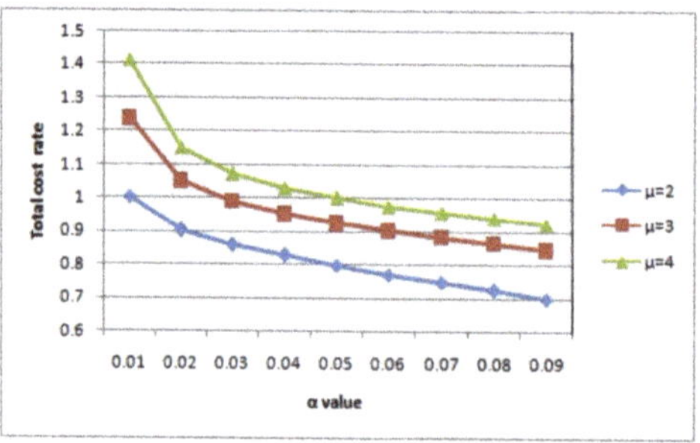

Figure 3. $TC(32,11)$ vs. α, $H = 8$; $\beta = 0.95$; N = 3; $\gamma = 0.6$; $C_C = 0.1$; $C_S = 0.8$; $C_B = 0.07$; $C_P = 0.05$; $C_R = 0.1$; $C_W = 0.1$.

Figure 4 grants the impact of the service rates (μ) on the total expected cost rate TC via four curves thatrelate to γ = 0.03,0.04,0.05,0.06. Because ofFigure 4, we perceive that the total cost value increases when the service rate (μ) increases and the perishable rate (γ) increases.

In Tables 4–9, we show the impact of the setup cost C_S, the carrying cost C_C, the balking cost C_B, the reneging cost C_R, and the waiting time cost C_W on the optimal values (S^*, s^*) and the corresponding total expected cost rate TC^*. Towards this end, we first fix the parameters and cost value as $H = 8$, $N = 5$, $\beta = 0.95$, $\mu = 1.04$, $\gamma = 0.6$, $\alpha = 0.35$, $\lambda = 0.8$, and $C_P = 0.05$.

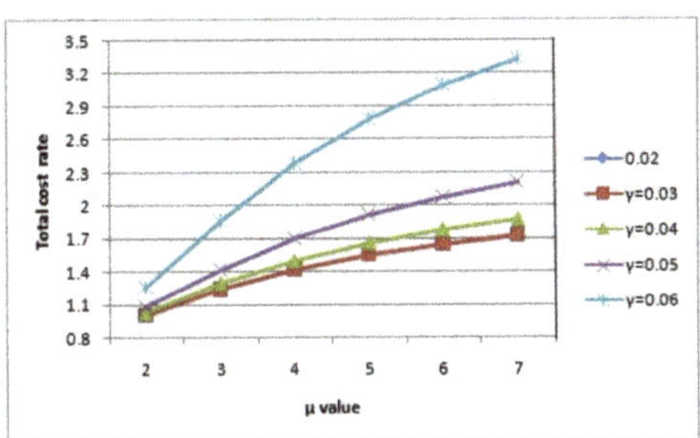

Figure 4. $TC(32,11)$ vs. μ, $H = 8$; $\beta = 0.95$; $N = 5$; $C_C = 0.1$; $C_S = 0.8$; $C_B = 0.07$; $C_P = 0.05$; $C_R = 0.1$; $C_W = 0.1$.

Table 4. Impact of C_C and C_S costs on the optimal value.

C_C/C_S	0.7	0.8	0.9	1.0	1.1
	32 11	32 11	32 11	32 11	32 11
0.09	0.664642	0.698524	0.732406	0.766289	0.800171
	32 11	32 11	32 11	32 11	32 11
0.10	0.694095	0.727977	0.761860	0.795742	0.829625
	32 11	32 11	32 11	32 11	32 11
0.11	0.723549	0.757431	0.791313	0.825196	0.859078
	32 11	32 11	32 11	32 11	32 11
0.12	0.753002	0.786885	0.820767	0.854649	0.888532
	32 11	32 11	32 11	31 10	31 10
0.13	0.782456	0.816338	0.850221	0.884103	0.917985

Table 5. Impact of C_W and C_B costs on the optimal value.

C_B/C_W	0.09	0.10	0.11	0.12	0.13
	32 11	32 11	32 11	32 11	32 11
0.07	0.719856	0.721000	0.722143	0.723287	0.724431
	32 11	32 11	32 11	32 11	32 11
0.08	0.786885	0.788028	0.789172	0.790316	0.791460
	32 11	32 11	32 11	32 11	32 11
0.09	0.853914	0.855057	0.856201	0.857345	0.858489
	32 11	32 11	32 11	32 11	32 11
0.10	0.920942	0.922086	0.923230	0.924374	0.925518
	32 12	32 12	32 12	33 12	33 12
0.11	0.980997	0.982192	0.983386	0.984581	0.985776

Table 6. Impact of C_W and C_R costs on the optimal value.

C_W/C_R	**0.09**	**0.10**	**0.11**	**0.12**	**0.13**
	32 11	32 11	32 11	32 11	32 11
0.09	0.696894	0.719856	0.742817	0.765779	0.788741
	32 11	32 11	32 11	32 11	32 11
0.10	0.763923	0.786885	0.809846	0.832808	0.855770
	32 11	32 11	32 11	32 11	32 11
0.11	0.830952	0.853914	0.876875	0.899837	0.922799
	32 12	32 11	32 11	32 11	32 11
0.12	0.897962	0.920942	0.943904	0.966866	0.989828
	32 12	32 12	32 12	32 12	33 12
0.13	0.956834	0.980997	1.005160	1.029322	1.053485

Table 7. Impact of C_W and C_S costs on the optimal value.

C_W/C_S	**0.7**	**0.8**	**0.9**	**1.0**	**1.1**
	32 11	32 11	32 11	32 11	32 11
0.09	0.685973	0.719856	0.753738	0.787620	0.821503
	32 11	32 11	32 11	32 11	32 11
0.10	0.753002	0.786885	0.820767	0.854649	0.888532
	32 11	32 11	32 11	32 11	32 11
0.11	0.820031	0.853914	0.887796	0.921678	0.955561
	32 11	32 11	32 11	32 11	32 11
0.12	0.885104	0.920942	0.954825	0.988707	1.022590
	33 11	33 11	33 11	33 11	33 11
0.13	0.943977	0.980997	1.018017	1.055037	1.089619

Table 8. Impact of C_W and C_C costs on the optimal value.

C_W/C_C	**0.09**	**0.10**	**0.11**	**0.12**	**0.13**
	32 11	32 11	32 11	32 11	32 11
0.09	0.631495	0.660949	0.690402	0.719856	0.749309
	32 11	32 11	32 11	32 11	32 11
0.10	0.698524	0.727977	0.757431	0.786885	0.816338
	32 11	32 11	32 11	32 11	32 11
0.11	0.765553	0.795006	0.824460	0.853914	0.883367
	32 11	32 11	32 11	32 11	32 11
0.12	0.832582	0.862035	0.891489	0.920942	0.946720
	31 11	31 11	31 11	31 10	31 10
0.13	0.899611	0.929064	0.956402	0.980997	1.005592

Table 9. Impact of C_S and C_R costs on the optimal value.

C_S/C_R	0.09		0.10		0.11		0.12		0.13	
	32	11	32	11	32	11	32	11	32	11
0.7	0.730040		0.753002		0.775964		0.798926		0.821888	
	32	11	32	11	32	11	32	11	32	11
0.8	0.763923		0.786885		0.809846		0.832808		0.855770	
	32	11	32	11	32	11	32	11	32	11
0.9	0.797805		0.820767		0.843729		0.866691		0.889652	
	32	11	32	11	32	11	32	11	32	11
1.0	0.831688		0.854649		0.877611		0.900573		0.923535	
	32	11	32	11	32	11	31	11	31	11
1.1	0.865570		0.888532		0.911493		0.934455		0.957417	

From Tables 4–9, we observe the below monotonic behavior of (S^*, s^*):

- The total expected cost rate increases when each of the setup cost C_S, the carrying cost C_C, the balking cost C_B, the reneging cost C_R, and the waiting time cost C_W increase.
- As is to be expected, (S^*, s^*) monotonically increase when C_W increases.
- (S^*, s^*) monotonically decrease when C_C and C_S increase.
- (S^*, s^*) monotonically increase when C_C and C_W increase.
- S^* increases with C_B and C_W increasing.

8. Conclusions

In this paper, we proposed a perishable inventory system model in which the demands arrive according to a MAP and the replenishment process is negatively exponential. The server provides service at least N number of customers in the system(i.e., N-policy). We investigated the effect of the N-policy on the arriving customers to the inventory system. The joint distribution is derived in the steady-state, and we analyzed some measures of system performance and obtained the total expected cost rate in the steady-state. Additionally, we presented the numerical illustration that describes the impact of impatient customers caused by the N-policy on the inventory system's total expected cost rate. From the sensitive analysis, we can see that the total expected cost value diminishes because of the impatient customer rate. The total expected cost value seriously diminishes when the customer requirements for service begin(i.e., N) with rate increments. The service rate building also did not assist with decreases in the effect on the total expected cost rate. Additionally, the total expected cost value decreases due to the customer loss cost by the impatient customer rate, which is greater than the total expected cost value decrease due to other cost and rate values. From these perceptions, we stated that the impatient customers due to N-policy have an enormous impact on the total expected cost of the system. Future work can investigate the way to reduce the increasing of impatient customers caused by the N-policy server in the inventory system by adding other concepts like vacation policy with the N-policy server.

Author Contributions: Conceptualization, R.S. and N.A.; Data curation, R.S.; Formal analysis, N.A.; Funding acquisition, L.N., N.A., and G.P.J.; Investigation, S.A. (Srijana Acharya); Methodology, S.A. (S. Amutha) and M.K.; Project administration, L.N. and S.A. (Srijana Acharya); Resources, L.N. and G.P.J.; Supervision, G.P.J., Validation, S.A. (S. Amutha); Visualization, M.K.; Writing—original draft, R.S.; Writing—review and editing, S.A. (Srijana Acharya) and G.P.J. All authors have read and agreed to the published version of the manuscript.

Funding: This research was funded by RUSA Phase 2.0(F 24-51/2014-U), DST-PURSE 2nd Phase programme (SR/PURSE Phase 2/38) and UGC-SAP (DRSI)(F.510/8/DRS-I/2016(SAP-I)), Govt. of India.

Institutional Review Board Statement: Not applicable.

Informed Consent Statement: Not applicable.

Data Availability Statement: Data sharing not applicable.

Conflicts of Interest: The authors declare no conflict of interest.

References

1. Nahmias, S. Perishable Inventory Theory: A Review. *Oper. Res.* **1982**, *30*, 680–708. [CrossRef] [PubMed]
2. Liu, A.; Zhu, Q.; Xu, L.; Lu, Q.; Fan, Y. Sustainable supply chain management for perishable products in emerging markets: An integrated location-inventory-routing model. *Transp. Res. E-Log.* **2021**, *150*, 102–319. [CrossRef]
3. Darestani, S.A.; Hemmati, M. Robust optimization of a bi-objective closed-loop supply chain network for perishable goods considering queue system. *Comput. Ind. Eng.* **2019**, *136*, 277–292. [CrossRef]
4. Ioannidis, S.; Jouini, O.; Economopoulos, A.A.; Kouikoglou, V.S. Control policies for single-stage production systems with perishable inventory and customer impatience. *Ann. Oper. Res.* **2013**, *209*, 115–138. [CrossRef]
5. Kalpakam, S.; Arivarignan, G. A continuous review perishable inventory model. *Stat. J. Theor. Appl. Stat.* **1988**, *19*, 389–398. [CrossRef]
6. Liu, L.; Lian, Z. (s, S) Continuous Review Models for Products with Fixed Lifetimes. *Oper. Res.* **1999**, *47*, 150–158. [CrossRef]
7. Ko, S.-S.; Kang, J.; Kwon, E.-Y. An (s,S) inventory model with level-dependent G/M/1-Type structure. *J. Ind. Manag. Optim.* **2016**, *12*, 609–624.
8. Weiss, H.J. Optimal Ordering Policies for Continuous Review Perishable Inventory Models. *Oper. Res.* **1980**, *28*, 365–374. [CrossRef]
9. Zhang, X.; Lam, J.S.L.; Iris, Ç. Cold chain shipping mode choice with environmental and financial perspectives. *Transp. Res. Part D Transp. Environ.* **2020**, *87*, 102–537. [CrossRef]
10. Berman, O.; Kaplan, E.H.; Shimshak, D.G. Deterministic approximations for inventory management at service facilities. *IIE Trans.* **1993**, *25*, 98–104. [CrossRef]
11. Amirthakodi, M.; Sivakumar, B. An inventory system with service facility and finite orbit size for feedback customers. *Opsearch* **2014**, *52*, 225–255. [CrossRef]
12. Arivarignan, G.; Elango, C.; Arumugam, N. A continuous review perishable inventory control system at service facilities. In *Advances in Stochastic Modelling*; Artalejo, J.R., Krishnamoorthy, A., Eds.; Notable Publications: NJ, USA, 2002; pp. 19–40.
13. Hamadi, H.M.; Sangeetha, N.; Sivakumar, B. Optimal control of service parameter for a perishable inventory system maintained at service facility with impatient customers. *Ann. Oper. Res.* **2015**, *233*, 3–23. [CrossRef]
14. Manuel, P.; Sivakumar, B.; Arivarignan, G. A perishable inventory system with service facilities, MAP arrivals and PH—Service times. *J. Syst. Sci. Syst. Eng.* **2007**, *16*, 62–73. [CrossRef]
15. Lawrence, A.S.; Sivakumar, B.; Arivarignan, G. A perishable inventory system with service facility and finite source. *Appl. Math. Model.* **2013**, *37*, 4771–4786. [CrossRef]
16. Yadin, M.; Naor, P. Queueing system with a removable service station. *Oper. Res.* **1963**, *14*, 393–405. [CrossRef]
17. Heyman, D. Optimal operating policies for M/G/1 queueing system. *Oper. Res.* **1968**, *16*, 362–382. [CrossRef]
18. Ke, J.-C. The control policy of an M[x]/G/1 queueing system with server startup and two vacation types. *Math. Methods Oper. Res.* **2001**, *54*, 471–490. [CrossRef]
19. Kella, O. The threshold policy in theM/G/1 queue with server vacations. *Nav. Res. Logist.* **1989**, *36*, 111–123. [CrossRef]
20. Wang, K.-H.; Ke, J.-C. Control Policies of an M/G/1 Queueing System with a Removable and Non-reliable Server. *Int. Trans. Oper. Res.* **2002**, *9*, 195–212. [CrossRef]
21. Krishnamoorthy, A.; Anbazhagan, N. Perishable inventory system at service facility with N policy. *Stoch. Anal. Appl.* **2008**, *26*, 1–17.
22. Jeganathan, K.; Anbazhagan, N.; Vigneshwaran, B. Perishable Inventory System with Server Interruptions, Multiple Server Vacations, and N Policy. *Int. J. Oper. Res. Inf. Syst.* **2015**, *6*, 32–52. [CrossRef]
23. Lucantoni, D.M.; Meier-Hellstern, K.S.; Neuts, M.F. A single server queue with server vacations and a class of non-renewal arrival processes. *Adv. Appl. Probab.* **1990**, *22*, 676–705. [CrossRef]
24. Latouche, G.; Ramaswami, V. *Introdution to Matrix Analytic Methods in Stochastic Modelling*; SIAM: Philadelphia, PA, USA, 1999.
25. Lee, G.; Jeon, J. A new approach to an N/G/1 queue. *Queueing Syst.* **2000**, *35*, 317–322. [CrossRef]
26. Chakravarthy, S.; Dudin, A. Analysis of a retrial queueing model with MAP arrivals and two types of ustomers. *Math. Comput. Model.* **2003**, *37*, 343–363. [CrossRef]
27. Ross, S.M. *Introduction to Probability Models*; Harcourt Asia: Singapore, 2000.

Article

A Knowledge-Based Hybrid Approach on Particle Swarm Optimization Using Hidden Markov Models

Mauricio Castillo [1], Ricardo Soto [1], Broderick Crawford [1], Carlos Castro [2] and Rodrigo Olivares [3,*]

1 Escuela de Ingeniería Informática, Pontificia Universidad Católica de Valparaíso, Valparaíso 2362807, Chile; mauricio.castillo.d@mail.pucv.cl (M.C.); ricardo.soto@pucv.cl (R.S.); broderick.crawford@pucv.cl (B.C.)
2 Departamento de Informática, Universidad Técnica Federico Santa María, Valparaíso 2390123, Chile; carlos.castro@inf.utfsm.cl
3 Escuela de Ingeniería Informática, Universidad de Valparaíso, Valparaíso 2362905, Chile
* Correspondence: rodrigo.olivares@uv.cl

Abstract: Bio-inspired computing is an engaging area of artificial intelligence which studies how natural phenomena provide a rich source of inspiration in the design of smart procedures able to become powerful algorithms. Many of these procedures have been successfully used in classification, prediction, and optimization problems. Swarm intelligence methods are a kind of bio-inspired algorithm that have been shown to be impressive optimization solvers for a long time. However, for these algorithms to reach their maximum performance, the proper setting of the initial parameters by an expert user is required. This task is extremely comprehensive and it must be done in a previous phase of the search process. Different online methods have been developed to support swarm intelligence techniques, however, this issue remains an open challenge. In this paper, we propose a hybrid approach that allows adjusting the parameters based on a state deducted by the swarm intelligence algorithm. The state deduction is determined by the classification of a chain of observations using the hidden Markov model. The results show that our proposal exhibits good performance compared to the original version.

Keywords: swarm intelligence method; parameter control; adaptive technique; hidden Markov model

Citation: Castillo, M.; Soto, R.; Crawford, B.; Castro, C.; Olivares, R. A Knowledge-Based Hybrid Approach on Particle Swarm Optimization Using Hidden Markov Models. *Mathematics* **2021**, *9*, 1417. https://doi.org/10.3390/math9121417

Academic Editor: Frank Werner

Received: 25 May 2021
Accepted: 15 June 2021
Published: 18 June 2021

1. Introduction

Swarm intelligence methods have attracted the attention of the scientific community in recent decades due to their impressive ability to adapt their methodology to complex problems [1]. These procedures are defined as bio-inspired computational processes observed in nature because they mimic the collective behavior of individuals when interacting in their environments [2]. Many of these procedures have become popular methods, such as genetic algorithms, differential evolution, ant colony system, particle swarm optimization, among several others, and they are still at the top of the main research in the optimization field [3]. Swarm intelligence methods work as a smart-flow using acquired knowledge in the iterative way to find near-optimal solutions [4]. The evolutionary strategy of these techniques mainly depends on the initial parameter configuration which is dramatically relevant for the efficient exploration of the search space, and therefore to the effective finding of high-quality solutions [5].

Finding the best value for a parameter is known as offline parameter setting, and it is done before executing the algorithm. This issue is treated even as an optimization problem in itself. On the other hand, the "online" parameters control is presented as a smart variation of the original version of the algorithm where the normal process is modified by new internal stimulants. Furthermore, according to the No Free Lunch theorem [6], there is no general optimal algorithm parameter setting. It is not obvious to define a priori which parameter setting should be used. The optimal values for the parameters mainly depend on the problem and even the instance to deal with and within the search time that the user wants to spend solving the problem. A universally optimal parameter value set for a given bio-inspired approximate method does not exist [7].

Using external techniques of the autonomous configuration, the swarm intelligence algorithms are able to adapt their internal processes during the run, based on performance metrics in order to be more efficient [8]. In this way, the user does not require any expert knowledge for reaching efficient solving processes. The adaptation process can be handled under two schemes: offline parameter tuning and online parameter control. In the online control, the parameters are handled and updated during the run of the algorithm, whereas in the offline parameter initialization, the values of different parameters are fixed before the run of the algorithm [6].

In this work, we tackle the online parameter definition problem with an autonomous search concept. This approach focuses on the algorithm parameters adjustment while the search process is performed as guided by the information obtained from the relation between the position of the solutions in the search space. This procedure allows the algorithm to change its functioning during the run, adapting to the particular conditions of the region discovered [9]. The main contribution of this work is to provide a bio-optimization solver with the ability to self-regulate their internal operation, without requiring advanced knowledge to efficiently calibrate the solving process. We propose a hybrid schema applying hidden Markov models to recompute the parameter values into a super-swarm optimization method. Hidden Markov models classify a chain of visible observations corresponding to the relationship between the distance of solutions given by the bio-inspired optimization algorithm. Recognizing when solutions are close to each other is a capability which is not part of all optimization algorithms. An external strategy that satisfies this problem is always valued for its potential use in complex engineering problems.

As a solver technique, we employ the popular particle swarm optimization (PSO) technique. The decision to use the PSO was based on two assumptions: (a) many bio-inspired methods that use the paradigm to generate solutions by updating velocity and position can be improved under this proposal; and (b) PSO is one of the most popular optimizer algorithms, so there is extensive literature that reports its excellent efficiency [10].

To evaluate this proposal, we solve a well-known optimization problem: the set covering problem. We treat a set of the hardest instances taken from the OR-Library [11] in order to demonstrate that the improved behavior of the particle swarm optimization exhibits a better yield than its original version and bio-inspired approximate methods of the state of the art.

The manuscript continues as follows: Section 2 presents a bibliographic search for related work in the field. The results of this search justify the proposal of this work; Section 3 explains offline and online parameter adjustment showing their differences; Section 4 describes the developed solution and the concepts used; Section 5 details the experimental setup, while Section 6 presents and discusses the main results obtained; finally, conclusions and future work are outlined in Section 7.

2. Related Work

Recent works show that swarm intelligence methods remain favorites in the optimization field [12–16], and their popularity has led them to be used in different application domains, such as resource planning, telecommunications, financial analysis, scheduling, space planning, energy distribution, molecular engineering, logistics, signal classification, and manufacturing, among others [17].

In [18], a recent survey on a new generation of nature-based optimization methods is detailed. This study presents metaheuristics as efficient solvers able to solve modern engineering problems in reduced time. Nevertheless, and despite this quality, these techniques present some complications inherent to the phenomenon that defines them. Among them, we find the adjustment and control of their input parameters that directly impact the exploration process of the search space [3,19]. This task is generally done when an external parameter is overcome by a non-deterministic move operator. The exploration phase operates with a set of agents that attempt to escape from local optima [20–22]. In this context, the autonomous search paradigm [23] describes how to empower metaheuristics to adjust its parameters during the resolution process, reacting to the information obtained. This technique accelerates the convergence of the algorithm, reducing the operational cost, and providing robustness to the process, making it react to the problem that is being solved.

By focusing on online parameter setting in optimization algorithms, we can find [24]. Here, the authors provide a perspective (from 30 years) explaining that an automatic tuning allows adapting the parameter values depending on the instance and the evolution of the algorithm. One of the first attempts using a controlled parameter variation in the particle swarm optimizer was proposed in [25]. A linear decreasing function was used for the inertia coefficient parameter w, with a start value of $w_s = 0.9$ and a final value of $w_f = 0.4$, testing this configuration for well-known benchmark functions. In [26], the parameter adjustment tries to make a transition between exploration and exploitation states, linearly decreasing and increasing the personal and social coefficients, respectively. In [27], an ideal velocity for the particles is defined, and the value of ω is updated to adjust the current average velocity to a value closer to the ideal. In [28,29], hybrid approaches to the PSO algorithm were developed. The first one includes a two-fold adaptive learning strategy to guarantee the exploration and exploitation phases of the algorithm. The second one proposes a learning strategy using a quasi-entropy index when local search works.

Related works that adjust the parameters of the PSO algorithm based on the fitness obtained along the iterations have been discussed. For example, in [30], two values that describe the state of the algorithm are defined: the evolutionary speed factor and the aggregation degree. Both values are used to update the ω values for each particle. In [31], the inertial and best solution acceleration coefficients are adjusted for each particle based on the relation between the current particle's fitness and the global best fitness. In [32], authors related the value of ω with the convergence factor and the diffusion factor to dynamically set its value. In [33], the value of ω for each particle in the swarm is computed based on the ratio of the personal best fitness value with the personal best fitness average, for all particles. In [34], the value of ω is calculated from the relation to fitness-based ranking for each particle and the problem dimension number. In [35], the value of ω is updated to accelerate the PSO convergence. In [36], a success rate for the PSO is defined based on the proportion of particles that improved their personal best position at iteration t. This method aims to increase the value of w when the proportion of particles that improved their personal best is high and decrease it otherwise. In [37], the inertial coefficient w is increased for the best particle and decreased for all others, based on the idea that the best particle is more confident on the direction of its movement. The inertial coefficient decreases linearly for the rest of particles.

Taking into account the fitness, but considering the relationship between the position of the particles in relation to the best, efforts have been made to adjust the value of the parameters. In [38], acceleration and inertial parameters are adjusted according to state deduction using a fuzzy classification system. The state classification depends on the calculation of the evolutionary factor. In [39], inertia weight and acceleration coefficients are adjusted using the gray relational analysis, proposed by [40], using the relation of the particle compared to the global best.

In recent years, the hybridization of metaheuristics with supervised and unsupervised methods has emerged as a promising field in approximation algorithms. In 2017, the term "Learnheuristic" was introduced in [41] to address the integration between metaheuristics and machine learning algorithms, and they provide a survey of the closest papers. In this research, a simple but robust idea is proposed. There are two work-groups: machine learning algorithms to enhance metaheuristics and metaheuristics to improve machine learning techniques. For the first group, is it possible to find (i) metaheuristics for improving clustering methods using the artificial bee colony algorithm [42], local search [43], particle swarm optimization [44] and ensemble-based metaheuristics [45]; (ii) metaheuristics to efficiently address the feature selection topic [46]; and (iii) metaheuristics for improving classification algorithms [47–49].

Recently, machine learning methods have also been used for the parameter control issue. An example of auto-tuning in a deep learning technique can be seen in [50]. Here, the authors provide an improvement to support and store a large number of parameters required by deep learning algorithms. Now, considering improvements that include machine learning in optimization algorithms, we can find in [51] that the PSOs parameters are adjusted by using an agent that chooses actions from a set in a probabilistic way, then measures the results and sends a reinforcement signal. In [52], a decision tree was used to perform the tuning of parameters. Hop-field neural networks were used in [53] to initiate solutions of a genetic algorithm applied to the economic dispatch problem. A mechanism for identifying and escaping from extreme points is punished in [54]. Here, the whale swarm algorithm includes new procedures to iteratively discard the attenuation coefficient and it enables the identification of extreme points during the run. An integration between the Gaussian mutation and an improved learning strategy were also proposed to boost a population-based method in [55]. New interactions between machine learning and optimization methods have recently been published in [56–58]. Moreover, improved machine learning techniques have been used for action recognition from collaborative learning networks [59], for the automatic recognition and classification of ECG and EEG signals [60–62], for complex processing on images [63], for health monitoring systems using IoT-based techniques [64], and several others works. In [65], a support vector machine is employed as a novel methodology to compute the genetic algorithm's fitness. A similar work can be seen in [66]. Clustering techniques were studied for the exploration of the search space [67] and for dynamic binarization strategies on combinatorial problems [68]. In [69], case-based reasoning techniques were investigated to the identify sub-spaces of searches to solve a combinatorial problem. In [70], an incremental learning technique was applied to constrained optimization problems. Finally, in [71], an alternative mechanism for the incorporation of negative learning on the ant colony optimization is proposed.

Finally, a few works explore the integration of hidden Markov models and optimization algorithms. In [72,73], a population-based method is proposed to train hidden Markov models. Another work which uses a bio-solver to optimize a hidden Markov model is presented in [74]. On the other hand, recent studies have studied how hidden Markov models improve the optimization algorithms. In [75], authors studied the relation of particle distances to determine the state of the particle swarm optimizer. The states were inspired by [38,76]. Parameters are updated according to the determined state. Similar work can be seen in [75,77,78].

3. Preliminaries

Parameter setting is known as a strategy for providing larger flexibility and robustness to the bio-inspired techniques, but requires an extremely careful initialization [7,19]. Indeed, the parameters of these procedures influence the efficiency and effectiveness of the search process [79]. To define a priori which parameter setting should be used is not an easy-task. The optimal values for the parameters mainly depend on the problem and even the instance to deal with and the search time within which the user wants to solve the problem.

This strategy is divided into two key approaches: the offline parameter tuning and the online parameter control (see Figure 1).

The adaptation process is called online when the performance information is obtained during solving, while the process is considered offline when a set of training instances is employed to gather the feedback [80]. The goal of parameter tuning is to obtain parameter values that could be useful over a wide range of problems. Such results require a large number of experimental evaluations and are generally based on empirical observations. Parameter control is divided into three branches according to the degree of autonomy of the strategies. Control is deterministic when parameters are changed according to a previously established schedule, adaptive when parameters are modified according to rules that take into account the state of the search, and self-adaptive when parameters are encoded into individuals in order to evolve conjointly with the other variables of the problem.

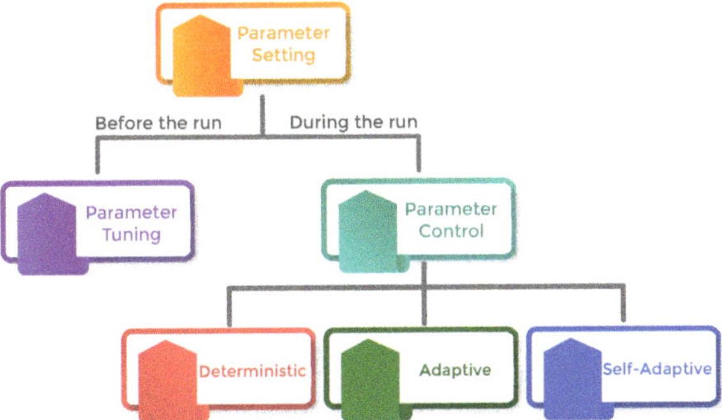

Figure 1. Scheme of the parameter adaptation process in bio-inspired methods.

The offline approaches require a high computational cost [19,81]. This cost increases when we use the offline approach for solving each input instance of a problem. Indeed, the optimal parameter values depend on the instances of the problem to be addressed. Furthermore, according to the No Free Lunch theorem, there is no generic optimal parameter setting. A universally optimal parameter value set for a given bio-inspired approximate method does not exist [6,82].

The great advantage of the online approach against the offline approach is that the effectiveness of the parameter control may change during the search process. That is, at different moments of the search, different optimal values are found for a given parameter. Hence, online approaches that change the parameter values during the search must be designed. Online approaches may be classified as follows [6]:

- Dynamic update: A random or deterministic updates the parameter value. This operation is performed without taking into account the search progress.
- Adaptive update: In this approach, parameter values evolve during the search progress. To change the parameter values, a function that mimics the behavior of the phenomenon is performed. For that, the memory of the search is mainly used. Hence, the parameters are associated with the representation and these are subject to updates in function of the problem's solution.

Online control is only recent but also interesting and challenging as the feedback is uniquely gathered during solving time with no prior knowledge from training phases and no user experts.

4. Developed Solution

Swarm intelligence methods have been developed for almost 30 years. The term swarm intelligence was coined in 1993 [83] and since its appearance, it has become a popular optimization method [84]. The distributed structure presents possible advantages over centralized methods, such as the simplicity of the search agents, the robustness provided by the redundancy of components, and the ability to escape local optimums [6]. This type of structure is typical in many biological systems, such as insect colonies, flocks of birds, and schools of fish. The synergy between the swarm members provides each of them with advantages that they could not achieve on their own, such as protection against predators and a more reliable supply of food [14,85].

Particle swarm optimization is a most popular population-based bio-inspired algorithm [86,87]. This method intelligently mimics the collaborative behavior of individuals or "particles" through two essential components: the position and the velocity. A set of particles (candidate solutions) forms the swarm that evolves during several iterations.

This procedure describes a powerful optimization method [88]. The technique operates by altering velocity through the search space and then updates its position according to its own experience and neighboring particles.

Particle swarm optimization can be identified as an intelligent system with two phases: (a) when the algorithm reaches large velocities in the initial phase, the current solutions focus more on diversification; (b) as velocities tend towards zero, the current solution focuses on intensification. The best reached solutions are memorized as *pBest*. The standard particle swarm optimization is governed by the movement of particles through two vectors: the velocity $V_i = \langle v_i^1, v_i^2, \ldots, v_i^D \rangle$ and the position $X_i = \langle x_i^1, x_i^2, \ldots, x_i^D \rangle$. First, the particles are randomly positioned in a D-dimensional heuristic space with random velocity values. During the evolution process, each particle updates its velocity via Equation (1) and position through Equation (2):

$$v_i^d = \omega v_i^d + c_1 \phi_1^d (pBest_i^d - x_i^d) + c_2 \phi_2^d (gBest^d - x_i^d) \tag{1}$$

$$x_i^d = x_i^d + v_i^d \tag{2}$$

where $d = \{1, 2, \ldots, D\}$ represents the size of the problem; the positive constants ω, c_1, and c_2 are acceleration coefficients; ϕ_1 and ϕ_2 are two uniformly distributed random numbers in the range $[0, 1]$; $pBest_i$ is the best position reached by ith particle; and $gBest$ is the global best position found by all particles during the resolution process.

4.1. Evolutionary Factor *f*

Diversity measures explain the distribution of a set of particles in the search space [89]. In this sense, ref. [38] proposes a measure derived from the distances between the PSO particles, known as the evolutionary factor f. This factor is computed in Equation (3):

$$f = \frac{d_p - d_w}{d_g - d_w} \in [0, 1] \tag{3}$$

where d_p represents the pBest fitness (position) reached by a particle until that moment. The d_w and d_g values describe the worst and gBest distance of the swarm, respectively.

Figure 2 shows the relationship between the position of the PSO particles and the state in which they will be classified: sub-figure (a) depicts the exploration state, where the particles are far away from each other; sub-figure (b) shows an exploitation/convergence state, where the particles are close to each other and the best particle appears in the center of the group; sub-figure (c) illustrates the jumping-out state, where the particles are close to each other, but the best particle has found a better zone and appears far from the others. To compute the difference among distances, the average of all distances d_{avg} is also required.

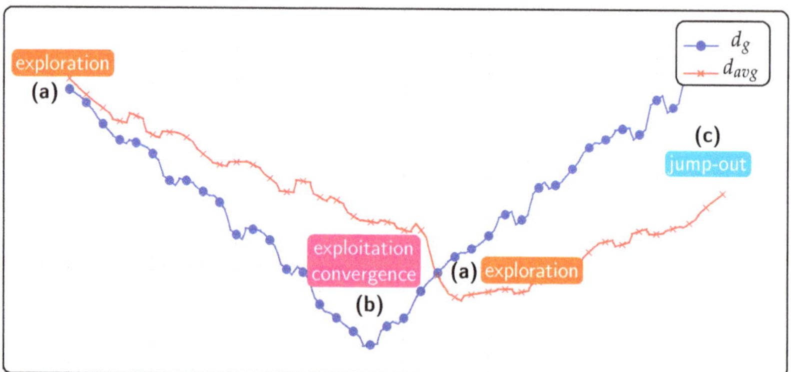

Figure 2. Example of PSO particle distribution: (**a**) $d_g \approx d_{avg}$ exploration; (**b**) $d_g \ll d_{avg}$ exploitation, convergence; and (**c**) $d_g \gg d_{avg}$ jump out.

4.2. Markov Models

This section defines the hidden Markov model used in this paper to identify the state (or inner-phase) of the PSO. For that, we detail how the states of a PSO can be modeled as a Markov chain, allowing the state to be inferred from the evolutionary factor calculated for the particle swarm optimizer.

4.2.1. Markov Chains

The Markov chain is a statistical model that defines the probability of transition from one state to another within a finite set of states. The Markov chain assumes that the transition to the next state only depends on the current state, regardless of previous states [90].

The states within a Markov chain are finite and the transitions between states are defined according to a probability. This probability must add up to 1, indicating all probable states that can be reached from the current state. Figure 3 shows a typical Markov chain.

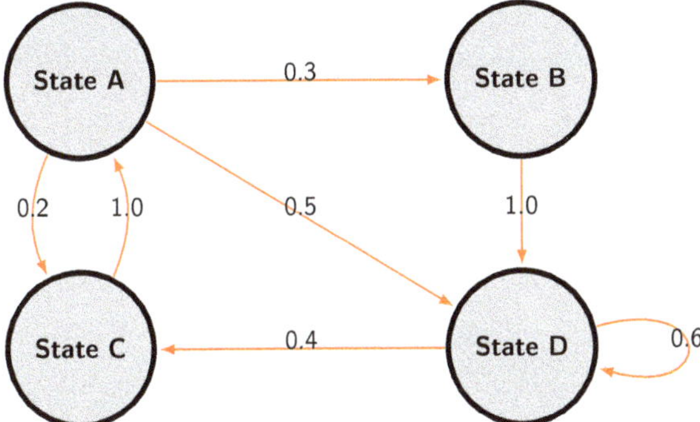

Figure 3. Markov chain. The arcs connecting the nodes/states indicate the transition probability between states.

For this Markov chain, we observe four states: A, B, C, and D. Arrows indicate which state can be accessed from a particular state and the number next to it indicates the probability that the transition occurs. These probabilities can be studied as a square matrix where the transition probability for all states is represented. The transition matrix M in the example above is defined as

$$M = \begin{bmatrix} 0 & 0.3 & 0.2 & 0.5 \\ 0 & 0 & 0 & 1 \\ 1 & 0 & 0 & 0 \\ 0 & 0 & 0.4 & 0.6 \end{bmatrix}$$

4.2.2. Hidden Markov Model

The hidden Markov model (HMM) is a framework that allows, through the observation of some visible state, to deduce elements of the Markov chain that are not directly visible, i.e., hidden [91,92]. The transition between states is assumed to be in the form of a Markov chain. The Markov chain can be defined by an initial probability vector π and a transition matrix A. Observable elements O are emitted according to some distribution in each hidden state, and they are noted in the emission matrix B.

There are three main tasks that an HMM solves:

1. Decoding.Given the parameters A, π, B, and the observed data O, estimate the optimal sequence of hidden states Q;
2. Likelihood. Given an HMM $\lambda = (A, B)$ and a sequence of observations O, determine the probability that those observations belong to the HMM, $P(O|\lambda)$;

3. Learning. Given a sequence of observations O and a set of states in the HMM, we learn its parameters A and B.

In our work, the decoding task will be used to determine the hidden state given a group of observations, obtained from a discretization of the evolutionary factor. The learning task will be used in each iteration to learn the value of the B emission matrix, specifically.

4.3. HMM-PSO Integration

In [38], four inner-phases (or states) through which PSO moves are defined: exploration, exploitation, convergence, and jumping-out. A previous work describes these states as a Markov chain [93] (detailed in Figure 4). This chain corresponds to the hidden chain that will be deduced using decoding and learning tasks.

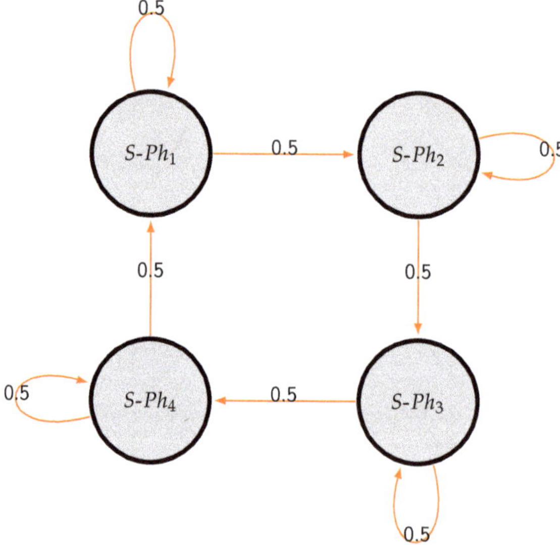

Figure 4. Evolutionary states defined for the Adaptive PSO algorithm: $S\text{-}Ph_{\{1,2,3,4\}}$ represent Exploration, Exploitation, Convergence, and Jump-out, respectively.

The HMM receives three input parameters: the first one is an initial probability vector π that computes a deterministic start in the exploration state: $\pi = [1, 0, 0, 0]$. The second parameter corresponds to the transition matrix between states A. As shown in figure, it is only possible to stay in the current state or to advance to the next state from left to right. As an initial value, all transitions have a probability of 0.5. The matrix A is defined as

$$A = \begin{bmatrix} 0.5 & 0.5 & 0 & 0 \\ 0 & 0.5 & 0.5 & 0 \\ 0 & 0 & 0.5 & 0.5 \\ 0.5 & 0 & 0 & 0.5 \end{bmatrix}$$

The type of hidden Markov model used in this work obtains its classifications from a group of observations belonging to a discrete alphabet. Therefore, we apply a discretization process to the evolutionary factor f of each inner-phase of the PSO. The discretization process used in this work is defined in [75] and corresponds to identifying the interval in which the calculated evolutionary factor belongs. The seven defined intervals are: $([0, 0.2], [0.2, 0.3], [0.3, 0.4], [0.4, 0.6], [0.6, 0.7], [0.7, 0.8], [0.8, 1])$.

The emission matrix B is the third parameter of the model. This matrix corresponds to the probability with which the elements of the alphabet of observations are emitted for each state. The emission matrix that we use in this work is defined in [75,94], and it is detailed as follows:

$$B = \begin{bmatrix} 0 & 0 & 0 & 0.5 & 0.25 & 0.25 & 0 \\ 0 & 0.25 & 0.25 & 0.5 & 0 & 0 & 0 \\ 2/3 & 1/3 & 0 & 0 & 0 & 0 & 0 \\ 0 & 0 & 0 & 0 & 0 & 1/3 & 2/3 \end{bmatrix}$$

The parameters π, A and B completely define the hmm model. Once defined, the model is capable of deducing hidden states—exploration, exploitation, convergence, jumping out—from the discretization of a chain of observations—evolutionary factor f—using the Viterbi algorithm. At each iteration, it is possible to adjust the parameters of the hmm model using task 2 with Baum–Welch's algorithm. For more details about the operation of both algorithms, please refer to [95].

5. Experimental Setup

In this section, we detail the proposal of integration of HMM in PSO, for the determination of the state and control of the parameters. This hybridization was tested on the set covering problem, which is a classic combinatorial optimization problem. One of the first works was proposed in [96], and it defines the Equation (4) as the formulation for the set covering problem:

$$\begin{aligned} &\texttt{minimize} \sum_{j=1}^{n} c_j x_j \\ &\text{subject to:} \\ &\sum_{j=1}^{n} a_{ij} x_j \geq 1 \quad \forall\, i \in M \\ &x_j \in \{0,1\} \quad \forall\, j \in N \end{aligned} \tag{4}$$

where c_j represents positive constants of the cost vector, and a_{ij} details binary values of the constraint matrix with M-rows and N-columns. If column j covers a row i, then $x_j = 1$. Otherwise, $x_j = 0$. We take the hardest instances of the set covering problem from the OR-Library [11].

To have an overview of the components involved in the search process, state identification and parameter control, Figures 5 and 6 show the flowchart of the algorithms.

Based on experimental analysis, the parameters are adjusted according to Table 1. Table 2 shows the initial parameter settings for the original PSO algorithm and our version. The initial values for the original PSO are the same as those used by the author. The initial values for the ω and np parameters on the hidden Markov model supporting the PSO algorithm (HPSO) come from [97].

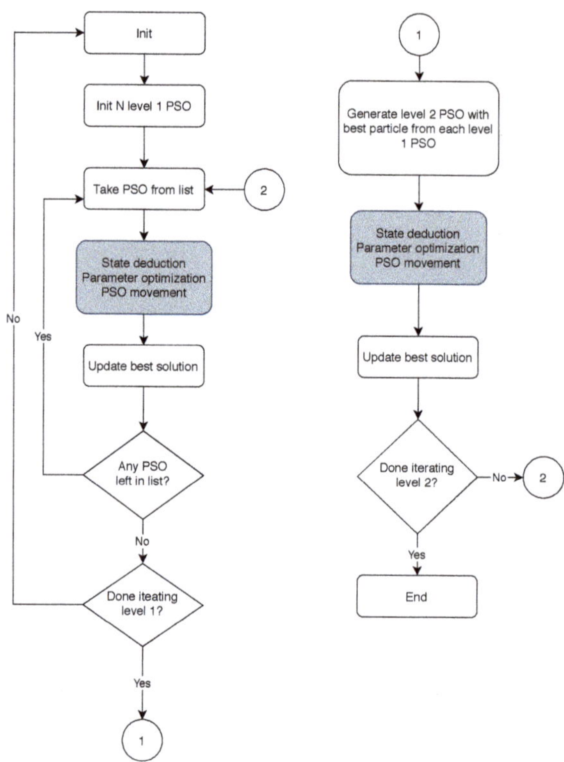

Figure 5. PSO algorithm state deduction integration.

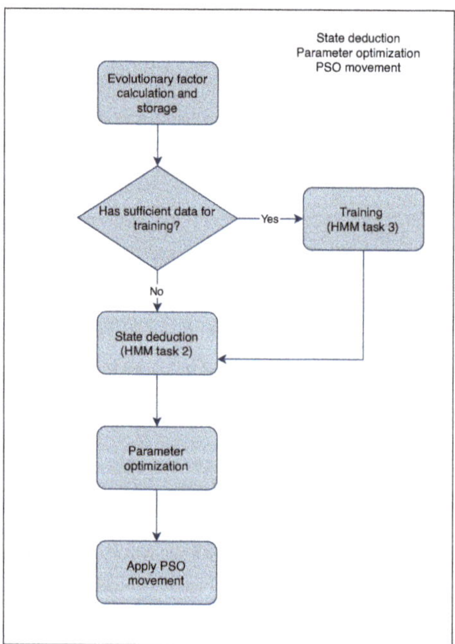

Figure 6. State deduction and parameter adjustment for PSO.

Table 1. Parameters update by identified state.

State (Inner-Phase)	Inertial Velocity w	Number of Particles np
Exploration	$\omega = \omega_{min} + (\omega_{max} - \omega_{min}) \cdot Rand(0,1)$	$np - 1$
Exploitation	$\omega = \frac{1}{1+\frac{3}{2}\exp^{-2.6f}}$	$np + 1$
Convergence	$\omega = \omega_{min}$	$np + 1$
Jump out	$\omega = \omega_{max}$	$np - 1$

Table 2. Initial configuration for input parameter values.

Orginal PSO Parameters		Proposed HPSO Parameters	
Parameter	**Value**	**Parameter**	**Value**
ω	$1 - \frac{k}{L+1}$	ω_{min}	0.4
		ω_{max}	0.9
np_{min}	5	np_{max}	30
np_{max}	50	np_{max}	30
$c1$	2.05 $rand(0,1)$	$c1$	2.05 $rand(0,1)$
$c2$	2.05 $rand(0,1)$	$c2$	2.05 $rand(0,1)$
iter. num.	50	iter. num.	50
iter. num.	250	iter. num.	250

6. Results and Discussion

In this section, we evaluated the functioning of our proposed HPSO. We compared our proposal against the original PSO. Then, we present a statistical comparison of the results obtained, and we illustrate the convergence of the search process and the percentages of exploration and exploitation.

Before integrating the adaptive approach, we analyzed the temporal complexity of the original PSO algorithm to evaluate that our proposal does not impact its performance. If we study each statement and expression, including control flows from the particle swarm optimization algorithm, we can state that time complexity is given by $(T \times np \times n)$, where T represents the maximum number of iterations, np stores is the number of particles (or solutions), and n is the dimension of each particle. In the worst case, the basic algorithm is upper bounded by $O(kn^2)$.

Then, performing a temporal analysis about our adaptive approach, we state that the temporal complexity of PSO is not altered. If we consider that: (a) this procedure operates in a determinate number of iterations (see Table 2); (b) it works with the same solutions; and (c) it runs in a way that is independent from the main algorithm, we can affirm that the upper bound is given by $(np \times n)$, which again, has an upper bound equal to $O(kn^2)$.

6.1. Original PSO Comparison

Table 3 shows the results obtained by our proposal and the original PSO in 11 hard instances of the set covering problem [68]. Each instance was executed 31 times and each run iterated 1000 cycles. These runs allow us to analyze the independence of the samples by determining the Z_{best}. The comparative includes the relative percentage distance (RPD). This value quantifies the deviation of the objective value Z_{best} from Z_{opt}, which is the minimal best-known value for each instance in our experiment, and it is computed as follows:

$$RDP = \left(\frac{Z_{best} - Z_{opt}}{Z_{opt}} \right) \quad (5)$$

Results show that the difference between both algorithms increases as the instance of the problem grows. The best results are highlighted with underline and maroon color.

For example, in the scp41, the best reached solution by HPSO overcomes than the classical PSO algorithm. The same strategy is used in all comparisons.

Table 3. Comparison of results between PSO and HPSO.

Instance	Optimum	Best HPSO	Best PSO	Avg. HPSO	Avg. PSO	RPD HPSO	RPD PSO
scp41	429	429	430	429.81	432.419	0	0.233
scp51	253	253	255	253.68	260.71	0	0.791
scp61	138	138	140	138.19	140.871	0	1.449
scpa1	253	253	256	254.32	258.097	0.395	1.186
scpb1	69	69	71	69	91.129	0	2.899
scpc1	227	227	234	228.36	238.258	0	3.084
scpd1	60	60	79	60.13	123.323	0	31.667
scpnre1	29	29	85	29	106.871	0	193.103
scpnrf1	14	14	39	14	49.29	0	178.571
scpnrg1	176	176	348	178.17	480.839	0.568	97.727
scpnrh1	63	65	277	65.25	349.452	1.587	339.683

Using the Wilcoxon–Mann–Whitney rank sum statistical test, we compare the results obtained by our proposal against the original PSO algorithm. It is valid to use this test because all runs are independent from each other and the results do not follow a normal distribution, since they are affected by pseudo-random numbers. Thirty-one samples of the obtained best fitness for 11 different instances of the set covering problem are compared. The test gives an p-value lower than 0.05 if it is possible to determine that one sample has statistically lower values than the other, and a value higher than 0.05 if not. Table 4 shows the comparison between the two algorithms.

Table 4. Statistical comparison.

Instance	HPSO $<$ PSO	PSO $<$ HPSO
scp41	0.728	0.277
scp51	0.002	0.998
scp61	0.000	1.000
scpa1	0.000	1.000
scpb1	0.000	1.000
scpc1	0.000	1.000
scpd1	0.000	1.000
scpnre1	0.000	1.000
scpnrf1	0.000	1.000
scpnrg1	0.000	1.000
scpnrh1	0.000	1.000

We can see that it was not possible to determine a statistical difference only for instance *scp41*, while for the other instances, the hypothesis that our algorithm improved the resolution process is confirmed.

6.2. Exploration/Exploitation Balance

In swarm intelligence methods, the population diversity is a measurement which evidences the performance of an algorithm, through the distribution of generated solutions [98,99]. This principle is significantly important to analyze the behavior of each solution in a swarm as well as the swarm as a whole. A recent work proposes a model based on the dimension-wise measurement to study the yield of algorithms [100]. The formulations that calculate this metric are defined in Equations (6) and (7):

$$Div(j) = \frac{1}{np} \sum_{i=1}^{np} |mean(x^j) - x_i^j| \tag{6}$$

$$Div = \frac{1}{D} \sum_{d=1}^{D} Div(d) \tag{7}$$

where $Div(j)$ describes the computed dimensional Hussain diversity over a solution x^j, $mean(x^j)$ represents the mean over each dimension j, np stores the number of solutions (population size). Finally, D saves the dimension size. After taking the dimension-wise distance of each swarm-individual i from the mean of the dimension j, we compute the average $Div(j)$ for all the individuals. Then, the average diversity of all dimensions is calculated in Div.

Using this fundamental, in [101], a model is proposed that allows to compute the evolution of the exploration and the exploitation effects obtained by the two algorithms, in each instance through all iterations. Resulting values represent percentages of exploration and exploitation on the population at iteration t. To calculate the exploration (XPL) balance, Equation (8) is applied, and to obtain the exploitation (XPLT) impacts, Equation (9) is employed:

$$XPL\% = \left(\frac{Div}{Div_{max}} \right) \times 100 \tag{8}$$

$$XPLT\% = \left(\frac{|Div - Div_{max}|}{Div_{max}} \right) \times 100 \tag{9}$$

For both equations, Div and Div_{max} represent the measures of diversity (distance) calculated over the population. Div represents the diversity of the full set of search agents through the aggregation of the diversity of each agent. However, Div_{max} represents the maximum value of diversity found. As can be intuited, the measurement of the percentage of exploration and exploitation varies depending on the measure of diversity used.

Figures 7 and 8 show the behavior of both algorithms. There are peaks, very noticeable in the original version and softer in our version. These peaks represent the change between inner-phases (the exploration and the exploitation processes).

6.3. Convergence Curves

We show plots for the convergence of the algorithms, PSO and HPSO, solving the set covering problem. Both algorithms reach a promising zone in the search space early on its execution. Convergence for both algorithms is very similar. Figures 9 and 10 shows the convergence achieved by both algorithms for their best execution on the instances scp41, scpa1, scpnre1, and scpnrh1.

6.4. Results Discussion

For the evaluation of the autonomous search method proposed, we used different measures that allowed us to evaluate the performance: a statistical comparison of the results of our algorithm against the original PSO, the variation of the evolutionary factor during the execution, the variation of the internal parameters of our algorithm, and the percentage of exploration and exploitation obtained in the search using the dimensional Hussain diversity measure.

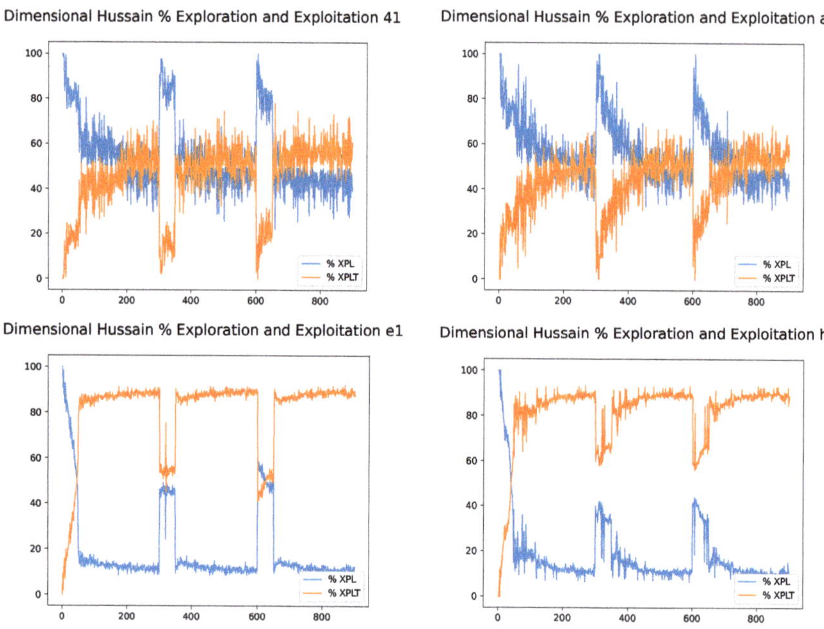

Figure 7. Exploration and exploitation percentage for original PSO. For small instances (scp41 and scpa1), the algorithm shows an exploitative behavior, for bigger instances (scpnre1 and scpnrh1), the algorithm shows an exploitative behavior. We can observe the transition between inner-phases at iterations 50, 300, and 600.

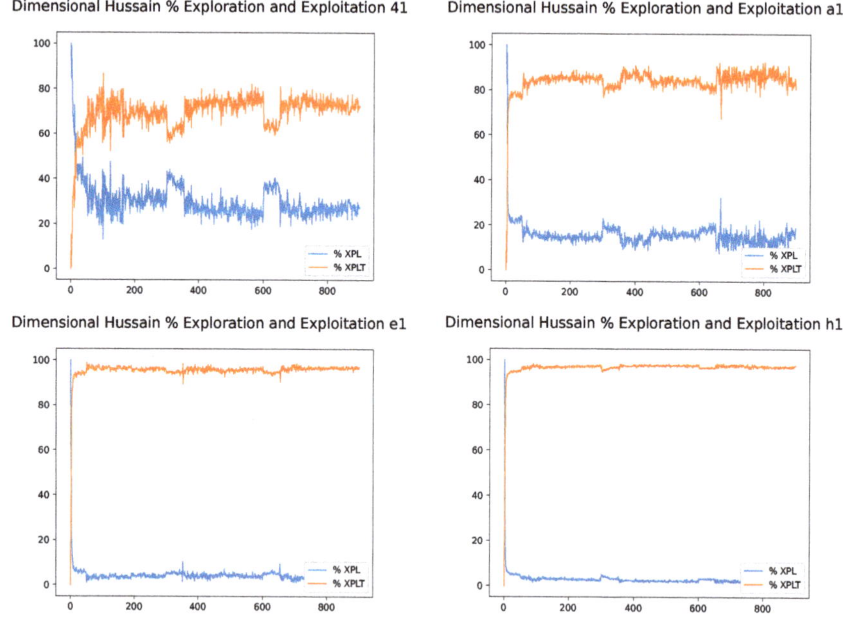

Figure 8. Exploration and exploitation percentage for HPSO. The algorithm shows a mostly exploitative behavior for small and big instances.

The statistical comparison of the results was conducted to determine whether there is an improvement to the original algorithm. Statistical tests confirm that there is a statistically distinguishable improvement when comparing the results of HPSO and PSO solving the combinatorial problem, for all tested instances.

The value of the evolutionary factor f shows a tendency to remain low, interrupted by sudden rises. Low values of f indicate that the PSO particles are close to each other and that the algorithm is converging. The higher value indicates that a particle found a solution with a better fit in an area far away from the group, which indicates that the PSO was able to avoid a local optimum.

The variation of the internal parameters shows an upward trend for the number of particles, and as the search progresses, new particles participate in the search. On the other hand, the inertia coefficient varies abruptly, going from the minimum value to the maximum value in a few iterations. This behavior did not affect the quality of the solutions; however, such an abrupt variation does not generate a recognizable pattern and the ω adjustment method must be reviewed.

The percentage of exploration and exploitation obtained shows that the algorithm maintains a mostly intensifying behavior, demonstrating that a promising area was found during the first iterations, maintaining the trend throughout the search. The transition between exploration and exploitation is more noticeable in smaller instances, which is explained by a smaller size of the search space. In general, the exploration and exploitation graphs show an efficient search.

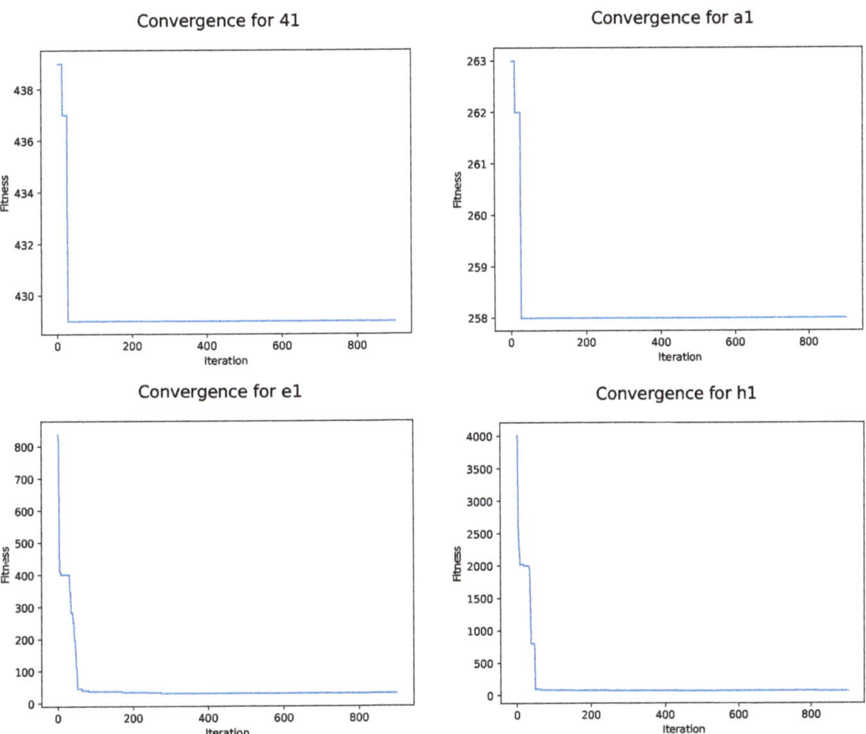

Figure 9. Convergence of PSO: The algorithm shows a premature convergence, with very few improvements after the first 50 iterations.

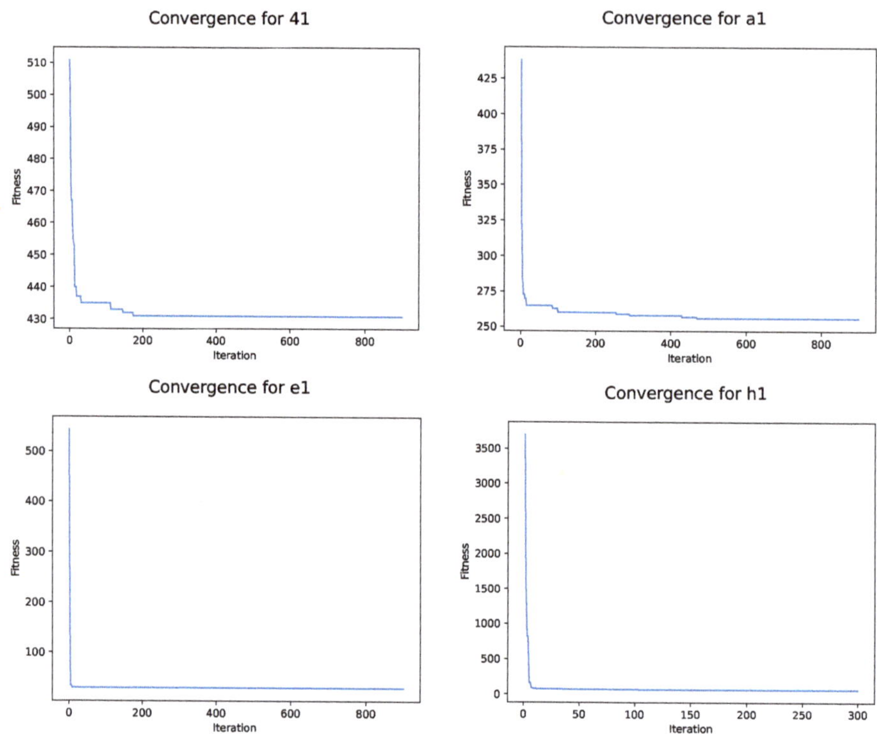

Figure 10. Convergence of HPSO. The algorithm shows improvements until approximately iteration 150, which represents 50% of the total iterations.

7. Conclusions and Future Work

In this work, we presented an autonomous search method for the PSO algorithm. This work was performed by hidden Markov models, which allow for the state identification of a PSO while the search process is running. The identification allows us to adjust PSO parameters based on a state deducted by the HMM. The deduction was made from the calculation of the evolutionary factor f metric, which gives information about the disposition of the particles inside PSO.

Different combinations of parameters to be adjusted for the PSO algorithm were evaluated, experimenting on a set of instances of the set covering problem and measuring the results. This experimentation showed that the combination of parameters w and np generates the best results. Then, the algorithm was compared against the original version of PSO without parameter control. The comparison of results was made using the Wilcoxom–Mann–Whitney statistical test, with the aim of testing the hypotheses posed for this work. The hypothesis was assumed and the parameter control shows a statistical difference in the quality of the solutions obtained. Moreover, we present figures that show the exploration and exploitation balance obtained by our proposal. If it is possible observe that the exploitation percentage increases compared to the original PSO. This behavior indicates that the HPSO was able to find better regions in the heuristic space, intensifying the search in those areas.

Future works consider verifying the impact on the classification of states when making changes to the transfer and binary functions in 0/1 optimization problems [102]. The discretizations made to the evolutionary factor f can also be adjusted, which will change the input data for the HMM model and its deductions. Finally, the PSO algorithm can be viewed as a framework for population-based metaheuristics, therefore testing with a different base algorithm is considered.

Author Contributions: Formal analysis, R.S., B.C.; investigation, R.S., B.C., R.O., and M.C.; methodology, R.S., R.O., and C.C.; resources, R.S. and B.C.; software, R.O. and M.C.; validation, B.C. and C.C.; writing—original draft, M.C., R.O.; writing—review and editing, R.S., and B.C., R.O., M.C., and C.C. All the authors of this paper are responsible for every part of this manuscript. All authors have read and agreed to the published version of the manuscript.

Funding: Ricardo Soto is supported by Grant CONICYT/FONDECYT/REGULAR/1190129. Broderick Crawford is supported by Grant ANID/FONDECYT/REGULAR/1210810.

Institutional Review Board Statement: Not applicable.

Informed Consent Statement: Not applicable.

Data Availability Statement: Not applicable.

Conflicts of Interest: The authors declare no conflict of interest. The founding sponsors had no role in the design of the study; in the collection, analyses, or interpretation of data; in the writing of the manuscript, and in the decision to publish the results.

References

1. Mavrovouniotis, M.; Li, C.; Yang, S. A survey of swarm intelligence for dynamic optimization: Algorithms and applications. *Swarm Evol. Comput.* **2017**, *33*, 1–17. [CrossRef]
2. Gill, S.S.; Buyya, R. Bio-inspired algorithms for big data analytics: A survey, taxonomy, and open challenges. In *Big Data Analytics for Intelligent Healthcare Management*; Elsevier: Amsterdam, The Netherlands, 2019; pp. 1–17.
3. Huang, C.; Li, Y.; Yao, X. A survey of automatic parameter tuning methods for metaheuristics. *IEEE Trans. Evol. Comput.* **2019**, *24*, 201–216. [CrossRef]
4. Nayyar, A.; Nguyen, N.G. Introduction to swarm intelligence. In *Advances in Swarm Intelligence for Optimizing Problems in Computer Science*; Chapman and Hall/CRC: Boca Raton, FL, USA, 2018; pp. 53–78.
5. Mejía-de Dios, J.A.; Mezura-Montes, E.; Quiroz-Castellanos, M. Automated parameter tuning as a bilevel optimization problem solved by a surrogate-assisted population-based approach. *Appl. Intell.* **2021**, 1–23. [CrossRef]
6. Talbi, E.G. *Metaheuristics: From Design to Implementation*; John Wiley & Sons: Hoboken, NJ, USA, 2009; Volume 74.
7. Stutzle, T.; Lopez-Ibanez, M.; Pellegrini, P.; Maur, M.; Montes de Oca, M.; Birattari, M.; Dorigo, M. Parameter adaptation in ant colony optimization. In *Autonomous Search*; Springer: Berlin/Heidelberg, Germany, 2011; pp. 191–215. [CrossRef]
8. Soto, R.; Crawford, B.; Aste Toledo, A.; Castro, C.; Paredes, F.; Olivares, R. Solving the manufacturing cell design problem through binary cat swarm optimization with dynamic mixture ratios. *Comput. Intell. Neurosci.* **2019**, *2019*, 4787856. [CrossRef]
9. Hamadi, Y. Autonomous search. In *Combinatorial Search: From Algorithms to Systems*; Springer: Berlin/Heidelberg, Germany, 2013; pp. 99–122.
10. Wang, D.; Tan, D.; Liu, L. Particle swarm optimization algorithm: An overview. *Soft Comput.* **2017**, *22*, 387–408. [CrossRef]
11. Beasley, J. OR-Library: Distributing test problems by electronic mail. *J. Oper. Res. Soc.* **1990**, *41*, 1069–1072. [CrossRef]
12. Darwish, A. Bio-inspired computing: Algorithms review, deep analysis, and the scope of applications. *Future Comput. Inform. J.* **2018**, *3*, 231–246. [CrossRef]
13. Nguyen, B.H.; Xue, B.; Zhang, M. A survey on swarm intelligence approaches to feature selection in data mining. *Swarm Evol. Comput.* **2020**, *54*, 100663. [CrossRef]
14. Khan, T.A.; Ling, S.H. A survey of the state-of-the-art swarm intelligence techniques and their application to an inverse design problem. *J. Comput. Electron.* **2020**, *19*, 1606–1628. [CrossRef]
15. Shaikh, P.W.; El Abd, M.; Khanafer, M.; Gao, K. A Review on Swarm Intelligence and Evolutionary Algorithms for Solving the Traffic Signal Control Problem. *IEEE Trans. Intell. Transp. Syst.* **2020**, 1–16. [CrossRef]
16. Tzanetos, A.; Dounias, G. A Comprehensive Survey on the Applications of Swarm Intelligence and Bio-Inspired Evolutionary Strategies. In *Learning and Analytics in Intelligent Systems*; Springer International Publishing: Berlin/Heidelberg, Germany, 2020; pp. 337–378. [CrossRef]
17. Gendreau, M.; Potvin, J.Y. *Handbook of Metaheuristics*; Springer: Berlin/Heidelberg, Germany, 2010; Volume 2.
18. Dokeroglu, T.; Sevinc, E.; Kucukyilmaz, T.; Cosar, A. A survey on new generation metaheuristic algorithms. *Comput. Ind. Eng.* **2019**, *137*, 106040. [CrossRef]
19. Eiben, Á.E.; Hinterding, R.; Michalewicz, Z. Parameter control in evolutionary algorithms. *IEEE Trans. Evol. Comput.* **1999**, *3*, 124–141. [CrossRef]
20. Mohamed, M.A.; Eltamaly, A.M.; Alolah, A.I. Swarm intelligence-based optimization of grid-dependent hybrid renewable energy systems. *Renew. Sustain. Energy Rev.* **2017**, *77*, 515–524. [CrossRef]

21. Oliveto, P.S.; Paixão, T.; Pérez, J.; Sudholt, D.; Trubenová, B. How to Escape Local Optima in Black Box Optimisation: When Non-elitism Outperforms Elitism. *Algorithmica* **2017**, *80*, 1604–1633. [CrossRef] [PubMed]
22. Qi, X.; Ju, G.; Xu, S. Efficient solution to the stagnation problem of the particle swarm optimization algorithm for phase diversity. *Appl. Opt.* **2018**, *57*, 2747. [CrossRef] [PubMed]
23. Hamadi, Y.; Monfroy, E.; Saubion, F. What is autonomous search? In *Hybrid Optimization*; Springer: Berlin/Heidelberg, Germany, 2011; pp. 357–391.
24. Jong, K.D. Parameter Setting in EAs: A 30 Year Perspective. In *Parameter Setting in Evolutionary Algorithms*; Springer: Berlin/Heidelberg, Germany, 2007; pp. 1–18. [CrossRef]
25. Shi, Y.; Eberhart, R.C. Empirical study of particle swarm optimization. In Proceedings of the 1999 Congress on Evolutionary Computation-CEC99 (Cat. No. 99TH8406), Washington, DC, USA, 6–9 July 1999; IEEE: Piscataway, NJ, USA, 1999; Volume 3, pp. 1945–1950.
26. Ratnaweera, A.; Halgamuge, S.K.; Watson, H.C. Self-organizing hierarchical particle swarm optimizer with time-varying acceleration coefficients. *IEEE Trans. Evol. Comput.* **2004**, *8*, 240–255. [CrossRef]
27. Xu, G. An adaptive parameter tuning of particle swarm optimization algorithm. *Appl. Math. Comput.* **2013**, *219*, 4560–4569. [CrossRef]
28. Wang, F.; Zhang, H.; Li, K.; Lin, Z.; Yang, J.; Shen, X.L. A hybrid particle swarm optimization algorithm using adaptive learning strategy. *Inf. Sci.* **2018**, *436–437*, 162–177. [CrossRef]
29. Cao, Y.; Zhang, H.; Li, W.; Zhou, M.; Zhang, Y.; Chaovalitwongse, W.A. Comprehensive Learning Particle Swarm Optimization Algorithm with Local Search for Multimodal Functions. *IEEE Trans. Evol. Comput.* **2019**, *23*, 718–731. [CrossRef]
30. Yang, X.; Yuan, J.; Yuan, J.; Mao, H. A modified particle swarm optimizer with dynamic adaptation. *Appl. Math. Comput.* **2007**, *189*, 1205–1213. [CrossRef]
31. Wu, Z.; Zhou, J. A self-adaptive particle swarm optimization algorithm with individual coefficients adjustment. In Proceedings of the 2007 International Conference on Computational Intelligence and Security (CIS 2007), Harbin, China, 15–19 December 2007; IEEE: Piscataway, NJ, USA, 2007; pp. 133–136.
32. Li, Z.; Tan, G. A self-adaptive mutation-particle swarm optimization algorithm. In Proceedings of the 2008 Fourth International Conference on Natural Computation, Jinan, China, 18–20 October 2008; IEEE: Piscataway, NJ, USA, 2008; Volume 1, pp. 30–34.
33. Li, X.; Fu, H.; Zhang, C. A self-adaptive particle swarm optimization algorithm. In Proceedings of the 2008 International Conference on Computer Science and Software Engineering, Wuhan, China, 12–14 December; IEEE: Piscataway, NJ, USA, 2008; Volume 5, pp. 186–189.
34. Dong, C.; Wang, G.; Chen, Z.; Yu, Z. A method of self-adaptive inertia weight for PSO. In Proceedings of the 2008 International Conference on Computer Science and Software Engineering, Wuhan, China, 12–14 December; IEEE: Piscataway, NJ, USA, 2008; Volume 1, pp. 1195–1198.
35. Chen, H.H.; Li, G.Q.; Liao, H.L. A self-adaptive improved particle swarm optimization algorithm and its application in available transfer capability calculation. In Proceedings of the 2009 Fifth International Conference on Natural Computation, Tianjin, China, 14–16 August 2009; IEEE: Piscataway, NJ, USA, 2009; Volume 3, pp. 200–205.
36. Nickabadi, A.; Ebadzadeh, M.M.; Safabakhsh, R. A novel particle swarm optimization algorithm with adaptive inertia weight. *Appl. Soft Comput.* **2011**, *11*, 3658–3670. [CrossRef]
37. Tanweer, M.R.; Suresh, S.; Sundararajan, N. Self regulating particle swarm optimization algorithm. *Inf. Sci.* **2015**, *294*, 182–202. [CrossRef]
38. Zhan, Z.H.; Zhang, J.; Li, Y.; Chung, H.S.H. Adaptive particle swarm optimization. *IEEE Trans. Syst. Ma, Cybern Part B (Cybern.)* **2009**, *39*, 1362–1381. [CrossRef] [PubMed]
39. Leu, M.S.; Yeh, M.F. Grey particle swarm optimization. *Appl. Soft Comput.* **2012**, *12*, 2985–2996. [CrossRef]
40. Julong, D. Introduction to grey system theory. *J. Grey Syst.* **1989**, *1*, 1–24.
41. Calvet, L.; Armas, J.; Masip, D.; Juan, A.A. Learnheuristics: Hybridizing metaheuristics with machine learning for optimization with dynamic inputs. *Open Math.* **2017**, *15*, 261–280. [CrossRef]
42. Singh, P.; Singh, S. Energy efficient clustering protocol based on improved metaheuristic in wireless sensor networks. *J. Netw. Comput. Appl.* **2017**, *83*, 40–52. [CrossRef]
43. Alvarenga, R.D.; Machado, A.M.; Ribeiro, G.M.; Mauri, G.R. A mathematical model and a Clustering Search metaheuristic for planning the helicopter transportation of employees to the production platforms of oil and gas. *Comput. Ind. Eng.* **2016**, *101*, 303–312. [CrossRef]
44. Kuo, R.J.; Kuo, P.H.; Chen, Y.R.; Zulvia, F.E. Application of metaheuristics-based clustering algorithm to item assignment in a synchronized zone order picking system. *Appl. Soft Comput.* **2016**, *46*, 143–150. [CrossRef]
45. Kuo, R.J.; Mei, C.H.; Zulvia, F.E.; Tsai, C.Y. An application of a metaheuristic algorithm-based clustering ensemble method to APP customer segmentation. *Neurocomputing* **2016**, *205*, 116–129. [CrossRef]
46. Fong, S.; Wong, R.; Vasilakos, A. Accelerated PSO Swarm Search Feature Selection for Data Stream Mining Big Data. *IEEE Trans. Serv. Comput.* **2015**, *9*, 33–45. [CrossRef]
47. Chou, J.S.; Putra, J.P. Metaheuristic optimization within machine learning-based classification system for early warnings related to geotechnical problems. *Autom. Constr.* **2016**, *68*, 65–80. [CrossRef]

48. Al-Obeidat, F.; Belacel, N.; Spencer, B. Combining Machine Learning and Metaheuristics Algorithms for Classification Method PROAFTN. In *Enhanced Living Environments;* Lecture Notes in Computer Science; Springer International Publishing: Berlin/Heidelberg, Germany, 2019; pp. 53–79. [CrossRef]
49. Chou, J.S.; Nguyen, T.K. Forward Forecast of Stock Price Using Sliding-Window Metaheuristic-Optimized Machine-Learning Regression. *IEEE Trans. Ind. Inform.* **2018**, *14*, 3132–3142. [CrossRef]
50. He, S.; Li, Z.; Tang, Y.; Liao, Z.; Li, F.; Lim, S.J. Parameters Compressing in Deep Learning. *Comput. Mater. Contin.* **2020**, *62*, 321–336. [CrossRef]
51. Hashemi, A.B.; Meybodi, M. Adaptive parameter selection scheme for PSO: A learning automata approach. In Proceedings of the 2009 14th International CSI Computer Conference, Tehran, Iran, 1–2 July 2009; IEEE: Piscataway, NJ, USA, 2009; pp. 403–411.
52. Ries, J.; Beullens, P. A semi-automated design of instance-based fuzzy parameter tuning for metaheuristics based on decision tree induction. *J. Oper. Res. Soc.* **2015**, *66*, 782–793. [CrossRef]
53. Salcedo-Sanz, S.; Yao, X. A Hybrid Hopfield Network-Genetic Algorithm Approach for the Terminal Assignment Problem. *IEEE Trans. Syst. Man Cybern. Part B (Cybern.)* **2004**, *34*, 2343–2353. [CrossRef] [PubMed]
54. Zeng, B.; Li, X.; Gao, L.; Zhang, Y.; Dong, H. Whale swarm algorithm with the mechanism of identifying and escaping from extreme points for multimodal function optimization. *Neural Comput. Appl.* **2019**, *32*, 5071–5091. [CrossRef]
55. Sun, Y.; Gao, Y. A Multi-Objective Particle Swarm Optimization Algorithm Based on Gaussian Mutation and an Improved Learning Strategy. *Mathematics* **2019**, *7*, 148. [CrossRef]
56. Olivares, R.; Munoz, R.; Soto, R.; Crawford, B.; Cárdenas, D.; Ponce, A.; Taramasco, C. An Optimized Brain-Based Algorithm for Classifying Parkinson's Disease. *Appl. Sci.* **2020**, *10*, 1827. [CrossRef]
57. Liu, J.; Wang, W.; Chen, J.; Sun, G.; Yang, A. Classification and Research of Skin Lesions Based on Machine Learning. *Comput. Mater. Contin.* **2020**, *62*, 1187–1200. [CrossRef]
58. Haoxiang, S.; Changxing, C.; Yunfei, L.; Mu, Y. Cooperative perception optimization based on self-checking machine learning. *Comput. Mater. Contin.* **2020**, *62*, 747–761. [CrossRef]
59. Zhou, S.; Chen, L.; Sugumaran, V. Hidden Two-Stream Collaborative Learning Network for Action Recognition. *Comput. Mater. Contin.* **2020**, *63*, 1545–1561. [CrossRef]
60. Zhou, S.; Tan, B. Electrocardiogram soft computing using hybrid deep learning CNN-ELM. *Appl. Soft Comput.* **2020**, *86*, 105778. [CrossRef]
61. Munoz, R.; Olivares, R.; Taramasco, C.; Villarroel, R.; Soto, R.; Alonso-Sánchez, M.F.; Merino, E.; de Albuquerque, V.H.C. A new EEG software that supports emotion recognition by using an autonomous approach. *Neural Comput. Appl.* **2018**, *32*, 11111–11127. [CrossRef]
62. Munoz, R.; Olivares, R.; Taramasco, C.; Villarroel, R.; Soto, R.; Barcelos, T.S.; Merino, E.; Alonso-Sánchez, M.F. Using Black Hole Algorithm to Improve EEG-Based Emotion Recognition. *Comput. Intell. Neurosci.* **2018**, *2018*, 3050214. [CrossRef]
63. Gui, Y.; Zeng, G. Joint learning of visual and spatial features for edit propagation from a single image. *Vis. Comput.* **2019**, *36*, 469–482. [CrossRef]
64. Santos, M.A.; Munoz, R.; Olivares, R.; Filho, P.P.R.; Ser, J.D.; de Albuquerque, V.H.C. Online heart monitoring systems on the internet of health things environments: A survey, a reference model and an outlook. *Inf. Fusion* **2020**, *53*, 222–239. [CrossRef]
65. Díaz, F.D.; Lasheras, F.S.; Moreno, V.; Moratalla-Navarro, F.; de la Torre, A.J.M.; Sánchez, V.M. GASVeM: A New Machine Learning Methodology for Multi-SNP Analysis of GWAS Data Based on Genetic Algorithms and Support Vector Machines. *Mathematics* **2021**, *9*, 654. [CrossRef]
66. Minonzio, J.G.; Cataldo, B.; Olivares, R.; Ramiandrisoa, D.; Soto, R.; Crawford, B.; Albuquerque, V.H.C.D.; Munoz, R. Automatic Classifying of Patients With Non-Traumatic Fractures Based on Ultrasonic Guided Wave Spectrum Image Using a Dynamic Support Vector Machine. *IEEE Access* **2020**, *8*, 194752–194764. [CrossRef]
67. Streichert, F.; Stein, G.; Ulmer, H.; Zell, A. A Clustering Based Niching Method for Evolutionary Algorithms. In *Genetic and Evolutionary Computation—GECCO 2003*; Springer: Berlin/Heidelberg, Germany, 2003; pp. 644–645. [CrossRef]
68. Valdivia, S.; Soto, R.; Crawford, B.; Caselli, N.; Paredes, F.; Castro, C.; Olivares, R. Clustering-Based Binarization Methods Applied to the Crow Search Algorithm for 0/1 Combinatorial Problems. *Mathematics* **2020**, *8*, 1070. [CrossRef]
69. Santos, H.G.; Ochi, L.S.; Marinho, E.H.; Drummond, L.M. Combining an evolutionary algorithm with data mining to solve a single-vehicle routing problem. *Neurocomputing* **2006**, *70*, 70–77. [CrossRef]
70. Jin, Y.; Qu, R.; Atkin, J. A Population-Based Incremental Learning Method for Constrained Portfolio Optimisation. In Proceedings of the 2014 16th International Symposium on Symbolic and Numeric Algorithms for Scientific Computing, Timisoara, Romania, 22–25 September 2014; IEEE: Piscataway, NJ, USA, 2014; p. 7031476. [CrossRef]
71. Nurcahyadi, T.; Blum, C. Adding Negative Learning to Ant Colony Optimization: A Comprehensive Study. *Mathematics* **2021**, *9*, 361. [CrossRef]
72. Rasmussen, T.K.; Krink, T. Improved Hidden Markov Model training for multiple sequence alignment by a particle swarm optimization—Evolutionary algorithm hybrid. *Biosystems* **2003**, *72*, 5–17. [CrossRef]
73. Prakash, A.; Chandrasekar, C. An Optimized Multiple Semi-Hidden Markov Model for Credit Card Fraud Detection. *Indian J. Sci. Technol.* **2015**, *8*, 165. [CrossRef]

74. Xue, L.; Yin, J.; Ji, Z.; Jiang, L. A particle swarm optimization for hidden Markov model training. In Proceedings of the 2006 8th International Conference on Signal Processing, Guilin, China, 16–20 November 2006; IEEE: Piscataway, NJ, USA, 2006; Volume 1, p. 345542.
75. Aoun, O.; Sarhani, M.; El Afia, A. Hidden markov model classifier for the adaptive particle swarm optimization. In *Recent Developments in Metaheuristics*; Operations Research/Computer Science Interfaces Series; Springer: Berlin/Heidelberg, Germany, 2018; pp. 1–15.
76. Wang, X.; Chen, H.; Heidari, A.A.; Zhang, X.; Xu, J.; Xu, Y.; Huang, H. Multi-population following behavior-driven fruit fly optimization: A Markov chain convergence proof and comprehensive analysis. *Knowl. Based Syst.* **2020**, *210*, 106437. [CrossRef]
77. Motiian, S.; Soltanian-Zadeh, H. Improved particle swarm optimization and applications to Hidden Markov Model and Ackley function. In Proceedings of the 2011 IEEE International Conference on Computational Intelligence for Measurement Systems and Applications (CIMSA) Proceedings, Ottawa, AB, Canada, 19–21 September 2011; IEEE: Piscataway, NJ, USA, 2011; p. 6045560. [CrossRef]
78. Sagayam, K.M.; Hemanth, D.J. ABC algorithm based optimization of 1-D hidden Markov model for hand gesture recognition applications. *Comput. Ind.* **2018**, *99*, 313–323. [CrossRef]
79. Trindade, Á.R.; Campelo, F. Tuning metaheuristics by sequential optimisation of regression models. *Appl. Soft Comput.* **2019**, *85*, 105829. [CrossRef]
80. Wei, Z.; Yong, Z.; Chen, L.; Lei, G.; Wenpei, Z. An improved particle swarm optimization algorithm and its application on distribution generation accessing to distribution network. In Proceedings of the IOP Conference Series: Earth and Environmental Science, Malang, Indonesia, 12–13 March 2019; Volume 342, p. 012011. [CrossRef]
81. Crawford, B.; Soto, R.; Monfroy, E.; Palma, W.; Castro, C.; Paredes, F. Parameter tuning of a choice-function based hyperheuristic using particle swarm optimization. *Expert Syst. Appl.* **2013**, *40*, 1690–1695. [CrossRef]
82. Pellegrini, P.; Stützle, T.; Birattari, M. A critical analysis of parameter adaptation in ant colony optimization. *Swarm Intell.* **2012**, *6*, 23–48. [CrossRef]
83. Beni, G.; Wang, J. Swarm intelligence in cellular robotic systems. In *Robots and Biological Systems: Towards a New Bionics?* Springer: Berlin/Heidelberg, Germany, 1993; pp. 703–712.
84. Beni, G. Swarm intelligence. In *Complex Social and Behavioral Systems: Game Theory and Agent-Based Models*; 2020; pp. 791–818. Available online: https://www.springer.com/gp/book/9781071603673 (accessed on 25 May 2021).
85. Zhu, H.; Wang, Y.; Ma, Z.; Li, X. A Comparative Study of Swarm Intelligence Algorithms for UCAV Path-Planning Problems. *Mathematics* **2021**, *9*, 171. [CrossRef]
86. Freitas, D.; Lopes, L.G.; Morgado-Dias, F. Particle Swarm Optimisation: A Historical Review Up to the Current Developments. *Entropy* **2020**, *22*, 362. [CrossRef]
87. Khare, A.; Rangnekar, S. A review of particle swarm optimization and its applications in Solar Photovoltaic system. *Appl. Soft Comput.* **2013**, *13*, 2997–3006. [CrossRef]
88. Kennedy, J.; Eberhart, R. Particle swarm optimization. In Proceedings of the ICNN'95-International Conference on Neural Networks, Perth, Australia, 27 November–1 December 1995; IEEE: Piscataway, NJ, USA, 1995; Volume 4, pp. 1942–1948.
89. Erwin, K.; Engelbrecht, A. Diversity Measures for Set-Based Meta-Heuristics. In Proceedings of the 2020 7th International Conference on Soft Computing & Machine Intelligence (ISCMI), Stockholm, Sweden, 14–15 November 2020; IEEE: Piscataway, NJ, USA, 2020; pp. 45–50.
90. Gavira-Durón, N.; Gutierrez-Vargas, O.; Cruz-Aké, S. Markov Chain K-Means Cluster Models and Their Use for Companies' Credit Quality and Default Probability Estimation. *Mathematics* **2021**, *9*, 879. [CrossRef]
91. Naranjo, L.; Esparza, L.J.R.; Pérez, C.J. A Hidden Markov Model to Address Measurement Errors in Ordinal Response Scale and Non-Decreasing Process. *Mathematics* **2020**, *8*, 622. [CrossRef]
92. Koike, T.; Hofert, M. Markov Chain Monte Carlo Methods for Estimating Systemic Risk Allocations. *Risks* **2020**, *8*, 6. [CrossRef]
93. El Afia, A.; Sarhani, M.; Aoun, O. Hidden markov model control of inertia weight adaptation for Particle swarm optimization. *IFAC-PapersOnLine* **2017**, *50*, 9997–10002. [CrossRef]
94. El Afia, A.; Aoun, O.; Garcia, S. Adaptive cooperation of multi-swarm particle swarm optimizer-based hidden Markov model. *Prog. Artif. Intell.* **2019**, *8*, 441–452. [CrossRef]
95. Rabiner, L.; Juang, B. An introduction to hidden Markov models. *IEEE ASSP Mag.* **1986**, *3*, 4–16. [CrossRef]
96. Beasley, J. An algorithm for set covering problem. *Eur. J. Oper. Res.* **1987**, *31*, 85–93. [CrossRef]
97. Harrison, K.R.; Engelbrecht, A.P.; Ombuki-Berman, B.M. Self-adaptive particle swarm optimization: A review and analysis of convergence. *Swarm Intell.* **2018**, *12*, 187–226. [CrossRef]
98. Hussain, K.; Salleh, M.N.M.; Cheng, S.; Shi, Y. On the exploration and exploitation in popular swarm-based metaheuristic algorithms. *Neural Comput. Appl.* **2018**, *31*, 7665–7683. [CrossRef]
99. Mattiussi, C.; Waibel, M.; Floreano, D. Measures of Diversity for Populations and Distances Between Individuals with Highly Reorganizable Genomes. *Evol. Comput.* **2004**, *12*, 495–515. [CrossRef]
100. Cheng, S.; Shi, Y.; Qin, Q.; Zhang, Q.; Bai, R. Population Diversity Maintenance In Brain Storm Optimization Algorithm. *J. Artif. Intell. Soft Comput. Res.* **2014**, *4*, 83–97. [CrossRef]

101. Morales-Castañeda, B.; Zaldivar, D.; Cuevas, E.; Fausto, F.; Rodríguez, A. A better balance in metaheuristic algorithms: Does it exist? *Swarm Evol. Comput.* **2020**, 54, 100671. [CrossRef]
102. Crawford, B.; Soto, R.; Astorga, G.; García, J.; Castro, C.; Paredes, F. Putting Continuous Metaheuristics to Work in Binary Search Spaces. *Complexity* **2017**, *2017*, 8404231. [CrossRef]

Article

Mixed-Integer Linear Programming Model and Heuristic for Short-Term Scheduling of Pressing Process in Multi-Layer Printed Circuit Board Manufacturing

Teeradech Laisupannawong [1], Boonyarit Intiyot [1,*] and Chawalit Jeenanunta [2]

[1] Department of Mathematics and Computer Science, Faculty of Science, Chulalongkorn University, Bangkok 10330, Thailand; teeradech.lai@gmail.com

[2] School of Management Technology, Sirindhorn International Institute of Technology (SIIT), Thammasat University, Pathum Thani 12120, Thailand; chawalit@siit.tu.ac.th

* Correspondence: boonyarit.i@chula.ac.th

Abstract: The main stages of printed circuit board (PCB) manufacturing are the design, fabrication, assembly, and testing. This paper focuses on the scheduling of the pressing process, which is a part of the fabrication process of a multi-layer PCB and is a new application since it has never been investigated in the literature. A novel mixed-integer linear programming (MILP) formulation for short-term scheduling of the pressing process is presented. The objective function is to minimize the makespan of the overall process. Moreover, a three-phase-PCB-pressing heuristic (3P-PCB-PH) for short-term scheduling of the pressing process is also presented. To illustrate the proposed MILP model and 3P-PCB-PH, the test problems generated from the real data acquired from a PCB company are solved. The results show that the proposed MILP model can find an optimal schedule for all small- and medium-sized problems but can do so only for some large-sized problems using the CPLEX solver within a time limit of 2 h. However, the proposed 3P-PCB-PH could find an optimal schedule for all problems that the MILP could find using much less computational time. Furthermore, it can also quickly find a near-optimal schedule for other large-sized problems that the MILP could not solved optimally.

Keywords: pressing process; printed circuit board; scheduling; mixed-integer linear programming; heuristic

Citation: Laisupannawong, T.; Intiyot, B.; Jeenanunta, C. Mixed-Integer Linear Programming Model and Heuristic for Short-Term Scheduling of Pressing Process in Multi-Layer Printed Circuit Board Manufacturing. *Mathematics* **2021**, *9*, 653. https://doi.org/10.3390/math9060653

Academic Editor: Frank Werner

Received: 9 February 2021
Accepted: 16 March 2021
Published: 18 March 2021

1. Introduction

A printed circuit board (PCB) is a major component in most electronics, such as televisions, mobile phones, digital cameras, computers, and medical devices. The manufacturing of PCBs has become a competitive industry due to the increased demand for electronic products. The PCBs can be classified into three types, according to the number of their layers, as single-layer PCBs, double-layer PCBs, and multi-layer PCBs.

According to Khandpur [1], PCB manufacturing consists of the design, fabrication, assembly, and testing. The PCB design is the process of creating a circuit schematic by PCB designers. Then, PCB fabrication is the process of constructing the PCB (bare board) before placing electronic components in the PCB assembly. The fabrication of each type of PCB is different. In this paper, we consider only the fabrication of multi-layer PCBs. As stated in Reference [1], the main materials used in multi-layer PCB fabrication include the copper-clad laminate sheets and prepregs. The fabrication of multi-layer PCBs can be summarized in the following five steps:

1. The laminate sheets are cut to the required size in the cutting process.
2. The circuit pattern is created on the cut laminate in the etching process.
3. A number of etched laminates (or cores) are stacked together with a prepreg inserted between each pair of them. The stack (or panel) is pressed using heat and pressure in the pressing process.

4. Holes will be drilled in the pressed board in the drilling process and the circuit pattern will be made on the outer surfaces.
5. The remaining steps are the quality control and labeling processes.

Figure 1 shows a schematic representation of the steps in multi-layer PCB fabrication. A major cost-consuming process in multi-layer PCB fabrication is the cutting process. Most PCB companies aim to cut the laminates so that the waste areas from cutting the laminates are minimized. This process could be formulated as a two-dimensional cutting stock problem (2DCSP). The drilling process is another time-consuming process in the multi-layer PCB fabrication. Most PCB companies aim to find an optimal path for drilling the holes in the designed positions in the circuit pattern so that the travel time or distance of the drilling device is minimized, and so the overall processing time is reduced. A mathematical problem that relates to the drilling process is the hole drill routing optimization problem (HDROP).

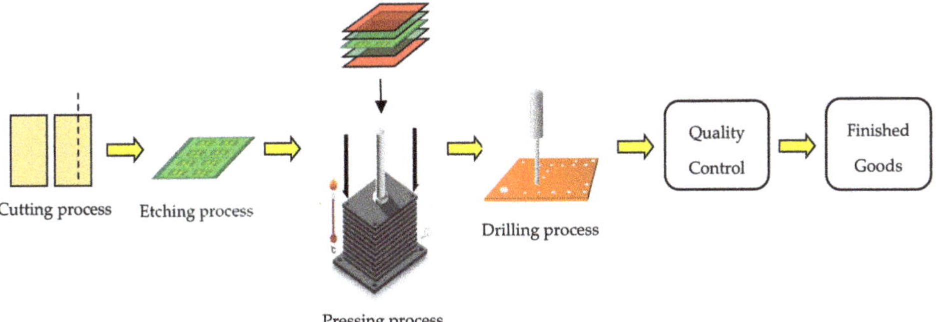

Figure 1. The steps of multi-layer printed circuit board (PCB) fabrication.

There have been many research studies reported on the 2DCSP and HDROP, where diverse techniques have been used to solve the 2DCSP, such as an integer linear programming model using a column generation technique [2], an exact arc-flow model [3], a branch-and-price algorithm [4], and heuristic algorithms based on column generation [5,6]. There are some reports on the cutting process that have used real data from PCB companies, such as in References [7,8]. As for the HDROP, numerous research studies have been developed to solve it, such as a particle swarm optimization (PSO) [9], an ant colony system [10], a cuckoo search algorithm [11], and a hybridized cuckoo search-genetic algorithm [12].

The PCB assembly is the process of placing electronic components, such as resistors, capacitors, and transistors, at the specified location on a bare board. In a PCB assembly line, there are many placement machines with different unit assembly times for the same component. A board is passed through all the machines to complete the component placement. Therefore, the components should be allocated to appropriate machines so that the assembly time is minimized. This leads to the problem of getting an optimal workload balance in the PCB assembly line [13–16]. The aim of this problem is to minimize the production cycle time of the assembly line for a given PCB type, which is the maximum time needed by one of the placement machines. Some techniques have been proposed to solve this problem, such as using a genetic algorithm [13] and a branch-and-bound-based optimization algorithm [14]. Some extended problems with additional constraints can be found in the literature, such as the use of feeder modules, precedence constraints between components, and feeder duplications [15], as well as an integrated workload balancing and single-machine optimization problem [16].

The testing is the process after assembling all components to the board. Environmental stress-screening chambers are commonly used to test PCBs to identify early fallouts before they are used in the field. The chamber can process multiple PCBs simultaneously, i.e., the PCBs are processed in batches. Therefore, the process of PCB testing can be considered as the batch-processing machine scheduling (BPMS) problem and has been addressed extensively in the literature. For example, a simulated annealing approach was proposed to minimize the makespan of a single BPMS problem [17]. A PSO algorithm was presented to minimize the makespan when scheduling non-identical parallel batch-processing machines [18]. A simulation-based intelligence optimization method was developed to minimize the makespan of a flow-shop scheduling problem with multiple heterogeneous batch-processing machines [19]. In addition, a PSO algorithm was presented to minimize the total weighted tardiness of non-identical parallel BPMS problems [20].

This paper focuses on the pressing process, which is also another time- and cost-consuming process in multi-layer PCB manufacturing. The pressing process, a stage in multi-layer PCB fabrication, consists of many phases that require a lot of materials and expensive machines. A good schedule is needed to reduce the production time and to increase the machine utilization, which requires effective assignment and scheduling. After extensively reviewing the literature on the scheduling problems that relate to PCB manufacturing, we have not found any studies linked to the scheduling of the pressing process. A similar mathematical problem in the literature is the flexible job shop scheduling problem (FJSP) [21–24]. The pressing process scheduling and FJSP have similar backgrounds, which are assignment and sequencing. In the FJSP, there are an operation-to-machine assignment and sequencing operation in each machine, but the pressing process scheduling has more than one stage of the assignment. In practice, most PCB companies manually schedule the pressing process, which may not yield the best resource utilization. Therefore, this paper aims to provide a mathematical model for scheduling the pressing process that maximizes the resource utilization. Furthermore, due to the complexity of the pressing process, an effective heuristic algorithm for solving this problem is also presented.

Novelties of the Paper

This paper investigates the pressing process scheduling, which is an application in real-world PCB industries, and, to the best of our knowledge, has never been investigated in the literature. Some PCB companies usually schedule the pressing process by dividing the planning horizon into fixed time intervals. Then, each time interval is assigned either to be in a cycle of a machine or to be vacant. However, this may not be the best way of scheduling the pressing process since in reality, the starting time and completion time of a cycle do not need to follow the fixed time intervals. It is more flexible if the starting and completion times of the cycles are considered as continuous variables. The contributions of this paper can be summarized as follows:

1. This paper proposes a novel mixed integer linear programming (MILP) model for the pressing process scheduling that can find an optimal schedule to meet the objective of maximizing the resource utilization, while the times are continuous values.
2. This paper presents a three-phase-PCB-pressing heuristic algorithm (3P-PCB-PH) for solving the pressing process scheduling, based on the proposed MILP, which can find a near-optimal solution within a reasonable computational time and is practical for real-life applications.

The remainder of this paper is organized as follows. In Section 2, the problem description of the pressing process scheduling is introduced. The proposed MILP model is presented in Section 3, while the 3P-PCB-PH algorithm is presented in Section 4. Numerical examples are shown in Section 5. The discussions and the conclusions are drawn in Sections 6 and 7, respectively.

2. Problem Description

This section explains the pressing process in multi-layer PCB manufacturing. The aim of the pressing process is to press the panel that consists of copper foils, prepregs, and core(s), and is shown schematically in Figure 2.

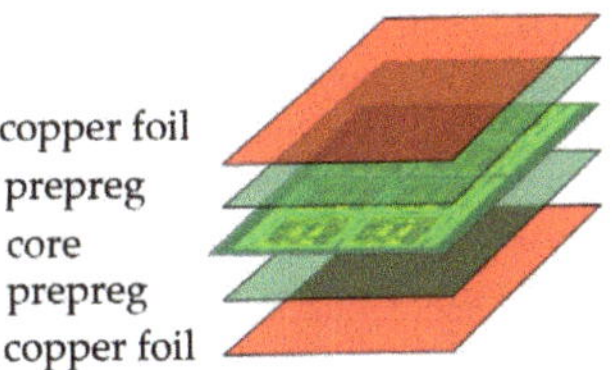

Figure 2. An example of a panel.

The overall processes of one cycle of a press machine are shown schematically in Figure 3. A single cycle of a press machine takes 360 min, which includes the following three phases:

1. Lay-up process phase: The panels are arranged on a selected stainless-steel template (SST), where the number of panels on the SST depends on the size of the SST, the gap between each panel on the SST, and the pattern layout of arrangement. The final arrangement of panels on a SST is called a book. Then, each book is loaded into slots (openings) of a press machine. The number of loaded books is equal to the number of openings of the press machine. This phase takes 120 min.
2. Pressing process phase: The press machine that is already loaded with books is sent into an oven, where the books are heated and pressed. After 120 min, the press machine is removed from the oven.
3. Cool-down process phase: The pressed books in the press machine are cooled down for 120 min. Finally, the books will be removed from the press machine to complete one cycle of the press machine.

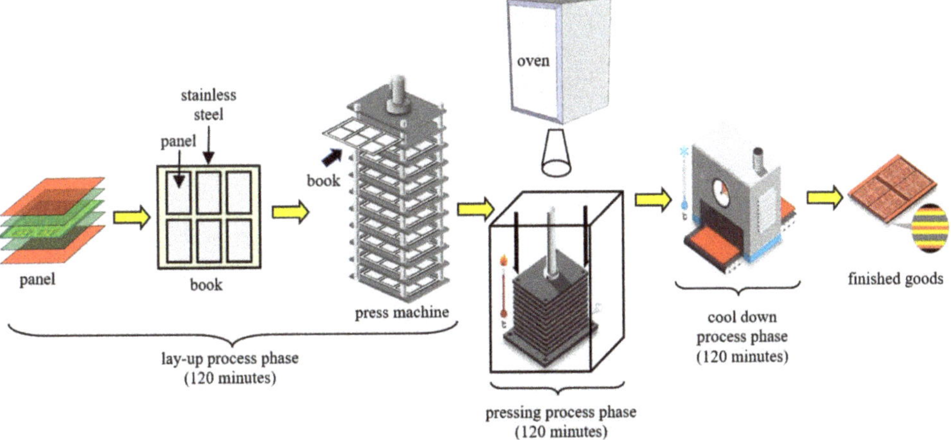

Figure 3. Schematic diagram showing one cycle of a press machine.

Note that, after a press machine has finished one cycle, it is immediately available for a new cycle. Similarly, an oven is immediately available for another press machine after finishing the pressing process phase. Moreover, the following assumptions are made:

- The three phases of a press machine cycle must be performed continuously (no idle time between phases).
- The number of press machines and ovens are known, and the number of ovens is less than the number of press machines. This is because the cost of an oven is very high, and hence the company usually has a small number of ovens.
- Each press machine has the same number of openings.

- There are many types of panels to press and the demand of each type of panel is given. The type of panel depends on the customer's design.
- All panels can be finished within the given due date and resources, i.e., the demands of panels, which are inputs from the customer, yield a feasible schedule.
- The maximum number of available cycles of each press machine to be operated within the due date is the same and this value is given. In practice, the production planning department can estimate this value from the order of the customers and the available resources.
- There are many sizes of SSTs, and each size is unlimitedly available.
- A layout is a pattern of arrangement of panels on a SST. In this study, there are eight layouts, as shown in Figure 4. For example, Figure 4a illustrates the layout with two horizontal sections and the panels are arranged vertically in each section.
- The inner gap is the minimal gap among two panels in a book and the outer gap is the minimal gap between each panel and the borders of the SST. The inner gap (g) and outer gap (G) of an arrangement of panels on a SST depend on the type of panel and these values are known.

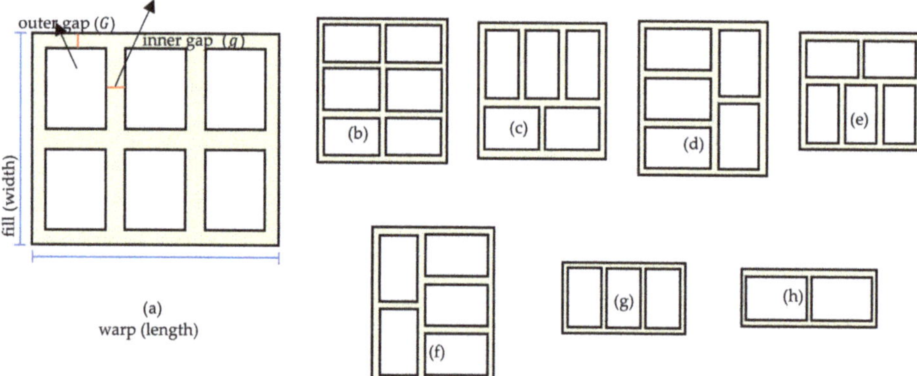

Figure 4. Illustration of the eight layouts (**a**–**h**) of the panel arrangement.

Note that Figure 4a–h are meant to show only the direction of the panel arrangement on a SST, and the number of panels in a book is not limited to those shown in the illustration. In fact, the actual number of panels on a SST using a given layout depends on the size of panel, the size of the SST, and the gaps. Normally, each PCB company may have its own formula for computing the number of panels on a SST with a layout.

The four principal constraints for the pressing process are as follows:

1. Only one type of panel can be arranged and pressed in a cycle of a press machine.
2. The books that are inserted in the same press machine must have the same layout and the same SST size.
3. Each oven can be used by only one press machine at a time to operate the pressing process phase.
4. The number of finished goods of each type of panel must be greater than or equal to the demand.

Constraints 1 and 2 are required so that the pressure from the press machine will be equally distributed to each panel. The objective of the process is to maximize utilization of all press machines and ovens.

3. Proposed Mathematical Model

This section presents a MILP model for scheduling the pressing process as described in Section 2. The indices, sets, parameters, and variables used in the proposed model are defined below.

Indices:

i The index of panel types.
k The index of SST sizes.
l The index of layouts.
p The index of press machines.
o The index of ovens.
t The index of cycles of a press machine.

Parameters:

I The number of types of panels.
K The number of all SST sizes.
L The number of layouts.
P The number of press machines.
O The number of ovens.
T The maximum number of available cycles of each press machine.
m The number of openings of each press machine.
n The processing time of each phase in the pressing process, i.e., the lay-up, pressing, and cool-down process phases. In our case, $n = 120$ min.
a_{ikl} The number of panels of type i per opening using stainless size k and layout l.
d_i Total demand of panel type i.
M A big positive number.

Sets:

$\hat{I}$ The set of all types of panels, $\hat{I} = \{1, 2, \ldots, I\}$.
$\hat{K}$ The set of all SST sizes, $\hat{K} = \{1, 2, \ldots, K\}$.
$\hat{L}$ The set of all layouts, $\hat{L} = \{1, 2, \ldots, L\}$.
$\hat{P}$ The set of all press machines, $\hat{P} = \{1, 2, \ldots, P\}$.
$\hat{O}$ The set of all ovens, $\hat{O} = \{1, 2, \ldots, O\}$.
$\hat{T}$ The set of all numbers of available cycles of each press machine, $\hat{T} = \{1, 2, \ldots, T\}$.

Decision variables:

x_{iklpt} 1, if panel type i is assigned with SST size k and layout l to press machine p at cycle t.
X_{pto} 1, if press machine p is put in oven o at cycle t.
$Y_{ptp't'o}$ 1, if cycle t of press machine p precedes cycle $t\prime$ of press machine $p\prime$ in oven o.
A_{pt} The starting time of the lay-up process phase in cycle t of press machine p.
B_{pto} The starting time of the pressing process phase in cycle t of press machine p in oven o.
C_{pt} The completion time of cycle t of press machine p.
D_{pto} The completion time of the pressing process phase in cycle t of press machine p in oven o.
C'_{pt} The auxiliary variable, which is equal to C_{pt} if there are a panel, a SST, and a layout assigned in press machine p at cycle t. Otherwise, it is equal to 0.
C_{max} The maximum completion time of the last cycle of all press machines which operate the pressing process, i.e., the makespan of the overall process.

In this model, the variable $Y_{ptp't'o}$ is a precedence binary variable that is only defined when $p \neq p\prime$. It is used to avoid the case where an oven operates the pressing process phase for more than one press machine at the same time. This variable is adapted from the precedence binary variable $Y_{iji'j'k}$ that is used to handle the sequencing operations on a machine in the mathematical model of the flexible job shop scheduling problem in Reference [21]. The proposed MILP model can be stated as follows:

$$Min \quad C_{max} \tag{1}$$

Subject to:

$$\sum_{i=1}^{I}\sum_{k=1}^{K}\sum_{l=1}^{L} x_{iklpt} \leq 1,\ \forall p \in \hat{P},\ \forall t \in \hat{T}, \tag{2}$$

$$x_{iklpt} \leq a_{ikl},\ \forall i \in \hat{I},\ \forall k \in \hat{K},\ \forall l \in \hat{L},\ \forall p \in \hat{P},\ \forall t \in \hat{T}, \tag{3}$$

$$\sum_{k=1}^{K}\sum_{l=1}^{L}\sum_{p=1}^{P}\sum_{t=1}^{T} x_{iklpt}(ma_{ikl}) \geq d_i,\ \forall i \in \hat{I}, \tag{4}$$

$$\sum_{i=1}^{I}\sum_{k=1}^{K}\sum_{l=1}^{L} x_{iklp(t-1)} \geq \sum_{i=1}^{I}\sum_{k=1}^{K}\sum_{l=1}^{L} x_{iklpt},\ \forall p \in \hat{P},\ \forall t \in \hat{T}-\{1\}, \tag{5}$$

$$\sum_{o=1}^{O} X_{pto} = 1,\ \forall p \in \hat{P},\ \forall t \in \hat{T}, \tag{6}$$

$$B_{pto} + D_{pto} \leq (X_{pto})M,\ \forall p \in \hat{P},\ \forall t \in \hat{T},\ \forall o \in \hat{O}, \tag{7}$$

$$A_{p,t} \geq C_{p,t-1},\ \forall p \in \hat{P},\ \forall t \in \hat{T}-\{1\}, \tag{8}$$

$$\sum_{o=1}^{O} B_{pto} = A_{pt} + n,\ \forall p \in \hat{P},\ \forall t \in \hat{T}, \tag{9}$$

$$C_{pt} = A_{pt} + 3n,\ \forall p \in \hat{P},\ \forall t \in \hat{T}, \tag{10}$$

$$(B_{pto} + n) - (1 - X_{pto})M \leq D_{pto},\ \forall p \in \hat{P},\ \forall t \in \hat{T},\ \forall o \in \hat{O}, \tag{11}$$

$$D_{pto} \leq (B_{pto} + n) + (1 - X_{pto})M,\ \forall p \in \hat{P},\ \forall t \in \hat{T},\ \forall o \in \hat{O}, \tag{12}$$

$$B_{pto} \geq D_{p't'o} - \left(Y_{ptp't'o}\right)M,\ \forall p, p' \in \hat{P},\ p \neq p',\ \forall t, t' \in \hat{T},\ \forall o \in \hat{O}, \tag{13}$$

$$B_{p't'o} \geq D_{pto} - \left(1 - Y_{ptp't'o}\right)M,\ \forall p, p' \in \hat{P},\ p \neq p',\ \forall t, t' \in \hat{T},\ \forall o \in \hat{O}, \tag{14}$$

$$C_{pt} - M\left[1 - \sum_{i=1}^{I}\sum_{k=1}^{K}\sum_{l=1}^{L} x_{iklpt}\right] \leq C'_{pt},\ \forall p \in \hat{P},\ \forall t \in \hat{T}, \tag{15}$$

$$C'_{pt} \leq C_{pt} + M\left[1 - \sum_{i=1}^{I}\sum_{k=1}^{K}\sum_{l=1}^{L} x_{iklpt}\right],\ \forall p \in \hat{P},\ \forall t \in \hat{T}, \tag{16}$$

$$C'_{pt} \leq M\left(\sum_{i=1}^{I}\sum_{k=1}^{K}\sum_{l=1}^{L} x_{iklpt}\right),\ \forall p \in \hat{P},\ \forall t \in \hat{T}, \tag{17}$$

$$C_{max} \geq C'_{pt},\ \forall p \in \hat{P},\ \forall t \in \hat{T}, \tag{18}$$

and,

$x_{iklpt} \in \{0,1\}$	$\forall i \in \hat{I},\ \forall k \in \hat{K},\ \forall l \in \hat{L},\ \forall p \in \hat{P},\ \forall t \in \hat{T},$
$X_{pto} \in \{0,1\}$	$\forall p \in \hat{P},\ \forall t \in \hat{T}, \forall o \in \hat{O},$
$Y_{ptp't'o} \in \{0,1\}$	$\forall p, p' \in \hat{P},\ p \neq p',\ \forall t, t' \in \hat{T},\ \forall o \in \hat{O},$
$A_{pt}, C_{pt} \geq 0$	$\forall p \in \hat{P},\ \forall t \in \hat{T},$
$B_{pto},\ D_{pto} \geq 0$	$\forall p \in \hat{P},\ \forall t \in \hat{T}, \forall o \in \hat{O},$
$C'_{pt} \geq 0$	$\forall p \in \hat{P},\ \forall t \in \hat{T},$
$C_{max} \geq 0$	

The objective function (1) is to minimize the makespan of the overall process. This can imply maximizing the utilization of all resources.

Constraint (2) is the panel-SST-layout assignment constraint. It is used to ensure that at most one panel type, one SST size, and one layout can be assigned in each cycle of each press machine. If there is an assignment of a panel type, a SST size, and a layout in a cycle of a press machine, it is assumed that these must be the same in all openings.

Constraint (3) is the panel-SST-size-layout compatibility constraint. If panel type i cannot use SST size k with layout l ($a_{ikl} = 0$), then constraint (3) ensures that this pattern cannot be assigned to any press machine p and any cycle t.

Constraint (4) is the demand constraint. It requires that the total outputs of each type of panel from all openings, all cycles, and all press machines must satisfy the demand.

Constraint (5) enforces that a panel type, a SST size, and a layout must be assigned in a press machine at cycle $t-1$ before cycle t. This helps push empty cycles (the cycles of a press machine with no panel assignment) to be after the cycles with a panel assignment.

Constraint (6) is the press machine assignment constraint. It is used to ensure that each cycle of each press machine must be assigned to one oven only.

Constraint (7) enforces that if cycle t of press machine p is assigned to oven o, then the starting time and completion time of the pressing process phase in cycle t of press machine p in oven o can be any non-negative value. Otherwise, these are set to be 0.

Constraint (8) makes sure that any cycle of a press machine can be started after the previous cycle has been finished.

Constraint (9) sets the starting time of the pressing process phase in cycle t of press machine p in its assigned oven to be equal to the starting time of this cycle of press machine p plus the processing time n that it takes in the lay-up process phase.

Constraint (10) sets the completion time of cycle t of press machine p to be equal to its starting time plus the processing time $3n$ (the processing time of one cycle).

Constraints (11) and (12) ensure that if $X_{pto} = 1$, the completion time of the pressing process phase in cycle t of press machine p in its assigned oven will be equal to its starting time plus the processing time n that it takes in the oven.

Constraints (13) and (14) take care of that the pressing process phase in cycle t of press machine p and the pressing process phase in cycle $t\prime$ of press machine $p\prime$, which are assigned in the same oven, cannot be done at the same time.

Constraints (15)–(17) require that if there is assignment of a panel, a SST size, and a layout in the press machine p at cycle t, then the variable C'_{pt} is equal to C_{pt}. Otherwise, it is equal to 0.

Constraint (18) determines the maximum completion time of the last cycle of press machines that has a panel assignment (non-empty cycles), which is the makespan of the overall process.

Note that for the cycle of the press machine that has no assignment of a panel, the proposed MILP model will still return its starting time (A_{pt}) and completion time (C_{pt}), which can be considered as it does not do any work (empty cycle). Also, note that the objective function (1) is to minimize the makespan of all the cycles of all the press machines that actually do the work (non-empty cycles). It means that the objective tries to minimize the makespan of all the cycles of all the press machines that are needed for the respective outputs to satisfy the demands.

The solution to the proposed MILP model provides information about the panel type, SST size, and layout that should be assigned in each cycle of a press machine. In addition, it also tells that each cycle of a press machine should be put into which oven, as well as its starting time and completion time. Hence, the proposed model can be an option to provide an optimized schedule in the pressing process of any PCB manufacturing industry.

4. Proposed 3P-PCB-PH Algorithm

Due to the complexity of the pressing process, using a mathematical programming model may not be suitable for solving a large-sized problem. This section presents a heuristic algorithm for scheduling the pressing process. The idea of this algorithm is to solve the proposed MILP model in three phases. Phase 1 consists of matching each panel type with a SST size and a layout and determining the number of cycles that is needed for the demands to be satisfied. Next, all cycles that are needed to be used are scheduled in Phase 2, which yields the number of non-empty cycles of each press machine and their starting and completion times. In Phase 3, each panel type with its selected SST size and layout from Phase 1 is assigned to a non-empty cycle of a press machine. The parameters that are used in the proposed 3P-PCB-PH algorithm are the same as described Section 3. The details of the designed 3P-PCB-PH algorithm include Steps 1–5, which are expressed below.

Step 1: Take the inputs $I,\ K,\ L,\ P,\ O,\ T, m, n,\ a_{ikl},\ \forall i \in \hat{I},\ k \in \hat{K},\ l \in \hat{L}$, and $d_i,\ \forall i \in \hat{I}$.

Step 2 (Phase 1): Selecting the SST size and layout.

In this phase, an appropriate SST size and a layout are chosen for each panel type. The inputs of Phase 1 include $I,\ K,\ L, m,\ d_i$, and $a_{ikl},\ \forall i \in \hat{I},\ k \in \hat{K},\ l \in \hat{L}$. For each panel type i, we select a SST size $\bar{k}_i$ and a layout $\bar{l}_i$ that give the maximum number of panels of type i, say $a_{i\bar{k}_i\bar{l}_i}$. Hence, the number of produced panels of this type per cycle of a press machine is $ma_{i\bar{k}_i\bar{l}_i}$. Next, the minimum number of cycles needed for pressing each panel of type $i \in \hat{I}$ can be computed from $d_{c_i} = \left\lceil \frac{d_i}{ma_{i\bar{k}_i\bar{l}_i}} \right\rceil$ (note that the notation $\lceil x \rceil$ is the smallest integer that is greater than or equal to x). Let d_c be the sum of these values of all panel types, which is the minimum number of total cycles that are needed to be used for pressing, so that the demands of all panel types are satisfied. Note that the value d_c does not exceed the number of all available cycles $P \times T$, since we have the assumption that the demands of panels (which are inputs from the customer) yield a feasible schedule. The flowchart of the algorithm for Phase 1 is shown in Figure 5.

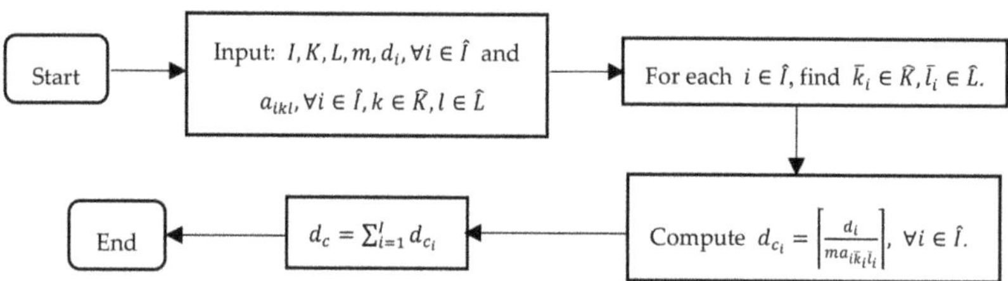

Figure 5. Flowchart for the 3P-PCB-PH algorithm for Phase 1.

Step 3 (Phase 2): Scheduling the press machines and ovens.

In this phase, all d_c cycles are distributed to all press machines to generate a schedule such that the makespan is minimized. The algorithm for Phase 2 is comprised of the following components.

1. $A = [A_{pt}]_{P\times T}$: the matrix that collects the starting time of cycle t of press machine p (the starting time of the lay-up process phase). Initially, A is set to be $[0]_{P\times T}$.
2. $C = [C_{pt}]_{P\times T}$: the matrix that collects the completion time of cycle t of press machine p. Initially, C is set to be $[0]_{P\times T}$.
3. *Can*: the candidate list represents the next earliest available cycle number to use each press machine. Initially, *Can* is set to be $[1]_{1\times P}$, i.e., for each press machine, the cycle that is ready to start is cycle 1.
4. (*start_time, end_time, press_machine, cycle*): A scheduled pressing job which collects the starting and end times of the pressing process phase of a press machine at a cycle, where the *start_time*, *end_time*, *press_machine*, and *cycle* are the starting time, end time, press machine number, and cycle number, respectively. For example, if we have a scheduled pressing job (240, 360, 1, 1), it means that the pressing job occurs from time 240 to 360 min and is the task of press machine 1 at cycle 1.
5. *Oven_Schedule_List*: the list of scheduled pressing jobs to use in each oven in a sequential order. Each element in the *Oven_Schedule_List* is also a list, which collects the scheduled pressing job tuples that are assigned in the corresponding oven. Figure 6 shows an example of an *Oven_Schedule_List* when the number of ovens (O) is three and the processing time of the pressing process phase (n) is 120 min. The first list in *Oven_Schedule_List* contains the scheduled pressing jobs that are already assigned to

oven 1. There are two pressing jobs in the first list. The first is (120, 240, 1, 1), which means oven 1 has to press from 120 to 240 min and is the task of press machine 1 at cycle 1, while the second is (480, 600, 1, 2) which means oven 1 has to press from 480 to 600 min and is the task of press machine 1 at cycle 2. Similarly, the list for oven 2 has only one job that is already assigned, and there is no job that is currently assigned to oven 3 since the third list is empty. Note that, initially, the list *Oven_Schedule_List* is set to be the list of O empty lists $[\,[\,]\,]_{1\times O}$. The algorithm for Phase 2 will later populate this list with suitable jobs.

6. *Oven_Idle_time_List*: the list of idle time intervals of each oven in a sequential order. Each element in the *Oven_Idle_time_List* is also a list which collects all the idle time intervals in the corresponding oven. Initially, each oven has only one idle time interval $[0, \infty)$, indicating that no task had been assigned to it yet.

$$Oven\_Schedule\_List = [[(120, 240, 1, 1), (480, 600, 1, 2)], [(120, 240, 2, 1)], [\ \ ]]$$

Jobs in oven 1 Jobs in oven 2 Jobs in oven 3

Figure 6. An example of an *Oven_Schedule_List*.

After introducing all the components, we proposed the algorithm for Phase 2 as follows. The inputs for the algorithm are P, O, n, and d_c, where d_c is used as the total number of iterations. For each iteration, a press machine with the minimum workload is selected, say $p\prime$. Next, we check whether $Can[p\prime]$, the next earliest available cycle of press machine $p\prime$, is the first cycle. If yes, the starting time of press machine $p\prime$ at cycle $Can[p\prime]$ is set to be 0. Otherwise, it is set to be the end time of the previous cycle. Let this starting time be *start_time_press_machine*. Note that this starting time is not yet a final starting time of the press machine since we need to check the feasibility with the assigned oven first. Then, the press machine $p\prime$ at cycle $Can[p\prime]$ will be assigned to the oven with the minimum workload, say $o\prime$, to operate the pressing process phase. Next, we check whether the oven $o\prime$ has been used yet. If not (i.e., the $Oven\_Idle\_time\_List[o\prime]$ has only one idle time interval $[0, \infty)$), the cycle $Can[p\prime]$ of press machine $p\prime$ can be started at *start_time_press_machine*, and sequentially, $p\prime$ is sent into the oven $o\prime$ at the time $start\_time\_press\_machine + n$. Otherwise, we consider all idle time intervals in the $Oven\_Idle\_time\_List[o']$. These intervals are examined from left to right to find the earliest time that the press machine $p\prime$ at cycle $Can[p\prime]$ can start the pressing process phase in the oven $o\prime$. An example is illustrated in Figure 7. Suppose that $o\prime$ is oven 1 that already has a task of cycle 1 from press machine 1 assigned before, and the processing time of the pressing process phase (n) is 120 min. Suppose $p\prime$ is press machine 2 and $Can[2]$ is cycle 1. Since this is the first cycle, the value *start_time_press_machine* is 0. However, since oven 1 has been used, we will examine the idle time intervals from left to right. From Figure 7, $Oven\_Idle\_time\_List[1] = [[0, 120], [240, \infty)]$. It is clear that the first interval [0, 120] is not feasible since the lay-up process phase has not been done. So, press machine 2 at cycle 1 can start the pressing process phase as early as possible in oven 1 at time 240 min in the second idle time interval [240, ∞). Let this time be *start_time_oven*. We can then find the time that the press machine $p\prime$ is removed from the oven $o\prime$ ($end\_time\_oven = start\_time\_oven + 120$) as well as the actual starting time (*start_time_press_machine*) and completion time of press machine $p\prime$ at cycle $Can[p\prime]$, which are the $start\_time\_oven - 120$ and $end\_time\_oven + 120$, respectively. We update these values in matrices A and C as well as update the list $Oven\_Schedule\_List[o\prime]$ and $Oven\_Idle\_time\_List[o\prime]$. Then, $Can[p\prime]$ is incremented by 1 so that the next cycle of the press machine $p\prime$ is a new candidate. The algorithm is repeated until all d_c cycles are scheduled. The flowchart of the algorithm for Phase 2 is shown in Figure 8.

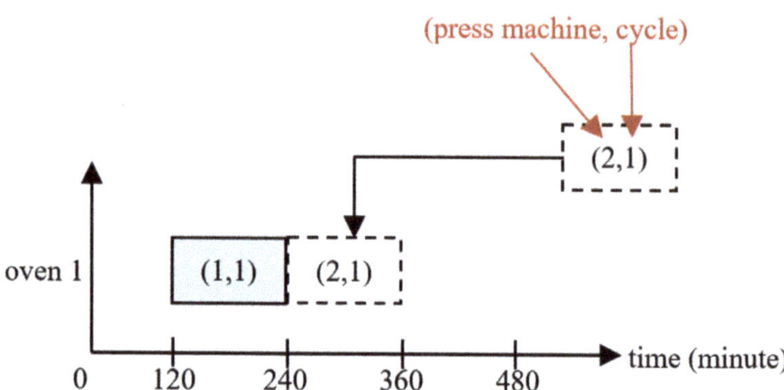

Figure 7. An example of finding *start_time_oven*.

Step 4 (Phase 3): Assigning the panel-SST-size-layout combinations to cycles of the press machines.

From Phase 2, the number of working cycles for each press machine is known. In Phase 3, each panel type with its selected SST size and layout will be assigned to a cycle of a press machine as follows. Recall that d_{c_i} is the minimum number of cycles needed to be used for pressing each panel of type $i \in \hat{I}$. The d_{c_1} cycles for the first panel type are chosen from the first cycles of all press machines such that the work is distributed among the press machines equally. The d_{c_2} cycles for the second panel type are then chosen from the next available cycles of all the press machines so that the work is distributed equally, and so on. As a result of this panel-cycle assignment, the panels of the same type are finished in a group, which is preferable in real-world situations. Figure 11 depicts an example of this assignment.

Step 5: Output the number of finished goods of each panel type $i \in \hat{I}$; A_{pt}, C_{pt}, $\forall p \in \hat{P}$, $t \in \hat{T}$, the schedule of press machines and ovens, x_{iklpt}, $\forall i \in \hat{I}$, $k \in \hat{K}$, $l \in \hat{L}, p \in \hat{P}$, $t \in \hat{T}$, and the makespan.

The total number of finished goods of each panel type i can be computed from $ma_{i\bar{k}_i\bar{l}_i} d_{c_i}, \forall i \in \hat{I}$. The value of A_{pt}, C_{pt}, $\forall p \in \hat{P}$, $t \in \hat{T}$ can be obtained from matrices A and C in Phase 2, and these values can then be used for creating the schedule of press machines. The schedule of ovens can be interpreted from the list *Oven_Schedule_List* in Phase 2. The makespan of the overall processes is the maximum element in C. The output $x_{iklpt}, \forall i \in \hat{I}$, $k \in \hat{K}$, $l \in \hat{L}, p \in \hat{P}$, $t \in \hat{T}$, which is equal to 1, can be obtained from Phase 3. From all three phases, the computational complexity of the proposed 3P-PCB-PH algorithm is $O\left(P^2T^2 + IKL\right)$.

Note that PCB manufacturing companies prefer to finish each PCB type in a group, since it is easier to prepare material and sequence the next work. The proposed MILP in the previous section can find an optimal schedule for a pressing process with the minimum makespan, but cycles of the same panel type may not be scheduled consecutively. This is a limitation of the proposed MILP model, whereas the proposed 3P-PCB-PH algorithm can handle this preference. Therefore, the proposed 3P-PCB-PH algorithm is more practical for real PCB manufacturing industries.

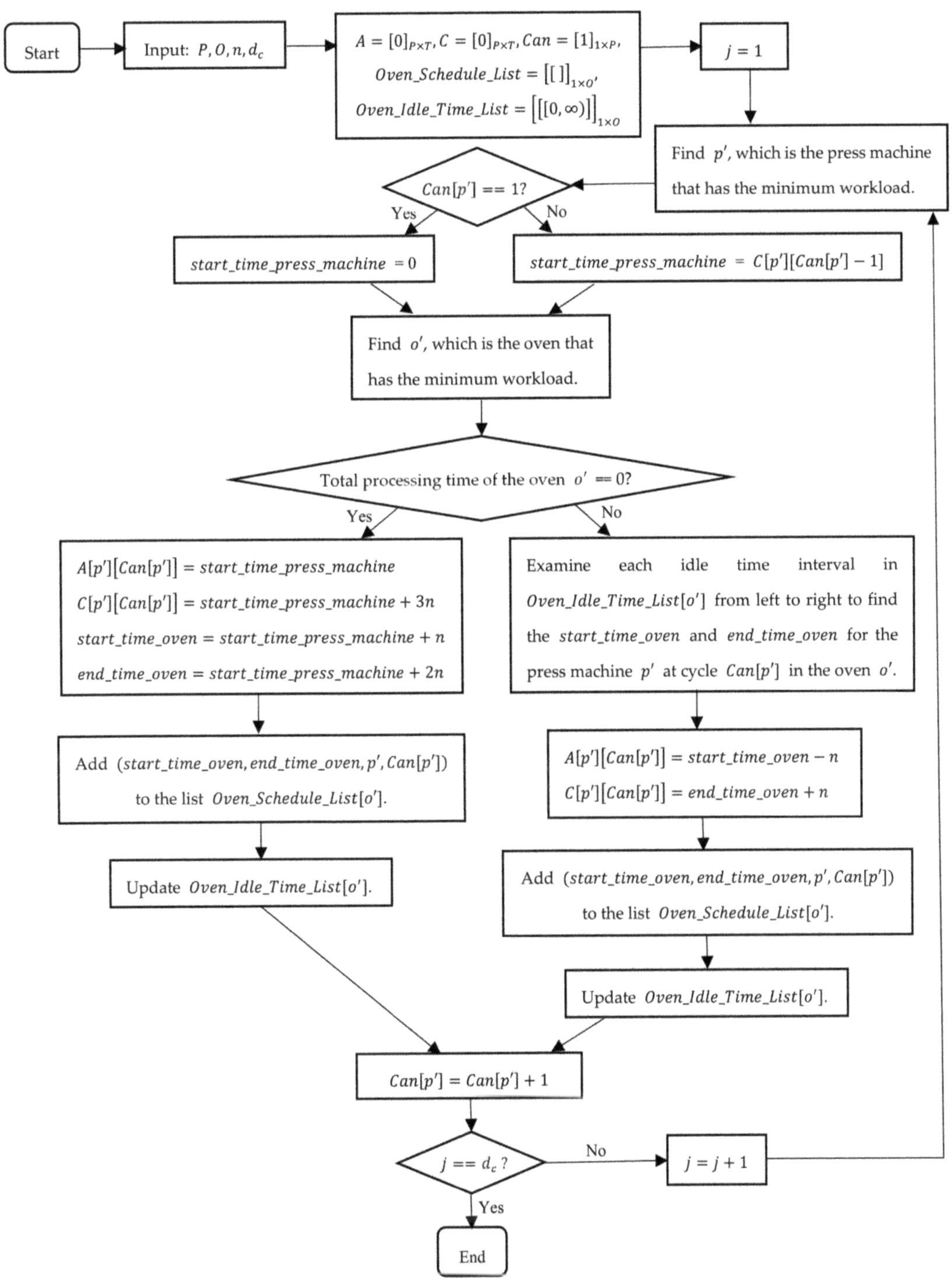

Figure 8. Flowchart of the 3P-PCB-PH algorithm for Phase 2.

5. Numerical Experiments

To demonstrate the proposed MILP model and 3P-PCB-PH algorithm, we used real-world data from a PCB company. The data and test problems are shown in Sections 5.1 and 5.2, respectively. The computational results from the proposed MILP model and heuristic algorithm are shown in Sections 5.3 and 5.4, respectively.

5.1. Data

The data acquired from an actual PCB company included seven panel types, six SST sizes, eight layouts, six press machines, each of which had 10 openings, and three ovens. We assumed that the processing time of each phase in the pressing process (the lay-up, pressing, and cool-down process phase) was 120 min, with a planning horizon of 3 days and a maximum number of available cycles of each press machine to be 12. This is because one cycle of a press machine takes 360 min (6 h). If a press machine works continuously, it can carry out up to 12 cycles of the pressing process in 3 days. We also considered a planning horizon of 2 and 1.5 days for the small problem, where the maximum number of available cycles of each press machine was eight and six cycles, respectively. The information of each type of panel, which consisted of warp (or length), fill (or width), inner gap, and outer gap, is shown in Table 1. The size of each SST is shown in Table 2.

Table 1. Sizes, inner gap, and outer gap of each panel type.

Panel	Warp (a)	Fill (b)	Inner Gap (g)	Outer Gap (G)
1	20.5	24	0.5	0.25
2	25.65	22.25	1	0.5
3	26	24	0.5	0.25
4	26.5	22.5	1	0.5
5	19	22.25	0.5	0.25
6	15	23.8	0.5	0.25
7	27.75	20.5	0.5	0.25

Table 2. Sizes of each SST.

Stainless-Steel	Warp (X)	Fill (Y)
1	50	44
2	50	53
3	50	56
4	50	58
5	43	25.5
6	43	27

The number of layouts was eight, as described in Figure 4 (in Section 2). The formulas for computing the number of panels (per book) based on the size of the SST and the layout are shown in Table 3. In the formulas, the values a, b, g, and G are the warp, fill, inner gap, and outer gap of panel type i, respectively. The values X and Y are the warp and fill of the SST size k, respectively. Note that the notation x is the greatest integer that is less than or equal to x.

5.2. Test Problems

According to Pan [25], the speed that mixed-integer linear programming problems can be solved at depends upon the number of binary variables, constraints, and continuous variables, where the most deciding factor is the number of binary variables. Therefore, the generated test problems are categorized to be 3 groups, i.e., small-, medium-, and large-sized test problems, depending on the number of binary variables.

Table 3. Formulas for computing the number of panels of type i per opening using SST size k and layout (a_{ikl}).

Layout (l)	a_{ikl}
1	$\left\lfloor \frac{X-2(G-\frac{g}{2})}{a+g} \right\rfloor \times \left\lfloor \frac{Y-2(G-\frac{g}{2})}{b+g} \right\rfloor$
2	$\left\lfloor \frac{X-2(G-\frac{g}{2})}{b+g} \right\rfloor \times \left\lfloor \frac{Y-2(G-\frac{g}{2})}{a+g} \right\rfloor$
3	$\left\lfloor \frac{X-2(G-\frac{g}{2})}{a+g} \right\rfloor + \left(\left\lfloor \frac{X-2(G-\frac{g}{2})}{b+g} \right\rfloor \times \left\lfloor \frac{Y-b-G-2(G-\frac{g}{2})}{a+g} \right\rfloor \right)$
4	$\left\lfloor \frac{Y-2(G-\frac{g}{2})}{a+g} \right\rfloor + \left(\left\lfloor \frac{Y-2(G-\frac{g}{2})}{b+g} \right\rfloor \times \left\lfloor \frac{X-b-G-2(G-\frac{g}{2})}{a+g} \right\rfloor \right)$
5	$\left\lfloor \frac{X-2(G-\frac{g}{2})}{b+g} \right\rfloor + \left(\left\lfloor \frac{X-2(G-\frac{g}{2})}{a+g} \right\rfloor \times \left\lfloor \frac{Y-a-G-2(G-\frac{g}{2})}{b+g} \right\rfloor \right)$
6	$\left\lfloor \frac{Y-2(G-\frac{g}{2})}{b+g} \right\rfloor + \left(\left\lfloor \frac{Y-2(G-\frac{g}{2})}{a+g} \right\rfloor \times \left\lfloor \frac{X-a-G-2(G-\frac{g}{2})}{b+g} \right\rfloor \right)$
7	$\left\lfloor \frac{X-2(G-\frac{g}{2})}{a+g} \right\rfloor$
8	$\left\lfloor \frac{X-2(G-\frac{g}{2})}{b+g} \right\rfloor$

5.2.1. Small-Sized Test Problems

The small-sized test problems were generated where the number of binary variables in each problem is less than 8500. The parameters in the small-sized test problems are as follows. The number of SST sizes (K) and the number of layouts (L) were six and eight respectively, as described in the previous subsection. The number of panel types (I) was three, which are the panel types 1–3 in Table 1. The number of press machines (P) and the number of ovens (O) were varied at three to four and two to three, respectively. The maximum number of available cycles of each press machine (T) was varied as six, eight, and 12 cycles, and the demand of each type of panel (d_i) was randomly generated. The details of the small-sized test problems are summarized in Table 4.

Table 4. The small-sized test problems for the proposed MILP model and 3P-PCB-PH algorithm.

No.	I	K	L	P	O	T	$d_i, i \in \{1,2,\ldots,I\}$
1	3	6	8	3	2	6	110, 150, 125
2	3	6	8	3	2	8	200, 220, 230
3	3	6	8	3	2	12	270, 250, 210
4	3	6	8	4	2	6	110, 150, 125
5	3	6	8	4	3	6	110, 150, 125

5.2.2. Medium-Sized Test Problems

The medium-sized test problems were generated where the number of binary variables in each problem is between 8500 to 30,000. The parameters in the medium-sized test problems are as follows. The number of panel types (I) was three to five, while the number of SST sizes (K), the number of layouts (L), the number of press machines (P), and the number of ovens (O) were six, eight, six, and three respectively, which are the real data from the previous subsection. The maximum number of available cycles (T) was varied at six, eight, or 12. The demands of each test problem were randomly generated. The details of the medium-sized test problems are shown in Table 5. In Problems 1–3, the number of types of panels was three, which included panel types 1–3 in Table 1. Problems 4–6 had panel types 1–4, and the other problems had panel types 1–5, as described in Table 1.

Table 5. The medium-sized test problems for the proposed MILP model and 3P-PCB-PH algorithm.

No.	I	K	L	P	O	T	d_i, $i \in \{1,2,\ldots,I\}$
1	3	6	8	6	3	6	300, 300, 300
2	3	6	8	6	3	8	450, 480, 500
3	3	6	8	6	3	12	720, 900, 600
4	4	6	8	6	3	6	200, 300, 400, 100
5	4	6	8	6	3	8	300, 400, 200, 500
6	4	6	8	6	3	12	500, 700, 700, 500
7	5	6	8	6	3	6	200, 250, 200, 250, 200
8	5	6	8	6	3	8	400, 300, 200, 250, 300

5.2.3. Large-Sized Test Problems

The large-sized test problems were generated where the number of binary variables in each problem is greater than 30,000. The parameters in the large-sized test problems are as follows. The number of panel types (I) was varied at five to seven. The number of SST sizes (K), the number of layouts (L), the number of press machines (P), and the number of ovens (O) were six, eight, six, and three respectively, which are the real data from the previous subsection. Furthermore, we also evaluated slightly larger-sized problems by increasing the number of press machines and ovens by one. The maximum number of available cycles (T) was 12 and the demand of each type of panel (d_i) was randomly generated. The details of the large-sized test problems are shown in Table 6. In Problems 1–3, the number of types of panels was five, which included panel types 1–5 in Table 1. Problems 4–6 had panel types 1–6, and the other problems had all seven panel types, as described in Table 1.

Table 6. The large-sized test problems for the proposed MILP model and 3P-PCB-PH algorithm.

No.	I	K	L	P	O	T	d_i, $i \in \{1,2,\ldots,I\}$
1	5	6	8	6	3	12	500, 500, 500, 500, 500
2	5	6	8	7	3	12	500, 500, 500, 500, 500
3	5	6	8	6	4	12	500, 500, 500, 500, 500
4	6	6	8	6	3	12	500, 360, 220, 180, 380, 720
5	6	6	8	7	3	12	500, 360, 220, 180, 380, 720
6	6	6	8	6	4	12	500, 360, 220, 180, 380, 720
7	7	6	8	6	3	12	300, 325, 290, 425, 450, 475, 200
8	7	6	8	7	3	12	300, 325, 290, 425, 450, 475, 200
9	7	6	8	6	4	12	300, 325, 290, 425, 450, 475, 200

5.3. Result of the Test Problems Using the Proposed MILP Model

In this section, all the test problems were solved using the proposed MILP model and the ILOG OPL CPLEX 12.6 software running on a personal computer with a core i7 2.20 GHz CPU and 8 GB RAM. The maximum running time was limited to 2 h.

5.3.1. Results of the Small-Sized Test Problems Using the Proposed MILP Model

The model size and computational results of each small-sized test problem using the proposed model are shown in Table 7. The model size consisted of the number of binary variables, continuous variables, and constraints. The results included the number of finished goods of each type of panel (outputs), CPU time, and the optimal makespan (C_{max}) of the overall process.

Table 7. Computational results of the small-sized test problems using the proposed MILP model.

No.	I	K	L	P	O	T	d_i	Model Size			Results		
								Binary	Continuous	Constraint	Outputs	CPU Time	C_{max}(min)
1	3	6	8	3	2	6	110, 150, 125	3060	128	3741	120, 160, 160	2.31 s	1440 [a]
2	3	6	8	3	2	8	200, 220, 230	4272	170	5373	200, 240, 240	2.48 s	2160 [a]
3	3	6	8	3	2	12	270, 250, 210	6984	254	9213	280, 280, 240	3.21 s	2520 [a]
4	3	6	8	4	2	6	110, 150, 125	4368	170	5563	120, 160, 160	8.15 s	1200 [a]
5	3	6	8	4	3	6	110, 150, 125	4824	218	6499	120, 160, 160	3.18 s	1080 [a]

[a] Optimal solution.

As shown in Table 7, all the small-sized test problems could be solved to an optimal solution within the 2 h time limit. The computational time of each problem is small. Note that Problems 1, 4, and 5 have the same demands. The results of Problem 4 indicate that if the number of press machines was increased by one from Problem 1, the pressing process of Problem 1 could be finished ahead of time for 240 min (i.e., the makespan was reduced from 1440 to 1200 min). However, the results of Problem 5 indicate that the pressing process of Problem 1 could be finished ahead of time for 360 min (i.e., the makespan was reduced from 1440 to 1080 min) if the number of press machines and ovens were increased by one from Problem 1. These show that the proposed MILP model can help in deciding which resources should be increased to reduce the production time.

5.3.2. Results of the Medium-Sized Test Problems Using the Proposed MILP Model

Table 8 shows the size and computational results of each medium-sized test problem using the proposed MILP model. The number of binary variables of each problem is between 8500 to 30,000. The results showed that all the medium-sized test problems could be solved to an optimal solution within the 2 h time limit. Note that the maximum computational time for solving the medium-sized test problems (9 min and 31 s in Problem 7 of the medium-sized test problems) increased significantly compared with the maximum computational time for solving the small-sized test problems, which is only around 8 s (in Problem 4 of the small-sized test problems).

5.3.3. Results of the Large-Sized Test Problems Using the Proposed MILP Model

The model size and computational results of each large-sized test problem using the proposed model are shown in Table 9. The results show that only Problems 1, 2, and 4 of the large-sized test problems could be solved to an optimal solution within the 2 h time limit, while the other problems could not, but we report the best feasible solution that could be found within the time limit. Note that the maximum computational time for solving the large-sized test problems to get an optimal solution (40 min and 11 s in Problem 4 of the large-sized test problems) increased significantly compared with the maximum computational time for solving the medium-sized test problems (9 min and 31 s in Problem 7 of the medium-sized test problems). In addition, an optimal solution could not be found for most large-sized test problems within the time limit of 2 h. This is common when solving large-sized mixed-integer linear programming problems. Since some practitioners can accept a promise solution within reasonable time instead of an optimal solution, this paper also presents a heuristic algorithm for solving the pressing process scheduling that could find a good solution within reasonable time, and the results of the proposed heuristic algorithm are presented in the next subsection.

Table 8. Computational results of the medium-sized test problems using the proposed MILP model.

No.	I	K	L	P	O	T	d_i	Model Size			Results		
								Binary	Continuous	Constraint	Outputs	CPU Time	C_{max}(min)
1	3	6	8	6	3	6	300, 300, 300	8532	326	12,339	320, 320, 320	32.81 s	1560 [a]
2	3	6	8	6	3	8	450, 480, 500	12,816	434	19,335	480, 480, 520	16.79 s	2520 [a]
3	3	6	8	6	3	12	720, 900, 600	23,544	650	37,647	720, 920, 600	1 min 28 s	3600 [a]
4	4	6	8	6	3	6	200, 300, 400, 100	10,260	326	14,068	200, 320, 400, 120	20.65 s	1800 [a]
5	4	6	8	6	3	8	300, 400, 200, 500	15,120	434	21,640	320, 400, 200, 520	3 min 59 s	2280 [a]
6	4	6	8	6	3	12	500, 700, 700, 500	27,000	650	41,104	520, 720, 720, 520	4 min 2 s	3960 [a]
7	5	6	8	6	3	6	200, 250, 200, 250, 200	11,988	326	15,797	200, 280, 200, 280, 200	9 min 31 s	1920 [a]
8	5	6	8	6	3	8	400, 300, 200, 250, 300	17,424	434	23,945	400, 320, 200, 280, 320	49.31 s	2520 [a]

[a] Optimal solution.

Table 9. Computational results of the large-sized test problems using the proposed MILP model.

No.	I	K	L	P	O	T	d_i	Model size			Results		
								Binary	Continuous	Constraint	Outputs	CPU Time	C_{max}(min)
1	5	6	8	6	3	12	500, 500, 500, 500, 500	30,456	650	44,561	520, 520, 520, 520, 520	44 min 38 s	4080 [a]
2	5	6	8	7	3	12	500, 500, 500, 500, 500	38,556	758	58,035	520, 520, 520, 520, 520	23 min 42 s	3600 [a]
3	5	6	8	6	4	12	500, 500, 500, 500, 500	34,848	794	53,417	520, 520, 520, 520, 520	2 h	4080 [b]
4	6	6	8	6	3	12	500, 360, 220, 180, 380, 720	33,912	650	48,018	520, 360, 240, 200, 400, 770	48 min 14 s	3360 [a]
5	6	6	8	7	3	12	500, 360, 220, 180, 380, 720	42,588	758	62,068	520, 360, 240, 200, 400, 770	2 h	3000 [b]
6	6	6	8	6	4	12	500, 360, 220, 180, 380, 720	38,304	794	56,874	520, 360, 240, 200, 400, 770	2 h	3360 [b]
7	7	6	8	6	3	12	300, 325, 290, 425, 450, 475, 200	37,368	650	51,475	320, 360, 320, 440, 480, 490, 200	2 h	3720 [b]
8	7	6	8	7	3	12	300, 325, 290, 425, 450, 475, 200	46,620	758	66,101	320, 360, 320, 440, 480, 490, 200	2 h	3360 [b]
9	7	6	8	6	4	12	300, 325, 290, 425, 450, 475, 200	41,760	794	60,331	320, 360, 320, 440, 480, 490, 200	2 h	3720 [b]

[a] Optimal solution. [b] The best-known solution from the proposed MILP model.

An example of an optimal solution from the proposed MILP model is described below. For the results of Problem 1 in Table 9, the number of outputs of panel types 1–5 that were obtained after the finishing pressing process was 520 each, which satisfied the demands. The variables x_{iklpt} and X_{pto}, which were equal to 1 in the optimal solution of Problem 1, are shown in Tables 10 and 11, respectively. The corresponding Gantt charts of the press machines and ovens are presented in Figures 9 and 10 respectively, where the same color represents the same panel type.

Table 10. List of non-zero x_{iklpt} values in the solution of Problem 1 of the large-sized problems using the proposed MILP model.

Press Machine	Non-Zero x_{iklpt}
1	$x_{13111}, x_{51212}, x_{24213}, x_{51214}, x_{33215}, x_{51216}, x_{51217}, x_{24218}, x_{43219}, x_{3221,10}, x_{4321,11}$
2	$x_{14621}, x_{53622}, x_{53223}, x_{54524}, x_{32225}, x_{12626}, x_{43227}, x_{44228}, x_{32229}, x_{1222,10}, x_{3222,11}$
3	$x_{23431}, x_{51532}, x_{44233}, x_{43234}, x_{43235}, x_{43236}, x_{12437}, x_{34238}, x_{12239}, x_{2343,10}, x_{4423,11}$
4	$x_{14141}, x_{43242}, x_{32243}, x_{24244}, x_{13445}, x_{32246}, x_{32247}, x_{32248}, x_{24649}, x_{5414,10}, x_{5134,11}$
5	$x_{12151}, x_{14352}, x_{54553}, x_{23454}, x_{32255}, x_{23656}, x_{42257}, x_{32258}, x_{43259}, x_{2365,10}$
6	$x_{51361}, x_{13362}, x_{24663}, x_{42264}, x_{24665}, x_{23466}, x_{24467}, x_{54468}, x_{11269}, x_{1126,10}, x_{3426,11}$

Table 11. List of non-zero X_{pto} values in the solution of Problem 1 of the large-sized problems using the proposed MILP model.

Press Machine	Non-Zero X_{pto}
1	$X_{111}, X_{122}, X_{131}, X_{142}, X_{151}, X_{161}, X_{171}, X_{181}, X_{191}, X_{1,10,3}, X_{1,11,3}$
2	$X_{212}, X_{223}, X_{233}, X_{241}, X_{253}, X_{263}, X_{272}, X_{282}, X_{292}, X_{2,10,1}, X_{2,11,1}$
3	$X_{311}, X_{322}, X_{331}, X_{341}, X_{353}, X_{363}, X_{373}, X_{383}, X_{393}, X_{3,10,2}, X_{3,11,2}$
4	$X_{412}, X_{421}, X_{433}, X_{442}, X_{451}, X_{461}, X_{472}, X_{482}, X_{491}, X_{4,10,1}, X_{4,11,3}$
5	$X_{513}, X_{522}, X_{531}, X_{543}, X_{553}, X_{563}, X_{573}, X_{581}, X_{593}, X_{5,10,2}$
6	$X_{613}, X_{623}, X_{632}, X_{643}, X_{652}, X_{662}, X_{673}, X_{681}, X_{692}, X_{6,10,3}, X_{6,11,1}$

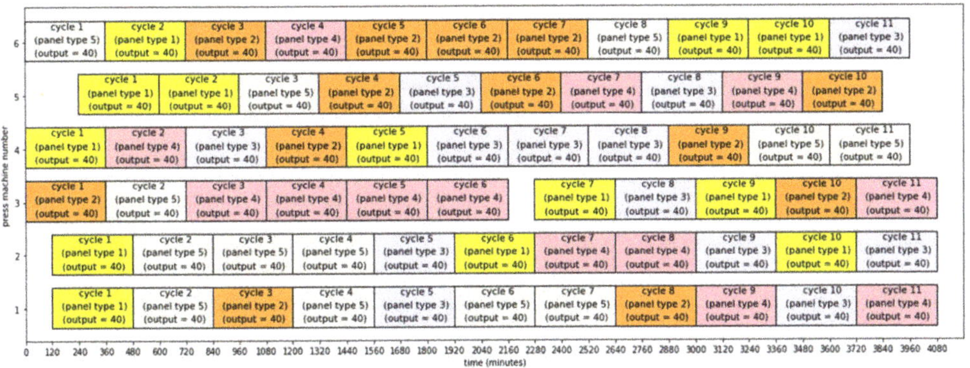

Figure 9. Gantt chart of the press machines for Problem 1 of the large-sized problems using the proposed MILP model.

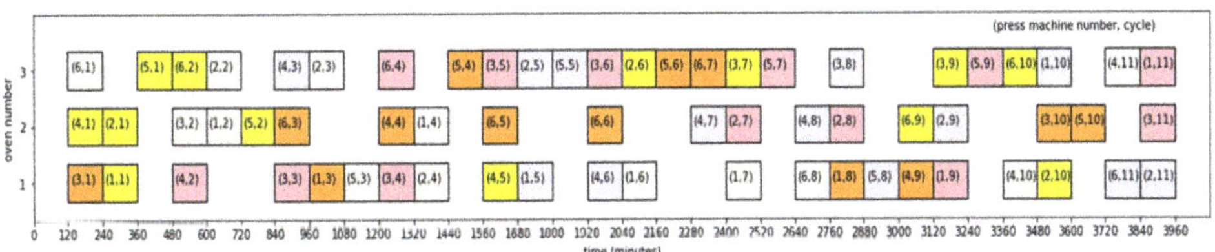

Figure 10. Gantt chart of the ovens for Problem 1 of the large-sized problems using the proposed MILP model.

From Table 10, the list of non-zero x_{iklpt} were sorted by cycle numbers (index t) in ascending order, while the list of non-zero X_{pto} values (Table 11) were also sorted in a similar manner.

Figure 9 shows the starting time and completion time of each cycle of each press machine. One cycle of the press machine takes 360 min, i.e., 120 min for each lay-up, pressing, and cool-down process phase. The time for the pressing process phase for each cycle of each press machine is depicted in Figure 10. For example, press machine 3 at cycle 1 had to lay up at 0–120 min (Figure 9), move into the oven 1 at 120–240 min (Figure 10), and cool-down at 240–360 min (Figure 9). The minimum makespan of the overall process was 4080 min (Figure 9).

5.4. Result of the Test Problems Using the Proposed 3P-PCB-PH Algorithm

In this section, all the test problems were solved using the proposed 3P-PCB-PH algorithm implemented in Python version 3.7.3 running under the same hardware environment as in the previous subsection. Each problem was run 10 times to capture the variation in the computational time. The results of each test problem when using the proposed heuristic algorithm were compared with the results from the proposed MILP model.

5.4.1. Results of the Small-Sized Test Problems Using the Proposed 3P-PCB-PH Algorithm

The results of the small-sized test problems from the heuristic algorithm and the proposed MILP model are compared in Table 12. The results included the number of finished goods of each type of panel (outputs), the average CPU time over 10 runs (Avg CPU time), and the makespan (C_{max}) of the overall process. The last column of Table 12 reports the percentage gap ($\%gap$) between the makespan from the proposed heuristic algorithm and the optimal makespan or best-known makespan from the proposed MILP model.

Table 12. Computational results of the small-sized test problems using the proposed 3P-PCB-PH algorithm.

No.	I	K	L	P	O	T	d_i	Results Using Proposed MILP Model			Results Using Proposed 3P-PCB-PH Algorithm			$\%gap$
								Outputs	**CPU Time**	C_{max} (min)	**Outputs**	**Avg CPU Time (SD)**	C_{max} (min)	
1	3	6	8	3	2	6	110, 150, 125	120, 160, 160	2.31 s	1440 [a]	120, 160, 160	0.00349 s (0.00085 s)	1440 [a]	0%
2	3	6	8	3	2	8	200, 220, 230	200, 240, 240	2.48 s	2160 [a]	200, 240, 240	0.00488 s (0.00246 s)	2160 [a]	0%
3	3	6	8	3	2	12	270, 250, 210	280, 280, 240	3.21 s	2520 [a]	280, 280, 240	0.00658 s (0.00346 s)	2520 [a]	0%
4	3	6	8	4	2	6	110, 150, 125	120, 160, 160	8.15 s	1200 [a]	120, 160, 160	0.00598 s (0.00342 s)	1200 [a]	0%
5	3	6	8	4	3	6	110, 150, 125	120, 160, 160	3.18 s	1080 [a]	120, 160, 160	0.00509 s (0.00371 s)	1080 [a]	0%

[a] Optimal solution.

As shown in Table 12, the proposed 3P-PCB-PH algorithm could solve all the small-sized test problems with an average and standard deviation (SD) computational time of less than 1 s respectively, for solving each problem. The makespans from the proposed heuristic algorithm were the same as the optimal makespans from the proposed MILP model ($\%gap = 0\%$), but the proposed heuristic algorithm used less computational times than the proposed MILP model. This shows that the proposed 3P-PCB-PH algorithm is very efficient and effective.

5.4.2. Results of the Medium-Sized Test Problems Using the Proposed 3P-PCB-PH Algorithm

Table 13 shows the computational results of each test problem when using the proposed heuristic algorithm compared with the results from the proposed MILP model. All the medium-sized test problems could still be solved to an optimal solution ($\%gap = 0\%$) by the proposed heuristic algorithm using only a very small average and SD computational time of less than 1 s each. This shows the efficiency and effectiveness of the proposed 3P-PCB-PH algorithm.

Table 13. Computational results of the medium-sized test problems using the proposed 3P-PCB-PH algorithm.

No.	I	K	L	P	O	T	d_i	Results Using Proposed MILP Model			Results Using Proposed 3P-PCB-PH Algorithm			$\%gap$
								Outputs	CPU Time	C_{max} (min)	Outputs	Avg CPU Time (SD)	C_{max} (min)	
1	3	6	8	6	3	6	300, 300, 300	320, 320, 320	32.81 s	1560 [a]	320, 320, 320	0.00469 s (0.00141 s)	1560 [a]	0%
2	3	6	8	6	3	8	450, 480, 500	480, 480, 520	16.79 s	2520 [a]	480, 480, 520	0.00519 s (0.00248 s)	2520 [a]	0%
3	3	6	8	6	3	12	720, 900, 600	720, 920, 600	1 min 28 s	3600 [a]	720, 920, 600	0.00658 s (0.00245 s)	3600 [a]	0%
4	4	6	8	6	3	6	200, 300, 400, 100	200, 320, 400, 120	20.65 s	1800 [a]	200, 320, 400, 120	0.00599 s (0.00266 s)	1800 [a]	0%
5	4	6	8	6	3	8	300, 400, 200, 500	320, 400, 200, 520	3 min 59 s	2280 [a]	320, 400, 200, 520	0.00768 s (0.00342 s)	2280 [a]	0%
6	4	6	8	6	3	12	500, 700, 700, 500	520, 720, 720, 520	4 min 2 s	3960 [a]	520, 720, 720, 520	0.00927 s (0.00509 s)	3960 [a]	0%
7	5	6	8	6	3	6	200, 250, 200, 250, 200	200, 280, 200, 280, 200	9 min 31 s	1920 [a]	200, 280, 200, 280, 200	0.00768 s (0.00282 s)	1920 [a]	0%
8	5	6	8	6	3	8	400, 300, 200, 250, 300	400, 320, 200, 280, 320	49.31 s	2520 [a]	400, 320, 200, 280, 320	0.00909 s (0.00331 s)	2520 [a]	0%

[a] Optimal solution.

5.4.3. Results of the Large-Sized Test Problems Using the Proposed 3P-PCB-PH Algorithm

The results of the large-sized test problems from the proposed heuristic algorithm and the proposed MILP model are compared in Table 14. Each problem was solved by the proposed heuristic algorithm using an average and SD computational time of less than 1 s each. For Problems 1, 2, and 4, the makespans from the proposed heuristic algorithm are the same as the optimal makespans from the proposed MILP model ($\%gap = 0\%$), but the proposed heuristic algorithm used much less computational time than the proposed MILP model. Furthermore, the proposed heuristic algorithm could find a near-optimal schedule with the same makespan as the best-known solution from the proposed MILP model for the other large-sized test problems using very small computational times. Note that the computational time of the proposed heuristic algorithm slightly increases when the size of problem is increased from small size to large size, which is different from the computational time of the proposed MILP model. These results show that the proposed 3P-PCB-PH algorithm is very efficient and effective for solving the pressing process scheduling.

Table 14. Computational results of the large-sized test problems using the proposed 3P-PCB-PH algorithm.

No.	I	K	L	P	O	T	d_i	Results Using Proposed MILP Model			Results Using Proposed 3P-PCB-PH Algorithm			%gap
								Outputs	CPU Time	C_{max} (min)	Outputs	Avg CPU Time (SD)	C_{max} (min)	
1	5	6	8	6	3	12	500, 500, 500, 500, 500	520, 520, 520, 520, 520	44 min 38 s	4080 [a]	520, 520, 520, 520, 520	0.00928 s (0.00346 s)	4080 [a]	0%
2	5	6	8	7	3	12	500, 500, 500, 500, 500	520, 520, 520, 520, 520	23 min 42 s	3600 [a]	520, 520, 520, 520, 520	0.00918 s (0.00297 s)	3600 [a]	0%
3	5	6	8	6	4	12	500, 500, 500, 500, 500	520, 520, 520, 520, 520	2 h	4080 [b]	520, 520, 520, 520, 520	0.00987 s (0.00447 s)	4080	0% [c]
4	6	6	8	6	3	12	500, 360, 220, 180, 380, 720	520, 360, 240, 200, 400, 770	48 min 14 s	3360 [a]	520, 360, 240, 200, 400, 770	0.00997 s (0.00326 s)	3360 [a]	0%
5	6	6	8	7	3	12	500, 360, 220, 180, 380, 720	520, 360, 240, 200, 400, 770	2 h	3000 [b]	520, 360, 240, 200, 400, 770	0.00908 s (0.00291 s)	3000	0% [c]
6	6	6	8	6	4	12	500, 360, 220, 180, 380, 720	520, 360, 240, 200, 400, 770	2 h	3360 [b]	520, 360, 240, 200, 400, 770	0.00993 s (0.00575 s)	3360	0% [c]
7	7	6	8	6	3	12	300, 325, 290, 425, 450, 475, 200	320, 360, 320, 440, 480, 490, 200	2 h	3720 [b]	320, 360, 320, 440, 480, 490, 200	0.01015 s (0.00319 s)	3720	0% [c]
8	7	6	8	7	3	12	300, 325, 290, 425, 450, 475, 200	320, 360, 320, 440, 480, 490, 200	2 h	3360 [b]	320, 360, 320, 440, 480, 490, 200	0.01250 s (0.00504 s)	3360	0% [c]
9	7	6	8	6	4	12	300, 325, 290, 425, 450, 475, 200	320, 360, 320, 440, 480, 490, 200	2 h	3720 [b]	320, 360, 320, 440, 480, 490, 200	0.01057 s (0.00566 s)	3720	0% [c]

[a] Optimal solution. [b] The best-known solution from the proposed MILP model. [c] The %gap between the solution from the heuristic algorithm and the best-known solution from the MILP model.

In addition, the results from the proposed heuristic algorithm can give valuable information. For example, from Problems 1–3, all parameters in the problems are the same except for the number of press machines and ovens. The results of Problem 2 indicate that if the number of press machines was increased by one from Problem 1, the pressing process of Problem 1 could be finished ahead of time for 360 min (i.e., the makespan was reduced from 4080 to 3600 min). However, the results of Problem 3 indicate that increasing the number of ovens by one from Problem 1 cannot reduce the makespan. The manager of the company should increase the number of press machines rather than the number of ovens if he/she wants to reduce the makespan of the pressing process of Problem 1. This is the same in Problems 4–6, and Problems 7–9. Note that if the number of press machines is increased, the number of cycles that is needed for the demands to be satisfied can be distributed to more press machines and, as a consequence, all demands can be finished faster. These show that the proposed 3P-PCB-PH algorithm can also help in deciding which resources should be increased to reduce the production time.

An example of a solution from the proposed heuristic algorithm is described below, where the results of Problem 1 are shown in Tables 15 and 16 for the variables x_{iklpt} and X_{pto}, which are equal to 1, and in Figures 11 and 12 for the Gantt charts of the press machines and ovens. These Gantt charts were different from the Gantt charts from the MILP model (Figures 9 and 10), and this shows that Problem 1 of the large-sized problems has an alternative optimal schedule. Note that, in Figure 11, each type of panel is finished as a group, which is preferable in the real manufacturing industry. The makespan of the overall process was 4080 min and the number of outputs of each panel type was 520, which satisfied the demand.

Table 15. List of non-zero x_{iklpt} values in the solution of Problem 1 of the large-sized problems using the proposed 3P-PCB-PH algorithm.

Press Machine	Non-Zero x_{iklpt}
1	$x_{11211}, x_{11212}, x_{11213}, x_{23214}, x_{23215}, x_{32216}, x_{32217}, x_{43218}, x_{43219}, x_{5121,10}, x_{5121,11}$
2	$x_{11221}, x_{11222}, x_{23223}, x_{23224}, x_{23225}, x_{32226}, x_{32227}, x_{43228}, x_{43229}, x_{5122,10}, x_{5122,11}$
3	$x_{11231}, x_{11232}, x_{23233}, x_{23234}, x_{32235}, x_{32236}, x_{32237}, x_{43238}, x_{43239}, x_{5123,10}, x_{5123,11}$
4	$x_{11241}, x_{11242}, x_{23243}, x_{23244}, x_{32245}, x_{32246}, x_{43247}, x_{43248}, x_{43249}, x_{5124,10}, x_{5124,11}$
5	$x_{11251}, x_{11252}, x_{23253}, x_{23254}, x_{32255}, x_{32256}, x_{43257}, x_{43258}, x_{51259}, x_{5125,10}, x_{5125,11}$
6	$x_{11261}, x_{11262}, x_{23263}, x_{23264}, x_{32265}, x_{32266}, x_{43267}, x_{43268}, x_{51269}, x_{5126,10}$

Table 16. List of non-zero X_{pto} values in the solution of Problem 1 of the large-sized problems using the proposed 3P-PCB-PH algorithm.

Press Machine	Non-Zero X_{pto}
1	$X_{111}, X_{121}, X_{131}, X_{141}, X_{151}, X_{161}, X_{171}, X_{181}, X_{191}, X_{1,10,1}, X_{1,11,1}$
2	$X_{212}, X_{222}, X_{232}, X_{242}, X_{252}, X_{262}, X_{272}, X_{282}, X_{292}, X_{2,10,2}, X_{2,11,2}$
3	$X_{313}, X_{323}, X_{333}, X_{343}, X_{353}, X_{363}, X_{373}, X_{383}, X_{393}, X_{3,10,3}, X_{3,11,3}$
4	$X_{411}, X_{421}, X_{431}, X_{441}, X_{451}, X_{461}, X_{471}, X_{481}, X_{491}, X_{4,10,1}, X_{4,11,1}$
5	$X_{512}, X_{522}, X_{532}, X_{542}, X_{552}, X_{562}, X_{572}, X_{582}, X_{592}, X_{5,10,2}, X_{5,11,2}$
6	$X_{613}, X_{623}, X_{633}, X_{643}, X_{653}, X_{663}, X_{673}, X_{683}, X_{693}, X_{6,10,3}$

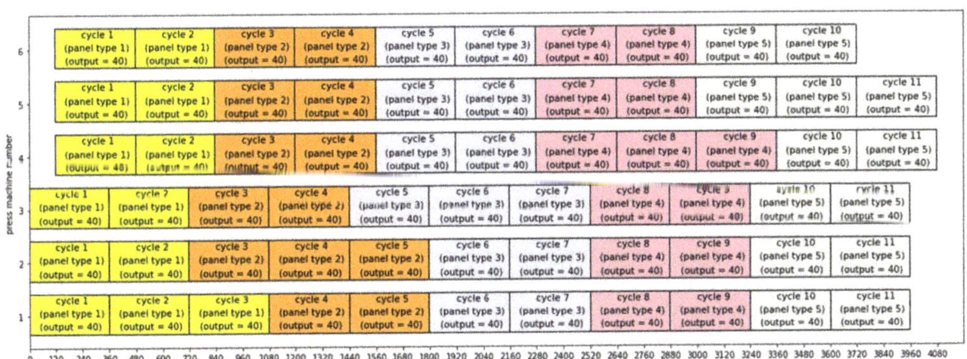

Figure 11. Gantt chart of the press machines for Problem 1 of the large-sized problems using the proposed 3P-PCB-PH algorithm.

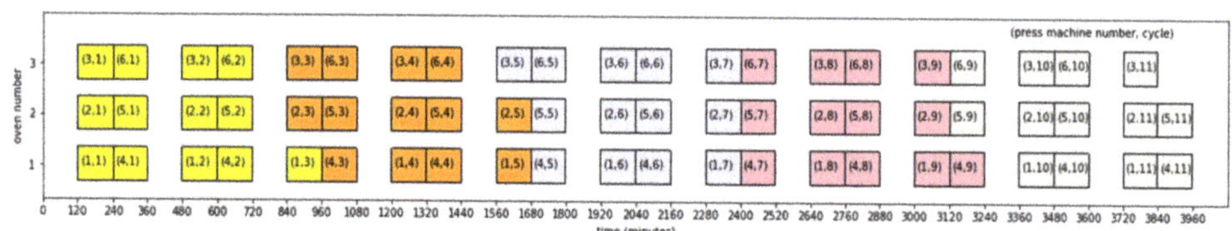

Figure 12. Gantt chart of the ovens for Problem 1 of the large-sized problems using the proposed 3P-PCB-PH algorithm.

6. Discussion

This paper presents a MILP model and a 3P-PCB-PH algorithm for solving the pressing process scheduling. From the numerical experiments, the proposed MILP model is suitable for the small-sized and medium-sized problems, where the number of binary variables is less than 30,000. The proposed MILP model tends to cause long computational times for solving the large-sized problems, where the number of binary variables is greater than 30,000. Furthermore, the running time was significantly increased as the size of the problem grows because there are a lot of feasible solutions to be verified for optimality due to many decision variables. However, the proposed MILP model has the benefit that it gives an optimal solution if one exists. On the other hand, the proposed 3P-PCB-PH algorithm is suitable for all sizes of problems. It could find an optimal solution for all problems that the proposed MILP model could find. It also can find the same best makespans as the proposed MILP model for all problems that the proposed MILP model could not find an optimal solution. The computational times of the proposed heuristic algorithm seem to be very fast and are not hugely increased when the size of the problem is increased from small size to large size. A benefit of the proposed heuristic algorithm is the saving in time to find a good solution since it used smaller computational times compared with the computational times of the proposed MILP model.

The proposed MILP model can also be easily extended to be more practical in the real-life application. For example, in the proposed model, the objective is to minimize the makespan of the overall process of the pressing process scheduling, where the demands must be satisfied. However, the surplus output of each panel type may be too large. If we also want to enforce that the surplus output of each panel type should not be too excessive with the main objective makespan, we can add the term $\varepsilon \sum_{i=1}^{I}\left[\sum_{k=1}^{K}\sum_{l=1}^{L}\sum_{p=1}^{P}\sum_{t=1}^{T} x_{iklpt}(ma_{ikl}) - d_i\right]$ to the objective function. The constant ε should be very small so that it has no effect on minimizing the main objective makespan.

7. Conclusions

This paper presented a new application of a mixed-integer linear programming to the scheduling of the pressing process in multi-layer PCB manufacturing. In the process, the panels are inserted into a press machine and then sent into an oven so that the panels are pressed and heated in the oven. The objective of the scheduling problem was to minimize the makespan of the overall pressing process. This objective can often imply increasing the utilization of available resources.

The goal of this study was to present two methods for solving the pressing process scheduling, i.e., a MILP model which is an exact method and a 3P-PCB-PH algorithm which is an approximation method. The first method illustrates a possible application of the integer linear programming that can handle a complicated problem from the real-world industry. The real data from a PCB company was used to generate the test problems. The computational results indicated that the proposed MILP model was suitable for small- and medium-size problems. The proposed MILP model could find an optimal solution for some large-sized problems and a good feasible solution for the other large-sized problems within the time limit. The MILP model has an advantage that it can guarantee to find an optimal solution if the problem can be solved optimally within the time limit. On the other hand, the proposed 3P-PCB-PH algorithm could find the optimal solutions and near optimal solutions within very small computational time. It is more suitable than the proposed MILP model when the size of the problem is large. Furthermore, the schedule from the proposed heuristic is preferable in real manufacturing than the schedule from the proposed MILP model since each type of panel is finished in a single group. Both the proposed MILP model and 3P-PCB-PH algorithm could be options to provide an optimal schedule for the pressing process in any PCB industries or could be adapted to other industrial applications with similar aspects of scheduling.

Some additional constraints can be introduced into the pressing process for further development. For example, the cycle time depends on each type of panel, one cycle of a press machine can press more than one type of panel, and some types of panels have a higher priority or different due date. Adding these factors to the problem would also be a very challenging task for the future research, but also increase the complexity of the problem.

The limitations of this paper are that the problem is assumed to have the same size of press machines and the same size of ovens. In reality, however, a PCB company may have several sizes of press machines or ovens.

Author Contributions: Conceptualization, T.L., B.I., and C.J.; methodology, T.L., B.I., and C.J.; software, T.L. and C.J.; validation, T.L., B.I., and C.J.; formal analysis, T.L., B.I., and C.J.; investigation, T.L., B.I., and C.J.; resources, C.J.; data curation, C.J.; writing—original draft preparation, T.L.; writing—review and editing, B.I. and C.J.; visualization, T.L.; supervision, B.I. and C.J. All authors have read and agreed to the published version of the manuscript.

Funding: This research received no external funding.

Acknowledgments: This work was supported in part by the Development and Promotion of Science and Technology Talented Project (DPST), Thailand, in part by Center of Excellence in Logistics and Supply Chain Systems Engineering and Technology (LogEn Tech), Sirindhorn International Institute of Technology, Thammasat University, Thailand, and in part by Department of Mathematics and Computer Science, Faculty of Science, Chulalongkorn University, Thailand.

Conflicts of Interest: The authors declare no conflict of interest.

References

1. Khandpur, R.S. *Printed Circuit Boards: Design, Fabrication, Assembly and Testing*; McGraw-Hill: New York, NY, USA, 2006.
2. Gilmore, P.C.; Gomory, R.E. Multistage cutting stock problems of two and more dimensions. *Oper. Res.* **1965**, *13*, 94–120. [CrossRef]
3. Macedo, R.; Alves, C.; de Carvalho, J.M.V. Arc-flow model for the two-dimensional guillotine cutting stock problem. *Comput. Oper. Res.* **2010**, *37*, 991–1001. [CrossRef]
4. Mrad, M.; Meftahi, I.; Haouari, M. A branch-and-price algorithm for the two-stage guillotine cutting stock problem. *J. Oper. Res. Soc.* **2013**, *64*, 629–637. [CrossRef]
5. Alvarez-Valdes, R.; Parajon, A.; Tamarit, J.M. A computational study of LP-based heuristic algorithms for two-dimensional guillotine cutting stock problems. *OR Spectr.* **2002**, *24*, 179–192. [CrossRef]
6. Furini, F.; Malaguti, E., Duran, R.M.; Persiani, A.; Toth, P. A column generation heuristic for the two-dimensional two-staged guillotine cutting stock problem with multiple stock size. *Eur. J. Oper. Res.* **2012**, *218*, 251–260. [CrossRef]
7. Tieng, K.; Sumetthapiwat, S.; Dumrongsiri, A.; Jeenanunta, C. Heuristics for two-dimensional rectangular guillotine cutting stock. *Thail. Stat.* **2016**, *14*, 147–164.
8. Sumetthapiwat, S.; Intiyot, B.; Jeenanunta, C. A column generation on two-dimensional cutting stock problem with fixed-size usable leftover and multiple stock sizes. *Int. J. Logist. Manag.* **2020**, *35*, 273–288. [CrossRef]
9. Onwubolu, G.C.; Clerc, M. Optimal path for automated drilling operations by a new heuristic approach using particle swarm optimization. *Int. J. Prod. Res.* **2004**, *42*, 473–491. [CrossRef]
10. Saealal, M.S.; Abidin, A.F.; Adam, A.; Mukred, J.; Khalil, K.; Yusof, Z.M.; Ibrahim, Z.; Nordin, N. An ant colony system for routing in PCB holes drilling process. *IJIMIP* **2013**, *4*, 50–56.
11. Lim, W.C.E.; Kanagaraj, G.; Ponnambalam, S.G. PCB drill path optimization by combinatorial cuckoo search algorithm. *Sci. World J.* **2014**, 264518. [CrossRef]

12. Kanagaraj, G.; Ponnambalam, S.G.; Lim, W.C.E. Application of a hybridized cuckoo search-genetic algorithm to path optimization for PCB holes drilling process. In Proceedings of the IEEE International Conference on Automation Science and Engineering, Taipei, Taiwan, 18–22 August 2014; pp. 18–22.
13. Ji, P.; Sze, M.T.; Lee, W.B. A genetic algorithm of determining cycle time for printed circuit board assembly lines. *Eur. J. Oper. Res.* **2001**, *128*, 175–184. [CrossRef]
14. Kodek, D.M.; Krisper, M. Optimal algorithm for minimizing production cycle time of a printed circuit board assembly line. *Int. J. Prod. Res.* **2004**, *42*, 5031–5048. [CrossRef]
15. Emet, S.; Knuutila, T.; Alhoniemi, E.; Maier, M.; Johnsson, M.; Nevalainen, O.S. Workload balancing in printed circuit board assembly. *Int. J. Adv. Manuf. Technol.* **2010**, *50*, 1175–1182. [CrossRef]
16. He, T.; Li, D.; Yoon, S.W. A heuristic algorithm to balance workloads of high-speed SMT machines in a PCB assembly line. *Procedia Manuf.* **2017**, *11*, 1790–1797. [CrossRef]
17. Damodaran, P.; Srihari, K.; Lam, S.S. Scheduling a capacitated batch-processing machine to minimize makespan. *Robot. Comput. Integr. Manuf.* **2007**, *23*, 208–216. [CrossRef]
18. Damodaran, P.; Diyadawagamage, D.A.; Ghrayeb, O.; Velez-Gallego, M.C. A particle swarm optimization algorithm for minimizing makespan of nonidentical parallel batch processing machines. *Int. J. Adv. Manuf. Technol.* **2012**, *58*, 1131–1140. [CrossRef]
19. Noroozi, A.; Mokhtari, H. Scheduling of printed circuit board (PCB) assembly systems with heterogeneous processors using simulation-based intelligent optimization methods. *Neural. Comput. Appl.* **2015**, *26*, 857–873. [CrossRef]
20. Hulett, M.; Damodaran, P.; Amouie, M. Scheduling non-identical parallel batch processing machines to minimize total weighted tardiness using particle swarm optimization. *Comput. Ind. Eng.* **2017**, *113*, 425–436. [CrossRef]
21. Ozguven, C.; Ozbakir, L.; Yavuz, Y. Mathematical models for job-shop scheduling problems with routing and process plan flexibility. *Appl. Math. Model.* **2010**, *34*, 1539–1548. [CrossRef]
22. Zhang, G.; Gao, L.; Shi, Y. An effective genetic algorithm for the flexible job-shop scheduling problem. *Expert Syst. Appl.* **2011**, *38*, 3563–3573. [CrossRef]
23. Li, X.; Gao, L. An effective hybrid genetic algorithm and tabu search for flexible job shop scheduling problem. *Int. J. Prod. Econ.* **2016**, *174*, 93–110. [CrossRef]
24. Luan, F.; Cai, Z.; Wu, S.; Liu, S.Q.; He, Y. Optimizing the low-carbon flexible job shop scheduling problem with discrete whale optimization algorithm. *Mathematics* **2019**, *7*, 688. [CrossRef]
25. Pan, C.H. A study of integer programming formulations for scheduling problems. *Int. J. Syst. Sci.* **1997**, *28*, 33–41. [CrossRef]

 mathematics

MDPI

Article

Branch Less, Cut More and Schedule Jobs with Release and Delivery Times on Uniform Machines

Nodari Vakhania [1] and Frank Werner [2,*]

[1] Centro de Investigacion en Ciences, Universidad Autónoma del Estado de Morelos, Cuernavaca 62209, Mexico; nodari@uaem.mx
[2] Faculty of Mathemaics, Otto-von-Guericke University Magdeburg, PSF 4120, 39016 Magdeburg, Germany
* Correspondence: frank.werner@ovgu.de; Tel.: +49-391-675-2025

Abstract: We consider the problem of scheduling n jobs with identical processing times and given release as well as delivery times on m uniform machines. The goal is to minimize the makespan, i.e., the maximum full completion time of any job. This problem is well-known to have an open complexity status even if the number of jobs is fixed. We present a polynomial-time algorithm for the problem which is based on the earlier introduced algorithmic framework blesscmore ("branch less and cut more"). We extend the analysis of the so-called behavior alternatives developed earlier for the version of the problem with identical parallel machines and show how the earlier used technique for identical machines can be extended to the uniform machine environment if a special condition on the job parameters is imposed. The time complexity of the proposed algorithm is $O(\gamma m^2 n \log n)$, where γ can be either n or the maximum job delivery time $q_{\max}$. This complexity can even be reduced further by using a smaller number $\kappa < n$ in the estimation describing the number of jobs of particular types. However, this number κ becomes only known when the algorithm has terminated.

Keywords: scheduling; uniform machines; release time; delivery time; time complexity; algorithm

Citation: Vakhania, N.; Werner, F. Branch Less, Cut More and Schedule Jobs with Release and Delivery Times on Uniform Machines. *Mathematics* **2021**, *9*, 633. https://doi.org/10.3390/math9060633

Academic Editor: Armin Fügenschuh

Received: 6 January 2021
Accepted: 8 March 2021
Published: 16 March 2021

1. Introduction

In this paper, we consider a basic optimization problem of scheduling jobs with release and delivery times on uniform machines with the objective to minimize the makespan. More precisely, n jobs from the set $J = \{1, 2, ..., n\}$ are to be processed by m parallel *uniform machines* (or *processors*) from the set $M = \{1, 2, ..., m\}$. Job $j \in J$ is available from its *release time* r_j; it needs a continuous (integer) *processing time* p, which is the time that it needs on a *slowest* machine. We assume that the machines in the set M are ordered by their speeds, the fastest machines first, i.e., $s_1 \geq s_2 \geq \cdots \geq s_m$ are the corresponding machine speeds, s_i being the speed of machine i. Without loss of generality, we assume that $s_m = 1$, and the processing time of job j on machine i is an integer p/s_i. Job j has one more parameter, the *delivery time* q_j, an integer number which represents the amount of additional time units which are necessary for the *full* completion of job j *after* it completes on the machine. Thus, notice that the delivery of job j consumes no machine time (the delivery is accomplished by an independent agent).

Now, we define a *feasible schedule* S as a function that assigns to each job j a starting time t_j^S and a machine i from the set M such that, for any job j, we have $t_j^S \geq r_j$ and $t_j^S \geq t_k^S + p/s_i$ holds for any job k scheduled before job j on the same machine. Note that the first inequality requires that a job cannot start its processing before before the given release time, and the second one describes the constraint that each machine can process only one job at any time. The *completion time* of job j in the schedule S is the time moment when the processing of job i is complete on the machine i to which it is assigned in the schedule S, i.e., $c_j^S = t_j^S + p/s_i$ and the *full completion time* of job j in the schedule S is $C_j^S = c_j^S + q_j$ (the full completion time of job j takes into account the delivery time of that job, whereas the completion time of job j does not depend on its delivery time). The goal is to determine an *optimal schedule* S being feasible and minimizing the maximum full job completion time of any job

$$C_{max}(S) = \max_j C_j$$

or the *makespan*.

The studied multiprocessor optimization problem, described below, is commonly abbreviated as $Q|p_j{=}p, r_j, q_j|C_{\max}$ (its version with identical parallel machines is abbreviated as $P|p_j{=}p, r_j, q_j|C_{\max}$, the first field specifies the machine environment, the second one the job parameters, and the third one the objective function).

It is well-known that there is an equivalent (perhaps more traditional) formulation of the above described problem, in which, instead of the delivery time q_j, every job j has its due-date d_j. The *lateness* of job j in the schedule S is $L_j^S = c_j^S - d_j$. Then, the objective becomes to minimize the maximum job lateness $L_{\max}$, i.e., find a feasible schedule $S_{\rm OPT}$ in which the maximum job lateness is not more than in any other feasible schedule, i.e., $S_{\rm OPT}$ is an *optimal* schedule. The equivalence is easily established by associating with each job delivery time a corresponding due-date, and, vice-versa, see e.g., Bratley et al. [1]). The version of the problem with due-dates with identical and uniform machine environments are commonly abbreviated as $P|r_j, d_j|L_{\max}$ and $Q|r_j, d_j|L_{\max}$, respectively.

For the problem considered, each machine from a group of parallel uniform machines is characterized by its own speed, independent from a particular job that can be assigned to it, unlike a machine from a group of unrelated machines whose speed is job-dependent. Because of the uniform speed characteristic, scheduling problems with uniform machines are essentially easier than scheduling problems with unrelated machines, whereas scheduling problems with identical machines are easier than those with uniform ones.

The general problem of scheduling jobs with release and delivery times on uniform machines is well-known to be strongly NP-hard as already the single-machine version is strongly NP-hard. However, if all jobs have equal processing times, the problem can be polynomially solved on identical machines. The version on uniform machines $Q|p_j{=}p, r_j, q_j|C_{\max}$ is a long-standing open problem even in case the number of machines m is fixed. In this paper, we present a polynomial-time algorithm for the uniform machine environment which finds an optimal solution to the problem if for any pair of jobs i and j with $q_i > q_j$ and $r_j > r_i$, we have

$$q_i - q_j \geq r_j - r_i \tag{1}$$

The proposed algorithm relies on the blesscmore ("branch less, cut more") framework for the identical machine case $P|p_j{=}p, r_j, q_j|C_{\max}$ from [2] (the blesscmore algorithmic concept was formally introduced later in [3]). A blesscmore algorithm generates a solution tree similar to a branch-and-bound algorithm, however, the branching and cutting criteria are based on a direct analysis of some structural properties of the problem under consideration without using lower bounds. The algorithmic framework, on which the blesscmore algorithm that we describe here is based, takes an advantage of some nice structural properties of specially created schedules which are analyzed in terms of the so-called behavior alternatives from [2]. The framework resulted in an $O(q_{\max} mn \log n + O(m\kappa n))$ time algorithm with $q_{\max}$ being the maximum delivery time of a job and $\kappa < n$ being a parameter which becomes known only after the algorithm has terminated. Each schedule is easily created by a well-known greedy algorithm commonly referred to as Largest Delivery Time heuristic (LDT-heuristic for short): iteratively, among all released jobs, it schedules one with the largest delivery time. The algorithm from [2] carries out the enumeration of LDT-schedules (ones created by the LDT-heuristic)—it is known that there is an optimal LDT-schedule. Based on the established properties, the set of LDT-schedules is reduced to a subset of polynomial size which yields a polynomial time overall performance. Although the LDT-heuristic applied to a problem instance with uniform machines does not provide the desirable properties, it can be modified to a similar method that takes into account the uniform speed characteristic. While scheduling identical machines, the minimum completion time of each next selected job is always achieved on the machine next to the one to which the previous job was assigned. With uniform machines, this is not necessarily the case, for example, the next machine can be much slower than the current one. Hence, the time moment at which the job will complete on each of the machines needs to be additionally determined and then this job can be assigned to a machine on which the above minimum is reached. In this paper, we use an adaptation of the LDT-heuristic, which will be referred to as the LDTC-heuristic, and a schedule created by the later heuristic will be referred to as an LDTC-schedule. Instead of enumerating the LDT-schedules (as in [2]), the algorithm proposed here enumerates LDTC-schedules. Some properties for the identical machine environment which do not immediately hold for uniform machines are reformulated in terms of uniform machines, which allows for maintaining the basic framework from [2], which, as suggested earlier, turned out to be sufficiently flexible.

Similarly to the existence of an optimal LDT-schedule for the identical machine environment, there exists an optimal LDTC-schedule for the uniform machine environment. The complete enumeration of the LDTC-schedules is avoided by the generalization of nice properties of LDT-schedules to LDTC-schedules for uniform machines. These properties are obtained via the analysis of the so-called behavior alternatives from [2] that are generalized for uniform machines. The algorithm presented in this paper in the worst case requires $O(\gamma m^2 n \log n)$ steps with γ being any of the number n of jobs or the maximum delivery time $q_{\max}$ of a job. In fact, n can be replaced by a smaller magnitude κ, the number of special types of jobs; this is the same parameter κ as for the algorithm from [2], which becomes known only when the algorithm halts. The running time of the proposed algorithm is worse than that of the one from [2], in part because of the cost of the LDTC-heuristic which is repeatedly used during the solution process.

The remainder of this paper is as follows: in Section 2, we give a brief literature review. Section 3 presents some necessary preliminaries. Then, the basic algorithmic framework is given in Section 4. Section 5 discusses the performance analysis of the developed blesscmore algorithm. Section 6 contains a final discussion and concluding remarks.

2. Literature Review

If the job processing times are arbitrary, then the problem is known to be strongly NP-hard, even if there is only a single machine $1|r_j, d_j|L_{\max}$ [4]. McMahon & Florian [5] gave an efficient branch and bound algorithm, and Carlier [6] later adopted it for the version with jobs delivery times $1|r_j, q_j|C_{\max}$ (a solution to the latter version can immediately be used for the calculation of lower bounds for a more general job shop scheduling problem). For the single machine case, Baptiste gave an $O(n^7)$ algorithm for the problem $1|r_j, p_j{=}p|\sum T_j$ [7] and also an algorithm of the same complexity for the problem $1|r_j, p_j{=}p|\sum w_j U_j$ [8] of minimizing the weighted number of late jobs. Chrobak et al. [9] have derived an algorithm of improved complexity $O(n^5)$ for the case of unit weights, i.e., for the problem $1|r_j, p_j{=}p|\sum U_j$. Later, Vakhania [10] gave an $O(n^2 \log n)$ blesscmore algorithm for this problem. Note that, for the problem $1|r_j, p_j, pmtn|\sum U_j$ with arbitrary processing times and allowed preemptions, Vakhania [11] derived an $O(n^3 \log n)$ blesscmore algorithm.

One may consider a slight relaxation of problems $1|r_j, d_j|L_{\max}$, $P|r_j, d_j|L_{\max}$ and $Q|r_j, d_j|L_{\max}$ in which one looks for a schedule in which no job completes after its due-date. Such a feasibility setting with a single machine was considered by Garey et al. [12]. They have proposed an $O(n^2 \log n)$ algorithm that has further been improved to an $O(n \log n)$ one by using a very sophisticated data structure. This paper uses the concept of a so-called forbidden region describing an interval in which it is forbidden to start any job in a feasible schedule. Later, Simons and Warmuth [13] have constructed an $O(n^2 m)$ time algorithm for the feasibility setting with the identical machine environment also using the concept of forbidden regions. (It can be mentioned that the minimization version of the problem can be solved by applying an algorithm for the feasibility problem by repeatedly increasing the due-dates of all jobs until a feasible schedule with the modified due dates is found. Using binary search makes such a reduction procedure more efficient and reduces the reduction cost to $O(\log(np/m))$.)

Dessousky et al. [14] considered scheduling problems on uniform machines with simultaneously released jobs (i.e., with $r_j = 0$ for every job j) and with different objective criteria. They proposed fast polynomial-time algorithms for these problems, in particular, for the criterion $L_{\max}$. In fact, the LDTC-heuristic is an adaptation of an optimal solution method that the authors in [14] constructed for the criterion $L_{\max}$.

For a uniform machine environment with allowed preemptions ($pmtn$), the problem $Q|r_j, pmtn|C_{max}$ is polynomially solvable even for arbitrary processing times [15], while a polynomial algorithm for the problem $Q|r_j, p_j{=}p, pmtn|\sum C_j$ with minimizing total weighted completion time in the case of equal processing times has been given in [16]. The case of unrelated machines is very hard. A polynomial algorithm exists for the problem $R|r_j, pmtn|L_{max}$ with allowed preemptions and minimizing maximum lateness, even for the case of arbitrary processing times [17]. If preemptions are forbidden, Vakhania et al. [18] gave a polynomial algorithm for the case of minimizing the makespan when only two processing times p and $2p$ are possible (i.e., for the problem $R|p \in \{p, 2p\}|C_{max}$. Note that the case of two arbitrary processing times p and q is known to be NP-hard [19]. For the special case of identical parallel machines, there exist several works for the same setting as considered in this paper but for more complicated objective functions regarding the complexity status. In particular, the problems $P|r_j, p_j{=}p|\sum w_j C_j$ of minimizing the weighted sum of completion times [20] and $P|r_j, p_j{=}p|\sum T_j$ of minimizing total tardiness [21] can be polynomially solved by a reduction to a linear programming problem. In [3], Vakhania presented an $O(n^3 \log n)$ blesscmore algorithm for the problem $P|r_j, p_j{=}p|\sum U_j$ of minimizing the number of late jobs. His blesscmore algorithm uses a solution tree, where the branching and cutting criteria are based on the analysis of behavior alternatives. Moreover, the problem $P|r_j, p_j{=}p|\sum f_j(C_j)$ can also be polynomially solved for the case that f_j is an arbitrary non-decreasing function such that the difference $f_i - f_j$ is monotonic for any indices i and j [22]. The authors also applied a linear programming approach. An overview of selected solution approaches for some related scheduling problems with equal processing times is given in Table 1. It can also be mentioned that a detailed survey on parallel machine scheduling problems with equal processing times has been given in [23].

Table 1. Overview of solution approaches for related problems with equal processing times.

Problem	Approach	Reference
$1\|r_j, p_j{=}p\|\sum T_j$	dynamic programming $O(n^7)$	Baptiste [7]
$1\|r_j, p_j{=}p\|\sum w_j U_j$	dynamic programming $O(n^7)$	Baptiste [8]
$1\|r_j, p_j{=}p\|\sum U_j$	blesscmore algorithm $O(n^2 \log n)$	Vakhania [10]
$P\|r_j, p_j{=}p\|\sum w_j C_j$	linear programming	Brucker & Kravchenko [20]
$P\|r_j, p_j{=}p\|\sum T_j$	linear programming	Brucker & Kravchenko [21]
$P\|r_j, p_j{=}p\|\sum T_j$	blesscmore algorithm $O(n^3 \log n)$, behavior alternatives	Vakhania [3]
$Q\|r_j, p_j{=}p, pmtn\|\sum C_j$	linear programming	Kravchenko & Werner [16]

3. Preliminaries

This section contains some useful properties, necessary terminology, and concepts, some of which were introduced in [2] for identical machines.

LDTC-heuristic. We first describe the LDTC-heuristic, an adaptation of the LDT-heuristic for uniform machines. As earlier briefly noted, unlike an LDT-schedule, an LDTC-schedule is not defined by a mere permutation of the given n jobs since the machine to which the next selected job is assigned depends on the machine speed. Starting from the minimal job release time, the current scheduling time is iteratively set as the minimum release time among all yet unscheduled jobs. Iteratively, among all jobs released by the current scheduling time, the LDTC-heuristic determines one with the largest delivery time (a most *urgent* one) and schedules it on the machine on which the earliest possible completion time of this job is attained (ties can be broken by selecting the machine with the minimum index):

Note that, in an LDTC-schedule S, a machine will contain an idle-time interval (a *gap*) if and only if there is no unscheduled job released by the current scheduling time. The running time of the modified heuristic is the same as that of the LDT-heuristic with an additional factor of m due to the machine selection at each iteration (which is not required for the uniform machine environment), which results in the time complexity $O(mn \log n)$.

Example 1. *We shall illustrate the basic notions and the algorithm described here on a small problem instance with 10 jobs and two uniform machines with* $s_1 = 2$ *and* $s_2 = 1$*. The processing time of all jobs (on machine 2) is 20. The rest of the parameters of these jobs are defined as follows:* $r_1 = r_2 = 0$, $r_3 = r_4 = 1$, $r_5 = r_6 = r_7 = 23$ *and* $r_8 = r_9 = r_{10} = 45$. $q_1 = q_2 = 0$, $q_3 = q_4 = 51$, $q_5 = q_6 = q_7 = 75$ *and* $q_8 = q_9 = q_{10} = 54$*. The LDTC-schedule obtained by the LDTC-heuristic for the problem instance of the above example is depicted in Figure 1. In general, we denote the LDTC-schedule obtained by the LDTC-heuristic for the initially given instance of problem* $Q|p_j{=}p, r_j, q_j|C_{\max}$ *by* σ *(as we will see in the following subsection, we may generate alternative LDTC-schedules by iteratively modifying the originally given problem instance).*

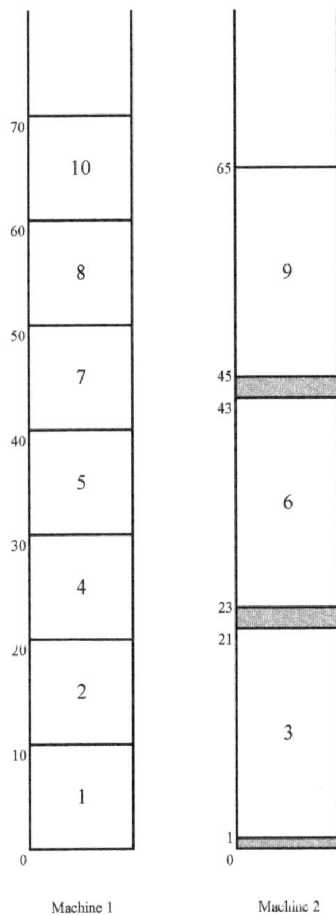

Figure 1. Initial LDTC schedule σ.

The next property of an LDTC-schedule easily follows from the definition of the LDTC-heuristic and the equality of the job processing times (job j is said to have the ordinal number i in schedule S if it is the ith scheduled job in that schedule S):

Property 1. *If in an LDTC-schedule S, job j is scheduled after job i, i.e., the ordinal number of job j in S is larger than that of job i, then $c_j^S \geq c_i^S$.*

Next, we give another easily seen important property of an LDTC-schedule S on which the proposed method essentially relies. Let A be a set of, say k jobs, all of which being released by time moment t, and let π be any permutation of k jobs all of them being also released by time t (recall that all the jobs have equal length). Let, further, $S(A)$ be a partial LDTC-schedule constructed for the jobs in the set A, and let $S(\pi)$ be a list schedule constructed for the permutation π (it includes the jobs according to the order in permutation π, leaves no avoidable gap, and assigns each job to the machine on which it will complete sooner breaking ties again by selecting the machine with the minimum index). The following property, roughly, states that both above schedules are indistinguishable if the delivery times of the jobs are ignored:

Property 2. *The completion time of every machine in both schedules $S(A)$ and $S(\pi)$ is the same. Moreover, the ith scheduled job in the schedule $S(A)$ starts and completes at the same time as the ith scheduled job on the corresponding machine in the schedule $S(\pi)$.*

The above property also holds for a group of identical machines and is helpful for the generalization of the earlier results for identical machines from [2] to uniform machines. Roughly, ignoring the job release times, the property states that two list schedules constructed for two different permutations with the same number of jobs have the same structure. Although the starting and completion times of the jobs scheduled in the same position on the corresponding machine are the same in both schedules, the full completion times will not necessarily be the same (this obviously depends on the delivery times of these jobs).

A block. A consecutive independent part in a schedule is commonly referred to as a *block* in the scheduling literature. We define a block in an LDTC-schedule S as the largest fragment (a consecutive sequence of jobs) of that schedule such that, for each two successively scheduled jobs i and j, job j starts no later than job i finishes (jobs i and j can be scheduled on the same or different machines according to the LDTC-heuristic). It follows that there is a single block that starts schedule S and there is also a single block that finishes this schedule. If these blocks coincide, then there is a single block in the schedule S; otherwise, each next block is "separated" from the previous one with gaps on each of the machines. Here, a zero length gap between jobs i and j will be distinguished in case job j is scheduled at time $r_j = c_i^S$ on the same machine as job i (it immediately succeeds job i on this machine). It is easily observed that the schedule given in Figure 1 consists of a single block (note that the order in which the jobs are included in this schedule coincides with the enumeration of these jobs).

A block B (with at least two elements) possesses the following property that will be used later. Suppose the ιth scheduled job j (not the last scheduled job of the block) is removed from that block and the LDTC-heuristic is applied to the remaining jobs of the block. As a consequence, in the resulting (partial) schedule, the processing interval of the job scheduled as the ιth one overlaps with the processing interval of job j in the block B earlier.

Some additional definitions are required to specify how the proposed algorithm creates and evaluates the feasible schedules. At any stage of the execution of the algorithm, independently whether a new LDTC-schedule will be generated or not, depends on specific properties of the LDTC-schedules already generated by that stage. The definitions below are helpful for the determination of these properties.

An overflow job. In an LDTC-schedule S, let o be a job realizing the maximum full completion time of a job, i.e.,

$$C_o(S) = C_{max}(S), \tag{2}$$

and let $B(S)$ be the *critical block* in the schedule S, i.e., the block containing the earliest scheduled job o satisfying Equation (2). The *overflow job* $o(S)$ in the schedule S is the last scheduled job in the block $B(S)$ satisfying Condition (2), i.e., one with the maximum $C_o(S)$. It can be easily verified that in the schedule in Figure 1, job 7 is the overflow job with $C_{\max} = C_7 = 50 + 75 = 125$ (the full completion time of the latest completed job 10 is $70 + 54 = 124$).

A kernel. Next, we define an important component in an LDTC-schedule S defined as its fragment containing the overflow job $o = o(S)$ such that the jobs scheduled before job o in the block $B(S)$ have a delivery time not smaller than q_o (we will write o instead of $o(S)$ when this causes no confusion). This sub-fragment of the block $B(S)$ is called its *kernel* and is denoted by $K(S)$. The kernel in the schedule in Figure 1 is the fragment of that schedule constituted by the jobs 5, 6, and 7.

Intuitively, on the one hand, the kernel $K(S)$ is a critical part in a schedule S, and, on the other hand, it is relatively easy to arrange the kernel jobs optimally. In fact, we will explore different LDTC-schedules identifying the kernel in each of these schedules. We will also relate this kernel to the kernels of the earlier generated LDTC-schedules. We need to introduce a few more definitions.

An emerging job. Suppose that a job j of kernel $K(S)$ is *pushed* by a non-kernel job e scheduled before that job in the block $B(S)$, that is, the LDTC-heuristic would schedule job j earlier if job i was forced to be scheduled after job j. If $q_e < q_o$, then job e is called a *regular emerging job* in the schedule S, and the latest scheduled (regular) emerging job (the one closest to job o) is called the *delaying* emerging job. The emerging jobs in the schedule in Figure 1 are jobs 1, 2, 3, and 4, and job 4 is the delaying emerging job (in general, there may exist a non-kernel non-emerging job scheduled before the kernel $K(S)$ in the block $B(S)$).

The following optimality condition can be established already in the initial LDTC-schedule σ.

Lemma 1. *If the initial LDTC-schedule σ contains a kernel K such that no job of that kernel is pushed by an emerging job, then this schedule is optimal.*

Proof. Using an interchange argument, we show that no reordering of the jobs of the kernel K can be beneficial. First, we note that the first job j of kernel K must be scheduled on machine 1 since otherwise, it would have been pushed by the corresponding job scheduled on that machine. But this job cannot exist since, by the condition of the lemma, it cannot be pushed by an emerging job (and if it is not an emerging job, it should have been a part of kernel K). Since machine 1 will finish job j at least as early as any other machine, the full completion of job j cannot be reduced.

Let i and j be two successively scheduled jobs from the kernel K, α and β be the machines to which jobs i and j are assigned, respectively, in the schedule σ. Without loss of generality, assume that $\alpha < \beta$ as otherwise it is easy to see that interchanging jobs i and j cannot give any benefit. We let σ' be the schedule obtained from schedule σ by interchanging jobs i and j.

Assume first that jobs i and j are among the first m (or less) scheduled jobs from the kernel K. By the condition of the lemma, both jobs start at their release time in the schedule σ. We show that interchanging jobs i and j cannot be beneficial by establishing that

$$\max\{C_i^{\sigma}, C_j^{\sigma}\} \leq \max\{C_i^{\sigma'}, C_j^{\sigma'}\}.$$

Since $C_i^{\sigma} < C_j^{\sigma}$, $\max\{C_i^{\sigma}, C_j^{\sigma}\} = C_j^{\sigma}$, hence we need to show that $\max\{C_i^{\sigma'}, C_j^{\sigma'}\} \geq C_j^{\sigma}$. Since $C_j^{\sigma'} \leq C_j^{\sigma}$, it will suffice to show that

$$C_i^{\sigma'} \geq C_j^{\sigma}. \tag{3}$$

We have

$$C_i^{\sigma'} = r_i + p/s_{\beta} + q_i$$

and

$$C_j^{\sigma} = r_j + p/s_{\beta} + q_j.$$

However, by Condition (1), $q_i - q_j \geq r_j - r_i$, which establishes Inequality (3).

Suppose now that jobs i and j are not among the first m scheduled jobs of the kernel K. If by the current scheduling time both jobs are released, then by a similar interchange argument Inequality (3) can easily be established (without using Condition (1)).

It remains to consider the case when job j is released within the execution interval of job i, hence $t_j^{\sigma} = r_j$. We have

$$C_i^{\sigma'} - C_j^{\sigma} = q_i - q_j + t_i^{\sigma'} - t_j^{\sigma} = (q_i - q_j) - (r_j - t_i^{\sigma'}).$$

Again, by Condition (1), $q_i - q_j \geq r_j - r_i \geq r_j - t_j^{\sigma'}$ and Inequality (3) again holds.

Applying repeatedly the above interchange argument to all pairs of jobs from the kernel K, we obtain that no rearrangement of the jobs of kernel K may result in a maximum full job completion time less than that of the overflow job $o(\sigma)$, i.e., the schedule σ is optimal. □

Constructing Alternative LDTC-Schedules

Due to Lemma 1, from here on, it is assumed that the condition in this lemma is not satisfied, i.e., there exists an emerging job e in the schedule S (note that $e \in B(S)$ as otherwise job e may not push a job of the kernel $K(S)$). Since job e is pushing a job of kernel $K(S)$, the removal of this job may potentially decrease the start and hence the full completion time of the overflow job $o(S)$. At the same time, note again that, by the definition of a block, the omission of a job not from the block $B(S)$ may not affect the starting time of any job from the block $B(S)$. This is why we restrict here our attention to the jobs of the block $B(S)$. (Here, we only mention that later we will also apply an alternative notion of a passive emerging job, and then the notion "emerging job" is used either for a regular or a passive emerging job; until then, we use "emerging job" for a "regular emerging job").

Clearly, no emerging job can actually be removed as the resultant schedule would be infeasible. Instead, to restart the jobs in the kernel $K(S)$ earlier, an emerging job e is *applied* to this kernel, i.e., it is forced to be rescheduled after all jobs of the kernel $K(S)$ whereas any job, scheduled after the kernel $K(S)$ in the schedule S, is maintained to be scheduled after that kernel. The LDTC-heuristic is newly applied with the restriction that the scheduling of job e and all jobs, scheduled after kernel $K(S)$ in schedule S is forbidden until all jobs of kernel $K(S)$ are scheduled. The resultant LDTC-schedule is denoted by S_e (the so-called *complementary schedule* or a *C-schedule*) (Such a schedule generation technique was originally suggested by McMahon and Florian [5] for the single-machine setting.)

By Lemma 1, the kernel $K(S)$, a fragment of the LDTC-schedule S considered as an independent LDTC-schedule is optimal if it possesses no emerging job. Otherwise, the jobs of the kernel $K(S)$ are pushed by the corresponding emerging jobs. Some of these emerging jobs can be scheduled after the kernel $K(S)$ in an optimal complete schedule S_{OPT}. Such a rescheduling is achieved by the creation of the corresponding C-schedules (as we will see in Lemma 2, it will suffice to consider only C-schedules, i.e., S_{OPT} is a C-schedule). In Figure 2, a complementary schedule σ_4 is depicted in which the delaying emerging job 4 is rescheduled after all jobs of kernel $K(\sigma)$ (where σ is the initial LDTC-schedule of Figure 1).

The application of an emerging job has two "opposite" effects. On the positive side, since the number of jobs scheduled before the kernel $K(S)$ in the schedule S_e is one less than that in the schedule S, the overflow job $o(S)$ in the schedule S_e will be completed earlier than it was completed in the schedule S; likewise, the completion time of that job, which is scheduled as the latest one of the kernel $K(S)$ in the schedule S_e, will be smaller than the completion time of the job $o(S)$ in the schedule S. Hence, the application of an emerging job gives a potential to improve schedule S. On the negative side, it creates a new gap within the former execution interval of job e or at a later time moment before kernel $K(S)$ (see Lemma 1 in [2] for a proof for the case of identical machines, the uniform machine case can be proved similarly). Such a gap may enforce a right-shift (delay) of the jobs included after job e in the schedule S_e; (for example, a new gap "[20–23]" that arises for machine 1 in the C-schedule σ_4 in Figure 2 enforces a right-shift of jobs 8 and 10 included behind job 4). Thus, roughly, the C-schedule S_e favors the kernel $K(S)$ but creates a potential conflict for later scheduled jobs (jobs 8 and 10 in the above example).

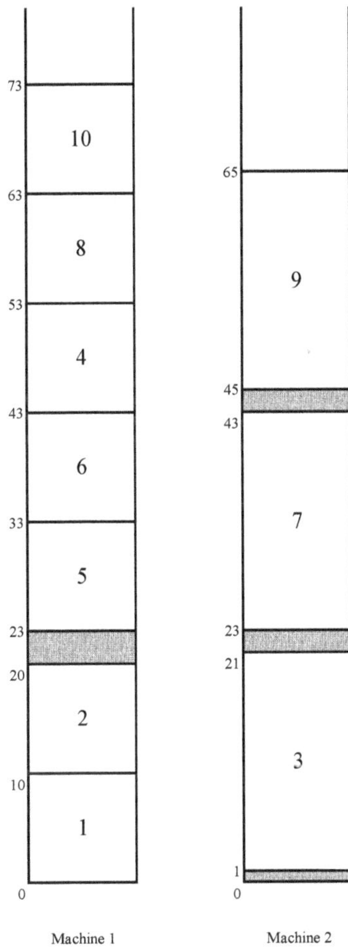

Figure 2. The C-schedule σ_4.

4. The Basic Algorithmic Framework

In this section, we give the basic skeleton of the algorithm in this paper and prove its correctness. The schedule $S_{\rm OPT}$ is characterized by a proper processing order of the emerging jobs scheduled in between the kernels. Starting with the initial LDTC-schedule σ, an emerging job in the current LDTC-schedule is applied and a new C-schedule is created; in this schedule, the kernel is again determined. The same operation is iteratively repeated as long as the established optimality conditions are not satisfied. As we will show later, it will suffice to enumerate all C-schedules to find an optimal solution to the problem.

We associate a complete feasible C-schedule with each node in a *solution tree* T, the initial LDTC-schedule being associated with the root. Aiming to avoid a brutal enumeration of all C-schedules, we carry out a deeper study of the structure of the problem and some additional useful properties of LDTC-schedules. In fact, our solution tree T consists of a single chain of C-schedules. We will refer to a node of the tree as a *stage* (since each node represents a particular stage in the algorithm with the corresponding LDTC-schedule). We let $T_h = (S^0, ..., S^h)$ be the sequence of C-schedules generated by stage h. Thus, S^0 is the initial LDTC-schedule, and the schedule S^h of stage $h > 0$, the immediate successor of schedule S^{h-1}, is obtained by one of the extension rules as described below.

In the schedule S^h, the overflow job $o(S^h)$, the delaying emerging job l, and the kernel $K(S^h)$ are determined. Using the *normal extension rule*, we let $S^{h+1} := S^h_l$, where l is the delaying emerging job in the schedule S^h (we may observe that the schedule σ_4 in Figure 2 is obtained from the schedule σ by the normal extension rule). Alternatively, the schedule S^{h+1} is constructed from the schedule S^h by the *emergency extension rule* as described in the following subsection.

4.1. Types of Emerging Jobs and the Extended Behavior Alternatives

A marched emerging job. An emerging job may be in different possible states. It is useful to distinguish these states and treat them accordingly. Suppose that e is an emerging job in the schedule S^g, and it is applied by stage h, $h > g$ (in a predecessor-schedule of schedule S^h). Then, job e is called *marched* in the schedule S^h if $e \in B(S^h)$ (job 4 is marched in the schedule σ_4 in Figure 2). Intuitively, the existence of a marched job in the schedule S^h indicates an "interference" of the kernel $K(S^h)$ with an earlier arranged part of the schedule preceding that kernel. In our example, it is easy to see that the kernel $K(\sigma_4)$ consists of jobs 8, 9, and 10, with $o(\sigma_4) = 10$ and with $C_{10} = 73 + 54 = 127$. Here, the marched job 4 has "provoked" the rise of the new kernel, where "the earlier arranged part" includes the kernel $K(\sigma)$ of the initial LDTC-schedule σ in Figure 1.

A stuck emerging job. Suppose that job e is marched in the schedule S^h and $E(S^h) = \emptyset$, where $B(S^h)$ is a non-primary block. Then, job e is called *stuck* in the schedule S^h if either it is scheduled before job $o(S^h)$ or $e = o(S^h)$ (observe that any job stuck in the schedule S^h belongs to the kernel $K(S^h)$). In Figure 3, the C-schedule $\sigma_{4,4}$ obtained from the C-schedule σ_4 by the application of the delaying emerging job 4 is depicted. Job 4 becomes the overflow job in the schedule $\sigma_{4,4}$ (with $C_4(\sigma_{4,4}) = 75 + 51 = 126$) and, hence, it is stuck in this schedule.

Block evolution in the solution tree T. Although $E(S^h) = \emptyset$, since $B(S)$ is a non-primary block, a "potential" regular emerging job might be "hidden" in some block preceding block $B(S^h)$, in the schedule S^h. In general, a block in the schedule S^h can be a part of a larger block from the schedule S^g for some $g < h$. Recall that the application of an emerging job e in a C-schedule S yields the raise of a new gap in the C-schedule S_e. As it can be straightforwardly seen, this can lead to a separation or to a *splitting* of the critical block $B(S)$ into two (or possibly even more) new blocks. Likewise, since job e may push the following jobs in the schedule S_e because of the forced right-shift of these jobs, two or more blocks may *merge* forming a bigger block consisting of the jobs from the former blocks.

We will refer to blocks from two different C-schedules as *congruent* if both of them are formed by the same set of jobs. A block in a C-schedule, which is congruent to a block from the initial LDTC-schedule, will be referred to as a *primary* block. Observe that a non-primary block may arise because of either block splitting or/and block merging and that all blocks in the initial LDTC-schedule are primary.

If the block B arises as a result of an application of an emerging job e, then this block is said to be a non-primary block *of* job e, and the latter job is said to be the *splitting* job of B. Note that, since the application of an emerging job does not necessarily lead to a splitting of a block, a non-primary block of job e may contain some other emerging jobs which have already been applied (recall that, before each application of an emerging job, the current release time of this job has to be increased accordingly; e.g., if job e is applied ι times, its release time is modified ι times).

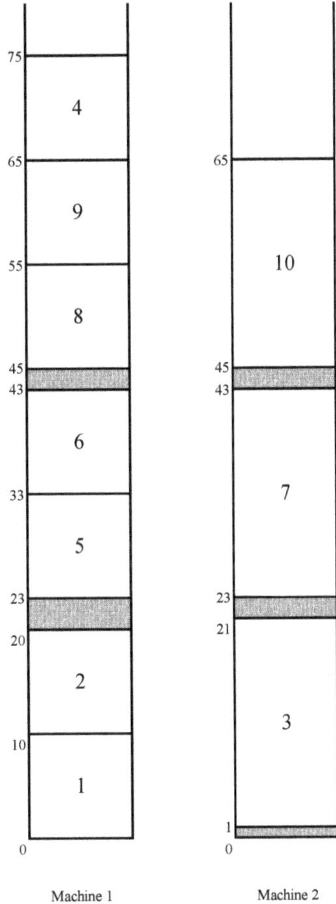

Figure 3. The C-schedule $\sigma_{4,4}$.

In the following, we call the blocks which arose after the splitting of one particular block as the *direct descendants* of this block (the latter block is called the *direct predecessor* of the former ones). Moreover, if the block $B \in S$ is the direct predecessor of the blocks $B_1, ..., B_k$, then the inclusion $B \subseteq B_1 \cup ... \cup B_k$ holds, but not the opposite one (due to a possible merging of blocks).

Recurrently, a *descendant* of a block is its direct descendant, since any descendant of a descendant of a block is obviously also a descendant of the latter block. Moreover, if a block is a descendant of another block, then the latter one is called a *predecessor* of the former block. Subsequently, we call two or more blocks *relative* if they possess at least one common predecessor block.

Returning to our example, the primary block from the schedule σ in Figure 1 is split into its two direct descendant blocks with the jobs 1, 2, 3 and 5, 6, 7, 4, 8, 9, 10, respectively, in the schedule σ_4 in Figure 2. The splitting job is the marched job 4. The first block in the schedule σ_4 is congruent to the first block in the schedule $\sigma_{4,4}$ in Figure 3. The second block in the schedule σ_4 is further split into two blocks in the schedule $\sigma_{4,4}$, and the splitting job is again job 4. The third block in the schedule $\sigma_{4,4}$ (consisting of the jobs 8, 9, 10, and 4 is a descendant of the primary block in the schedule σ.

A passive emerging job. A *passive emerging job e* in the schedule S^h is a ("hidden") regular emerging job from the schedule S^g, i.e., job e belongs to a block from the schedule S^g, relative to block $B(S^h)$ (preceding this block), such that $q_e < q_{o(S^h)}$. For example, in the schedule σ_4 in Figure 2, the passive emerging jobs are 1, 2, and 3, and in the schedule $\sigma_{4,4}$ in Figure 3, the passive emerging jobs are 1 and 2.

Extended behavior alternatives. Now, we define two extended behavior alternatives, which, together with the five basic behavior alternatives, were introduced earlier in [2] (Section 2.3). Suppose that there exists no regular emerging job in the C-schedule S (i.e., the block $B(S)$ starts with the kernel $K(S)$), and there is no stuck job in this schedule. Then, we say that an *exhaustive instance of alternative (a)* occurs in the schedule S which we abbreviate by EIA(a). If now there exists a stuck job in the schedule S, then an *extended instance of alternative (b)* with this job in the schedule S (abbreviated EIA(b)) is said to occur (there may exist more than one job stuck in the schedule S^h). We easily observe that in the schedule $\sigma_{4,4}$ in Figure 3, an EIA(b) with job 4 arises.

The first above behavior alternative immediately yields an optimal solution, and the second one indicates that some rearrangement of the already applied emerging jobs might be required. The first and the second behavior alternatives, respectively, are treated in the following lemma and in the next subsection, respectively.

Lemma 2. *A C-schedule S^h is optimal if an EIA(a) in it occurs.*

Proof. By the condition, there exists no stuck job in the schedule S^h. This implies that none of the jobs of the kernel $K(S^h)$ can be scheduled at some earlier time moment without causing a forced delay of a more urgent job from this kernel, and the lemma can easily be proved by an interchange argument. □

4.2. Emergency Extension Rule

Throughout this subsection, assume that there arises an EIA(b) with job $e \in B(S^h)$ in the schedule S^h (this job is stuck in that schedule) and there exists a passive emerging job in schedule S^h. We let e be the latest applied job stuck in the schedule S^h, and let l be the latest scheduled passive emerging job in that schedule. By the definition of job l, there is a schedule S^g, a predecessor of schedule S^h in the solution tree T such that $e \in B(S^g)$ and $l \in B(S^g)$. Although jobs l and e belong to different blocks in the schedule S^h, the corresponding blocks can be merged by reverting the application(s) of job e. This can clearly be accomplished by restoring the corresponding earlier release time of job e (recall that the release time of an emerging job is increased each time it is applied). Once these blocks are merged, job l becomes a regular emerging job and, hence, it can be applied.

Denote by r the release time of an emerging job e before it is applied to a kernel K. Then, we say that job e is *revised* (for the kernel K) if it is sequenced back before the kernel K; this means that we reassign the value r to its release time and apply the LDTC-heuristic.

In more detail, let B be a block relative to $B(S^h)$ in the schedule S^h containing job l. Now, B and $B(S^h)$ are different blocks and, in addition, there also might exist a chain $B_1, ..., B_{k-1}$ of succeeding (relevant) blocks between the two blocks B and $B(S^h)$ in this schedule S^h. Let $B_0 = B$ and $B_k = B(S^h)$. First, the blocks B_{k-1} and B_k are merged by reverting the application of the corresponding emerging job. Then, the resulting block is similarly merged with block B_{k-2}, and so on. In general, to merge the block B' with its successive (relative) block B'', the corresponding release time of one of the currently applied jobs scheduled in block B'' (which are scheduled between the jobs of the two blocks B' and B'' before the merging is applied) is restored, i.e., this job is *revised*.

According to our definition, the revision of the splitting job of the two blocks B' and B'' will lead to a merging of these two blocks. Note also that the revision of any other applied emerging job, which is scheduled in the block B'', will also lead to this effect. Among all such jobs with the largest delivery time, the latest scheduled one in block B'' will be referred to as the *active splitting* job for the blocks B' and B''.

The blocks B_{k-1} and B_k are merged by the revision of the active splitting job of these two blocks which is scheduled in block B_k. In a similar way, the active splitting job of the block B_{k-2} is revised in order to merge the block B_{k-2} with the block obtained earlier and so on, and this process continues until all blocks from the chain $B_0, B_1, ..., B_{k-1}, B_k$ are merged.

Observe that the active splitting job in the schedule $\sigma_{4,4}$ in Figure 3 is job 4. Its revision yields the merging of the three blocks from this schedule into a single primary block of the schedule σ of Figure 1.

We denote the resultant merged block by $\mathcal{B}(l)$ (this block, ending with the jobs from block $B(S^h)$, can, in general, be non-primary), and we will refer to the above described procedure as the *chain of revisions* for the passive emerging job l. Note that the chain of revisions is accomplished only if, besides a passive emerging job l, there exists a stuck job e. In addition, observe that, although this procedure somewhat resembles the traditional backtracking, it is still different as it keeps untouched the "intermediate" applications that could have been earlier carried out between the reverted applications.

Let $Rev_{l,e}(S^h)$ be the C-schedule, obtained from schedule S^h by the chain of revisions for job l (here e is the corresponding stuck job). Observe that job l changes its status from a passive to a regular emerging job in this schedule, i.e., it is *activated* in the C-schedule $Rev_{l,e}(S^h)$. The emergency extension rule applies job l in the schedule $Rev_{l,e}(S^h)$ to the kernel $K(Rev_{l,e}(S^h))$, setting $S^{h+1} := (Rev_{l,e}(S^h))_l$.

In the schedule $\sigma_{4,4}$ in Figure 3, we have $l = 2$ and $e = 4$; the C-schedule $Rev_{2,4}(\sigma_{4,4})_2$ is represented in Figure 4 (which turns out to be an optimal schedule for the problem instance of our example.

4.3. The Description of the Algorithm and Its Correctness

We give the following Algorithm 1 and prove its correctness:

Algorithm 1 Blesscmore Algorithm

Step 1: Set $h := 0$ and $S^h := \sigma$.

Step 2: If the condition of Lemma 1 holds, then return the schedule σ and stop.

Step 3: Set $h := h + 1$.

Step 4: { *iterative stopping rules* } If in the schedule S^h: either (i) there exists no regular emerging job and no job from block $B(S^h)$ is stuck, or (ii) there is neither regular nor passive emerging job (there may exist a stuck job in block $B(S^h)$), or (iii) there occurs an EIA(a), then return a schedule from the tree T with the minimum makespan and stop;

Step 5: { *normal extension rule* }: if in the schedule S^h, there occurs no EIA(b), then $S^{h+1} := S^h_l$, where l is the regular delaying emerging job else
{ *emergency extension rule* } if in the schedule S^h there occurs an EIA(b), then $S^{h+1} := Rev_{l,e}(S^h))_l$, where l is the latest scheduled passive emerging job, and e is the latest applied stuck job in schedule S^h.

Step 6: goto Step 3.

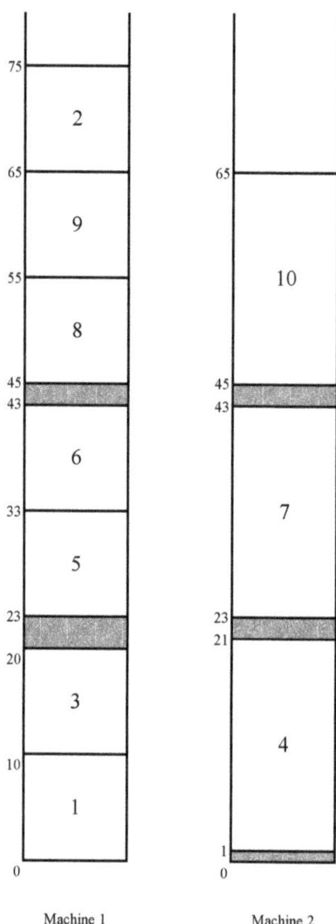

Figure 4. An optimal C-schedule $Rev_{2,4}(\sigma_{4,4})_2$.

We give the final illustration of the algorithm for our example. At Step 1, we have $S^0 = \sigma$ (Figure 1). Since the condition at Step 2 is not satisfied, $h := 1$, and the normal extension rule is used to generate the C-schedule $S^1 := \sigma_4$ in Figure 2. Then, similarly, the normal extension rule is used to generate the C-schedule $S^2 = \sigma_{4,4}$ in Figure 3. Since, in the latter schedule S^2, an EIA(b) with job 4 occurs, this job is revised and an alternative C-schedule $S^3 := Rev_{2,4}(\sigma_{4,4})_2$ after the chain of revisions for the passive emerging job 2 is created. The kernel $K(S^3)$ of this schedule consists of jobs 8, 9, and 10 (as that of schedule S^1) with $o(S^3) = 10$ and $L_{10}(S^3) = 65 + 54 = 119$. This is the optimal makespan: There is neither a regular emerging job nor a stuck job in the C-schedule S^3. Hence, the stopping rule (i) applies, and the algorithm stops with an optimal solution.

Now, we prove the following theorem:

Theorem 1. *For some stage h, the C-schedule S^h is an optimal schedule S_{OPT}.*

Proof. Suppose that the optimality condition in Lemma 1 is not satisfied for the schedule $S^0 = \sigma$, and let us consider C-schedule S^h of an iteration $h > 1$. Assume schedule S^h is not optimal, and assume first that there is no stuck job in schedule S^h. Then, in any feasible schedule S with a better makespan, the number of jobs scheduled before the kernel $K(S^h)$ in block $B(S^h)$ must be one less than in the latter schedule as otherwise, due to Property 2 and the fact that there is no stuck job in schedule S^h, a job from the kernel $K(S^h)$ cannot have a smaller full completion time in the schedule S' than job $o(S^h)$ in the schedule S^h (as the jobs of the kernel $K(S^h)$ are already included in an optimal sequence, see Lemma 1).

Thus, some job $l \in B(S^h)$ included before the kernel $K(S^h)$ in the schedule S^h must be scheduled after that kernel in the schedule S. We claim that job l is to be a regular emerging job. Indeed, if it is not, then $q_l \geq q_o$, $o = o(S^h)$ or/and $l \notin B(S^h)$. If $q_l \geq q_o$, then, due to inequality $t_l^{(S^h)_l} \geq t_o^{S^h}$ and Proposition 2, $|(S^h)_l| \geq |S^h|$. Hence, the makespan of any feasible schedule in which job l is scheduled after the kernel $K(S^h)$ cannot be less than that of schedule S^h. Suppose now $l \notin B(S^h)$. If l is not a passive emerging job, then obviously the above reasoning applies again. Suppose l is a passive emerging job, and suppose first that there is no marched job in the block $B(S^h)$. Then, since the block $B(S^h)$ starts with the kernel $K(S^h)$ (there exists no regular emerging job schedule S^h), the full completion time of the overflow job is a lower bound on the optimum schedule makespan (this can be seen similarly to Lemma 1). Obviously, the same reasoning applies in case there are marched jobs in the block $B(S^h)$, but none of them is stuck in schedule S^h.

The above proves the validity of the stopping rule (i) from Algorithm 1. Suppose now that there is neither a regular nor passive emerging job, and there is a stuck job in schedule S^h. Clearly, the full completion time of the overflow job $o \in B(S^h)$ cannot be decreased unless such stuck job $e \in B(S^h)$ is revised. Note that the corresponding C-schedule coincides with an earlier generated C-schedule S^g, for some $g < h$ from the solution tree T. Furthermore, in any feasible schedule having a makespan less than that of the schedule S^h, another job l with $q_l < q_o$ is to be applied instead of job e. Moreover, job l should belong to a block, relative to the block $B(S^h)$, as otherwise the time interval released by the removal of that job from its current execution interval may not yield a right-shift of any job from the block $B(S^h)$. It follows that job l is a passive emerging job. This proves the stopping rule (ii). The stopping rule (iii) follows from Lemma 2.

It remains to show that the search in the space of the C-schedules is correctly organized. There are two extension rules. The normal extension rule is used at stage h if there exists a regular emerging job in the schedule S^h. In this case, the delaying emerging job l is applied, i.e., $S^{h+1} := S_l^h$. Consider an alternative feasible schedule S_j^h, where j is another emerging job (above, we have shown that only emerging jobs need to be considered). It is easy to see that the left-shift of the kernel jobs in the schedule S_j^h cannot be more than that in the schedule S_l^h, and the forced right-shift for the jobs scheduled after job j in the schedule S_j^h cannot be less than that of the jobs scheduled after job l in the schedule S_l^h (recall that $p_j = p_l$). Hence, the schedule S_j^h is dominated by the schedule S_l^h unless job l gets stuck at a later stage $h' > h$. In the latter case, the emergency extension rule revises first job l. In the resultant C-schedule, the passive delaying job converts into a regular delaying emerging job. Then, the emergency extension rule applies this (converted) regular regular delaying emerging job. We complete the proof by repeatedly applying the above reasoning for the normal extension rule. □

5. Performance Analysis

It is not difficult to see that the direct application of Algorithm 1 of the previous section may yield the generation of some redundant C-schedules: the jobs from the same kernel K including the overflow job o may be forced to be right-shifted after they are already "arranged" (i.e., the corresponding emerging job(s) are already applied to that kernel) due to the arrangement accomplished for the kernel K' preceding kernel K. As a result (because of the application of an emerging job for the kernel K'), one or more redundant C-schedules in which a job from the kernel K repeatedly becomes an overflow job might be created. Such an unnecessary rearrangement of the portion of a C-schedule between the kernels K' and K is avoided by restricting the number of jobs that are allowed to be scheduled in that portion. This number becomes well-defined after the first disturbance of this portion caused by the application of an emerging job for the kernel K'. This issue was studied in detail for the case of identical machines in [2] (see Section 4.1). It can be readily verified that the basic estimations for the case of identical machines similarly hold for uniform machines. In particular, the number of the enumerated C-schedules remains the same for uniform machines. A complete time complexity analysis requires a number of additional concepts and definitions from [2] and would basically repeat the arguments for identical machines.

Recall that we use a different schedule generation mechanism for identical and uniform machines: the LDT-heuristic applied for the schedule generation in the identical machine environment is replaced by the LDTC-heuristic for the uniform machine environment. LDT-schedules possess a number of nice properties used in the algorithm from [2]. LDTC-schedules also possess such necessary useful properties (Properties 1 and 2) that allowed us to use the basic framework from [2]. While generating an LDT-schedule for identical machines, every next job is scheduled on the next available machine (the next to the last machine m being machine 1) and the starting and completion time of each next scheduled job is not smaller than that of all previously scheduled ones. In some sense, the generalization of these properties are Properties 1 and 2, which still assure that the structural pattern of the generated schedules is kept, and it does not depend on which particular jobs are being scheduled in a particular time interval (note that this would not be the case for an unrelated machine environment). This allowed us to adopt the blesscmore framework from [2] for the uniform machine environment (for instance, intuitively, while restricting the number of the scheduled jobs between two successive kernels, no matter which particular jobs are being scheduled between these kernels).

Another "redundancy issue" occurs when a series of emerging jobs are successively applied to the same kernel K without reaching the desired result, i.e., the applied emerging jobs become new overflow jobs in the corresponding C-schedules (see Section 4.2 in [2]). In Lemma 7 from the latter reference, it is shown that this yields an additional factor of p in the running time of the algorithm. This result also holds for the uniform machine environment. The magnitude p remains valid for the uniform machine environment as the difference between the completion times of two successively scheduled jobs on the same machine cannot be more than p (recall that p is the processing time of any job on the slowest machine m). The desired result follows since the delivery time of each emerging job next applied to kernel K is strictly less than that of the previously applied one (see the proof of Lemma 7 in [2]).

The above results yield the same bound $O(\gamma m)$ on the number of the enumerated C-schedules as in [2]), where γ can be either n or $q_{\max}$ (see Lemma 8 from Section 4.3 and Theorem 2 from Section 6.1, in [2]). In fact, γ is the total number of the applied emerging jobs, a magnitude that can be essentially smaller than n. This yields the overall cost $O(\gamma m^2 n \log n)$ due to the cost $O(mn \log n)$ of the LDTC-heuristic (instead of $O(n \log n)$ for the LDT-heuristic). A further refinement of the overall time complexity accomplished in [2] is not possible for the uniform machine environment. In the algorithm from [2], while generating every next C-schedule, instead of applying the LDT-heuristic to the whole set of jobs, it is only applied to the jobs from a small part of the current C-schedule, the so-called critical segment (a specially determined part of the latter schedule containing its kernel), and the remaining jobs are scheduled in linear time just by right-shifting the jobs following the critical segment by the required amount of time units (conserving their current processing order). This is not possible for uniform machines as such an obtained schedule will not necessarily remain an LDTC-schedule, i.e., a linear time rescheduling will not provide the desired structure.

6. Discussion and Concluding Remarks

We showed that the earlier developed technique for scheduling identical machines can be extended to the uniform machine environment if Condition (1) on the job parameters is satisfied, thus making a step towards the settlement of the complexity status of this long-standing open problem. In particular, the imposed condition reflects potential conflicts that arise in the uniform machine environment but do not arise in the identical machine environment. It is a challenging question whether the removal of Condition (1) results in an NP-hard problem or if it can still be solved in polynomial time, at least, for a fixed number of machines. Although the LDTC-heuristic would not give the desired results if

Condition (1) is not satisfied, it might still be possible to develop a more intelligent heuristic that can successfully be combined with the blesscmore framework and the analysis of the behavior alternatives from [2]. This approach may have some limitations though. As we have mentioned earlier, it is unlikely that it can be applied to the unrelated machine environment, mainly because the structural pattern of the generated schedules will depend on, which particular jobs are scheduled in a particular time interval on each machine from a group of unrelated machines, which makes the analysis of the behavior alternatives much more complicated. At the same time, the approach might be extensible to shop scheduling problems. It is a challenging question whether it can be extended to the case where there are two allowable job processing times (this turned out to be possible for the single-machine environment, see the blesscmore algorithm in [24]) and for a much more general setting with mutually divisible job processing times for identical and uniform machine environments (this turned out to be also possible for the single-machine environment—a maximal polynomially solvable special case of (a strongly NP-hard) problem $1|r_j, d_j|L_{\max}$ with mutually divisible job processing times was dealt with recently in [25]). Finally, we note that the algorithm presented here can also be used as an approximate one for non-equal job processing times, as it is often the case in practical applications.

Author Contributions: Conceptualization, N.V.; investigation, N.V. and F.W.; writing—original draft preparation, N.V. and F.W. All authors have read and agreed to the published version of the manuscript.

Funding: The first author is grateful for the support of the DAAD grant 57507438 for Research Stays for University Academics and Scientists.

Institutional Review Board Statement: Not applicable.

Informed Consent Statement: Not applicable.

Data Availability Statement: Not applicable.

Acknowledgments: The authors would like to thank the anonymous referees for useful suggestions and express very special thanks to referee 2 whose very careful comments helped greatly to correct the inconsistencies from the original submission.

Conflicts of Interest: The authors declare no conflict of interest.

References

1. Bratley, P.; Florian, M.; Robillard, P. On sequencing with earliest start times and due-dates with application to computing bounds for ($n/m/G/F_{max}$) problem. *Nav. Res. Logist. Quart.* **1973**, *20*, 57–67. [CrossRef]
2. Vakhania, N. A better algorithm for sequencing with release and delivery times on identical processors. *J. Algorithms* **2003**, *48*, 273–293. [CrossRef]
3. Vakhania, N. Branch less, cut more and minimize the number of late equal-length jobs on identical machines. *Theor. Comput. Sci.* **2012**, *465*, 49–60. [CrossRef]
4. Garey, M.R.; Johnson, D.S. *Computers and Intractability: A Guide to the Theory of NP-Completeness*; Freeman: San Francisco, CA, USA, 1979.
5. McMahon, C.; Florian, M. On scheduling with ready times and due dates to minimize maximum lateness. *Operations* **1975**, *23*, 475–482. [CrossRef]
6. Carlier, J. The one-machine sequencing problem. *Eur. J. Oper. Res.* **1982**, *11*, 42–47. [CrossRef]
7. Baptiste, P. Scheduling equal-length jobs on identical parallel machines. *Discret. Appl. Math.* **2000**, *103*, 21–32. [CrossRef]
8. Baptiste, P. Polynomial time algorithms for minimizing the weighted number of late jobs on a single machine with equal processing times. *J. Sched.* **1999**, *2*, 245–252. [CrossRef]
9. Chrobak, M.; Dürr, C.; Jawor, W.; Kowalik, L.; Kurowski, M. A note on scheduling equal-length jobs to maximize throughput. *J. Sched.* **2006**, *9*, 71–73. [CrossRef]
10. Vakhania, N. A study of single-machine scheduling problem to maximize throughput. *J. Sched.* **2013**, *16*, 395–403. [CrossRef]
11. Vakhania, N. Scheduling jobs with release times premptively on a single machine to minimize the number of late jobs. *Oper. Res. Lett.* **2009**, *37*, 405–410. [CrossRef]
12. Garey, M.R.; Johnson, D.S.; Simons, B.B.; Tarjan, R.E. Scheduling unit-time tasks with arbitrary release times and deadlines. *SIAM J. Comput.* **1981**, *10*, 256–269. [CrossRef]

13. Simons, B.; Warmuth, M.A. fast algorithm for multiprocessor scheduling of unit-length jobs. *SIAM J. Comput.* **1989**, *18*, 690–710. [CrossRef]
14. Dessouky, M.; Lageweg, B.J.; Lenstra, J.K.; van de Velde, S.L. Scheduling identical jobs on uniform parallel machines. *Stat. Neerl.* **1990**, *44*, 115–123. [CrossRef]
15. Labetoulle, J.; Lawler, E.L.; Lenstra, J.K.; Rinnooy Kan, A.H.G. *Preemptive Scheduling of Uniform Machines Subject to Release Dates*; Pulleyblank, H.R., Ed.; Progress in Combinatorial Optimization; Academic Press: New York, NY, USA, 1984; pp. 245–261.
16. Kravchenko, S.; Werner, F. Preemptive scheduling on uniform machines to minimize mean flow time. *Comput. Oper. Res.* **2009**, *36*, 2816–2821. [CrossRef]
17. Lawler, E.L.; Labetoulle, J. On preemptive scheduling of unrelated parallel processors by linear programming. *J. Assoc. Comput. Mach.* **1978**, *25*, 612–619. [CrossRef]
18. Vakhania, N.; Hernandez, J.; Werner, F. Scheduling unrelated machines with two types of jobs. *Int. J. Prod. Res.* **2014**, *52*, 3793–3801. [CrossRef]
19. Lenstra, J.K.; Shmoys, D.B.; Tardos, E. Approximation algorithms for scheduling unrelated machines. *Math. Program.* **1990**, *46*, 259–271. [CrossRef]
20. Brucker, P.; Kravchenko, S. *Scheduling Jobs with Release Times on Parallel Machines to Minimize Total Tardiness*; OSM Reihe P; Heft 257; Universität Osnabrück: Osnabrück, Germany, 2005.
21. Brucker, P.; Kravchenko, S. Scheduling jobs with equal processing times and time windows on identical parallel machines. *J. Sched.* **2008**, *11*, 229–237. [CrossRef]
22. Kravchenko, S.; Werner, F. On a parallel machine scheduling problem with equal processing times. *Discret. Appl. Math.* **2009**, *157*, 848–852. [CrossRef]
23. Kravchenko, S.; Werner, F. Parallel machine problems with equal processing times: A survey. *J. Sched.* **2011**, *14*, 435–444. [CrossRef]
24. Vakhania, N. Single-Machine Scheduling with Release Times and Tails. *Ann. Oper. Res.* **2004**, *129*, 253–271 . [CrossRef]
25. Vakhania, N. Dynamic Restructuring Framework for Scheduling with Release Times and Due-Dates. *Mathematics* **2019**, *7*, 1104. [CrossRef]

 *mathematics*

Article

A Numerical Comparison of the Sensitivity of the Geometric Mean Method, Eigenvalue Method, and Best–Worst Method

Jiří Mazurek *, Radomír Perzina, Jaroslav Ramík and David Bartl

Department of Informatics and Mathematics, School of Business Administration in Karviná, Silesian University in Opava, Univerzitní Náměstí 1934/3, 733 40 Karviná, Czech Republic; perzina@opf.slu.cz (R.P.); ramik@opf.slu.cz (J.R.); bartl@opf.slu.cz (D.B.)
* Correspondence: mazurek@opf.slu.cz

Abstract: In this paper, we compare three methods for deriving a priority vector in the theoretical framework of pairwise comparisons—the Geometric Mean Method (GMM), Eigenvalue Method (EVM) and Best–Worst Method (BWM)—with respect to two features: sensitivity and order violation. As the research method, we apply One-Factor-At-a-Time (OFAT) sensitivity analysis via Monte Carlo simulations; the number of compared objects ranges from 3 to 8, and the comparison scale coincides with Saaty's fundamental scale from 1 to 9 with reciprocals. Our findings suggest that the BWM is, on average, significantly more sensitive statistically (and thus less robust) and more susceptible to order violation than the GMM and EVM for every examined matrix (vector) size, even after adjustment for the different numbers of pairwise comparisons required by each method. On the other hand, differences in sensitivity and order violation between the GMM and EMM were found to be mostly statistically insignificant.

Keywords: Best–Worst Method; Eigenvalue Method; Geometric Mean Method; Monte Carlo simulations; pairwise comparisons; sensitivity

Citation: Mazurek, J.; Perzina, R.; Ramík, J.; Bartl, D. A Numerical Comparison of the Sensitivity of the Geometric Mean Method, Eigenvalue Method, and Best–Worst Method. *Mathematics* **2021**, *9*, 554. https://doi.org/10.3390/math9050554

Academic Editors: Frank Werner and Ioannis Stratis

Received: 4 February 2021
Accepted: 2 March 2021
Published: 5 March 2021

1. Introduction

Pairwise comparisons constitute a fundamental part of sophisticated multiple-criteria decision-making frameworks, such as the analytic hierarchy/network process (AHP/ANP), PROMETHEE, ELECTRE, PAPRIKA, and many others [1–9]. In addition, hundreds of successful applications of pairwise comparisons in almost all domains of human activity have been published in the literature [10].

The objective of a pairwise comparison method is to assign weights to compared objects corresponding to their preference/importance and to rank objects from the most preferred/important to the last. The Geometric Mean Method (GMM) proposed by Crawford [11] and the Eigenvalue (eigenvector) Method (EVM) proposed by Saaty [7] are the most popular methods for deriving weights from pairwise comparisons arranged in the form of a pairwise comparison (PC) matrix. The Best–Worst Method (BWM) proposed by Jafar Rezaei in 2015 (see Rezaei [12]) is one of the latest contributions to the field of pairwise comparisons. It is based on the pairwise comparisons of all objects (in the original paper, the objects were criteria) with the best object and the worst object (known a priori) only. Therefore, it belongs in the family of pairwise comparison methods with missing elements and/or incomplete pairwise comparison matrices (with additional information). Since its introduction, the BWM has attracted the attention of many researchers and practitioners and has been applied to various problems in areas such as waste management, tourism, sustainability or biochemistry [13–18].

The appeal of the BWM lies in its obvious simplicity; however, until now, numerical comparisons of the BWM with other methods, the GMM and EVM (AHP) in particular, have rarely been covered in the literature. The studies of Ajrina et al. [19] and Haseli et al. [20] compared the BWM and the AHP via one and two numerical examples, respectively.

The original paper on the BWM [12] provided a comparison of the results of the BWM with the AHP in only one particular example based on an experiment with 46 respondents (university undergraduate students) and 322 PC matrices (and pairs of vectors) of the order $n = 4$. The work came to the conclusion that the BWM performs better than the AHP, and the weights derived by the BWM are highly reliable. Other comparison studies of the BWM and AHP/GMM are not known to the authors.

Therefore, the aim of this paper is to bridge this gap and provide a comprehensive numerical comparison of the BWM, EVM (AHP), and GMM with respect to two crucial method properties: sensitivity and the violation of the order of preferences (order violation in short). Sensitivity analysis is a well-established tool for assessing how the input of a model/method affects the output (or vice-versa) and is widely used in natural sciences, in particular in climatology [21–24]. To assess sensitivity, we apply One-Factor-At-a-Time (OFAT) methodology [25]. The second feature we focus on, order violation, describes how often a unit change in the input leads to a change in the final ranking (ordering) of compared objects, therefore providing useful information on the robustness of rankings. As a research method, we apply Monte Carlo simulations, where pairwise comparisons are selected from Saaty's fundamental scale from 1 to 9 (with reciprocals) and where the number n of compared objects ranges from 3 to 8, as real-world multiple criteria problems usually do not involve large numbers of criteria. Then, we perform a statistical analysis of the acquired results, enabling a final comparison of all three methods.

This paper is organized as follows: preliminaries on pairwise comparison methods, prioritization, sensitivity, and order violation are provided in Section 2, while Monte Carlo simulations are described in Section 3 followed by a discussion in Section 4. Conclusions close the article.

2. Preliminaries

The input data for the *PC* method is a *PC* matrix $C = [c_{ij}]$, where $c_{ij} \in \mathbb{R}_+$ and $i, j \in \{1, \ldots, n\}$. The values of c_{ij} and c_{ji} indicate the relative importance (or preference) of the objects i and j.

In the context of the BWM, the compared objects are criteria. The set of n criteria to be compared and ranked is denoted as $F = \{F_1, \ldots, F_n\}$.

Definition 1. *The matrix $C = [c_{ij}]$ is said to be reciprocal if $\forall i, j \in \{1, \ldots, n\}: \; c_{ij} = c_{ji}^{-1}$ and $C = [c_{ij}]$ is said to be consistent if $\forall i, j, k \in \{1, \ldots, n\}: \; c_{ij} \cdot c_{jk} \cdot c_{ki} = 1$.*

Note that if $C = [c_{ij}]$ is consistent, then it is also reciprocal, but not vice versa. In this paper, it is assumed that a PC matrix is always reciprocal. The reciprocity condition seems to be natural in many decision-making situations. For instance, if an element c_{ij} of a PCM $C = [c_{ij}]$ expresses that the i-th criterion is c_{ij} times more important than the j-th criterion, then it is evident that the j-th criterion is $1/c_{ij}$ times more important than the i-th one; thus, $c_{ij} = 1/c_{ji}$. In particular, for each criterion i, we obtain $c_{ii} = 1$, which corresponds to the fact that the importance of each criterion with respect to itself is equal to one.

2.1. The Eigenvalue Method and the Geometric Mean Method

The result of a pairwise comparison method is a priority vector (vector of weights) w.

According to one of the most popular prioritization methods, the EVM (the Eigenvalue Method) proposed by Saaty [7], the vector w is determined as the rescaled principal eigenvector of the matrix C. Thus, assuming that $C\bar{w} = \lambda_{\max}\bar{w}$, the priority vector w is $w = [w(F_1), \ldots, w(F_n)]^T = \gamma[\bar{w}_1, \ldots, \bar{w}_n]^T$, where γ is a scaling factor, $\gamma = \left[\sum_{i=1}^{n} \bar{w}_i\right]^{-1}$, so that $\|w\| = 1$.

In the Geometric Mean Method (GMM) (see Crawford [11]) the weight of the ith alternative is given by the geometric mean of the ith row of the matrix $C = [c_{ij}]$. Thus, the priority vector is given as follows:

$$w = \left[w(F_1), \dots, w(F_n)\right]^{\mathrm{T}}$$
$$= \gamma\left[\left(\prod_{r=1}^{n} c_{1r}\right)^{\frac{1}{n}}, \dots, \left(\prod_{r=1}^{n} c_{nr}\right)^{\frac{1}{n}}\right]^{\mathrm{T}} \tag{1}$$

where $\gamma = \left[\sum_{i=1}^{n} \prod_{r=1}^{n} c_{ir}\right]^{-1}$ is the scaling factor again.

Several aspects of these methods are discussed in more detail, e.g., in Ramík [26].

2.2. The Best–Worst Method

In the Best–Worst method (see Rezaei [12]), each criterion is pairwise compared only with the best criterion and the worst criterion.

The Best–Worst method proceeds as follows [12]:

Step 1. A set of criteria is determined.
Step 2. The decision maker identifies the best (most desirable, most important) criterion and the worst (least desirable, least important) criterion.
Step 3. Preferences of the best criterion with respect to all other criteria are determined on a scale from 1 (equal importance) to 9 (absolute preference).
Step 4. Preferences of all other criteria with respect to the worst criterion are determined onathe scale from 1 to 9.
Step 5. bOptimal weights of all criteria are found by solving a corresponding non-linear programming problem; see Equation (2).

Let c_{Bj} denote the preference of the best criterion (B) over the criterion F_j, and let c_{iW} denote the preference of the criterion F_i over the worst criterion (W). Let w_B and w_W be the weights of the best and worst criterion, respectively. The goal is to find the vector of criteria weights (a priority vector) $w = (w_1, w_2, \dots, w_n)$.

Rezaei [12] suggested finding the priority vector by solving the following optimization problem:

$$\min\left(\max_j\left\{\left|\frac{w_B}{w_j} - c_{Bj}\right|, \left|\frac{w_j}{w_W} - c_{jW}\right|\right\}\right) \tag{2}$$

s.t.

$$\sum_{j=1}^{n} w_j = 1 \tag{3}$$

$$w_j \geq 0, \qquad \forall j = 1, \dots, n. \tag{4}$$

The problem can equivalently be stated as follows:

$$\min \xi \tag{5}$$

s.t.

$$\left|\frac{w_B}{w_j} - c_{Bj}\right| \leq \xi, \qquad \forall j = 1, \dots, n, \tag{6}$$

$$\left|\frac{w_j}{w_W} - c_{jW}\right| \leq \xi, \qquad \forall j = 1, \dots, n, \tag{7}$$

$$\sum_{j=1}^{n} w_j = 1, \tag{8}$$

$$w_j \geq 0, \qquad \forall j = 1, \dots, n. \tag{9}$$

Further, it is assumed that for all j, the following inequalities hold:

$$c_{BW} \geq c_{jW} \geq 1; \qquad c_{BW} \geq c_{Bj} \geq 1. \tag{10}$$

A linear version of the BWM was introduced by Brunelli & Rezaei [27] and Rezaei [28], where the letter "L" denotes linear:

$$\min \xi_L \tag{11}$$

s.t.

$$\left|w_B - c_{Bj}w_j\right| \leq \xi_L, \qquad \forall j = 1,\ldots,n, \tag{12}$$

$$\left|w_j - c_{jW}w_W\right| \leq \xi_L, \qquad \forall j = 1,\ldots,n, \tag{13}$$

$$\sum_{j=1}^{n} w_j = 1, \tag{14}$$

$$w_j \geq 0, \qquad \forall j = 1,\ldots,n. \tag{15}$$

Notice that the solution to the linear version of the BWM differs from the solution to the non-linear version in general. In addition, in this case, the value of ξ_L^* should not be divided by CI.

When comparing n objects (criteria, alternatives, etc.) pairwise, the EVM and GMM require $n(n-1)/2$ comparisons to be made. The BWM requires only comparisons (of criteria) with the best and worst criterion, and the reduced number of comparisons amounts to $2n - 3$. This reduction might be very important when dealing with a large number of compared objects.

2.3. Order Violation

First, let us explain the concept of order violation.

Definition 2. *Let $C = [c_{ij}]$ be a pairwise comparison matrix of n objects, let $c_{ij} \in \left\{\frac{1}{9}, \frac{1}{8}, \ldots, 1, \ldots, 8, 9\right\}$, and let $w = (w_1, \ldots, w_n)$ be a corresponding vector of weights (a priority vector). The order violation occurs when for some pair of objects (i,j), $i,j \in \{1,\ldots,n\}$, with the weights w_i and w_j, respectively, it holds that after a change of one element c_{kl}, $k,l \in \{1,\ldots,n\}$, by one unit of the scale, the relation $w_i \geq w_j$ changes into $w_i < w_j$.*

Remark 1. *By a unit change in Definition 2, we mean the change to an adjacent point of a given discrete scale; e.g., a change from 6 to 7 (and reciprocal values change as well), or from $\frac{1}{6}$ to $\frac{1}{7}$ (and reciprocal values also change again). In addition, other scales than Saaty's fundamental scale can be used for comparisons, but we decided to adhere to the scale of the original study of Rezaei [12].*

Order violation means that after a change of only one pairwise comparison by just one unit (which is a minimal possible change) the order (ranking) of compared objects provided by the given PC method changes as well. Thus, it can be considered an undesirable feature, as it indicates that the order of objects is unstable and a minimal change or error is sufficient to disturb it.

2.4. Sensitivity

To evaluate the sensitivity of the BW, EV, and GM methods, we applied One-Factor-At-a-Time (OFAT) methodology. We define sensitivity as a change in the priority vector (output) when one preference (input) is changed by one unit:

Definition 3. *Let $w = (w_1, \ldots, w_n)$ be the vector of weights obtained by a generic prioritization method PM. Let $w^* = (w_1^*, \ldots, w_n^*)$ be the vector of weights after one preference was changed by one unit. Then, the sensitivity $\triangle w$ is defined as follows:*

$$\triangle\, w_{(n)}^{(PM)} = \frac{100}{n} \sum_{i=1}^{n} |w_i - w_i^*|. \tag{16}$$

The sensitivity $\triangle w$ expresses, in a percentage, a per-weight change in the original priority vector w. If, for instance, $\triangle w_{(n)}^{(GMM)} = 2$, this means that each component of the original weight vector $w = (w_1, ..., w_n)$ changed by 2% on average when the GMM was applied.

The following example illustrates the use of order violation and sensitivity.

Example 1 (Best Worst Method [29]). *Consider buying a car according to five criteria: quality, price, comfort, safety, and style. The best criterion is price, and the worst criterion is style. The buyer provides the following pairwise comparisons:*

Best to all preferences: $(2, 1, 4, 3, 8)$. *All to worst preferences:* $(4, 8, 4, 2, 1)$.

By using the linear BWM model and the MS Excel Solver from Best Worst Method [29], we obtain the following weights of criteria:

$$w = (0.246, 0.431, 0.123, 0.154, 0.046).$$

Now, we change one preference by one unit (in bold) in the best to all preferences: $(2, 1, \mathbf{3}, 3, 8)$. *The weights of criteria are now*

$$w^* = (0.233, 0.427, 0.155, 0.136, 0.048).$$

The sensitivity $\triangle w_{(5)}^{(BWM)} is therefore \frac{100}{5} \sum_{i=1}^{5} |w_i - w_i^*| = 1.4$.

Thus, the weights of the criteria changed, on average, by 1.4%.

As for the order violation, initially the criterion of comfort'was ranked fourth and the criterion of safety was ranked third. After the unit change in one PC comparison, the criterion of comfort was ranked third and the criterion of safety was ranked fourth. Therefore, an order violation occurred.

3. Monte Carlo Simulations

Simulations of the EVM and GMM were performed in C#; simulations of the BWM were carried out via the MS Excel Solver [29].

The procedure for a full PC matrix, the GMM (EVM), and Saaty's scale was as follows:

Step 1. A random PC (reciprocal) matrix C of the order n with entries from Saaty's scale (from 1 to 9) was generated.
Step 2. The priority vector w was derived by the GMM (EVM).
Step 3. A randomly chosen element c_{ij} of a PC matrix C was randomly changed by one unit of a scale up or down, and the reciprocal element c_{ji} was changed accordingly.
Step 4. The priority vector w^* was derived by the GMM (EVM).
Step 5. Sensitivity (16) was calculated and order violation was checked.
Step 6. The procedure was repeated 500–1000 times for each matrix size $n \in \{3, 4, 5, 6, 7, 8\}$.

The EVM and GMM were performed on the same set of random PC matrices.

The procedure for the BWM and Saaty's scale was as follows:

Step 1. Pairwise comparisons of all n objects with respect to the best and the worst object (in the form of two vectors) were randomly generated with the use of Saaty's scale while preserving relations (10).
Step 2. The priority vector w was calculated by the linear version of the BWM according to Equation (11).
Step 3. A randomly chosen element from one of the two vectors generated in Step 1 was changed by one unit (up or down, again randomly), while preserving Saaty's scale.
Step 4. The priority vector w^* was derived by the linear BWM according to Equation (11).
Step 5. Sensitivity (16) was calculated and order violation was checked.
Step 6. The procedure was repeated 500–1000 times for each matrix size $n \in \{3, 4, 5, 6, 7, 8\}$

Table 1 provides the average values of sensitivity along with standard deviations and the frequency of the order violation. As can be seen, the least sensitive (thus, the most robust) method was the GMM followed by the EVM. The Best–Worst method performed significantly worse. In the case of order violation, again, the GMM performed best and the BWM worst.

Table 1. Simulation results: mean sensitivity and average order violation occurrence (for each method separately). BWM: Best–Worst Method; GMM: Geometric Mean Method; EVM: Eigenvalue Method.

Method	n	Mean Sensitivity (st. dev.)	Order Violation (%)
BWM	3	2.801 (2.7771)	0
	4	2.161 (2.170)	31.4
	5	1.621 (1.687)	38.9
	6	1.273 (1.276)	50.8
	7	0.966 (1.031)	53.8
	8	0.741 (0.831)	97.2
GMM	3	1.942 (1.475)	10.6
	4	0.919 (0.681)	12.2
	5	0.465 (0.364)	12.3
	6	0.263 (0.212)	15.2
	7	0.185 (0.148)	19.1
	8	0.113 (0.081)	61.0
EVM	3	1.790 (1.444)	10.9
	4	0.932 (0.619)	14.8
	5	0.474 (0.314)	16.6
	6	0.304 (0.178)	22.0
	7	0.196 (0.094)	24.3
	8	0.134 (0.059)	64.7

Figures 1 and 2 provide graphical illustrations of the sensitivity and order violation for all examined matrix sizes.

Figure 3 shows a comparison of the GMM and EVM with the BWM adjusted for the lower number of pairwise comparisons.

Figures 4–6 show the sensitivity distribution of all three methods for $n \in \{4, 6, 8\}$.

Figures 7–9 provide frequency diagrams of the sensitivity for all three methods and $n = 6$.

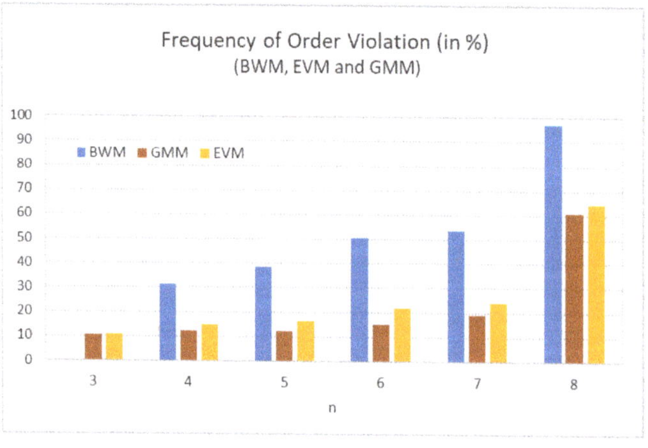

Figure 1. Average order violation: BWM, EVM, and GMM.

A discussion of the results is provided in the next section. The data are available in the Mendeley repository [30].

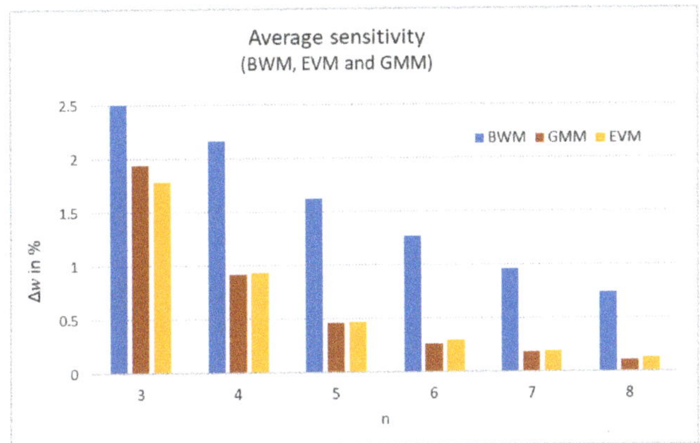

Figure 2. Mean sensitivity: BWM, EVM, and GMM.

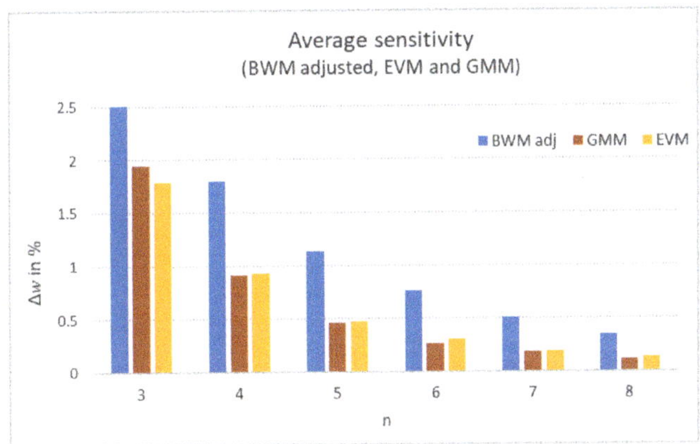

Figure 3. Mean sensitivity: BWM adjusted, EVM, and GMM.

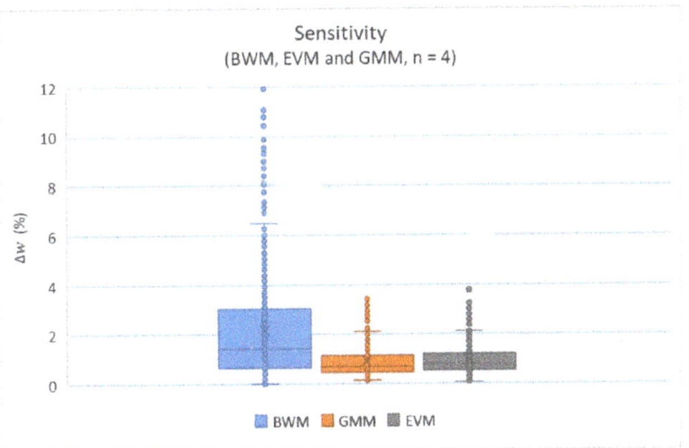

Figure 4. Sensitivity distribution: all methods, $n = 4$.

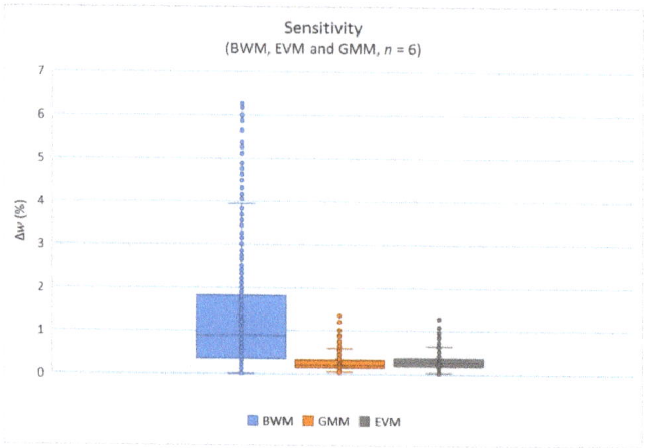

Figure 5. Sensitivity distribution: all methods, $n = 6$.

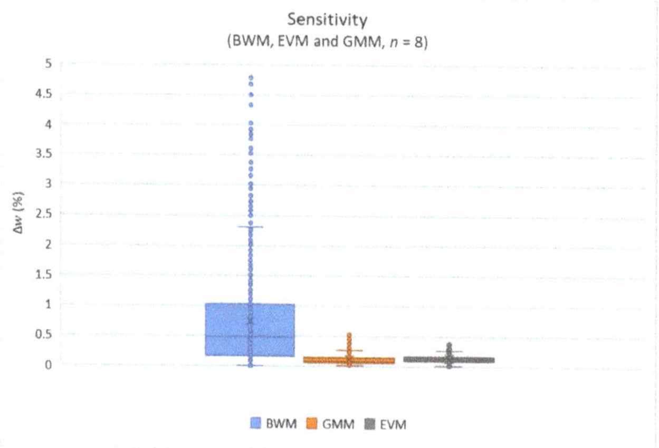

Figure 6. Sensitivity distribution: all methods, $n = 8$.

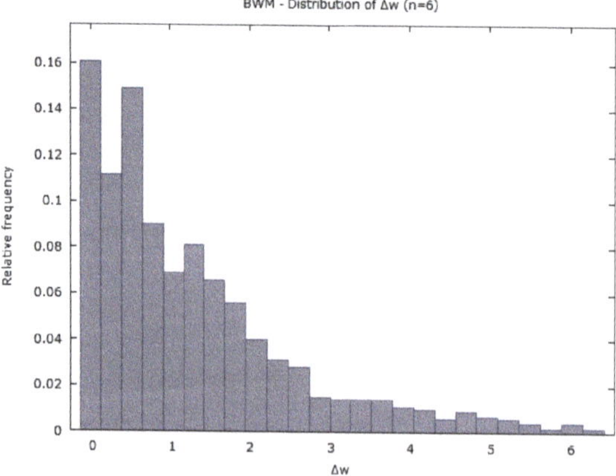

Figure 7. Sensitivity of the BWM: frequency distribution, $n = 6$.

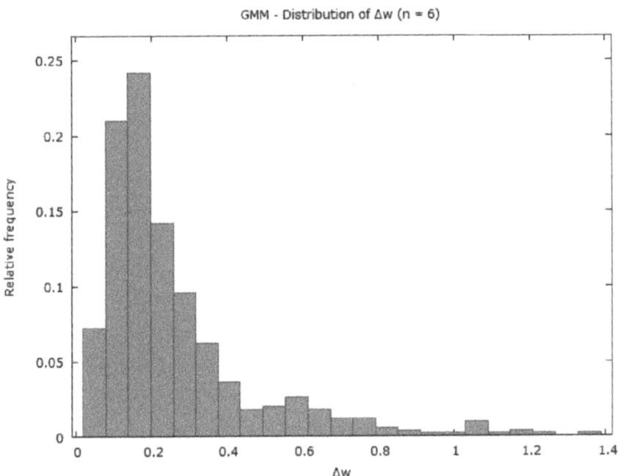

Figure 8. Sensitivity of the GMM: frequency distribution, $n = 6$.

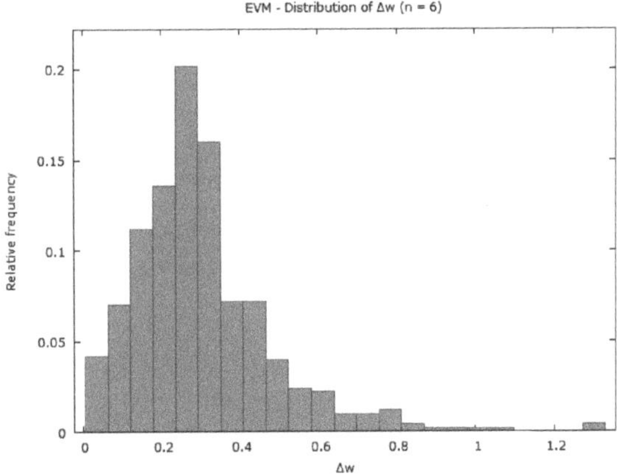

Figure 9. Sensitivity of the EVM: frequency distribution, $n = 6$.

4. Discussion

Our results, summarized in Table 1, indicate that the mean sensitivity decreases with the matrix size for all three prioritization methods; however, the sensitivity of the BWM is markedly larger than that of the GMM and EVM. In Figures 7–9, it can be seen that while the sensitivity of more than 1% is rare for the GMM and EVM (for $n = 6$), it is quite common for the BWM. A standard statistical tool for testing the equality of three or more means is the ANOVA (analysis of variance). However, the ANOVA requires an equality of variances, and from Table 1 and Figures 4–6, it is clear that this assumption was violated. Bartlett's test confirmed that variances in the sensitivity of all three methods were not equal (at $p = 0$). Therefore, we applied Welch's test instead, and the null hypothesis of sensitivity equality was rejected at the $p = 0$ level (p was so small that MS Excel rounded the value to 0) for all $3 \leq n \leq 8$. Since the BWM is "handicapped" by the lower number of pairwise comparisons required, we also performed Welch's test for sensitivity equality of all three methods, where the sensitivity of the BWM was adjusted (decreased) by the factor $\frac{2n-3}{n(n-1)/2}$, corresponding to the number of pairwise comparisons required by the BWM and EVM/GMM, respectively; see also Figure 3. Even after the adjustment, the null hypothesis of equal sensitivity was rejected at $p = 0$ for all examined n again. Interestingly, the two-sample t-test revealed that differences in the sensitivity of the GMM and EVM were statistically significant at the $p = 0.01$ level only for $n \in \{6, 8\}$.

As for order violation, its occurrence increased for all three methods as the matrix size n increased. In the case of $n = 8$, the order violation occurrence was more likely (above 50%) than not, and in the case of the BWM, it was almost certain to happen (above 97%). This result for $n = 8$ means that even the smallest possible deviation (or error) in one pairwise comparison on the input leads to a different ranking of objects in more than 50% cases. It is likely that for greater matrix sizes, this phenomenon will happen even more often; this therefore implies that the robustness of rankings of objects derived from pairwise comparisons by the BWM, EVM, and GMM is rather low, and the decision maker should take this into account. Differences in order violation occurrence (considered to be a binomial variable where either order violation happened or not) between all three methods were again tested for statistical significance. The null hypothesis that there is no difference between the BWM and GMM, and the BWM and EVM, was rejected at $p < 10^{-10}$ for all n. Differences between the GMM and EVM were statistically significant at the $p = 0.01$ level for $n \in \{5, 6, 7\}$, but not statistically significant at $p = 0.01$ for $n \in \{3, 4, 8\}$.

5. Illustrative Application of Our Approach to Order Violation Evaluation

The study of Zabihi et al. [31] focused on developing a global information system (GIS)-based multiple-criteria decision making model for a citrus land suitability assessment. The authors selected five relevant criteria: elevation, maximum temperature, minimum temperature, slope angle, and rainfall. To determine the importance of criteria, the authors pairwise compared all criteria with the Saaty's scale. The resulting pairwise comparison matrix is shown in Table 2. By using the EVM, we obtained the vector of the weights of all criteria, and we ranked them from the most important to the least important as follows: elevation (weight 0.497), minimum temperature (0.242), rainfall (0.132), maximum temperature (0.087), and slope angle (0.041). (Notice that the weights in parentheses slightly differ from the study of Zabihi et al. [31], perhaps due to numerical errors in [31].)

Table 2. A pairwise comparison matrix for the calculation of criteria weights for the citrus site selection (Table 4, Zabihi et al. [31]).

Sustainability Criterion	Elevation	Maximum Temperature	Minimum Temperature	Slope Angle	Rainfall
elevation	1	5	3	7	5
maximum temperature	1/5	1	1/3	3	1/2
minimum temperature	1/3	3	1	5	3
slope angle	1/7	1/3	1/5	1	1/5
rainfall	1/5	2	1/3	5	1

Since measurements or judgments are usually associated with errors, the question arises as to how stable the obtained ranking of the criteria is, or, in other words, is there an element (the so called critical element) in the PC matrix in Table 2 such that the minimal change of this matrix element leads to an order violation; i.e., a change in the ranking of all criteria?

Without loss of generality, we can examine all matrix elements in Table 2 (we denote the matrix as $C = [c_{ij}]$) larger than or equal to 1 (with the exception of diagonal elements):

- Let us start with $c_{12} = 5$. We change the value of 5 down by 1 scale unit (to 4), apply the EVM to find the priority vector and the ranking of all five criteria, and we get the following result: the ranking of elevation, minimum temperature, rainfall, maximum temperature, and slope angle remains unchanged. Next, we change the value of

$c_{12} = 5$ up by 1 unit (to 6), repeat the procedure, and find that the final ranking is unchanged again.

- We take another matrix element larger than or equal to 1, namely $c_{13} = 3$, and change it by 1 scale unit up and down. Again, the final ranking obtained by the EVM is unchanged in both cases.
- We proceed with the remaining elements larger than or equal to 1, namely c_{14}, c_{15}, c_{24}, c_{32}, c_{34}, c_{35}, c_{52}, and c_{54}, and change each of them by 1 scale unit up and down. The final ranking obtained by the EVM remains unchanged in all cases.

We thus conclude that the pairwise comparison matrix given in Table 2 is "robust" in the sense that the ranking of the alternatives (elevation, minimum temperature, rainfall, maximum temperature, and slope angle) remains the same if one element of the matrix is changed by 1 scale unit up or down.

The pairwise comparisons presented in Table 2 were obtained by an expert team, and the EVM yielded the priority vector $w = (0.497, 0.087, 0.242, 0.041, 0.132)$. Taking these weights of the criteria into account, another expert team may pairwise compare the criteria's importance as presented in Table 3. Notice that Saaty's consistency ratio of the PC matrix presented in Tables 2 and 3 is $CR = 0.055$ and $CR = 0.007$, respectively; i.e., the PC matrix presented in Table 3 appears to be much more consistent than the original PC matrix presented in Table 2.

Table 3. A pairwise comparison matrix for the calculation of criteria weights for the citrus site selection by another expert team.

Sustainability Criterion	Elevation	Maximum Temperature	Minimum Temperature	Slope Angle	Rainfall
elevation	1	6	2	9	4
maximum temperature	1/6	1	1/3	2	1
minimum temperature	1/2	3	1	6	2
slope angle	1/9	1/2	1/6	1	1/3
rainfall	1/4	1	1/2	3	1

By applying the EVM to the PC matrix found in Table 3, we obtain the vector of the weights of the criteria. The ranking of the criteria is the same as above: elevation (0.486), minimum temperature (0.256), rainfall (0.119), maximum temperature (0.093), and slope angle (0.045).

Although the new pairwise comparison matrix, denoted as $C' = [c'_{ij}]$, shown in Table 3 is more consistent, a critical element can be found in it. In particular, by changing the value of the element $c'_{25} = 1$ by 1 scale unit up (to 2) and by using the EVM, we obtain the following weights and ranking of the criteria: elevation (0.483), minimum temperature (0.255), maximum temperature (0.112), rainfall (0.106), and slope angle (0.045). We can see that the change of the critical element c'_{25} by 1 scale unit, which is the smallest possible change of a PC matrix, caused order violation; i.e., it led to a different ranking of the criteria for citrus land suitability assessment. The sensitivity $\Delta w^{(EVM)}_{(5)} = 0.749$ was above the mean value 0.474 (see Table 1) in this case. Knowing this, the decision maker should pay special attention to this particular PC comparison (or a measurement in general) to ensure its accuracy and to avoid a distortion of the final ranking. Our Monte Carlo simulations revealed that the frequency occurrence of critical elements increased with the increasing matrix size for all three priority deriving methods; see Table 1.

6. Conclusions

The aim of this paper was to provide a comparison of the sensitivity and order violation of three popular prioritization methods in pairwise comparisons: the Geometric Mean Method, the Eigenvalue Method, and the Best–Worst Method.

Our results suggest that the Best–Worst Method is statistically significantly more sensitive and more susceptible to order violation than the Geometric Mean Method and the Eigenvalue Method for $3 \leq n \leq 8$ compared objects. On the other hand, the difference in sensitivity of the Geometric Mean Method and the Eigenvalue Method was found to be statistically insignificant in most cases.

Since both the GMM and EVM outperformed the BWM, and the differences in the GMM and EVM were rather small, both "standard" methods can equally be recommended as a suitable prioritization method with regard to sensitivity and order violation.

Further, we demonstrated how our approach can be used in practice for the evaluation of the stability of a ranking obtained by a given PC method. We came across a surprising finding that, with the increasing size of the PC matrix, the relative frequency of critical elements also increases.

Future research may focus on a comparison of extensions of the aforementioned methods aiming at interval pairwise comparisons or fuzzy pairwise comparisons.

Author Contributions: Conceptualization, J.M. and J.R.; methodology, J.M., J.R. and D.B.; software, R.P.; validation, J.M., D.B. and J.R.; formal analysis, J.R. and D.B.; investigation, R.P. and J.M.; data curation, R.P.; writing—original draft preparation, J.M.; writing—review and editing, D.B.; visualization, R.P. and J.M.; supervision, J.R.; project administration, J.R.; funding acquisition, J.R. All authors have read and agreed to the published version of the manuscript.

Funding: This work was supported by the Czech Science Foundation, grant number GAČR 18-01246S and the Ministry of Education, Youth and Sports Czech Republic within the Institutional Support for Long-term Development of a Research Organization in 2021.

Data Availability Statement: The data used in this study are available in the Mendeley repository [30].

Conflicts of Interest: The authors declare no conflict of interest. The funder had no role in the design of the study; in the collection, analyses, or interpretation of data; in the writing of the manuscript, or in the decision to publish the results.

Abbreviations

The following abbreviations are used in this manuscript:

AHP	Analytic Hierarchy Process
BWM	Best–Worst Method
EVM	Eigenvalue (eigenvector) Method
GMM	Geometric Mean Method
PC	Pairwise Comparison

References

1. Brans, J.P.; Vincke, P. A preference ranking organisation method: The PROMETHEE method for MCDM. *Manag. Sci.* **1985**, *31*, 647–656. [CrossRef]
2. Figueira, J.; Greco, S.; Ehrgott, M. *Multiple Criteria Decision Analysis: State of the Art Surveys*; Springer Science + Business Media, Inc.: New York, NY, USA, 2005.
3. Hansen, P.; Ombler, F. A new method for scoring additive multi-attribute value models using pairwise rankings of alternatives. *J. Multi-Criteria Decis. Anal.* **2008**, *15*, 87–107. [CrossRef]
4. Koksalan, M.M.; Wallenius, J.; Zionts, S. *Multiple Criteria Decision Making: From Early History to the 21st Century*; World Scientific Publishing: Singapore, 2011.
5. Mardani, A.; Jusoh, A.; Nor, K.M.D.; Khalifah, Z.; Zakwan, N.; Valipour, A. Multiple criteria decision-making techniques and their applications–a review of the literature from 2000 to 2014. *Econ. Res.-Ekon. Istraz.* **2015**, *28*, 516–571. [CrossRef]
6. Roy, B. Classement et choix en presence de points de vue multiples (la methode ELECTRE). *Rev. D'Inform. Rech. Oper. (RIRO)* **1968**, *8*, 57–75.

7. Saaty, T.L. A Scaling Method for Priorities in Hierarchical Structures. *J. Math. Psychol.* **1977**, *15*, 234–281. [CrossRef]
8. Saaty, T.L. *Analytic Hierarchy Process*; McGraw-Hill: New York, NY, USA, 1980.
9. Saaty, T.L. *Fundamentals of Decision Making*; RWS Publications: Pittsburgh, PA, USA, 1994.
10. Vaidya, O.; Kumar, S. Analytic Hierarchy Process: An Overview of Applications. *Eur. J. Oper. Res.* **2006**, *169*, 1–29. [CrossRef]
11. Crawford, G.B. The geometric mean procedure for estimating the scale of a judgement matrix. *Math. Model.* **1987**, *9*, 327–334. [CrossRef]
12. Rezaei, J. Best-worst multi-criteria decision-making method. *Omega* **2015**, *53*, 40–57. [CrossRef]
13. Abadia, F.A.; Sahebib, I.G.; Arabc, A.; Alavid, A.; Karachi, H. Application of best-worst method in evaluation of medical tourism development strategy. *Decis. Sci. Lett.* **2018**, *7*, 77–86. [CrossRef]
14. Ahmadi, A.B.; Kusi-Sarpong, S.; Rezaei, J. Assessing the social sustainability of supply chains using Best Worst Method. *Resour. Conserv. Recycl.* **2017**, *126*, 99–106. [CrossRef]
15. Chang, M-H.; Liou, J.J.H.; Lo, H.-W. A Hybrid MCDM Model for Evaluating Strategic Alliance Partners in the Green Biopharmaceutical Industry. *Sustainability* **2019**, *11*, 4065. [CrossRef]
16. Gupta, H.; Barua, M.K. Supplier selection among SMEs on the basis of their green innovation ability using BWM and fuzzy TOPSIS. *J. Clean. Prod.* **2017**, *152*, 242–258. [CrossRef]
17. Rezaei, J.; Nispeling, T.; Sarkis, J.; Tavasszy, L. A supplier selection life cycle approach integrating traditional and environmental criteria using the best worst method. *J. Clean. Prod.* **2016**, *135*, 577–588. [CrossRef]
18. Setyono, R.P.; Sarno, R. Vendor Track Record Selection Using Best Worst Method. In Proceedings of the 2018 International Seminar on Application for Technology of Information and Communication, Semarang, Indonesia, 21–22 September 2018; pp. 41–48.
19. Ajrina, A.S.; Sarno, R.; Hari Ginardi, R.V. Comparison of AHP and BWM Methods Based on Geographic Information System for Determining Potential Zone of Pasir Batu Mining. In Proceedings of the 2018 International Seminar on Application for Technology of Information and Communication, Semarang, Indonesia, 21–22 September 2018; pp. 453–457.
20. Haseli, G.; Sheikh, R.; Sankar Sana, S. Base-criterion on multi-criteria decision-making method and its applications. *Int. J. Manag. Sci. Eng. Manag.* **2020**, *15*, 79–88. [CrossRef]
21. Cacuci, D.G.; Ionescu-Bujor, M.; Navon, M. *Sensitivity and Uncertainty Analysis, Vol. II: Applications to Large-Scale Systems*; Chapman & Hall/CRC: Boca Raton, FL, USA, 2005.
22. Helton, J.C.; Johnson, J.D.; Salaberry, C.J.; Storlie, C.B. Survey of sampling based methods for uncertainty and sensitivity analysis. *Reliab. Eng. Syst. Saf.* **2006**, *91*, 1175–1209. [CrossRef]
23. Saltelli, A.; Chan, K.; Scott, M. (Eds). *Sensitivity Analysis*; Wiley Series in Probability and Statistics; John Wiley and Sons: New York, NY, USA, 2000.
24. Saltelli, A. Sensitivity Analysis for Importance Assessment. *Risk Anal.* **2002**, *22*, 579–590. [CrossRef] [PubMed]
25. Czitrom, V. One-Factor-at-a-Time Versus Designed Experiments. *Am. Stat.* **1999**, *53*, 126–131.
26. Ramík, J. Ranking Alternatives by Pairwise Comparisons Matrix and Priority Vector. *Sci. Ann. Econ. Bus.* **2017**, *64*, 85–95. [CrossRef]
27. Brunelli, M.; Rezaei, J. A multiplicative best-worst method for multi-criteria decision making. *Oper. Res. Lett.* **2019**, *47*, 12–15. [CrossRef]
28. Rezaei, J. Best-worst multi-criteria decision-making method: Some properties and a linear model. *Omega* **2016**, *64*, 126–130. [CrossRef]
29. Best Worst Method: BWM Solvers. Available online: https://bestworstmethod.com/software/ (accessed on 31 January 2021).
30. Perzina, R.; Mazurek, J.; Ramík, J.; Bartl, D. Data for 'A Numerical Comparison of the Geometric Mean Method, the Eigenvalue Method, and the Best Worst Method', Mendeley Data, v1 (2021). Available online: https://data.mendeley.com/datasets/svdjzz2nk6/draft?a=a1fbe712-5d57-44da-b79d-0d4d56eaad2d (accessed on 31 January 2021)
31. Zabihi, H.; Alizadeh, M.; Kibet Langat, P.; Karami, M.; Shahabi, H.; Ahmad, A.; Nor Said, M.; Lee, S. GIS Multi-Criteria Analysis by Ordered Weighted Averaging (OWA): Toward an Integrated Citrus Management Strategy. *Sustainability* **2019**, *11*, 1009. [CrossRef]

Article

Properties of the Global Total k-Domination Number

Frank A. Hernández Mira [1], Ernesto Parra Inza [2], José M. Sigarreta Almira [3,*] and Nodari Vakhania [2]

1 Regional Development Sciences Center, Autonomous University of Guerrero, Los Pinos s/n, Suburb El Roble, Acapulco, Guerrero 39070, Mexico; fmira8906@gmail.com
2 Science Research Center, Autonomous University of Morelos, Cuernavaca 62209, Mexico; eparrainza@gmail.com (E.P.I.); nodari@uaem.mx (N.V.)
3 Faculty of Mathematics, Autonomous University of Guerrero, Carlos E. Adame 5, Col. La Garita, Acapulco, Guerrero 39070, Mexico
* Correspondence: josemariasigarretaalmira@hotmail.com

Abstract: A nonempty subset $D \subset V$ of vertices of a graph $G = (V, E)$ is a dominating set if every vertex of this graph is adjacent to at least one vertex from this set except the vertices which belong to this set itself. $D \subseteq V$ is a total k-dominating set if there are at least k vertices in set D adjacent to every vertex $v \in V$, and it is a global total k-dominating set if D is a total k-dominating set of both G and $\overline{G}$. The global total k-domination number of G, denoted by $\gamma^g_{kt}(G)$, is the minimum cardinality of a global total k-dominating set of G, GTkD-set. Here we derive upper and lower bounds of $\gamma^g_{kt}(G)$, and develop a method that generates a GTkD-set from a GT$(k-1)$D-set for the successively increasing values of k. Based on this method, we establish a relationship between $\gamma^g_{(k-1)t}(G)$ and $\gamma^g_{kt}(G)$, which, in turn, provides another upper bound on $\gamma^g_{kt}(G)$.

Keywords: global total domination; total k-domination number

Citation: Hernández Mira, F.A.; Parra Inza, E.; Sigarreta Almira, J.M.; Vakhania, N. Properties of the Global Total k-Domination Number. *Mathematics* **2021**, *9*, 480. https://doi.org/10.3390/math9050480

Academic Editors: Christos G. Massouros, Ricardo Lopez-Ruiz and Dumitru Baleanu

Received: 11 January 2021
Accepted: 22 February 2021
Published: 26 February 2021

Publisher's Note: MDPI stays neutral with regard to jurisdictional claims in published maps and institutional affiliations.

1. Introduction

We start by introducing the basic notation. Suppose we are given a simple graph $G = (V, E)$ with $|V| = n$ (n is called the order of graph G) and $|E| = m$ (m is called the size of graph G). Given $D \subseteq V$ ($D \neq \emptyset$) and vertex $v \in V$, let $N_D(v)$ be the set of all vertices from set D, adjacent to vertex v (also called the neighbors of vertex v from set D); we will use $\overline{N}_D(v)$ for the set of vertices in set D which are not neighbors of vertex v ($\overline{N}_D[v] = \overline{N}_D(v) \cup \{v\}$). We let $N_D[v] = N_D(v) \cup \{v\}$, and we call $\delta_D(v) = |N_D(v)|$ the degree of vertex v in set D. We denote by $\overline{\delta}_D(v)$ the cardinality of set $\overline{N}_D(v)$ ($\overline{\delta}_D(v) = |\overline{N}_D(v)|$). We will use more compact notation $N(v)$, $N[v]$, $\delta(v)$, $\overline{N}(v)$ and $\overline{N}[v]$ instead of $N_G(v)$, $N_G[v]$, $\delta_G(v)$, $\overline{N}_G(v)$ and $\overline{N}_G[v]$, respectively, when this will cause no confusion. The minimum (the maximum, respectively) degree in graph G is traditionally denoted by δ (Δ, respectively). $G[S]$ and $\overline{G}$, respectively, will stand for the subgraph of graph G induced by $S \subseteq V$ and the complement of graph G, respectively.

Let X and Y be subsets of set V. We denote by $E(X, Y)$ the set of all the edges in graph G joining a vertex $x \in X$ with a vertex $y \in Y$. Let u and v be vertices from set V. Then the distance between these two vertices $d(u, v)$ is the length (the number of edges) of a minimum $u - v$-path. The length of the longest $u - v$ path, for any u and v, is called the diameter of graph G, denoted by $diam(G)$. The girth of graph G is the length of the shortest cycle in that graph and is denoted by $g(G)$.

Let $D \subseteq V$ be a nonempty subset of set V. Then D is called a total k-dominating set for graph G if there are at least k vertices in set D adjacent to every vertex $v \in V$ (we will also say that vertex v is totally k-dominated by set D). The cardinality of a total k-dominating set in graph G with the minimum cardinality is called the total k-domination number of graph G and is denoted by $\gamma_{kt}(G)$. We will refer to a total k-dominating set with cardinality $\gamma_{kt}(G)$ as a $\gamma_{kt}(G)$-set. A total 1-dominating set is normally referred to as a total dominating set, and the total 1-domination number is referred to as the total domination number, denoted by $\gamma_t(G)$. We refer the reader to [1–9] for more detail on these definitions.

Given again a non-empty set $D \subseteq V$, D is called a global total k-dominating set of graph G (GTkD set for short) if D is a total k-dominating set of both graphs G and $\overline{G}$. The global total k-domination number of G, denoted by $\gamma^g_{kt}(G)$, is the cardinality of a global total k-dominating set with the minimum cardinality. A global total k-dominating set of cardinality $\gamma^g_{kt}(G)$ will be referred to as a $\gamma^g_{kt}(G)$-set. Again, if $k = 1$, a global total 1-dominating set is a global total dominating set (see [10,11]).

As it is well-known and also easily be seen,

$$2k + 1 \leq \gamma^g_{kt}(G) \leq n,$$

for any graph G with order n. Here we shall exclusively deal with the connected graphs due to a known fact that if $G_1, G_2, \ldots, G_r$ ($r \geq 2$) are the connected components in graph G, then

$$\gamma^g_{kt}(G) = \sum_{i=1}^{r} \gamma_{kt}(G_i)$$

(see [12]).

The main goal of this paper is to complete the current study of global total k-domination number in graphs. First, we give upper and lower bounds on $\gamma^g_{kt}(G)$, and then we develop a method that generates a GTkD-set from a GT$(k-1)$D-set for the successively increasing values of k. Based on this method, we establish a relationship between $\gamma^g_{(k-1)t}(G)$ and $\gamma^g_{kt}(G)$, which, in turn, provides another upper bound on $\gamma^g_{kt}(G)$.

The rest of the paper is organized as follows. In the next section, we present known results and give some remarks. In Sections 3 and 4, we derive upper and lower bounds, respectively, for global total k-domination number. In the Section 5, we provide our method that obtains a global total $(k+1)$-dominating set from a global total k-dominating set.

2. Relations between $\gamma^g_{kt}(G)$ and $\gamma_{kt}(G)$

Clearly, the definition of a GTkD set gives us an implicit lower bound for the parameter $\gamma^g_{kt}(G)$:

Observation 1. *Let G be a graph; then* $\gamma^g_{kt}(G) \geq \max\{\gamma_{kt}(G), \gamma_{kt}(\overline{G})\}$.

The above lower bound is not necessarily attainable, as we illustrate in the following figure: we depict graph G and its complement $\overline{G}$, and the corresponding minimum total 2 dominating set in both graphs (black vertices); see Figure 1.

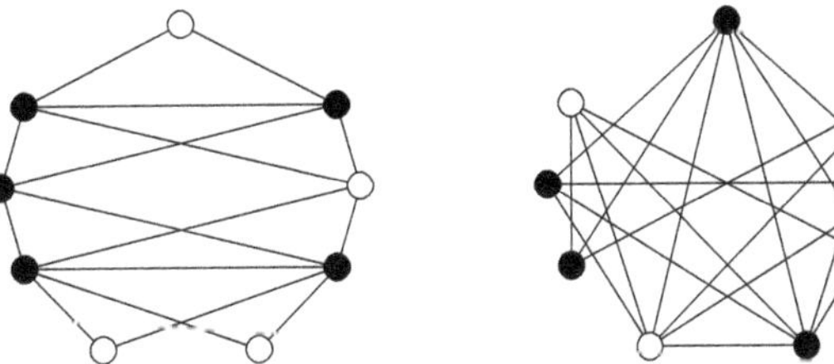

Figure 1. Graph G and its complement $\overline{G}$, which satisfy $\gamma_{2t}(G) = 5$, $\gamma_{2t}(\overline{G}) = 5$ and $\gamma^g_{2t}(G) = 6$.

The following proposition was proved in [12].

Proposition 1. *Let G be a graph,*

(i) *If* $\gamma_{kt}(G) > \Delta(G) + k$, *then* $\gamma^g_{kt}(G) = \gamma_{kt}(G)$.

(ii) *If* $\gamma_{kt}(G) \leq \Delta(G) + k$, *then* $\gamma^g_{kt}(G) \leq \Delta(G) + k + 1$.

Corollary 1. *Let G be a graph with maximum degree* Δ*. Then,* $\gamma^g_{kt}(G) \leq \max\{\gamma_{kt}(G), \Delta + k + 1\}$.

Proposition 2. *Let G be a graph with order n and maximum degree* Δ*. If* $n > \frac{\Delta(\Delta+k)}{k}$*, then* $\gamma^g_{kt}(G) = \gamma_{kt}(G)$.

Proof. If $n > \frac{\Delta(\Delta+k)}{k}$, then $\Delta + k < \frac{kn}{\Delta} \leq \gamma_{kt}(G)$; consequently, $\Delta + k + 1 \leq \gamma_{kt}(G)$. By Corollary 1 we have $\gamma^g_{kt}(G) = \gamma_{kt}(G)$. □

Theorem 1. *For any graph G,* $\gamma^g_{kt}(G) = \gamma_{kt}(G)$ *if and only if there exists a minimum total k-dominating set D such that any subset D′ of D with* $|D| - k + 1$ *vertices is not included in any star in the graph—that is, and only if there is not a vertex* $v \in V$ *such that* $D' \subseteq N[v]$.

Proof. Let D be a minimum total k-dominating set which is also a global total k-dominating set, and let D' be a subset of D with cardinality $|D| - k + 1$. If there exists a vertex $v \in V$ such that $D' \subseteq N[v]$, then $v \in D'$ and it is adjacent to $|D| - k$ vertices in D', so v has less than k non-adjacent vertices in D, or $v \notin D'$, and it is adjacent to $|D| - k + 1$ vertices in D', so v has less than k non-adjacent vertices in D. In both cases we have a contradiction with the fact that D is a global total k-dominating set.

On the other hand, we take a minimum total k-dominating set D such that for any subset D' of D with $|D| - k + 1$ vertices and every vertex $v \in V$, we have $D' \not\subseteq N[v]$. Then, for any vertex $v \in D$ we have $|N(v)| < |D| - k$, so v has, at least, k non-neighbors in D. If $v \in V \setminus D$ we have $|N(v)| < |D| - k + 1$, so v has, at least, k non-neighbors in D. Therefore, D is a global total k-dominating set. □

3. Upper Bounds for the Global Total k-Domination Number

In this section, we obtain some upper bounds for the global total k-domination number in a graph. Bermudo et al. in [12] showed a characterization when the global total k-domination number is equal to the order of the graph, but we give here that characterization in a more specific way. To do that, in the following proposition we give a condition to guarantee that the global total k-domination number is less than n.

Proposition 3. *Let G be a graph with order n, minimum degree* δ *and maximum degree* Δ*. If* $k < \min\{\delta, n - \Delta - 1\}$*, then* $\gamma^g_{kt}(G) \leq n - 1$.

Proof. Let us see that, for any $v \in V$, the set $D = V \setminus \{v\}$ is a GTkD set of G. We have that $\delta_D(v) = \delta(v) \geq \delta > k$ and $\overline{\delta}_D(v) = n - 1 - \delta(v) \geq n - 1 - \Delta > k$. For every $u \in D$ we have $\delta_D(u) \geq \delta(u) - 1 \geq \delta - 1 \geq k$ and $\overline{\delta}_D(u) \geq n - 1 - \delta(u) - 1 \geq n - 2 - \Delta \geq k$. Therefore, D is a GTkD set of G. □

Proposition 3 is not an equivalence, as we can see if we consider a triangle and we add a leaf to every vertex of the triangle. In such a case $\gamma^g_{1t}(G) \leq n - 1 = 5$ and $k = 1 = \min\{\delta, n - \Delta - 1\}$.

Now, in order to present the characterization of all graphs having a global total k-domination number equal to the number of vertices, we need to define the following set. Given a graph G and an integer i, let $T_i(G) = \{v \in V(G) : \delta(v) = i\}$ (i.e., the set of vertices in graph G with the degree i).

Theorem 2. *Given graph G with order n and the minimum and the maximum degrees* δ *and* Δ*,* $\gamma^g_{kt}(G) = n$ *if and only if one of the conditions (a)–(c) below hold*

(a) $k = \delta < n - \Delta - 1$ *and* $V = \bigcup_{v \in T_\delta(G)} N(v)$.

(b) $k = n - \Delta - 1 < \delta$ *and* $V = \bigcup_{w \in T_\Delta(G)} (V \setminus N[w])$.

(c) $k = \delta = n - \Delta - 1$ *and* $V = \left(\bigcup_{v \in T_\delta(G)} N(v) \right) \cup \left(\bigcup_{w \in T_\Delta(G)} (V \setminus N[w]) \right).$

Proof. (a) If $k = \delta < n - \Delta - 1$ and $V = \bigcup_{v \in T_\delta(G)} N(v)$, we consider $D = V \setminus \{u\}$ for any $u \in V$. We note that there exists $v \in N(u)$ such that $\delta(v) = k$; this implies that $\delta_D(v) < k$. Thus, D is not a GTkD set of G. Hence, $\gamma_{kt}^g(G) = n$.

(b) If $k = n - \Delta - 1 < \delta$ and $V = \bigcup_{w \in T_\Delta(G)} (V \setminus N[w])$, for any $u \in V$ there exists $w \in V$ such that $\delta(w) = \Delta$ and $u \notin N[w]$. If we consider $D = V \setminus \{u\}$, then $\overline{\delta}_D(w) \leq n - \Delta - 2 < k$; thus, D is not a GTkD set of G. Therefore, $\gamma_{kt}^g(G) = n$.

(c) If $k = \delta = n - \Delta - 1$ and $V = \left(\bigcup_{v \in T_\delta(G)} N(v) \right) \cup \left(\bigcup_{w \in T_\Delta(G)} (V \setminus N[w]) \right)$, using (a) or (b), we obtain that $V \setminus \{u\}$ is not a GTkD set of G, for any $u \in V$. Consequently, $\gamma_{kt}^g(G) = n$.

Finally, if we assume that $\gamma_{kt}^g(G) = n$, by Proposition 3 we have that $k \in \{\delta, n - \Delta - 1\}$. For every vertex $v \in V$, we note that $D = V \setminus \{v\}$ is not a GTkD set of G, so there exists $u \in D$ such that $\delta_D(u) < k$ or $\overline{\delta}_D(u) < k$. If $k = \delta < n - \Delta - 1$, since $\overline{\delta}_D(u) \geq n - 2 - \delta(u) \geq n - 2 - \Delta \geq k$, then we have that $\delta_D(u) < k = \delta$; this implies that $u \in T_\delta(G)$ and $v \in N(u)$. If $k = n - \Delta - 1 < \delta$, since $\delta_D(u) \geq \delta(u) - 1 \geq \delta - 1 \geq k$, then we have that $n - 2 - \delta(u) \leq \overline{\delta}_D(u) < k = n - \Delta - 1$; that is, $n - 2 - \delta(u) = \overline{\delta}_D(u) = n - \Delta - 2$, so $u \in T_\Delta(G)$ and $v \in V \setminus N[u]$. If $k = \delta = n - \Delta - 1$, since $\delta_D(u) < k$ or $\overline{\delta}_D(u) < k$, we have that $u \in T_\delta(G)$ and $v \in N(u)$, or $u \in T_\Delta(G)$ and $v \in V \setminus N[u]$. □

The following corollary was directly obtained from Theorem 2.

Corollary 2. *Let G be a graph with minimum degree δ, maximum degree Δ and order $n \neq \Delta + \delta + 1$. Then $\gamma_{kt}^g(G) = n$ if and only if one of the following condition holds.*

(a) $k = \delta < n - \Delta - 1$ *and* $\gamma_{kt}(G) = n$.

(b) $k = n - \Delta - 1 < \delta$ *and* $\gamma_{kt}(\overline{G}) = n$.

Corollary 3. *Let G be a graph of order n, minimum degree δ and maximum degree Δ. If one of the following conditions holds:*

(a) $k = \delta < n - \Delta - 1$ *and* $|T_\delta(G)| \geq n - \delta$.

(b) $k = n - \Delta - 1 < \delta$ *and* $|T_\Delta(G)| \geq \Delta + 1$

(c) $k = \delta = n - \Delta - 1$ *and* $|T_\delta(G)| \geq n - \delta$ *or* $|T_\Delta(G)| \geq \Delta + 1$,

then $\gamma_{kt}^g(G) = n$.

Proof. Since $\gamma_{kt}^g(G) = \gamma_{kt}^g(\overline{G})$, $\overline{\Delta} = n - \delta - 1$, $T_{\overline{\Delta}}(\overline{G}) = T_\delta(G)$ and $V \setminus N_{\overline{G}}[w] = N(w)$, it is enough to check that $|T_\Delta(G)| \geq \Delta + 1$ implies $V = \bigcup_{w \in T_\Delta(G)} (V \setminus N[w])$. However, for any vertex $v \in V$, if $|T_\Delta(G)| \geq \Delta + 1$, then there exists a vertex $w \in T_\Delta(G)$ which is not a neighbor of v, so $v \in \bigcup_{w \in T_\Delta(G)} (V \setminus N[w])$. □

It was proved in [12] that $\gamma_{kt}^g(G) \leq \min\{\gamma_{kt}(G) + \Delta, \gamma_{kt}(G) + \gamma_{kt}(\overline{G})\}$. It would be convenient to characterize the graphs G such that $\gamma_{kt}^g(G) = \gamma_{kt}(G) + \Delta$, and the graphs G such that $\gamma_{kt}^g(G) = \gamma_{kt}(G) + \gamma_{kt}(\overline{G})$. On the other hand, the invariants of a graph are important when characterizing them; below we use some of them such as diameter and girth. The following proofs use the ideas showed in [11].

Theorem 3. *If G is a graph such that* $diam(G) \geq 5$*, every total k-dominating set is a GTkD set of G.*

Proof. Let D be a total k-dominating set and $u, v \in V$ such that $d(u,v) \geq 5$. Since $\delta_D(u) \geq k$ and $\delta_D(v) \geq k$, there exist $\{u_1, \ldots, u_k\} \subseteq D \cap N(u)$ and $\{v_1, \ldots, v_k\} \subseteq D \cap N(v)$. For any vertex $w \in V$ we know that $\delta_D(w) \geq k$. If $u_i \in N(w)$ for some $i \in \{1, \ldots, k\}$, then $w \notin \bigcup_{i=1}^{k} N[v_i]$; that means, $\overline{\delta}_D(w) \geq k$. Therefore, D is a GTkD set of G. □

Corollary 4. *If G is a graph such that* $diam(G) \geq 5$*, then* $\gamma^g_{kt}(G) = \gamma_{kt}(G)$.

According to the idea given in [11], we obtain the following result.

Proposition 4. *If G is a graph such that* $diam(G) = 4$ *and there exist* $\{u, v_1, \ldots, v_k\} \subseteq V$ *such that* $dist(u, v_j) = 4$ *for every* $j \in \{1, \ldots, k\}$*, then* $\gamma^g_{kt}(G) \leq \gamma_{kt}(G) + k$.

Proof. Let D be a minimum total k-dominating set of a graph; then there exists the vertex set $\{u_1, \ldots, u_k\} \subseteq D$ such that $\{u_1, \ldots, u_k\} \subseteq N(u)$. For any vertex $w \in V$ and $i \in \{1, \ldots, k\}$, w cannot be adjacent to both u_i and v_i, so $D \cup \{v_1, \ldots, v_k\}$ is a global total k-dominating set. □

In Figure 2 we can see an example where the equality in Proposition 4 for $k = 2$ is attained. Taking into account that any neighbor of a vertex of degree 2 must belong to any total 2-dominating set (grey vertices), we show in that figure the minimum total 2-dominating set (b) and the minimum global total 2-dominating set (c).

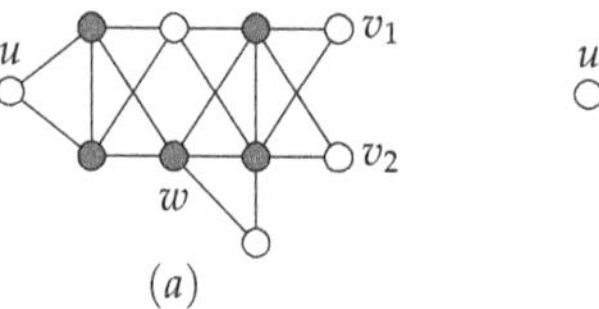

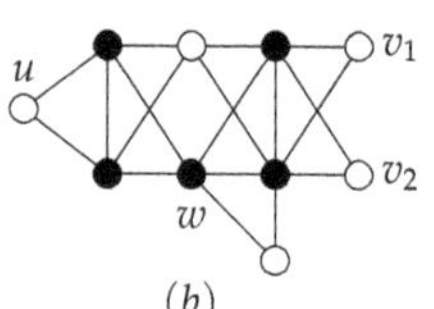

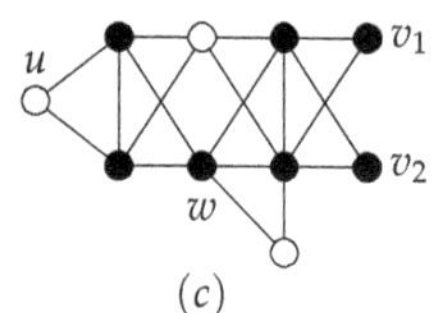

Figure 2. (**a**) Grey vertices are neighbors of vertices of degree 2. (**b**) Minimum total 2-dominating set and (**c**) minimum global total 2-dominating set.

For a graph G, we let $\delta^*(G) = min\{\delta(G), \delta(\overline{G})\}$.

Proposition 5. *Let G be a graph of order n and minimum degree* δ*; then* $\gamma^g_{kt}(G) \leq n - \delta^*(G) + k$.

Proof. Let us see that every set $D \subseteq V$ such that $|D| \geq n - \delta^*(G) + k$ is a global total k-dominating set. Since $|D| \geq n - \delta + k$, every vertex v satisfies $\delta_{V \setminus D}(v) \leq \delta - k$, $\delta_D(v) \geq k$. Since $|D| \geq n - \overline{\delta} + k$, every vertex v satisfies $\overline{\delta}_{V \setminus D}(v) \leq \overline{\delta} - k$, so $\overline{\delta}_D(v) \geq k$. □

4. Lower Bounds for the Global Total k-Domination Number

We know that any graph G satisfies $\gamma^g_{kt}(G) \geq 2k + 1$, and a characterization for graphs satisfying the equality was given in [12]. Additionally, in that work the authors showed the following inequality.

Remark 1. *Let G be a graph with order n, minimum degree* δ *and maximum degree* Δ*. Then,*

$$\gamma^g_{kt}(G) \geq \max\left\{\frac{kn}{\Delta}, \frac{kn}{n - \delta - 1}\right\}$$

For example, the lower bound given above can be reached in the graph shown in Figure 3.

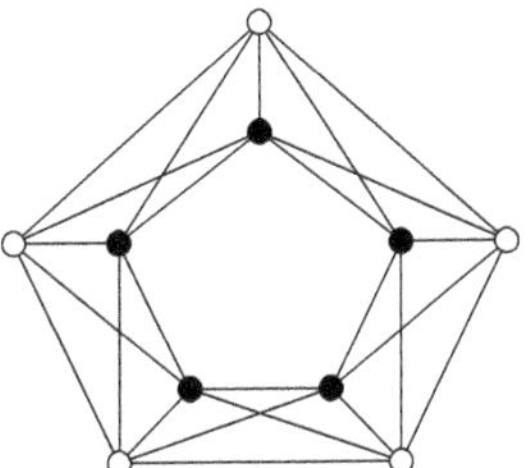

Figure 3. A graph G with order $n = 10$, $\delta = 5$ and $\gamma^g_{2t}(G) = \frac{2n}{n-\delta-1}$.

Theorem 4. *Let G be a graph of order n, maximum degree Δ and size m. Then*

$$\gamma^g_{kt}(G) \geq \frac{2m + n(2k-\Delta) + (2k+1)^2}{n+2k}.$$

Proof. Let D be a $\gamma_{kt}(G)$-set. Since every vertex in $V \setminus D$ cannot have more that $|D| - k$ neighbors in D, we have $E(D, V \setminus D) \leq (n - |D|)(|D| - k)$, so

$$\begin{aligned} m &= E(D,D) + E(D, V\setminus D) + E(V \setminus D, V \setminus D) \\ &\leq \frac{|D|\Delta(G) - E(D, V\setminus D)}{2} + E(D, V \setminus D) + \frac{(\Delta - k)(n-|D|)}{2} \\ &\leq \frac{|D|\Delta + (n-|D|)(|D|-k) + (\Delta-k)(n-|D|)}{2} \\ &= \frac{|D|\Delta + (n-|D|)(|D| - 2k + \Delta)}{2} \\ &= \frac{-|D|^2 + (n+2k)|D| + n\Delta - 2kn}{2}, \end{aligned}$$

which implies that

$$(2k+1)^2 + 2m \leq |D|^2 + 2m \leq (n+2k)|D| + n\Delta - 2kn,$$

then

$$|D| \geq \frac{2m + n(2k-\Delta) + (2k+1)^2}{n+2k}.$$

□

Theorem 5. *Let G be a graph with order n, maximum degree Δ and size m. Then,*

$$\gamma^g_{kt}(G) \geq \frac{2m + n(\Delta - 2k)}{n+k-\Delta}.$$

Proof. We suppose that D is a $\gamma_{kt}(G)$-set and $|D| \geq 2r+1$ for some $r \geq 2$, and $|D| \geq 2k+2$. Since D is minimal, for any vertex $v_1 \in D$ there exists a vertex w_{v_1} such that one of the following conditions holds.

(1) $w_{v_1} \in D, v_1 \in N(w_{v_1})$ and $\delta_D(w_{v_1}) = k$,

(2) $w_{v_1} \in D, v_1 \notin N(w_{v_1})$ and $\delta_D(w_{v_1}) = |D| - k - 1$,

(3) $w_{v_1} \in V \setminus D, v_1 \in N(w_{v_1})$ and $\delta_D(w_{v_1}) = k$,

(4) $w_{v_1} \in V \setminus D, v_1 \notin N(w_{v_1})$ and $\delta_D(w_{v_1}) = |D| - k$.

Now, in cases (1) and (3), we take $v_2 \in D \setminus N(w_{v_1})$, and in cases (2) and (4), we take $v_2 \in D \cap N(w_{v_1})$, and we know that there exists a vertex $w_{v_2} \neq w_{v_1}$ such that one of the above conditions holds. Since $|D| \geq 2r+1$ we can obtain $w_{v_1}, \ldots, w_{v_r}$ vertices satisfying one of the conditions above. We suppose that there exist i, $j-i$, s and $r-j-s$ vertices satisfying (1), (2), (3) and (4), respectively. Then,

$$\begin{aligned} E(D,D) &\le \frac{ik+(j-i)(|D|-k-1)+(|D|-j)(|D|-k-1)}{2} \\ &= \frac{ik-i(|D|-k-1)+|D|(|D|-k-1)}{2} \\ &= \frac{i(2k-|D|+1)+|D|(|D|-k-1)}{2}, \end{aligned}$$

$$\begin{aligned} E(D,V\setminus D) &\le \frac{sk+(r-j-s)(|D|-k)+(n-|D|-r+j)(|D|-k)}{2} \\ &= \frac{sk-s(|D|-k)+(n-|D|)(|D|-k)}{2} \\ &= \frac{(n-|D|)(|D|-k)+s(2k-|D|)}{2}, \end{aligned}$$

and

$$\begin{aligned} E(V\setminus D,V\setminus D) &\le \frac{s(\Delta-k)+(r-j-s)(\Delta-|D|+k)}{2} \\ &+ \frac{(n-|D|-r+j)(\Delta-k)}{2} \\ &= \frac{s(\Delta-k)+(r-j-s)(\Delta-k-|D|+2k)}{2} \\ &+ \frac{(n-|D|-r+j)(\Delta-k)}{2} \\ &= \frac{(\Delta-k)(n-|D|)+(r-j-s)(2k-|D|)}{2}; \end{aligned}$$

therefore,

$$\begin{aligned} m &\le E(D,D)+E(D,V\setminus D)+E(V\setminus D,V\setminus D) \\ &\le \frac{i(2k-|D|+1)+|D|(|D|-k-1)}{2}+\frac{(n-|D|)(|D|-k)+s(2k-|D|)}{2} \\ &\quad +\frac{(\Delta-k)(n-|D|)+(r-j-s)(2k-|D|)}{2} \\ &= \frac{i(2k-|D|+1)+|D|(|D|-k-1)}{2} \\ &+ \frac{(n-|D|)(|D|-2k+\Delta)+(r-j)(2k-|D|)}{2} \\ &= \frac{|D|(n+k-\Delta)+n(-2k+\Delta)+(i+r-j)(2k-|D|)+i}{2} \\ &\le \frac{|D|(n+k-\Delta)+n(-2k+\Delta)}{2}; \end{aligned}$$

then

$$|D| \ge \frac{2m+n(\Delta-2k)}{n+k-\Delta}.$$

□

Let us see another lower bound using the algebraic connectivity. Given a graph G, its adjacency matrix A and the diagonal matrix D whose entries are the degrees of all vertices in the graph, the Laplacian matrix is defined as $L = A - D$. The algebraic connectivity of G, denoted by μ is the second smallest eigenvalue of the Laplacian matrix.

The algebraic connectivity of $G = (V, E)$ with order n satisfies the following equality given by Fielder [13].

$$\mu = 2n \min\left\{ \frac{\sum_{v_i v_j \in E}(w_i - w_j)^2}{\sum_{v_i \in V}\sum_{v_j \in V}(w_i - w_j)^2} : w \neq \alpha \mathbf{j} \text{ for } \alpha \in \mathbb{R} \right\},$$

where $\mathbf{j} = (1, 1, \ldots, 1)$ and $w \in \mathbb{R}^n$.

Theorem 6. *Let G be a graph with order n and algebraic connectivity μ. Then,*

$$\gamma^g_{kt}(G) \geq \frac{kn}{n - \mu}.$$

Proof. Let D be a $\gamma_{kt}(G)$-set. It can be found that if we take

$$w = \begin{cases} 1 \text{ if } v \in D \\ 0 \text{ if } v \notin D \end{cases}$$

in the set given above, since μ is the minimum, we have

$$\mu \leq \frac{n\sum_{v \in D}\delta_{\overline{D}}(v)}{|D|(n - |D|)} \leq \frac{n(n - |D|)(|D| - k)}{|D|(n - |D|)} = \frac{n(|D| - k)}{|D|};$$

therefore, $|D| \geq \frac{kn}{n-\mu}$. □

Theorem 7. *Let G be a graph of order n and maximum degree Δ. If $k \geq \min\left\{\frac{\Delta}{2}, \frac{n-\delta-1}{2}\right\}$, then*

$$\gamma^g_{kt}(G) \geq \frac{\sqrt{4kn + 1} + 1}{2}.$$

Proof. Let D be a $\gamma_{kt}(G)$-set. For every $v \in D$, if we suppose that $k \geq \frac{\Delta}{2}$, we have $\delta_D(v) \geq \delta_{\overline{D}}(v)$, then

$$|D|(|D| - k - 1) \geq \sum_{v \in D}\delta_D(v) \geq \sum_{v \in D}\delta_{\overline{D}}(v) \geq (n - |D|)k,$$

which implies that $|D|^2 - |D| \geq kn$, or equivalently, that $\left(|D| - \frac{1}{2}\right)^2 \geq kn + \frac{1}{4}$; that is, $|D| \geq \frac{\sqrt{4kn+1}+1}{2}$.

If $\frac{n-\delta-1}{2} \leq k < \frac{\Delta}{2}$, since $\gamma^g_{kt}(G) = \gamma^g_{kt}(\overline{G})$ and $\overline{\Delta} = n - \delta - 1$, we can obtain the same result. □

The lower bound given in Theorem 7 is attained, for instance, in the graph given in Figure 4.

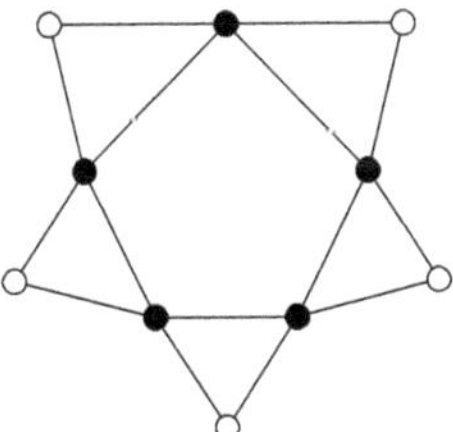

Figure 4. A graph G such that $\gamma^g_{2t}(G) \geq \frac{\sqrt{8n+1}+1}{2}$.

In graph theory, it is common to analyze graphs obtained by some transformation from an originally given graph. An example of such a transformation is the elimination of one or more edges of the graph. Given a graph G, it is natural to think about what happens if you add or delete edges on the graph. We note that removing an edge in G is equivalent to adding an edge to graph $\overline{G}$. Therefore, it suffices to study just one of these cases.

Proposition 6. *Let G be a graph with order n, minimum degree δ and maximum degree Δ, and let $k < min\{\delta, n-\Delta-1\}$. Then the following inequalities are satisfied (for an edge e):*

$$\gamma^g_{kt}(G-e) \le \gamma^g_{kt}(G)+2,$$
$$\gamma^g_{kt}(G+e) \le \gamma^g_{kt}(G)+2.$$

Proof. Let G be a graph and D be a $\gamma^g_{kt}(G)$-set, and we consider $e \in E$. Notice that $e \in E(V \setminus D, V \setminus D)$, $e \in E(D, V \setminus D)$ or $e \in E(D,D)$; we will divide the proof into three cases and we denote $G' = G - e$.

Case 1: If $e \in E(V \setminus D, V \setminus D)$. Note that every vertex in $V(G')$ has at least k neighbors and k non-neighbors in D. Therefore, $\gamma^g_{kt}(G') \le |D| = \gamma^g_{kt}(G) < \gamma^g_{kt}(G)+2$.

Case 2: If $e \in E(D, V \setminus D)$. Let $e = uv$, where $u \in D$ and $v \in V \setminus D$. We note that for every $w \in V(G) - \{v\}$, $\delta_D(w) \ge k$ and $\overline{\delta}_D(w) \ge k$. On the other hand, note that $\overline{\delta}_D(v) > k$ in G', and if $\delta_D(v) \ge k$ in G', then $\gamma^g_{kt}(G') \le |D| = \gamma^g_{kt}(G) < \gamma^g_{kt}(G)+2$. Now, if $\delta_D(v) = k-1$ in G', then there exists $w \in V(G') \setminus D$ such that $w \in N_{G'}(v)$. Therefore, $D \cup \{w\}$ is a GTkD set of G', so $\gamma^g_{kt}(G') \le |D \cup \{w\}| = \gamma^g_{kt}(G)+1 < \gamma^g_{kt}(G)+2$.

Case 3: If $e \in E(D,D)$. Let $e = uv$ where $u, v \in D$. We note that for every $w \in V(G) - \{u,v\}$, $\delta_D(w) \ge k$ and $\overline{\delta}_D(w) \ge k$. In the worst case $\delta_D(u) < k$ and $\delta_D(v) < k$; the others cases are solved as the above; there exists $w, p \in V(G') \setminus D$ such that $w \in N_{G'}(u)$ and $p \in N_{G'}(v)$. Now, if $w = p$ then $D \cup \{w\}$ is a GTkD set of G' and $\gamma^g_{kt}(G') \le |D \cup \{w\}| = \gamma^g_{kt}(G)+1 < \gamma^g_{kt}(G)+2$; otherwise, $w \ne p$ and then $D \cup \{w,p\}$ is a GTkD set of G'; hence $\gamma^g_{kt}(G') \le |D \cup \{w,p\}| = \gamma^g_{kt}(G)+2$.

Thus, the first inequality is satisfied: $\gamma^g_{kt}(G-e) \le \gamma^g_{kt}(G)+2$. Now, as we say above for this problem, removing an edge in G is analogous to adding an edge in $\overline{G}$. Since $G-e$ and $\overline{G}+e$ are complementary graphs and it is known that $\gamma^g_{kt}(\overline{G}) = \gamma^g_{kt}(G)$, it is verified that $\gamma^g_{kt}(G-e) = \gamma^g_{kt}(\overline{G}+e)$. Hence, by the first inequality $\gamma^g_{kt}(\overline{G}+e) = \gamma^g_{kt}(G-e) \le \gamma^g_{kt}(G)+2 = \gamma^g_{kt}(\overline{G})+2$. So, $\gamma^g_{kt}(G+e) \le \gamma^g_{kt}(G)+2$. □

Let S be a subset of set V such that the maximum degree of the subgraph induced by the vertices from set S is no more than $k-1$. Then set S will be referred to as a k-independent set of vertices. The cardinality of a k-independent set of the maximum cardinality will be referred to as the k-independence number in graph G and will be denoted by $\beta_k(G)$. The lower k-independence number $i_k(G)$ is the minimum cardinality of a maximal k-independent set in graph G.

Proposition 7. *Let D be a global total k-dominating set in G and let $V \setminus D$ be a maximum $(\Delta-k)$-independent. Then,*

$$n - \beta_{\Delta-k}(G) \le |D| \le \min\{n-\gamma(G), n - i_{\Delta-k}(G)\}.$$

Proof. Since $V \setminus D$ is a maximal $(\Delta-k)$-independent set, $V \setminus D$ is a dominating set; thus, $n - |D| \ge \gamma(G)$. Moreover, $i_{\Delta-k}(G) \le n - |D| \le \beta_{\Delta-k}(G)$. □

5. Deriving Upper Bounds for $\gamma^g_{(k+1)t}(G)$ from $\gamma^g_{kt}(G)$

It is intuitively clear that the greater k is, the more difficult is to find a global total k-dominating set of graph $G = (V,E)$ with the minimum cardinality. In particular, the following relationship is easy to see: $\gamma^g_{1t}(G) \le \gamma^g_{2t}(G) \le \gamma^g_{3t}(G) \le \ldots \le \gamma^g_{kt}(G)$, for every

$k \leq min\{\delta, n - \Delta - 1\}$. Ideally, one would wish to have a method that obtains a GT$(k+1)$D set of minimum cardinality from a GTkD set with the minimum cardinality. It is clear that this is not an easy task. In this next section we develop a method that generates a GT$(k+1)$D set from a GTkD, based on which we establish a relationship between minimum cardinality GTkD and GT$(k+1)$D sets—more precisely, between $\gamma^g_{kt}(G)$ and $\gamma^g_{(k+1)t}(G)$, which, in turn, provides upper bounds for $\gamma^g_{(k+1)t}(G)$.

We first need to introduce some necessary definitions. Given $D \subseteq V$, a subset of the set of vertices V, let $N(D)$ be the set of vertices from $V \setminus D$ having at least one neighbor in D; that is, $N(D) = \{x \in V \setminus D \mid \exists\, y \in D$ such that $x \in N_G(y)\}$. Similarly, we denote by $\overline{N}(D)$ the set of vertices from $V \setminus D$ having at least one non-neighbor in D.

Now let A and B be subsets of set V. We will say that a subset $D \subseteq A$ is a relative dominating set of B from set A if for every $x \in B$ there exists at least one vertex $v \in D$ such that $v \in N(x)$ or $v \in B$. Correspondingly, we call the minimum cardinality of such a relative dominating set the relative domination number of set B from set A and denote it by $\gamma'(A, B)$. We abbreviate by $\gamma'(A, B)$-set a relative dominating set of B from set A of cardinality $\gamma'(A, B)$.

Finally, $\gamma'(\overline{A, B})$ is the relative domination number of B from set A in graph $\overline{G}$ and $\gamma'(\overline{A, B})$-set is a relative dominating set of B from set A in graph $\overline{G}$ with cardinality $\gamma'(\overline{A, B})$; see an example in Figure 5.

Lemma 1. *Let G be a graph with diam(G) = 2 and g(G) = 4, and let S be an induced subgraph isomorphic to C_4. Let $B = V \setminus (N(S) \cup S)$ and $A = N(B)$. Then $\gamma^g_{1t}(G) \leq \gamma'(A, B) + 4$.*

Proof. Let D' be a $\gamma'(A, B)$-set, $D = S \cup D'$ and $v \in V$. Note that since $diam(G) = 2$, $D' \subseteq A \subseteq N(S)$. Thus, we can see that $v \in N(S)$, $v \in B$ or $v \in S$. If $v \in N(S)$, then it has at least one neighbor in S and hence also in D. On the other hand, if $v \in B$, then v must have at least one neighbor in D' and hence also in D. If $v \in S$, then v has at least one neighbor in S, and hence also in D. Therefore, D is a total 1-dominating set of G.

If $v \in S$, then there exists one non-neighbor vertex of v in S, and hence also in D. If $v \in B$, then the four vertices in S are non-neighbors of v, and hence vertex v has at least one non-neighbor in set D. If $v \in N(S)$, since $g(G) = 4$, v it has at most two neighbors in S; thus, it has at least two non-neighbors in S and hence also in D. Therefore, D is a global 1-dominating set of G. Finally, D is a global total 1-dominating set of G, so $\gamma^g_{1t}(G) \leq \gamma'(A, B) + |S| = \gamma'(A, B) + 4$. □

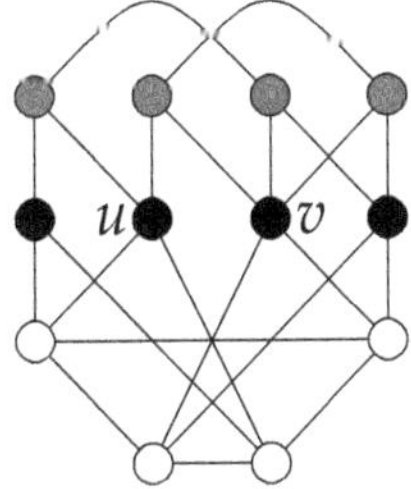

Figure 5. In the depicted graph G, the set S is formed by the white vertices, set A is formed by the black vertices and set B is formed by the gray vertices. Note that $\gamma'(A, B) = 2$ (the set $\{u, v\}$ is a $\gamma'(A, B)$-set) and $\gamma^g_{1t}(G) = 6$.

Corollary 5. *Let G be a graph with $diam(G) = 2$ and $g(G) = 4$; let S be an induced subgraph isomorphic to C_4, $B = V \setminus (N(S) \cup S)$ and $A = N(B)$. Then the following conditions hold.*

- *If $B = \emptyset$, then $\gamma^g_{1t}(G) = 4$.*
- *Since $\gamma'(A, B) \leq |B|$, $\gamma^g_{1t}(G) \leq |B| + 4$.*

- *If* $|N(x) \cap S| = 2, \forall x \in A$, *then* $\gamma^g_{2t}(G) \leq 2|B| + 4$.

Let k be a positive integer with $1 \leq k < min\{\delta, n - \Delta - 1\}$, and D be a $\gamma^g_{kt}(G)$-set for graph G. Below we define special sets of vertices that will be used in future derivations.

- $H = V(G) \setminus D$.
- $Z = \{x \in H \mid \delta_D(x) \geq k+1 \text{ and } \overline{\delta}_D(x) \geq k+1\}$ are all vertices in H which are global total $(k+1)$-dominated.
- $X = T_k(G[D])$ are all vertices in D with only k neighbors.
- $Y = T_{|D|-k-1}(G[D])$ are all vertices in D with only k non-neighbors.
- $X' = N(X) \cap H$ are all the vertices in H which have at least one neighbor in set X.
- $N = \gamma'(X', X)$-set, a relative dominating set of X from set X'.
- $Y' = \overline{N}(Y) \cap H$ are all the vertices in set H which have at least one non-neighbor in set Y.
- $R = \gamma'(\overline{Y', Y})$-set, a relative dominating set of X from set X' in $\overline{G}$.
- $P = H \setminus Z$ are all the vertices in set H which are not yet global total $(k+1)$-dominated.
- $M = \gamma'(H, P)$-set $\cup \, \gamma'(\overline{H, P})$-set;
- $S = D \cup N \cup R \cup M$;

Now we show that the set S obtained as above is a global total $(k+1)$-dominating set given a $\gamma^g_{kt}(G)$-set D.

Theorem 8. *Let G be a graph and D be an arbitrary $\gamma^g_{kt}(G)$-set. Then the set S obtained as above is a global total $(k+1)$-dominating set of graph G.*

Proof. Let D be an arbitrary $\gamma^g_{kt}(G)$-set, $H = V \setminus D$, $Z = \{x \in H\colon \delta_D(x) \geq k+1$ and $\overline{\delta}_D(x) \geq k+1\}$, $X = T_k(G[D])$ and $Y = T_{|D|-k-1}(G[D])$. Further, let $P = H \setminus Z$, E be a $\gamma'(H, P)$-set, F be a $\gamma'(\overline{H, P})$-set and $M = E \cup F$ (all these sets being constructed as above specified). If $X = \emptyset$ and $Y = \emptyset$, then every vertex from $D \cup Z$ has at least $k+1$ adjacent and $k+1$ non-adjacent vertices in set D. Besides, note that every vertex $v \in P$ has at least $k+1$ adjacent and $k+1$ non-adjacent vertices in set $D \cup M$. Additionally, since $V = D \cup Z \cup P$, $D \cup M$ is a global total $(k+1)$-dominating set of graph G.

Assume now that $X \neq \emptyset$ and $Y = \emptyset$, and let $X' = N(X) \cap H$ and N be a $\gamma'(X', X)$-set (notice that by the construction of the set X', there always exists the set N). Observe that every vertex from set $D \cup Z$ has at least $k+1$ adjacent and $k+1$ non-adjacent vertices in set $D \cup N$. Besides, every vertex $v \in P$ has at least $k+1$ adjacent and $k+1$ non-adjacent vertices in set $D \cup M$. Since $V = D \cup Z \cup P$, $D \cup N \cup M$ is a global total $(k+1)$-dominating set of G.

The case $X = \emptyset$ and $Y \neq \emptyset$ is analogous to the above case. We obtain that $D \cup R \cup M$ is a global total $(k+1)$-dominating set of G, where $Y' = \overline{N}(Y) \cap H$ and R is a $\gamma'(\overline{Y', Y})$-set.

Finally, assume that $X \neq \emptyset$ and $Y \neq \emptyset$. Let $X' = N(X) \cap H$, $Y' = \overline{N}(Y) \cap H$, N be a $\gamma'(X', X)$-set and R be a $\gamma'(\overline{Y', Y})$-set. Using a similar arguments as above, we again obtain that S is a global total $(k+1)$-dominating set of graph G. □

In the next proposition we derive an upper bound on the cardinality of the global total $(k+1)$-domination number. In the same lemma, we give a necessary condition when the global total $(k+1)$-domination number is equal to the total $(k+1)$-domination number.

Proposition 8. *Let G be a graph with $\delta \geq k$ and D be a $\gamma^g_{kt}(G)$-set. Then the following conditions hold:*

(a) $\gamma^g_{(k+1)t}(G) \leq \gamma^g_{kt}(G) + |N \cup R \cup M|$.

(b) *If* $|N \cup M| > \Delta + k - \gamma^g_{kt}(G)$, *then* $\gamma^g_{(k+1)t}(G) = \gamma_{(k+1)t}(G)$.

Proof. (a) By Theorem 8, S is a global total $(k+1)$-dominating set of G; hence, the bound trivially holds.

(b) Recall that $|S| = \gamma^g_{kt}(G) + |N \cup R \cup M|$. Additionally, it is easy to see that $S \setminus R$ is a total $(k+1)$-dominating set of G. In [12] it is proved that if $\gamma_{kt}(G) > \Delta + k$, then $\gamma^g_{kt}(G) = \gamma_{kt}(G)$ (see Proposition 2.10). Hence, if $|S| \geq \gamma^g_{kt}(G) + |N \cup M| \geq \gamma_{(k+1)t}(G) > \Delta + k + 1$, then $\gamma^g_{(k+1)t}(G) = \gamma_{(k+1)t}(G)$. Hence, if $|N \cup M| > \Delta + k + 1 - \gamma^g_{kt}(G)$ then $\gamma^g_{(k+1)t}(G) = \gamma_{(k+1)t}(G)$. □

Using the definition of the above introduced sets and Theorem 8 and Proposition 8, we can obtain a global total k-domination set for any $k = 2, \ldots, \min\{\delta, n - \Delta - 1\}$. As a side-result, we also obtain the corresponding upper bounds to a global total k-domination number. Finally, we note that this procedure provides a global total k-dominating set of minimum cardinality, $2 \leq k \leq min\{\delta, n - \Delta - 1\}$, for some graphs; see Figure 6.

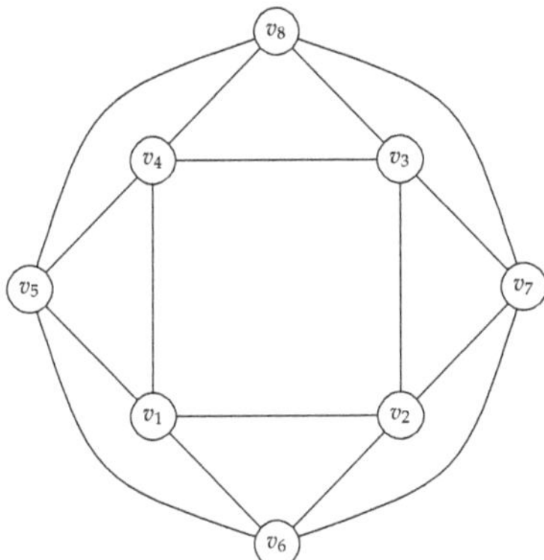

Figure 6. A graph G with $\gamma^g_{1t}(G) = 4$, $\gamma^g_{2t}(G) = 6$ and $\gamma^g_{3t}(G) = 8$. Note that if $D = \{v_1, v_2, v_3, v_4\}$ a $\gamma^g_{1t}(G)$-set, then $S = \{v_1, v_2, v_3, v_4, v_5, v_7\}$ which is a $\gamma^g_{2t}(G)$-set. Likewise, from S we construct $S' = \{v_1, v_2, v_3, v_4, v_5, v_6, v_7, v_8\}$ which is a $\gamma^g_{3t}(G)$-set.

6. Conclusions

We studied the global total k-domination number in general graphs. In particular, we presented new upper and lower bounds using the algebraic connectivity in graphs. We also established a relationship between the global total k-domination numbers of the originally given graph G and the transformed ones. Then we derived an explicit relationship between a $\gamma^g_{kt}(G)$-set and a $\gamma^g_{(k+1)t}(G)$-set, which allowed us to obtain another upper bound for the global total k-domination number in a recurrent fashion, starting from $k = 1$. We gave an example of a graph G for which a $\gamma^g_{kt}(G)$-set, for every $k = 2, \ldots, \min\{\delta, n - \Delta - 1\}$ is provided. For future work, the global total k-domination number could be studied on unitary operations in graphs, such as edge subdivision, edge contraction, path contraction and removal of a vertex. It would be a challenging task to adopt the proposed method as such and also extend it for a wider class of graphs.

Author Contributions: The authors contributed equally to this research. Investigation, F.A.H.M., E.P.I., J.M.S.A. and N.V.; writing—review and editing, F.A.H.M., E.P.I., J.M.S.A. and N.V. All authors have read and agreed to the published version of the manuscript.

Funding: J.M.S.A was supported by a grant from Agencia Estatal de Investigación (PID2019-106433GBI00/AEI/10.13039/501100011033), Spain. This work was partially supported by SEP PRODEP publication grant. The fourth author was supported by SEP PRODEP 511/6 grant and CONACyT 2020-000019-01NACV-00008 grant.

Institutional Review Board Statement: Not applicable.

Informed Consent Statement: Not applicable.

Conflicts of Interest: The authors declare no conflict of interest.

References

1. Bermudo, S.; Hernández-Gómez, J.C.; Sigarreta, J.M. On the total k-domination in graphs. *Discuss. Graph Theory* **2018**, *38*, 301–317. [CrossRef]
2. Bermudo, S.; Jalemskaya, D.L.; Sigarreta, J.M. Total 2-domination in Grid graphs. *Util. Math.* **2019**, *210*, 151–173.
3. Bermudo, S.; Sánchez, J.L.; Sigarreta, J.M. Total k-domination in Cartesian product graphs. *Period. Math. Hung.* **2017**, *75*, 255–267. [CrossRef]
4. Fernau, H.; Rodríguez-Velázquez, J.A.; Sigarreta, J.M. Global powerful r-alliances and total k-domination in graphs. *Util. Math.* **2015**, *98*, 127–147.
5. He, J.; Liang, H. Complexity of Total k-Domination and Related Problems. In Proceedings of the Joint Conference of FAW 2011 (the 5th International Frontiers of Algorithmics Workshop) and AAIM 2011 (the 7th International Conference on Algorithmic Aspects of Information and Management), Jinhua, China, 28–31 May 2011; pp. 147–155.
6. Kulli, V.R. On n-total domination number in graphs. In *Graph Theory, Combinatorics, Algorithms and Applications*; SIAM: Philadelphia, PA, USA, 1991; pp. 319–324.
7. Cabrera Martínez, A.; Hernández-Gómez, J.C.; Inza, E.P.; Sigarreta, J.M. On the Total Outer k-Independent Domination Number of Graphs. *Mathematics* **2020**, *8*, 194. [CrossRef]
8. Desormeaux, W.J.; Haynes, T.W.; Henning, M.A.; Yeo, A. Total domination in graphs with diameter 2. *J. Graph Theory* **2014**, *75*, 91–103. [CrossRef]
9. Goddard, W.; Henning, M.A. Domination in planar graphs with small diameter. *J. Graph Theory* **2002**, *40*, 1–25. [CrossRef]
10. Akhbari, M.H.; Eslahchiz, C.; Rad, N.J.; Hasni, R. Some Remarks on Global Total Domination in Graphs. *Appl. Math. E-Notes* **2015**, *15*, 22–28.
11. Kulli, V.R.; Janakiram, B. The total global domination number of a graph. *Indian J. Pure Appl. Math.* **1996**, *27*, 537–542.
12. Bermudo, S.; Martínez, A.C.; Mira, F.A.H.; Sigarreta, J.M. On the global total k-domination number of graphs. *Discret. Math.* **2019**, *263*, 42–50. [CrossRef]
13. Fiedler, M. A property of eigenvectors of nonnegative symmetric matrices and its application to graph theory. *Czechoslov. Math. J.* **1975**, *25*, 619–633. [CrossRef]

MDPI

Article

Unified Polynomial Dynamic Programming Algorithms for P-Center Variants in a 2D Pareto Front †

Nicolas Dupin [1,*], Frank Nielsen [2] and El-Ghazali Talbi [3]

1 Université Paris-Saclay, CNRS, Laboratoire Interdisciplinaire des Sciences du Numérique, 91400 Orsay, France
2 Sony Computer Science Laboratories Inc., Tokyo 141-0022, Japan; Frank.Nielsen@acm.org
3 CNRS UMR 9189-CRIStAL-Centre de Recherche en Informatique Signal et Automatique de Lille, Université Lille, F-59000 Lille, France; el-ghazali.talbi@univ-lille.fr
* Correspondence: nicolas.dupin@universite-paris-saclay.fr
† This paper is an extended version of our paper published in OLA 2020, International Conference in Optimization and Learning, Cadiz, Spain, 17–19 February 2020.

Abstract: With many efficient solutions for a multi-objective optimization problem, this paper aims to cluster the Pareto Front in a given number of clusters K and to detect isolated points. K-center problems and variants are investigated with a unified formulation considering the discrete and continuous versions, partial K-center problems, and their min-sum-K-radii variants. In dimension three (or upper), this induces NP-hard complexities. In the planar case, common optimality property is proven: non-nested optimal solutions exist. This induces a common dynamic programming algorithm running in polynomial time. Specific improvements hold for some variants, such as K-center problems and min-sum K-radii on a line. When applied to N points and allowing to uncover $M < N$ points, K-center and min-sum-K-radii variants are, respectively, solvable in $O(K(M+1)N\log N)$ and $O(K(M+1)N^2)$ time. Such complexity of results allows an efficient straightforward implementation. Parallel implementations can also be designed for a practical speed-up. Their application inside multi-objective heuristics is discussed to archive partial Pareto fronts, with a special interest in partial clustering variants.

Keywords: discrete optimization; operational research; computational geometry; complexity; algorithms; dynamic programming; clustering; k-center; p-center; sum-radii clustering; sum-diameter clustering; bi-objective optimization; Pareto Front; parallel programming

Citation: Dupin, N.; Nielsen, F.; Talbi, E.-G. Unified Polynomial Dynamic Programming Algorithms for P-Center Variants in a 2D Pareto Front. *Mathematics* **2021**, *9*, 453. https://doi.org/10.3390/math9040453

Academic Editor: Frank Werner

Received: 21 December 2020
Accepted: 16 February 2021
Published: 23 February 2021

1. Introduction

This paper is motivated by real-world applications of multi-objective optimization (MOO). Some optimization problems are driven by more than one objective function, with conflicting optimization directions. For example, one may minimize financial costs, while maximizing the robustness to uncertainties or minimizing the environmental impact [1,2]. In such cases, higher levels of robustness or sustainability are likely to induce financial over-costs. Pareto dominance, preferring one solution to another if it is better for all the objectives, is a weak dominance rule. With conflicting objectives, several non-dominated points in the objective space can be generated, defining efficient solutions, which are the best compromises. A Pareto front (PF) is the projection in the objective space of the efficient solutions [3]. MOO approaches may generate large sets of efficient solutions using Pareto dominance [3]. Summarizing the shape of a PF may be required for presentation to decision makers. In such a context, clustering problems are useful to support decision making to present a view of a PF in clusters, the density of points in the cluster, or to select the most central cluster points as representative points. Note than similar problems are of interest for population MOO heuristics such as evolutionary algorithms to archive representative points of a partial Pareto fronts, or in selecting diversified efficient solutions to process mutation or cross-over operators [4,5].

With N points in a PF, one wishes to define $K \ll N$ clusters while minimizing the measure of dissimilarity. The K-center problems, both in the discrete and continuous versions, define the cluster costs in this paper, covering the PF with K identical balls while minimizing the radius of the balls used. By definition, the ball centers belong to the PF for the discrete K-center version, whereas the continuous version is similar to geometric covering problems, without any constraint for the localization of centers. Furthermore, sum-radii or sum-diameter are min-sum clustering variants, where the covering balls are not necessarily identical. For each variant, one can also consider partial clustering variants, where a given percentage (or number) of points can be ignored in the covering constraints, which is useful when modelling outliers in the data.

The K-center problems are NP-hard in the general case, [6] but also for the specific case in $\mathbb{R}^2$ using the Euclidean distance [7]. This implies that K-center problems in three-dimensional (3D) PF are also NP-hard, with the planar case being equivalent to an affine 3D PF. We consider the case of two-dimensional (2D) PF in this paper, focusing on the polynomial complexity results. It as an application to bi-objective optimization, the 3D PF and upper dimensions are shown as perspectives after this work. Note that 2D PF are a generalization of one-dimensional (1D) cases, where polynomial complexity results are known [8,9]. A preliminary work proved that K-center clustering variants in a 2D PF are solvable in polynomial time using a Dynamic Programming (DP) algorithm [10]. This paper improves these algorithms for these variants, with an extension to min-sum clustering variants, partial clustering, and Chebyshev and Minkowski distances. The properties of the DP algorithms are discussed for efficient implementation, including parallelization.

This paper is organized as follows. The considered problems are defined formally with unified notations in Section 2. In Section 3, related state-of-the-art elements are discussed. In Sections 4 and 5, intermediate results and specific complexity results for sub-cases are presented. In Section 6, a unified DP algorithm with a proven polynomial complexity is designed. In Section 7, specific improvements are presented. In Section 8, the implications and applications of the results of Sections 5–7 are discussed. In Section 9, our contributions are summarized, with a discussion of future research directions.

2. Problem Statement and Notation

In this paper, integer intervals are denoted as $[\![a,b]\!] = [a,b] \cap \mathbb{Z}$. Let $E = \{x_1, \dots, x_N\} = \{x_i\}_{i \in [\![1,N]\!]}$ a set of N elements of $\mathbb{R}^2$, such that for all $i \neq j$, $x_i \; \mathcal{I} \; x_j$ defining the binary relations $\mathcal{I}, \prec$ for all $y = (y^1, y^2), z = (z^1, z^2) \in \mathbb{R}^2$ with

$$y \prec z \iff y^1 < z^1 \text{ and } y^2 > z^2 \quad (1)$$

$$y \preccurlyeq z \iff y \prec z \text{ or } y = z \quad (2)$$

$$y \; \mathcal{I} \; z \iff y \prec z \text{ or } z \prec y \quad (3)$$

These hypotheses on E define 2D PF considering the minimization of two objectives [3,11]. Such configuration is illustrated by Figure 1. Without loss of generality, transforming objectives to maximize f into $-f$ allows for the consideration of the minimization of two objectives. This assumption impacts the sense of the inequalities of $\mathcal{I}, \prec$. A PF can also be seen as a Skyline operator [12]. A 2D PF can be extracted from any subset of $\mathbb{R}^2$ using an output-sensitive algorithm [13], or using any MOO approach [3,14].

The results of this paper will be given using the Chebyshev and Minkowski distances, generically denoting $d(y,z)$ the l_∞ and l_m norm-induced distance, respectively. For a given $m > 0$, the Minkowski distance is denoted d_m, and given by the formula

$$\forall y = (y^1, y^2), z = (z^1, z^2) \in \mathbb{R}^2, \quad d_m(y,z) = \sqrt[m]{|y^1 - z^1|^m + |y^2 - z^2|^m} \quad (4)$$

The case $m = 2$ corresponds to the Euclidean distance; it is a usual case for our application. The limit with $m \to \infty$ defines the Chebyshev distance, denoted d_∞ and given by the formula

$$\forall y = (y^1, y^2), z = (z^1, z^2) \in \mathbb{R}^2, \quad d_\infty(y,z) = \max\left(\left|y^1 - z^1\right|, \left|y^2 - z^2\right|\right) \tag{5}$$

Once a distance d is defined, a dissimilarity among a subset of points $E' \subset E$ is defined using the radius of the minimal enclosing ball containing E'. Numerically, this dissimilarity function, denoted as f_C, can be written as

$$\forall E' \subset E, \quad f_C(E') = \min_{y \in \mathbb{R}^2} \max_{x \in E'} d(x,y) \tag{6}$$

A discrete variant considers enclosing balls with one of the given points as the center. Numerically, this dissimilarity function, denoted f_D, can be written as

$$\forall E' \subset E, \quad f_D(E') = \min_{y \in E'} \max_{x \in E'} d(x,y) \tag{7}$$

For the sake of having unified notations for common results and proofs, we define $\gamma \in \{0,1\}$ to indicate which version of the dissimilarity function is considered. $\gamma = 0$ (respectively, 1) indicates that the continuous (respectively, discrete) version is used, f_γ, thus denoting $f_1 = f_C$ (respectively, $f_0 = f_D$). Note that $\gamma \in \{0,1\}$ will be related to complexity results which motivated such a notation choice.

For each a subset of points $E' \subset E$ and integer $K \geqslant 1$, we define $\Pi_K(E)$, as the set of all the possible partitions of E' in K subsets. Continuous and discrete K-center are optimization problems with $\Pi_K(E)$ as a set of feasible solutions, covering E with K identical balls while minimizing the radius of the balls used

$$\min_{\pi \in \Pi_K(E)} \max_{P \in \pi} f_\gamma(P) \tag{8}$$

The continuous and discrete K-center problems in the 2D PF are denoted K-γ-CP2DPF. Another covering variant, denoted min-sum-K-radii problems, covers the points with non-identical balls, while minimizing the sum of the radius of the balls. We consider the following extension of min-sum-K-radii problems, with $\alpha > 0$ being a real number

$$\min_{\pi \in \Pi_K(E)} \sum_{P \in \pi} f_\gamma(P)^\alpha \tag{9}$$

$\alpha = 1$ corresponds to the standard min-sum-K-radii problem. $\alpha = 2$ with the standard Euclidean distance is equivalent to the minimization of the area defined by the covering disks. For the sake of unifying notations for results and proofs, we define a generic operator $\oplus \in \{+, \max\}$ to denote, respectively, sum-clustering and max-clustering. This defines the generic optimization problems

$$\min_{\pi \in \Pi_K(E)} \bigoplus_{P \in \pi} f_\gamma(P)^\alpha \tag{10}$$

Lastly, we consider a partial clustering extension of problems (10), similarly to the partial p-center [15]. The covering with balls mainly concerns the extreme points, which make the results highly dependent on outliers. One may consider that a certain number $M < N$ of the points may be considered outliers, and that M points can be removed in the evaluation. This can be written as

$$\min_{E' \subset E : |E \setminus E'| \leqslant M} \ \min_{\pi \in \Pi_K(E')} \bigoplus_{P \in \pi} f_\gamma(P)^\alpha \tag{11}$$

Problem (11) is denoted K-M-$\oplus$-(α, γ)-BC2DPF. Sometimes, the partial covering is defined by a maximal percentage of outliers. In this case, if M is much smaller than N, we have $M = \Theta(N)$, which we have to keep in mind for the complexity results. K-center problems, K-γ-CP2DPF, are K-M-max-(α, γ)-BC2DPF problems for all $\alpha > 0$; the value of α does not matter for max-clustering, defining the same optimal solutions as $\alpha = 1$. The standard min-sum-k-radii problem, equivalent to the min-sum diameter problem, corresponds to k-0-+-$(1, \gamma)$-BC2DPF problems for discrete and continuous versions, k-M-+-$(1, \gamma)$-BC2DPF problems consider partial covering for min-sum-k-radii problems.

3. Related Works

This section describes works related to our contributions, presenting the state-of-the-art for p-center problems and clustering points in a PF. For more detailed survey on the results for the p-center problems, we refer to [16].

3.1. Solving P-Center Problems and Complexity Results

Generally, the p-center problem consists of locating p facilities among possible locations and assigning n clients, called $c_1, c_2, \ldots, c_n$, to the facilities in order to minimize the maximum distance between a client and the facility to which it is allocated. The continuous p-center problem assumes that any place of location can be chosen, whereas the discrete p-center problem considers a subset o m potential sites, denoted $f_1, f_2, \ldots, f_m$, and distances $d_{i,j}$ for all $i \in [\![1,n]\!]$ and $j \in [\![1,m]\!]$. Discrete p-center problems can be formulated with bipartite graphs, modeling that si unfeasibile for some assignments. In the discrete p-center problem defined in Section 2, points $f_1, f_2, \ldots, f_m$ are exactly $c_1, c_2, \ldots, c_n$, and the distances are defined using a norm, so that triangle inequality holds for such variants.

P-center problems are NP-hard [6,17]. Furthermore, for all $\alpha < 2$, any α-approximation for the discrete p-center problem with triangle inequality is NP-hard [18]. Two approximations were provided for the discrete p-center problem running in $O(pn \log n)$ time and in $O(np)$ time, respecitvely [19,20]. The discrete p-center problem in $\mathbb{R}^2$ with a Euclidean distance is also NP-hard [17]. Defining binary variables $x_{i,j} \in \{0,1\}$ and $y_j \in \{0,1\}$ with $x_{i,j} = 1$ if and only if the customer i is assigned to the depot j, and $y_j = 1$ if and only if location f_j is chosen as a depot, the following Integer Linear Programming (ILP) formulation models the discrete p-center problem [21]

$$\min_{x,y,z} \quad z \tag{12a}$$

$$s.t: \quad \sum_{j=1}^{n} d_{i,j} x_{i,j} \leqslant z \qquad \forall i \in [\![1,n]\!] \tag{12b}$$

$$\sum_{j=1}^{m} y_j = p \tag{12c}$$

$$\sum_{j=1}^{m} x_{i,j} = 1 \qquad \forall i \in [\![1,n]\!] \tag{12d}$$

$$x_{i,j} \leqslant y_j \qquad \forall (i,j) \in [\![1,N]\!] \times [\![1,n]\!], \tag{12e}$$

$$x_{i,j}, y_j \in \{0,1\} \quad \forall i,j \in [\![1,n]\!] \times [\![1,m]\!], \tag{12f}$$

Constraints (12b) are implied by a standard linearization of the min–max original objective function. Constraint (12c) fixes the number of open facilities to p. Constraints (12d) assign each client to exactly one facility. Constraints (12e) are necessary to induce that any considered assignment $x_{i,j} = 1$ implies that facility j is open with $y_j = 1$. Tighter ILP formulations than (12) were proposed, with efficient exact algorithms relying on the IP models [22,23]. Exponential exact algorithms were also designed for the continuous p-center problem [24,25]. An $n^{O(\sqrt{p})}$-time algorithm was provided for the continuous Euclidean p-center problem in the plane [26]. An $n^{O(p^{1-1/d})}$-time algorithm is available for the continuous p-center problem in $\mathbb{R}^d$ under Euclidean and L_∞ -metric [27].

Specific cases of p-center problems are solvable in polynomial time. The continuous 1-center problem is exactly the minimum covering ball problem, which has a linear complexity in $\mathbb{R}^2$. Indeed, a "prune and search" algorithm finds the optimum bounding sphere and runs in linear time if the dimension is fixed as a constant [28]. In dimension d, its complexity is in $O((d+1)(d+1)!n)$ time, which is impractical for high-dimensional applications [28]. The discrete 1-center problem is solved in time $O(n \log n)$, using furthest-neighbor Voronoi diagrams [29]. The continuous and planar 2-center problem is solved in randomized expected $O(n \log^2 n)$ time [30,31]. The discrete and planar 2-center problem is solvable in $O(n^{4/3} \log^5 n)$ time [32].

1D p-center problems, or those with equivalent points that are located in a line, have specific complexity results with polynomial DP algorithms. The discrete 1D k-center problem is solvable in $O(n)$ time [33]. The continuous and planar k-centers on a line, finding k disks with centers on a given line l, are solvable in polynomial time, in $O(n^2 \log^2 n)$ time in the first algorithm by [29], and in $O(nk \log n)$ time and $O(n)$ space in the improved version provided by [34]. An intensively studied extension of the 1D sub-cases is the p-center in a tree structure. The continuous p-center problem is solvable in $O(n \log^3 n)$ time in a tree structure [7]. The discrete p-center problem is solvable in $O(n \log n)$ time in a tree structure [35].

Rectilinear p-center problems, using the Chebyshev distances, were less studied. Such distance is useful for complexity results; however, it has fewer applications than Euclidean or Minkowski norms. For the planar and rectangular 1-center and 2-center problems, $O(n)$ algorithms are available for the 1-center problem, and such 3-center problems can be solved in $O(n \log n)$ time [36]. In a general dimension d, continuous and discrete versions of rectangular p-center problems are solvable in $O(n)$ and $O(n \log^{d-2} n \log \log n + n \log n)$ running time, respectively. Specific complexity results for rectangular 2-center problems are also available [37].

3.2. *Solving Variants of P-Center Problems and Complexity Results*

Variants of p-center problems were studied less intensively than the standard p-center problems. The partial variants were introduced in 1999 by [15], whereas a preliminary work in 1981 considered a partial weighted one-center variant and a DP algorithm to solve it running in $O(n^2 \log n)$ time [38]. The partial discrete p-center can formulated as an ILP starting from the formulation provided by [21] as written in (12). Indeed, considering that n_0 points can be uncovered, constraints (12.4) become inequalities $\sum_{j=1}^{m} x_{i,j} \leqslant 1$ for all i, j and the maximal number of unassigned points is set to n_0, adding one constraint $\sum_{j=i}^{n} \sum_{j=1}^{m} x_{i,j} \geqslant n - n_0$. Similarly, the sum-clustering variants K-M-+-(α, γ)-BC2DPF can be written as the following ILP

$$
\begin{aligned}
&\min_{z,r\geqslant 0} && \sum_{n=1}^{N} r_n && && \text{(13a)}\\
&s.t: && \sum_{n=1}^{N} d(x_n, x_{n'})^{\alpha} z_{n,n'} \leqslant r_{n'} && \forall n' \in [\![1,N]\!] && \text{(13b)}\\
& && \sum_{n'=1}^{N} z_{n',n'} = K && && \text{(13c)}\\
& && \sum_{n'=1}^{N} z_{n,n'} \leqslant 1 && \forall n \in [\![1,N]\!] && \text{(13d)}\\
& && \sum_{n=1}^{N} \sum_{n'=1}^{N} z_{n,n'} \geqslant N - M && && \text{(13e)}\\
& && z_{n,n'} \leqslant z_{n',n'} && \forall (n,n') \in [\![1,N]\!]^2, && \text{(13f)}\\
& && z_{n,n'} \in \{0,1\} && \forall (n,n') \in [\![1,N]\!]^2, && \text{(13g)}\\
& && r_n \geqslant 0 && \forall n \in [\![1,N]\!], && \text{(13h)}
\end{aligned}
$$

In this ILP formulation, binary variables $z_{n,n'} \in \{0,1\}$ are defined such that $z_{n,n'} = 1$ if and only if the points x_n and $x_{n'}$ are assigned in the same cluster, with $x_{n'}$ being the discrete center. Continuous variables $r_n \geqslant 0$ denote the powered radius of the ball centered in x_n, if x_n is chosen as a center, and $r_n = 0$ otherwise. Constraint (13b) is a standard linearization of the non-linear objective function. $z_{n,n}$ indicates that if point x_n is chosen as the center, then this implies with (13c) that K such variables are nonzero, and with (13f) that a nonzero variable $z_{n,n'}$ implies that the corresponding $z_{n',n'}$ is not null. (13d) and (13e) allow the extension with partial variants, as discussed before.

Min-sum radii or diameter problems were rarely studied. However, such objective functions are useful for meta-heuristics to break some "plateau" effects [39]. Min-sum diameter clustering is NP-hard in the general case and polynomial within a tree structure [40]. The NP-hardness is also proven, even in metrics induced by weighted planar graphs [41]. Approximation algorithms were studied for min-sum diameter clustering. A logarithmic approximation with a constant factor blowup in the number of clusters was provided by [42]. In the planar case with Euclidean distances, a polynomial time approximation scheme was designed [43].

3.3. Clustering/Selecting Points in Pareto Frontiers

Here, we summarize the results related to the selection or the clustering of points in PF, with applications for MOO algorithms. Polynomial complexity resulting in the use of 2D PF structures is an interesting property; clustering problems have a NP-hard complexity in general [17,44,45].

To the best of our knowledge, no specific work focused on PF sub-cases of k-center problems and variants before our preliminary work [10]. A Distance-Based Representative Skyline with similarities to the discrete p-center problem in a 2D PF may not be fully available in the Skyline application, which makes a significant difference [46,47]. The preliminary results proved that K-γ-CP2DPF is solvable in $O(KN \log^{\gamma} N)$ time using $O(N)$ additional memory space [10]. Partial extensions and min-sum-k-radii variants were not considered for 2D PF. We note that the 2D PF case is an extension of the 1D case, with 1D cases being equivalent to the cases of an affine 2D PF. In the study of complexity results, a tree structure is usually a more studied extension of 1D cases. The discrete k-center problem on a tree structure, and thus the 1D sub-case, is solvable in $O(N)$ time [33]. 3F PF cases are NP-complete, as already mentioned in the introduction, this being a consequence of the NP-hardness of the general planar case.

Maximization of the quality of discrete representations of Pareto sets was studied with the hypervolume measure in the Hypervolume Subset Selection (HSS) problem [48,49]. The HSS problem is known to be NP-hard in dimension 3 (and greater dimensions) [50]. HSS is solvable with an exact algorithm in $N^{O(\sqrt{K})}$ and a polynomial-time approximation scheme for any constant dimension d [50]. The 2D case is solvable in polynomial time with a DP algorithm with a complexity in $O(KN^2)$ time and $O(KN)$ space [49]. The time complexity of the DP algorithm was improved in $O(KN + N \log N)$ by [51], and in $O(K(N-K) + N \log N)$ by [52].

The selection of points in a 2D PF, maximizing the diversity, can also be formulated using p-dispersion problems. Max–Min and Max-Sum p-dispersion problems are NP-hard problems [53,54]. Max–Min and Max-Sum p-dispersion problems are still NP-hard problems when distances fulfill the triangle inequality [53,54]. The planar (2D) Max–Min p-dispersion problem is also NP-hard [9]. The one-dimensional (1D) cases of Max–Min and Max-Sum p-dispersion problems are solvable in polynomial time, with a similar DP algorithm running in $O(\max\{pN, N \log N\})$ time [8,9]. Max–Min p-dispersion was proven to be solvable in polynomial time, with a DP algorithm running in $O(pN \log N)$ time and $O(N)$ space [55]. Other variants of p-dispersion problems were also proven to be solvable in polynomial time using DP algorithms [55].

Similar results exist for k-means, k-medoid and k-median clustering. K-means is NP-hard for 2D cases, and thus for 3D PF [44]. K-median and K-medoid problems are known to be NP hard in dimension 2, since [17], where the specific case of 2D PF was proven to be solvable in $O(N^3)$ time with DP algorithms [11,56]. The restriction of *k*-means to 2D PF would be also solvable in $O(N^3)$ time with a DP algorithm if a conjecture was proven [57]. We note that an affine 2D PF is a line in $\mathbb{R}^2$, where clustering is equivalent to 1D cases. 1D k-means were proven to be solvable in polynomial time with a DP algorithm in $O(KN^2)$ time and $O(KN)$ space. This complexity was improved for a DP algorithm in $O(KN)$ time and $O(N)$ space [58]. This is thus the complexity of K-means in an affine 2D PF.

4. Intermediate Results

4.1. Indexation and Distances in a 2D PF

Lemma 1. *$\preccurlyeq$ is an order relation, and $\prec$ is a transitive relation*

$$\forall x, y, z \in \mathbb{R}^2, \ \ x \prec y \ \ and \ \ y \prec z \Longrightarrow x \prec z \tag{14}$$

Proposition 1 implies an order among the points of E, for a re-indexation in $O(N \log N)$ time

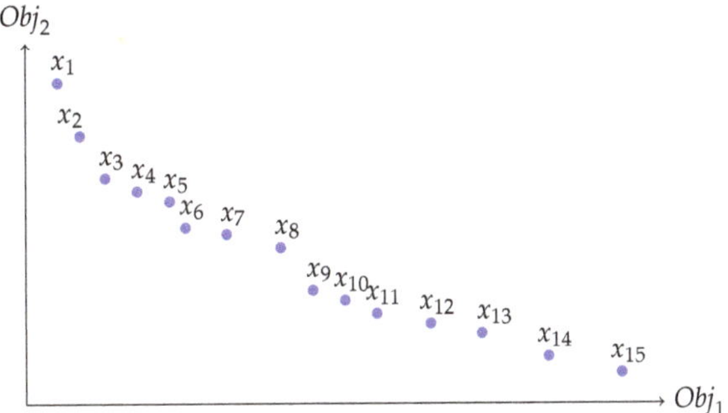

Figure 1. Illustration of a 2D Pareto Front (PF) with 15 points and the indexation implied by Proposition 1.

Proposition 1 (Total order). *Points (x_i) can be re-indexed in $O(N \log N)$ time, such that*

$$\forall (i_1, i_2) \in [\![1; N]\!]^2, \ \ i_1 < i_2 \Longrightarrow \ \ x_{i_1} \prec x_{i_2} \tag{15}$$

$$\forall (i_1, i_2) \in [\![1; N]\!]^2, \ \ i_1 \leqslant i_2 \Longrightarrow \ \ x_{i_1} \preccurlyeq x_{i_2} \tag{16}$$

Proof. We index E such that the first coordinate is increasing. This sorting procedure runs in $O(N \log N)$ time. Let $(i_1, i_2) \in [\![1; N]\!]^2$, with $i_1 < i_2$. We thus have $x^1_{i_1} < x^1_{i_2}$. Having $x_{i_1} \mathcal{I} x_{i_2}$ implies that $x^2_{i_1} > x^2_{i_2}$. $x^1_{i_1} < x^1_{i_2}$ and $x^2_{i_1} > x^2_{i_2}$ is by definition $x_{i_1} \prec x_{i_2}$. □

The re-indexation also implies monotonic relations among distances of the 2D PF

Lemma 2. *We suppose that E is re-indexed as in Proposition 1. Letting d be a Minkowski, Euclidean or Chebyshev distance, we obtain the following monotonicity relations*

$$\forall (i_1, i_2, i_3) \in [\![1; N]\!]^3, \ \ i_1 \leqslant i_2 < i_3 \Longrightarrow d(x_{i_1}, x_{i_2}) < d(x_{i_1}, x_{i_3}) \tag{17}$$

$$\forall (i_1, i_2, i_3) \in [\![1; N]\!]^3, \ \ i_1 < i_2 \leqslant i_3 \Longrightarrow d(x_{i_2}, x_{i_3}) < d(x_{i_1}, x_{i_3}) \tag{18}$$

Proof. We first note that the equality cases are trivial, so we can suppose that $i_1 < i_2 < i_3$ in the following proof. We prove the propriety (17); the proof of (18) is analogous.

Let $i_1 < i_2 < i_3$. We note that $x_{i_1} = (x^1_{i_1}, x^2_{i_1})$, $x_{i_2} = (x^1_{i_2}, x^2_{i_2})$ and $x_{i_3} = (x^1_{i_3}, x^2_{i_3})$. Proposition 1 re-indexation ensures $x^1_{i_1} < x^1_{i_2} < x^1_{i_3}$ and $x^2_{i_1} > x^2_{i_2} > x^2_{i_3}$. With $x^1_{i_3} - x^1_{i_1} > x^1_{i_2} - x^1_{i_1} > 0$, $|x^1_{i_1} - x^1_{i_2}| < |x^1_{i_1} - x^1_{i_3}|$

With $x^2_{i_3} - x^2_{i_1} < x^2_{i_2} - x^2_{i_1} < 0$, $|x^2_{i_1} - x^2_{i_2}| < |x^2_{i_1} - x^2_{i_3}|$

Thus, for any $m > 0$, $d_m(x_{i_1}, x_{i_2}) < |x^1_{i_1} - x^1_{i_3}|^m + |x^2_{i_1} - x^2_{i_3}|^m = d_m(x_{i_1}, x_{i_3})$ and also

$d_\infty(x_{i_1}, x_{i_2}) = \max(|x^1_{i_1} - x^1_{i_2}|, |x^2_{i_1} - x^2_{i_2}|) < \max(|x^1_{i_1} - x^1_{i_3}|, |x^2_{i_1} - x^2_{i_3}|) = d_\infty(x_{i_1}, x_{i_3})$. Hence, the result is proven for Euclidean, Minkowski and Chebyshev distances. □

4.2. Lemmas Related to Cluster Costs

This section provides the relations needed to compute or compare cluster costs. Firstly, one notes that the computation of cluster costs is easy in a 2D PF in the continuous clustering case.

Lemma 3. *Let $P \subset E$, such that $card(P) \geqslant 1$. Let i (resp i') be the minimal (respective maximal) index of points of P with the indexation of Proposition 1. Then, $f_C(P)$ can be computed with $f_C(P) = \frac{1}{2}d(x_i, x_{i'})$.*

To prove the Lemma 3, we use the Lemmas 4 and 5.

Lemma 4. *Let $P \subset E$, such that $card(P) \geqslant 1$. Let i (resp i') the minimal (resp maximal) index of points of P with the indexation of Proposition 1. We denote with $O = \frac{x_i + x_{i'}}{2}$ the midpoint of $x_i, x_{i'}$. Then, using a Minkowski or Chebyshev distance d, we have for all $x \in P$: $d(x, O) \leqslant d(x_i, O) = d(x_{i'}, O)$.*

Proof of Lemma 4: We denote with $r = d(x_i, O) = d(x_{i'}, O) = \frac{1}{2}d(x_i, x_{i'})$, with the equality being trivial as points $O, x_i, x_{i'}$ are on a line and d is a distance. Let $x \in P$. We calculate the distances using a new system of coordinates, translating the original coordinates such that O, is a new origin (which is compatible with the definition of Pareto optimality). x_i and $x_{i'}$ have coordinates $(-a, b)$ and $(a, -b)$ in the new coordinate system, with $a, b > 0$ and $a^m + b^m = r^m$ if a Minkowski distance is used, otherwise it is $\max(a, b) = r$ for the Chebyshev distance. We use (a', b') to denote the coordinates of x. $x_i \prec x \prec x_{i'}$ implies that $-a \leqslant a' \leqslant a$ and $-b \leqslant b' \leqslant b$, i.e., $|a'| \leqslant a$ and $|b'| \leqslant b$, which implies $d(x, O) \leqslant r$, using Minkowski or Chebyshev distances. □

Lemma 5. *Let $P \subset E$ such that $card(P) \geqslant 1$. Let i (respective i') be the minimal (respective maximal) index of points of P with the indexation of Proposition 1. We denote, using $O = \frac{x_i + x_{i'}}{2}$, the midpoint of $x_i, x_{i'}$. Then, using a Minkowski or Chebyshev distance d, we have for all $y \in \mathbb{R}^2$: $d(x_i, O) = d(x_{i'}, O) \leqslant \max(d(x_i, y), d(x_{i'}, y))$.*

Proof of Lemma 5: As previously noted, let $r = d(x_i, O) = d(x_{i'}, O) = \frac{1}{2}d(x_i, x_{i'})$. Let $y \in \mathbb{R}^2$. We have to prove that $d(x_{i'}, y) \geqslant r$ or $d(x_i, y) \geqslant r$. If we suppose that $d(x_i, y) < r$, this implies that $y \prec O$. Then, having $y \prec O \prec x_{i'}$ implies $d(x_{i'}, y) > d(x_{i'}, 0) = r$ with Lemma 2. □

Proof of Lemma 3: We first note that $f_C(P) = \min_{y \in \mathbb{R}^2} \max_{x \in P} d(x, y) \leqslant \max_{x \in P} d(x, O)$, using the particular point $O = \frac{x_i + x_{i'}}{2}$. Using Lemma 4, $\max_{x \in P} d(x, O) \leqslant r$, and thus $f_C(P) \leqslant r$ with $r = d(x_i, O) = d(x_{i'}, O) = \frac{1}{2}d(x_i, x_{i'})$. Reciprocally, for all $y \in \mathbb{R}^2$, $r \leqslant \max(d(x_i, y), d(x_{i'}, y))$ using Lemma 5, and thus $r \leqslant \max_{x \in P} d(x, y)$. This implies that $r \leqslant \min_{y \in \mathbb{R}^2} \max_{x \in P} d(x, y) = f_C(P)$. □

Lemma 6. *Let $P \subset E$ such that $card(P) \geqslant 3$. Let i (respective i') the minimal (respective maximal) index of points of P.*

$$f_{\mathcal{D}}(P) = \min_{j \in [\![i+1, i'-1]\!], x_j \in P} \max(d(x_j, x_i), d(x_j, x_{i'})) \tag{19}$$

Proof. Let $y \in P - \{x_i, x_{i'}\}$. We denote $j \in [\![i, i']\!]$, such that $y = x_j$. Applying Lemma 2 to $i < j < i'$, for all $k \in [\![i, i']\!]$,, we have $d(x_j, x_k) \leqslant \max(d(x_j, x_i), d(x_j, x_i))$. Then

$$\begin{aligned}
f_{\mathcal{D}}(P) &= \min_{y = x_j \in P} \max_{x \in P} d(x, y) \\
f_{\mathcal{D}}(P) &= \min_{j \in [\![i, i']\!], x_j \in P} \max\left(\max(d(x_j, x_i), d(x_j, x_i)), \max_{k \in [\![i, i']\!]} d(x_j, x_k) \right) \\
f_{\mathcal{D}}(P) &= \min_{j \in [\![i, i']\!], x_j \in P} \max(d(x_j, x_i), d(x_j, x_{i'}))
\end{aligned}$$

Lastly, we notice that extreme points are not optimal centers. Indeed, $\max(d(x_i, x_i), d(x_i, x_{i'})) = d(x_i, x_{i'}) > \max(d(x_{i+1}, x_{i'}), d(x_{i+1}, x_i))$ with Proposition 2, i.e., i, is not optimal in the last minimization, dominated by $i + 1$. Similarly, i' is dominated by $i' - 1$. □

Lemma 7. *Let $\gamma \in \{0, 1\}$. Let $P \subset P' \subset E$. We have $f_\gamma(P) \leqslant f_\gamma(P')$.*

Proof. Using the order of Proposition 1, let i (respectively, i') the minimal index of points of P (respectively, P') and let j (respectively, j') the maximal indexes of points of P (respectively, P'). $f_{\mathcal{C}}(P) \leqslant f_{\mathcal{C}}(P')$ is trivial using Lemmas 2 and 3. To prove $f_{\mathcal{D}}(P) \leqslant f_{\mathcal{D}}(P')$, we use $i \leqslant i' \leqslant j' \leqslant j$, and Lemmas 2 and 6

$$\begin{aligned}
f_{\mathcal{D}}(P) &= \min_{k \in [\![i, j]\!], x_k \in P} \max(d(x_k, x_i), d(x_j, x_k)) \\
&\leqslant \min_{k \in [\![i', j']\!], x_k \in P} \max(d(x_k, x_i), d(x_j, x_k)) \\
&\leqslant \min_{k \in [\![i', j']\!], x_k \in P'} \max\left(d(x_k, x_{i'}), d\left(x_{j'}, x_k\right)\right) = f_{\mathcal{D}}(P')
\end{aligned}$$

□

Lemma 8. *Let $\gamma \in \{0, 1\}$. Let $P \subset E$, such that $card(P) \geqslant 1$. Let i (respectively, i') the minimal (respectively, maximal) index of points of P. For all $P' \subset P$, such that $x_i, x_{i'} \in P'$, we have $f_\gamma(P) = f_\gamma(P')$*

Proof. Let $P' \subset P$ such that $x_i, x_{i'} \in P'$. With Lemma 7, we have $f_\gamma(P') \leqslant f_\gamma(P)$. $f_{\mathcal{C}}(P') = f_{\mathcal{C}}(P)$ is trivial using Lemma 3, so that we have to prove $f_{\mathcal{D}}(P) \leqslant f_{\mathcal{D}}(P')$.
$f_{\mathcal{D}}(P) = \min_{k \in [\![i, i']\!], x_k \in P} \max(d(x_k, x_i), d(x_i - x_{i'})) \leqslant \min_{k \in [\![i', j']\!], x_k \in P'} \max(d(x_k, x_i), d(x_k, x_{i'})) = f_{\mathcal{D}}(P')$ □

4.3. Optimality of Non-Nested Clustering

In this section, we prove that non-nested clustering property, the extension of interval clustering from 1D to 2D PF, allows the computation of optimal solutions, which will be a key element for a DP algorithm. For (partial) p-center problems, i.e., K-M-max-(α, γ)-BC2DPF, optimal solutions may exist without fulfilling the non-nested property, whereas for K-M-$+$-$(\alpha, 0)$-BC2DPF problems, the nested property is a necessary condition to obtaining an optimal solution.

Lemma 9. *Let $\gamma \in \{0,1\}$; let $M > 0$. There is an optimal solution of 1-M-$\oplus$-(α,γ)-BC2DPF on the shape $\mathcal{C}_{i,i'} = \{x_j\}_{j \in [\![i,i']\!]} = \{x \in E \mid \exists j \in [\![i,i']\!],\ x = x_j\}$, with $|i'-i| \geqslant N-M$.*

Proof. Let $\pi \in \Pi_K(E)$ represent an optimal solution of 1-M-$\oplus$-(α,γ)-BC2DPF, let OPT be the optimal cost, and $\mathcal{C} \subset E$ with $|\mathcal{C}| \geqslant N-M$ and $f_\gamma(\mathcal{C}) = OPT$. Let i (respectively, i') be the minimal (respectively maximal) index of $\mathcal{C}$ using order of Proposition 1. $\mathcal{C} \subset \mathcal{C}_{i,i'}$, so Lemma 8 applies and $f_\gamma(\mathcal{C}_{i,i'}) = f_\gamma(\mathcal{C}) = OPT$. $|\mathcal{C}_{i,i'}| \geqslant |\mathcal{C}| \geqslant N-M$, thus $\mathcal{C}_{i,i'}$ defines an optimal solution of 1-M-$\oplus$-(α,γ)-BC2DPF. □

Proposition 2. *Let $E = (x_i)$ be a 2D PF, re-indexed with Proposition 1. There are optimal solutions of K-M-$\oplus$-(α,γ)-BC2DPF using only clusters on the shape $\mathcal{C}_{i,i'} = \{x_j\}_{j \in [\![i,i']\!]} = \{x \in E \mid \exists j \in [\![i,i']\!],\ x = x_j\}$.*

Proof. We prove the results by the induction on $K \in \mathbb{N}^*$. For $K = 1$, Lemma 9 gives the initialization.

Let us suppose that $K > 1$ and the Induction Hypothesis (IH) that Proposition 2 is true for K-M-$\oplus$-(α,γ)-BC2DPF. Let $\pi \in \Pi_K(E)$ be an optimal solution of K-M-$\oplus$-(α,γ)-BC2DPF; let OPT be the optimal cost. Let $X \subset E$ be the subset of the non-selected points, $|X| \leqslant M$, and $\mathcal{C}_1, \ldots, \mathcal{C}_K$ be the K subsets defining the costs, so that $X, \mathcal{C}_1, \ldots, \mathcal{C}_K$ is a partition of E and $\bigoplus_{k=1}^K f_\gamma(\mathcal{C}_k)^\alpha = OPT$. Let N' be the maximal index, such that $x_{N'} \notin X$, which is, necessarily, $N' \geqslant N-M$. We reindex the clusters $\mathcal{C}_k$, such that $x_{N'} \in \mathcal{C}_K$. Let i be the minimal index such that $x_i \in \mathcal{C}_K$.

We consider the subsets $\mathcal{C}'_K = \{x_j\}_{j \in [\![i,N']\!]}$, $X' = X \cap [\![1,i-1]\!]$ and $\mathcal{C}'_k = \mathcal{C}_k \cap \{x_j\}_{j \in [\![1,i-1]\!]}$ for all $k \in [\![1,K-1]\!]$. It is clear that $X', \mathcal{C}'_1, \ldots, \mathcal{C}'_{K-1}$ is a partition of $\{x_j\}_{j \in [\![1,i-1]\!]}$, and $X', \mathcal{C}'_1, \ldots, \mathcal{C}'_K$ is a partition of E. For all $k \in [\![1,K-1]\!]$, $\mathcal{C}'_k \subset \mathcal{C}_k$, so that $f_\gamma(\mathcal{C}'_k) \leqslant f_\gamma(\mathcal{C}_k)$ (Lemma 7).

$X', \mathcal{C}'_1, \ldots, \mathcal{C}'_K$ is a partition of E, and $\bigoplus_{k=1}^K f_\gamma(\mathcal{C}'_k)^\alpha \leqslant OPT$. $\mathcal{C}'_1, \ldots, \mathcal{C}'_K$ is an optimal solution of K-$|X'|$-$\oplus$-(α,γ)-BC2DPF. $\mathcal{C}'_1, \ldots, \mathcal{C}'_{K-1}$ is an optimal solution of $(K-1)$-$|X'|$-$\oplus$-(α,γ)-BC2DPF, applied to points $E' = \cup_{k=1}^{K-1} \mathcal{C}'_1 \cup X'$. Letting OPT' be the optimal cost of $(K-1)$-$|X'|$-$\oplus$-(α,γ)-BC2DPF, we have $OPT = OPT' \oplus f_\gamma(\mathcal{C}'_K)^\alpha$. Applying IH for of $(K-1)$-$|X'|$-$\oplus$-(α,γ)-BC2DPF to points E', we have $\mathcal{C}''_1, \ldots, \mathcal{C}''_{K-1}$ an optimal solution of $(K-1)$-$|X'|$-$\oplus$-(α,γ)-BC2DPF among E' on the shape $\mathcal{C}_{i,i'} = \{x_j\}_{j \in [\![i,i']\!]} = \{x \in E' \mid \exists j \in [\![i,i']\!],\ x = x_j\}$. $\bigoplus_{k=1}^K f_\gamma(\mathcal{C}''_k)^\alpha = OPT'$, and thus $\bigoplus_{k=1}^K f_\gamma(\mathcal{C}''_k)^\alpha \oplus f_\gamma(\mathcal{C}'_K)^\alpha = OPT$. $\mathcal{C}''_1, \ldots, \mathcal{C}''_{K-1}, \mathcal{C}'_K$ is an optimal solution of K-M-$\oplus$-(α,γ)-BC2DPF in E using only clusters $\mathcal{C}_{i,i'}$. Hence, the result is proven by induction. □

Proposition 3. *There is an optimal solution of K-M-$\oplus$-$(\alpha,0)$-BC2DPF, removing exactly M points in the partial clustering.*

Proof. Starting with an optimal solution of K-M-$+$-$(\alpha,0)$-BC2DPF, let OPT be the optimal cost, and let $X \subset E$ be the subset of the non-selected points, $|X| \leqslant M$, and $\mathcal{C}_1, \ldots, \mathcal{C}_K$, the K subsets defining the costs, so that $X, \mathcal{C}_1, \ldots, \mathcal{C}_K$ is a partition of E. Removing random $M - |X|$ points in $\mathcal{C}_1, \ldots, \mathcal{C}_K$, we have clusters $\mathcal{C}'_1, \ldots, \mathcal{C}'_K$ such that, for all $k \in [\![1,K-1]\!]$, $\mathcal{C}'_k \subset \mathcal{C}_k$, and thus $f_\gamma(\mathcal{C}'_k) \leqslant f_\gamma(\mathcal{C}_k)$ (Lemma 7). This implies $\bigoplus_{k=1}^K f_\gamma(\mathcal{C}'_k)^\alpha \leqslant \bigoplus_{k=1}^K f_\gamma(\mathcal{C}_k)^\alpha = OPT$, and thus the clusters $\mathcal{C}'_1, \ldots, \mathcal{C}'_K$ and outliers $X' = E \setminus \cup_k \mathcal{C}'_k$ define and provide the optimal solution of K-M-$\oplus$-$(\alpha,0)$-BC2DPF with exactly M outliers. □

Reciprocally, one may investigate if the conditions of optimality in Propositions 2 and 3 are necessary. The conditions are not necessary in general. For instance, with $E = \{(3,1);(2,2);(1,3)\}$, $K = M = 1$ and the discrete function F_D, ie $\gamma = 1$, the selection of each pair of points defines an optimal solution, with the same cost as the selection of the three points, which do not fulfill the property of Proposition 3. Having an optimal solution with the two extreme points also does not fulfill the property of Proposition 2. The optimality conditions are necessary in the case of sum-clustering, using the continuous measure of the enclosing disk.

Proposition 4. *Let an optimal solution of K-M-+-(α, 0)-BC2DPF be defined with $X \subset E$ as the subset of outliers, with $|X| \leqslant M$, and $\mathcal{C}_1, \dots, \mathcal{C}_K$ as the K subsets defining the optimal cost. We therefore have*

(i) $\left|\bigcup_{k=1}^{K} \mathcal{C}_k\right| = M$, in other words, exactly M points are not selected in π.

(ii) For each $k \in [\![1, K]\!]$, defining $i_k = \min\{i \in [\![1, N]\!] | x_i \in \mathcal{C}_k\}$ and $j_k = \max\{i \in [\![1, N]\!] | x_i \in \mathcal{C}_k\}$, we have $\mathcal{C}_k = \{x_i\}_{i \in [\![i_k, j_k]\!]}$.

Proof. Starting with an optimal solution of *K-M-+-*(α, 0)-BC2DPF, let *OPT* be the optimal cost, and let $X \subset E$ be the subset of the non-selected points, $|X| \leqslant M$, and $\mathcal{C}_1, \dots, \mathcal{C}_K$ be the K subsets defining the costs, so that $X, \mathcal{C}_1, \dots, \mathcal{C}_K$ is a partition of E. We prove (i) and (ii) ad absurdum.

If $|X| < M$, one may remove one extreme point of the cluster $\mathcal{C}_1$, defining $\mathcal{C}'_1$. With Lemmas 2 and 3, we have $f_\mathcal{C}(\mathcal{C}'_1) < f_\mathcal{C}(\mathcal{C}_1)$, and $f_\mathcal{C}(\mathcal{C}'_1)^\alpha + \sum_{k=1}^{K} f_\mathcal{C}(\mathcal{C}_k)^\alpha < f_\mathcal{C}(\mathcal{C}_1)^\alpha + \sum_{k=1}^{K} f_\mathcal{C}(\mathcal{C}_k)^\alpha = OPT$. This is in contraction with the optimality of $\mathcal{C}_1, \dots, \mathcal{C}_K, \mathcal{C}'_1, \mathcal{C}_2 \dots, \mathcal{C}_K$, defining a strictly better solution for *K-M-+-*(α, 0)-BC2DPF. (i) is thus proven ad absurdum.

If (ii) is not fulfilled by a cluster $\mathcal{C}_k$, there is $x_i \notin \mathcal{C}_k$ with $i \in [\![i_k, j_k]\!]$. If $x_i \in X$, we have a better solution than the optimal one with $X' = X \cup \{x_{i_k}\} \setminus \{x_i\}$ and $\mathcal{C}'_k = \mathcal{C}_k \cup \{x_i\} \setminus \{x_{i_k}\}$. If $x_i \in \mathcal{C}_l$ with $l \neq k$, we have nested clusters $\mathcal{C}_l$ and $\mathcal{C}_k$. We suppose that $i_k < i_l$ (otherwise, reasoning is symmetrical). We define a better solution than the optimal one with $\mathcal{C}'_l = \mathcal{C}_k \cup \{x_i\} \setminus \{x_{i_l}\}$ and $\mathcal{C}'_k = \mathcal{C}_k \cup \{x_{i_l}\} \setminus \{x_i\}$. (ii) is thus proven ad absurdum. □

4.4. Computation of Cluster Costs

Using Proposition 2, only cluster costs $\mathcal{C}_{i,i'}$ are computed. This section allows the efficient computation of such cluster costs. Once points are sorted using Proposition 1, cluster costs $f_\mathcal{C}(\mathcal{C}_{i,i'})$ can be computed in $O(1)$ using Lemma 3. This makes a time complexity in $O(N^2)$ to compute all the cluster costs $f_\mathcal{C}(\mathcal{C}_{i,i'})$ for $1 \leqslant i \leqslant i' \leqslant N$.

Equation (19) ensures that cluster costs $f_\mathcal{D}(\mathcal{C}_{i,i'})$ can be computed in $O(i' - i)$ for all $i < i'$. Actually, Algorithm 1 and Proposition 5 allow for computations in $O(\log(i' - i))$ once points are sorted following Proposition 1, with a dichotomic and logarithmic search.

Lemma 10. *Letting (i, i') with $i < i'$. $f_{i,i'} : j \in [\![i, i']\!] \longmapsto \max(d(x_j - x_i), d(x_j - x_{i'}))$ decreases before reaching a minimum $f_{i,i'}(l)$, $f_{i,i'}(l+1) \geqslant f_{i,i'}(l)$, and then increases for $j \in [\![l+1, i']\!]$.*

Proof : We define $g_{i,i',j}, h_{i,i',j}$ with $g_{i,i'} : j \subset [\![i, i']\!] \longmapsto d(x_j - x_i)$ and $h_{i,i'} : j \in [\![i, i']\!] \longmapsto d(x_j - x_{i'})$.

Let $i < i'$. Proposition 2, applied to i and any $j, j+1$ with $j \geqslant i$ and $j < i'$, ensures that g is decreasing. Similarly, Proposition 2, applied to i' and any $j, j+1$, ensures that h is increasing.

Let $A = \{j \in [\![i, i']\!] | \forall m \in [\![i, j]\!] g_{i,i'}(m) < h_{i,i'}(m)\}. g_{i,i'}(i) = 0$ and $h_{i,i'}(i) = d(x_{i'} - x_i) > 0$, so that $i \in A$. A is a non-empty and bounded subset of $\mathbb{N}$, so that A has a maximum. We note that $l = \max A. h_{i,i'}(i') = 0$ and $g_{i,i'}(i') = d(x_{i'} - x_i) > 0$, so that $i' \notin A$ and $l < i'$.

Let $j \in [\![i, l-1]\!]$. $g_{i,i'}(j) < g_{i,i'}(j+1)$ and $h_{i,i',j}(j+1) < h_{i,i'}(j)$, using the monotony of $g_{i,i'}$ and $h_{i,i'}. f_{i,i'}(j+1) = \max(g_{i,i'}(j+1), h_{i,i'}(j+1)) = h_{i,i}(j+1)$ and $f_{i,i'}(j) = \max(g_{i,i'}(j), h_{i,i'}(j)) = h_{i,i}(j)$ as $j, j+1 \in A$. Hence, $f_{i,i'}(j+1) = h_{i,i'}(j+1) < h_{i,i'}(j) = f_{i,i'}(j)$. This proves that $f_{i,i'}$ is decreasing in $[\![i, l]\!]$.

$l+1 \notin A$ and $g_{i,i'}(l+1) > h_{i,i'}(l+1)$ have to be coherent with the fact that $l = \max A$.

Let $j \in [\![l+1, i'-1]\!]$. $j+1 > j \geqslant l+1$, so $g_{i,i'}(j+1) > g_{i,i'}(j) \geqslant g_{i,i'}(l+1) > h_{i,i'}(l+1) \geqslant h_{i,i'}(j) > h_{i,i'}(j+1)$ using the monotony of $g_{i,i'}$ and $h_{i,i'}$. This proves that $f_{i,i'}$ is increasing in $[\![l+1, i']\!]$.

Lastly, the minimum of f can be reached in l or in $l+1$, depending on the sign of $f_{i,i'}(l+1) - f_{i,i'}(l)$. If $f_{i,i'}(l+1) = f_{i,i'}(l)$, there are two minimums $l, l+1$. Otherwise, there is a unique minimum $l_0 \in \{l, l+1\}$, $f_{i,i'}$, which decreases before increasing.

Algorithm 1: Computation of $f_{\mathcal{D}}(\mathcal{C}_{i,i'})$

input: indexes $i < i'$, a distance d
output: the cost $f_{\mathcal{D}}(\mathcal{C}_{i,i'})$

define $\underline{i} := i$, $\underline{v} := d(x_i - x_{i'})$, $\overline{i} := i'$, $\overline{v} := d(x_i - x_{i'})$,
while $\overline{i} - \underline{i} \geqslant 2$
$i'' := \left\lfloor \frac{\overline{i}+\underline{i}}{2} \right\rfloor$
if $d(x_i - x_{i''}) < d(x_{i'} - x_{i''})$ **then** $\underline{i} := i''$ and $\underline{v} := d(x_{i'} - x_{i''})$
else $\overline{i} := i''$ and $\overline{v} := d(x_i - x_{i''})$
end while
return $\min(\underline{v}, \overline{v})$

Proposition 5. *Let $E = \{x_1, \ldots, x_N\}$ be N points of $\mathbb{R}^2$, such that for all $i < j$, $x_i \prec x_j$. The computing cost $f_{\mathcal{D}}(\mathcal{C}_{i,i'})$ for any cluster $\mathcal{C}_{i,i'}$ has a complexity in $O(\log(i'-i))$ time, using $O(1)$ additional memory space.*

Proof. Let $i < i'$. Let us prove the correctness and complexity of Algorithm 1. Algorithm 1 is a dichotomic and logarithmic search; it iterates $O(\log(i'-i))$ times, with each iteration running in $O(1)$ time. The correctness and complexity of Algorithm 1 is a consequence of Lemma 10 and the loop invariant, which exists as a of minimum of $f_{i,i'}$, $f_{i,i'}(j^*)$ with $\underline{i} \leqslant j^* \leqslant \overline{i}$, also having $\underline{v} = f_{i,i'}(\underline{i})$ and $\overline{v} = f_{i,i'}(\overline{i})$. By construction in Algorithm 1, we have $d(x_i - x_{\underline{i}}) < d(x_{i'} - x_{\underline{i}})$, and thus $f_{i,i'}(\underline{i}) = d(x_{i'} - x_{\underline{i}})$. This implies that $f_{i,i'}(\underline{i}-1) = d(x_{i'} - x_{\underline{i}-1}) > f_{i,i'}(\underline{i})$, and thus $\underline{i} \leqslant j^*$, using Lemma 10. Similarly, we always obtain $d(x_i - x_{\underline{i}}) \geqslant d(x_{i'} - x_{\underline{i}})$, and thus $f_{i,i'}(\overline{i}) = d(x_i - x_{\overline{i}})$, $f_{i,i'}(\overline{i}+1) = d(x_i - x_{\overline{i}+1}) > f_{i,i'}(\overline{i})$, so that $\overline{i} \geqslant j^*$ with Lemma 10. At the convergence of the dichotomic search, $\overline{i} - \underline{i} = 1$ and j^* is $\underline{i}$ or $\overline{i}$; therefore, the optimal value is $f_{\mathcal{D}}(\mathcal{C}_{i,i'}) = f_{i,i'}(j^*) = \min(\underline{v}, \overline{v})$. □

Remark 1. *Algorithm 1 improves the previously proposed binary search algorithm [10]. If it has the same logarithmic complexity, this leads to two times fewer calls of the distance function. Indeed, in the previous version, the dichotomic algorithm is computed at each iteration $f_{i,i'}(i'')$ and $f_{i,i'}(i''+1)$ to determine if i'' is in the increasing or decreasing phase of $f_{i,i'}$. In Algorithm 1, the computations that are provided for each iteration are equivalent to the evaluation of only $f_{i,i'}(i'')$, computing $d(x_i - x_{i''})$ and $d(x_{i'} - x_{i''})$.*

Proposition 5 can compute $f_{\mathcal{D}}(\mathcal{C}_{i,i'})$ for all $i < i'$ in $O(N^2 \log N)$. Now, we prove that the costs $f_{\mathcal{D}}(\mathcal{C}_{i,i'})$ of all $i < i'$ can be computed in $O(N^2)$ time instead of $O(N^2 \log N)$ using $O(N^2)$-independent computations. Two schemes are proposed, computing the lines of the cost matrix in $O(N)$ time, computing $f_{\mathcal{D}}(\mathcal{C}_{j,j'})^\alpha$ for all $j' \in [\![j; N]\!]$ for a given $j \in [\![1; N]\!]$ in Algorithm 2, and computing $f_{\mathcal{D}}(\mathcal{C}_{j',j})^\alpha$ for all $j' \in [\![1; j]\!]$ for a given $j \in [\![1; N]\!]$ in Algorithm 3.

Lemma 11. *Let $i, i' \in [\![1, N]\!]$, with $i+1 < i'$. Let $c \in [\![i+1, i'-1]\!]$, such that $f_{i,i'}(c) = f_{\mathcal{D}}(\mathcal{C}_{i,i'})$.*
(i) If $i' < N$, then there is c', such that $c \leqslant c' \leqslant i'$, with $f_{i,i'+1}(c') = \min_{l \in [\![i+1,i'-1]\!]} f_{i,i'}(l) = f_{\mathcal{D}}(\mathcal{C}_{i,i'+1})$.
(ii) If $i > 1$, then there is c'', such that $i \leqslant c' \leqslant c$, with $f_{i-1,i'}(c') = \min_{l \in [\![i+1,i'-1]\!]} f_{i-1,i'}(l) = f_{\mathcal{D}}(\mathcal{C}_{i-1,i'})$.

Proof. We prove (i); we suppose that $i' < N$ and we prove that, for all $c' < c$ $f_{i,i'+1}(c) \leqslant f_{i,i'+1}(c')$, so that either c is an argmin of the minimization, and the superior minimum to

c. (ii) is similarly proven. Let $c' < c$. $f_{i,i'}(c) = f_{\mathcal{D}}(\mathcal{C}_{i,i'})$, which implies $f_{i,i'}(c) \leqslant f_{i,i'}(c')$ and, with Lemma 10, $f_{i,i'}$ is decreasing in $[\![c', c]\!]$, i.e., $f_{i,i'}(c'') = d(x_{c''}, x_{i'})$ for all $c'' \in [\![c', c]\!]$ We thus have $d(x_{c'}, x_{i'}) \geqslant d(x_{c'}, x_i)$, and, with lemma 2, $d(x_{c'}, x_{i'+1}) \geqslant d(x_{c'}, x_{i'})$. Thus, $f_{i,i'+1}(c'') = d(x_{c'}, x_{i'+1}$. With lemma 2, $d(x_{c'}, x_{i'+1} \geqslant d(x_c, x_{i'+1}$. $f_{i,i'}(c) = d(x_c, x_{i'})$ implies that $d(x_c, x_{i'}) \geqslant d(x_c, x_i)$, and then $d(x_c, x_i) \leqslant d(x_c, x_{i'}) \leqslant d(x_c, x_{i'+1})$. Thus $f_{i,i'+1}(c) = d(x_{c'}, x_{i'+1}$, and $f_{i,i'}(c) \leqslant f_{i,i'}(c')$. □

Proposition 6. *, Let $E = \{x_1, \ldots, x_N\}$ be N points of $\mathbb{R}^2$, such that for all $i < j$, $x_i \prec x_j$. Algorithm 2 computes $f_{\mathcal{D}}(\mathcal{C}_{j,j'})^\alpha$ for all $j' \in [\![j; N]\!]$ for a given $j \in [\![1; N]\!]$ in $O(N)$ time using $O(N)$ memory space.*

Proof. The validity of Algorithm 2 is based on Lemmas 10 and 11: once a discrete center c is known for a $f_{\mathcal{D}}(\mathcal{C}_{j,j'})^\alpha$, we can find a center c' of $f_{\mathcal{D}}(\mathcal{C}_{j,j'+1})^\alpha$ with $c' \geqslant c$, and Lemma 10 gives the stopping criterion to prove a discrete center. Let us prove the time complexity; the space complexity is obviously within $O(N)$ memory space. In Algorithm 2, each computation $f_{j',j}(curCtr)$ is in $O(1)$ time; we have to count the number of calls for this function. In each loop in j', one computation is used for the initialization; the total number of calls for this initialization is $N - j \leqslant N$. Then, denoting, with $c_N \leqslant N$, the center found for $\mathcal{C}_{j,N}$, we note that the number of loops is $c_N - j \leqslant N$. Lastly, there are less that $2N$ computations calls $f_{j',j}(curCtr)$; Algorithm 2 runs in $O(N)$ time. □

Algorithm 2: Computing $f_{\mathcal{D}}(\mathcal{C}_{j,j'})^\alpha$ for all $j' \in [\![j; N]\!]$ for a given $j \in [\![1; N]\!]$

Input: $E = \{x_1, \ldots, x_N\}$ indexed with Proposition 1, $j \in [\![1; N]\!]$, $\alpha > 0$, N points of $\mathbb{R}^2$,
Output: for all $j' \in [\![1; j]\!]$, $v_{j'} = f^\alpha_{\mathcal{D}}(\mathcal{C}_{j',j})$

```
define vector v with v_j' := 0 for all j' ∈ [[j; N]]
define curCtr := j + 1, curCost := 0
for j' := j + 1 to N
    curCost := f_j',j(curCtr)
    while curCost ⩽ f_j',j(curCtr + 1)
        curCtr := curCtr + 1
        curCost := f_j',j(curCtr)
    end while
    v_j' := curCost^α
  end for
return vector v
```

Proposition 7. *Let $E = \{x_1, \ldots, x_N\}$ be N points of $\mathbb{R}^2$, such that for all $i < j$, $x_i \prec x_j$ Algorithm 3 computes $f_{\mathcal{D}}(\mathcal{C}_{j',j})^\alpha$ for all $j' \in [\![1; j]\!]$ for a given $j \in [\![1; N]\!]$ in $O(N)$ time, using $O(N)$ memory space.*

Algorithm 3: Computing $f_{\mathcal{D}}(\mathcal{C}_{j',j})^\alpha$ for all $j' \in [\![1;j]\!]$ for a given $j \in [\![1;N]\!]$

Input: $E = \{x_1, \ldots, x_N\}$ indexed with Proposition 1, $j \in [\![1;N]\!]$, $\alpha > 0$, N points of $\mathbb{R}^2$,
Output: for all $j' \in [\![1;j]\!]$, $v_{j'} = f_{\mathcal{D}}(\mathcal{C}_{j',j})^\alpha$

```
define vector v with v_j' := 0 for all j' ∈ [[1; j]]
define curCtr := j − 1, curCost := 0
for j' := j − 1 to 1 with increment j' := j' − 1
    curCost := f_j',j(curCtr)
    while curCost ⩽ f_j',j(curCtr − 1)
        curCtr := curCtr − 1
        curCost := f_j',j(curCtr)
    end while
    v_j' := curCost^α
end for
return vector v
```

Proof. The proof is analogous with Proposition 6, applied to Algorithm 2. □

5. Particular Sub-Cases

Some particular sub-cases have specific complexity results, which are presented in this section.

5.1. Sub-Cases with $K = 1$

We first note that sub-cases $K = 1$ show no difference between 1-0-+-(α,γ)-BC2DPF and 1-0-max-$(1,\gamma)$-BC2DPF problems, defining the continuous or the discrete version of 1-center problems. Similarly, 1-M-+-(α,γ)-BC2DPF and 1-M-max-$(1,\gamma)$-BC2DPF problems define the continuous or the discrete version of partial 1-center problems. 1-center optimization problems have a trivial solution; the unique partition of E in one subset is E. To solve the 1-center problem, it is necessary to compute the radius of the minimum enclosing disk covering all the points of E (centered in one point of E for the discrete version). Once the points are re-indexed with Proposition 1, the cost computation is in $O(1)$ time for the continuous version using Proposition 3, and in $O(\log N)$ time for the discrete version using Proposition 5. The cost of the re-indexation in $O(N \log N)$ forms the overall complexity time with such an approach. One may improve this complexity without re-indexing E.

Proposition 8. *Let $E = \{x_1, \ldots, x_N\}$, a subset of N points of $\mathbb{R}^2$, such that for all $i \neq j$, $x_i \mathcal{I} x_j$. 1-0-⊕-(α,γ)-BC2DPF problems are solvable in $O(N)$ time using $O(1)$ additional memory space.*

Proof. Using Lemma 3 or Lemma 6, computations of f_γ are, at most, in $O(N)$ once the extreme elements following the order $\prec$ have been computed. Computation of the extreme points is also seen in $O(N)$, with one traversal of the elements of E, storing only the current minimal and maximal element with the order relation $\prec$. Finally, the complexity of one-center problems is in linear time. □

Proposition 9. *Let $M \in \mathbb{N}^*$, let $E = \{x_1, \ldots, x_N\}$ a subset of N points of $\mathbb{R}^2$, such that for all $i \neq j$, $x_i \mathcal{I} x_j$. The continuous partial 1-center, i.e., 1-M-⊕-$(\alpha,0)$-BC2DPF problems, is solvable in $O(N \min(M, \log N))$ time. The discrete partial 1-center, i.e., 1-M-⊕-$(\alpha,1)$-BC2DPF problems, is solvable in $O(N \log N)$ time.*

Proof. Using Proposition 2, *one*-center problems are computed equivalently: $\min_{m \in [\![0;M]\!]} f_\gamma(\mathcal{C}_{1+m,N-m})^\alpha$.

For the continuous and the discrete case, re-indexing the whole PF with Proposition 1 runs in $O(N \log N)$ time, leading to M computations in $O(1)$ or $O(\log(N-M))$ time, which are dominated by the complexity of re-indexing. The time complexity for both cases are highest in $O(N \log N)$. In the continuous case, i.e., $\gamma = 0$, one requires only the M minimal and maximal points with the total order $\prec$ to compute the cluster costs using. If $M < \log N$, one may use one traversal of E, storing the current m minimal and extreme points, which has a complexity in $O(MN)$. Choosing among the two possible algorithms, the time complexity is in $O(N \min(M, \log N))$. □

5.2. Sub-Cases with K = 2

Specific cases with $K = 2$ define two clusters, and one separation as defined in Proposition 2. For these cases, specific complexity results are provided, enumerating all the possible separations.

Proposition 10. *Let N points of $\mathbb{R}^2$, $E = \{x_1, \ldots, x_N\}$, such that for all $i \neq j$, $x_i \mathcal{I} x_j$. 2-0-⊕-(α,γ)-BC2DPF problems are solvable in $O(N \log N)$ time, using $O(N^\gamma)$ additional memory space.*

Proof. Using Proposition 2, optimal solutions exist, considering two clusters: $\mathcal{C}_{1,i}$ and $\mathcal{C}_{i+1,N}$. One enumerates the possible separations $i \in [\![1;N]\!]$. First, the re-indexation phase runs in $O(N \log N)$ time, which will be the bottleneck for the time complexity. Enumerating the (N-1) values $f_\gamma(\mathcal{C}_{1,i})^\alpha \oplus f_\gamma(\mathcal{C}_{i+1,N})^\alpha$ and storing the minimal value induces (N-1) computations in $O(1)$ time for the continuous case $\gamma = 0$, and uses $O(1)$ additional memory space: the current best value and the corresponding index. Considering the discrete case, one uses $O(N)$ additional memory space $f_\gamma(\mathcal{C}_{1,i})^\alpha, f_\gamma(\mathcal{C}_{i+1,N})^\alpha$ to maintain the time complexity result. □

One can extend the previous complexity results with the partial covering extension.

Proposition 11. *Let $E = \{x_1, \ldots, x_N\}$ be a subset of N points of $\mathbb{R}^2$, such that for all $i \neq j$, $x_i \mathcal{I} x_j$. 2-M-⊕-(α,γ)-BC2DPF problems are solvable in $O(N((M+1)^2 + \log N))$ time and $O(N^\gamma)$ additional memory space, or in $O(N((M+1)^2 \log^\gamma N) + \log N))$ time and $O(1)$ additional memory space.*

Proof. After the re-indexation phase running in $O(\log N)$ time), Proposition 2 ensures that there is an optimal solution for 2-M-⊕-(α,γ)-BC2DP, removing the $m_1 \geqslant 0$ first indexes, the $m_3 \geqslant 0$ last indexes, and $m_2 \geqslant 0$ points between the two selected clusters, with $m_1 + m_2 + m_3 \leqslant M$. Using Proposition 3, there is an optimal solution, exactly defining the M outliers, so that we can consider that $m_1 + m_2 + m_3 = M$. Denoting i as the last index of the first cluster, the first selected cluster is $\mathcal{C}_{1+m_1,i}$; the second one is $\mathcal{C}_{i+m_2+1,N-M+m_1+m_2}$. We have $i \geqslant m_1 + 1$ and $i + m_2 + 1 \leqslant N - M + m_1 + m_2$ i.e., $i \leqslant N - M + m_1$. We denote, with X, the following feasible i, m_1, m_2

$$X = \{(i, m_1, m_2) \in [\![1;N]\!] \times [\![0;M]\!]^2,\ 0 \leqslant m_1 + m_2 \leqslant M \text{ and } m_1 + 1 \leqslant i \leqslant N - M + m_1\}$$

Computing an optimal solution for 2-M-⊕-(α,γ)-BC2DP brings the following enumeration

$$OPT = \min_{(i,m_1,m_2) \in X} f_\gamma(\mathcal{C}_{1+m_1,i})^\alpha + f_\gamma(\mathcal{C}_{i+m_2+1,N-M+m_1+m_2})^\alpha \tag{20}$$

In the continuous case (ie $\gamma = 0$), we use $O((M+1)^2)$ computations to enumerate the possible m_1, m_2, and $O(N)$ computations to enumerate the possible i once m_1, m_2 are defined. With cost computations running in $O(1)$ time, the computation of (20) by enumeration runs in $O(N(M+1)^2)$ time, after the re-indexation in $O(N \log N)$ time. This induces the time complexity announced for $\gamma = 0$. This computation uses $O(1)$ additional memory space, storing only the best current solution $(i, m_1, m_2) \in X$ and its cost; this is also the announced memory complexity .

In the discrete case (i.e., $\gamma = 1$), we use $O((M+1)^2)$ computations to enumerate the possible m_1, m_2, and $O(N)$ computations to enumerate the possible i once m_1, m_2 are fixed. This uses $O(1)$ additional memory space, and the total time complexity is $O(N \log N((M+1)^2)$. To decrease the time complexity, one can use two vectors of size N to store a given m_1, m_2, for which the cluster costs $f_\gamma(\mathcal{C}_{1+m_1,i})^\alpha$ and $f_\gamma(\mathcal{C}_{i+m_2+1,N-M+m_1+m_2})^\alpha$ are given by Algorithms 2 and 3, so that the total time complexity remains in $O(N((M+1)^2 + \log N))$ with an $O(N)$ additional memory space. These two variants, using $O(1)$ or $O(N)$ additional memory space, induce the time complexity announced in Proposition 11. □

5.3. Continuous Min-Sum K-Radii on A Line

To the best of our knowledge, the 1D continuous min-sum k-radii and the min-sum diameter problems were not previously studied. The specific properties hold, as proven in Lemma 12. This allows a time complexity of $O(N \log N)$.

Lemma 12. *Let $E = \{x_1, \ldots, x_N\}$ be N points in a line of $\mathbb{R}^2$, indexed such that for all $i < j$, $x_i \prec x_j$. The min-sum k-radii in a line, K-0-+-(1,0)-BC2DPF, is equivalent to selecting the $K-1$ highest values of the distance among consecutive points, with the extremity of such segments defining the extreme points of the disks.*

Proof. Let a feasible and non-nested solution of K-0-+-(1,0)-BC2DPF be defined with clusters $\mathcal{C}_{a_1,b_1}, \mathcal{C}_{a_2,b_2}, \ldots, \mathcal{C}_{a_K,b_K}$ such that $1 = a_1 \leqslant b_1 < a_2 \leqslant b_2 < \cdots < a_K \leqslant b_K = N$. Using the alignment property, we can obtain

$$d(x_1, x_N) = \sum_{i=1}^{n-1} d(x_i, x_{i+1}) = \sum_{k=1}^{K} d(x_{a_k}, x_{b_k}) + \sum_{k=2}^{K} d(x_{b_{k-1}}, x_{a_k}) = \sum_{k=1}^{K} f_0(\mathcal{C}_{a_k,b_k}) + \sum_{k=2}^{K} d(x_{b_{k-1}}, x_{a_k})$$

Reciprocally, this is equivalent to considering K-0-+-(1,0)-BC2DPF or the maximization of the sum of $K-1$ sa a different distance among consecutive points. The latter problem is just computing the $K-1$ highest distances among consecutive points. □

Proposition 12. *Let $E = \{x_1, \ldots, x_N\}$ be a subset of N points of $\mathbb{R}^2$ on a line. K-0-+-(1,0)-BC2DPF, the continuous min-sum-k-radii, is solvable in $O(N \log N)$ time and $O(N)$ memory space.*

Proof. Lemma 12 ensures the validity of Algorithm 4, determining the $K-1$ highest values of the distance among consecutive points. The additional memory space in Algorithm 4 is in O(N), computing the list of consecutive distances. Sorting the distances and the re-indexation both have a time complexity in $O(N \log N)$. □

Algorithm 4: Continuous min-sum K-radii on a line

Input: $K \in \mathbb{N}^*$, N points of $\mathbb{R}^2$ on a line $E = \{x_1, \ldots, x_N\}$

re-index E using Proposition 1
initialize vector v with $v_i := (i, d(x_{i+1}) - d(x_i))$ for $i \in [\![1; N-1]\!]$
initialize vector w with $v_j := 0$ for $j \in [\![1; K-1]\!]$
sort vector v with $d(x_{i+1}) - d(x_i)$ increasing
for the $K-1$ elements of v with the maximal value $d(x_{i+1}) - d(x_i)$, store the indexes i in w
sort w in the increasing order
initialize $\mathcal{P} = \emptyset$, $\underline{i} = \overline{i} = 1$, $OPT = 0$.
for $j \in [\![1; K-1]\!]$ in the increasing order
  $\overline{i} := w_j$
  add $\mathcal{C}_{\underline{i},\overline{i}}$ in $\mathcal{P}$
  $OPT := OPT + f_C(\mathcal{C}_{\underline{i},\overline{i}})$
  $\underline{i} := \overline{i} + 1$
end for
add $\mathcal{C}_{\underline{i},N}$ in $\mathcal{P}$
$OPT := OPT + f_C(\mathcal{C}_{\underline{i},N})$

return OPT the optimal cost and the partition of selected clusters $\mathcal{P}$

6. Unified DP Algorithm and Complexity Results

Proposition 2 allows the design of a common DP algorithm for p-center problems and variants, and to prove polynomial complexities. The key element is to design Bellman equations.

Proposition 13 (Bellman equations). *Defining $O_{i,k,m}$ as the optimal cost of k-m-$\oplus$-(α,γ)-BC2DPF among points $[\![1,i]\!]$ for all $i \in [\![1,N]\!]$, $k \in [\![1,K]\!]$ and $m \in [\![0,M]\!]$, we have the following induction relations*

$$\forall i \in [\![1,N]\!], \; O_{i,1,0} = f_\gamma(\mathcal{C}_{1,i})^\alpha \tag{21}$$

$$\forall m \in [\![1,M]\!], \forall k \in [\![1,K]\!], \forall i \in [\![1,m+k]\!], \; O_{i,k,m} = 0 \tag{22}$$

$$\forall m \in [\![1,M]\!], \forall i \in [\![m+2,N]\!], \; O_{i,1,m} = \min(O_{i-1,1,m-1}, f_\gamma(\mathcal{C}_{1+m,i})^\alpha) \tag{23}$$

$$\forall k \in [\![2,K]\!], \forall i \in [\![k+1,N]\!], \; O_{i,k,0} = \min_{j \in [\![k,i]\!]} \left(O_{j-1,k-1,0} \oplus f_\gamma(\mathcal{C}_{j,i})^\alpha \right) \tag{24}$$

$\forall m \in [\![1,M]\!], \forall k \in [\![2,K]\!], \forall i \in [\![k+m+1,N]\!]$,

$$O_{i,k,m} = \min\left(O_{i-1,k,m-1}, \min_{j \in [\![k+m,i]\!]} \left(O_{j-1,k-1,m} \oplus f_\gamma(\mathcal{C}_{j,i})^\alpha \right) \right) \tag{25}$$

Proof. (21) is the standard 1-center case. (22) is a trivial case, where it is possible to fill the clusters with singletons, with a null and optimal cost. (23) is a recursion formula among the partial 1-center cases, an optimal solution of 1-m-$\oplus$-(α,γ)-BC2DPF among points $[\![1,i]\!]$, selecting the point x_i, and the optimal solution is cluster $\mathcal{C}_{1+m,i}$ with Proposition 3, with a cost $f_\gamma(\mathcal{C}_{1+m,i})^\alpha$ or an optimal solution of 1-$m-1$-$\oplus$-(α,γ)-BC2DPF if the point x_i is not selected. (24) is a recursion formula among the k-0-$\oplus$-(α,γ)-BC2DPF cases among points $[\![1,i]\!]$; when generalizing the ones from [10] for the powered sum-radii cases, the proof is similar. Let $k \in [\![2,K]\!]$ and $i \in [\![k+1,N]\!]$. Let $j \in [\![k,i]\!]$, when selecting an optimal solution of k-0-$\oplus$-(α,γ)-BC2DPF among points indexed in $[\![1,j-1]\!]$, and adding cluster $\mathcal{C}_{j,i}$, a feasible solution is obtained for k-0-$\oplus$-(α,γ)-BC2DPF among the points indexed in

$[\![1, i]\!]$ with a cost $O_{j-1,k-1,0} \oplus f_\gamma(\mathcal{C}_{j,i})^\alpha$. This last cost is lower than the optimal cost, thus $O_{i,k,0} \leqslant O_{j-1,k-1,0} \oplus f_\gamma(\mathcal{C}_{j,i})^\alpha$. Such inequalities are valid for all $j \in [\![k, i]\!]$; this implies

$$O_{i,k,0} \leqslant \min_{j \in [\![k,i]\!]} \left(O_{j-1,k-1,0} \oplus f_\gamma(\mathcal{C}_{j,i})^\alpha \right) \tag{26}$$

Let $j_1 < j_2 < \cdots < j_{k-1} < j_k = i$ indexes, such that $\mathcal{C}_{1,j_1}, \mathcal{C}_{j_1+1,j_2}, \ldots, \mathcal{C}_{j_{k-1}+1,i}$ defines the optimal solution of k-0-$\oplus$-(α,γ)-BC2DPF among the points indexed in $[\![1, i]\!]$; its cost is $O_{i,k,0}$. Necessarily, $\mathcal{C}_{1,j_1}, \mathcal{C}_{j_1+1,j_2}, \ldots, \mathcal{C}_{j_{k-2}+1,j_{k-1}}$ defines the optimal solution of $k-1$-0-$\oplus$-(α,γ)-BC2DPF among the points indexed in $[\![1, j_{k-1}]\!]$. On the contrary, a better solution for $O_{i,k,0}$ would be constructed, adding the cluster $\mathcal{C}_{j_{k-1}+1,i}$. We thus have $O_{i,k,0} = O_{j_{k-1},k-1,0} \oplus f_\gamma(\mathcal{C}_{j_{k-1}+1,i})^\alpha$. Combined with (26), this proves $O_{i,k,0} = \min_{j \in [\![k,i]\!]} \left(O_{j-1,k-1,0} \oplus f_\gamma(\mathcal{C}_{j,i})^\alpha \right)$.

Lastly, we prove (25). Let $m \in [\![1, M]\!], k \in [\![2, K]\!], i \in [\![k+m+1, N]\!]$. $O_{i,k,m} \leqslant O_{i-1,k,m-1}$; each solution of k-$m-1$-$\oplus$-(α,γ)-BC2DPF among the points indexed in $[\![1, i-1]\!]$ defines a solution of k-m-$\oplus$-(α,γ)-BC2DPF among the points indexed in $[\![1, i]\!]$, with the selecting point x_i as an outlier. Let $O'_{i,k,m}$, with the cost of k-m-$\oplus$-(α,γ)-BC2DPF among the points indexed in $[\![1, i]\!]$; necessarily selecting the point i, we obtain $O_{i,k,m} \leqslant O'_{i,k,m}$. $O'_{i,k,m}$ is defined by a cluster $\mathcal{C}_{j,i}$ and an optimal solution of k-m-$\oplus$-(α,γ)-BC2DPF among the points indexed in $[\![1, j-1]\!]$, so that $O'_{i,k,m} = \min_{j \in [\![k+m,i]\!]} \left(O_{j-1,k-1,m} \oplus f_\gamma(\mathcal{C}_{j,i})^\alpha \right)$. We thus have

$$O_{i,k,m} \leqslant \min \left(O_{i-1,k,m-1}, \min_{j \in [\![k+m,i]\!]} \left(O_{j-1,k-1,m} \oplus f_\gamma(\mathcal{C}_{j,i})^\alpha \right) \right) \tag{27}$$

Reciprocally, let $1 = a_1 < b_1 < a_2 < b_2 < \cdots < a_k < b_k$ indexes, such that $\mathcal{C}_{a_1,b_1}, \mathcal{C}_{a_2,b_2}, \ldots, \mathcal{C}_{a_k,b_k}$ defines an optimal solution of k-m-$\oplus$-(α,γ)-BC2DPF among the points indexed in $[\![1, i]\!]$; its cost is $O_{i,k,m}$. If $b_k = i$, then $O_{i,k,m} = O'_{i,k,m}$ and (27) is an equality. If $b_k < i$, then $\mathcal{C}_{a_1,b_1}, \mathcal{C}_{a_2,b_2}, \ldots, \mathcal{C}_{a_k,b_k}$ defines an optimal solution of k-$m-1$-$\oplus$-(α,γ)-BC2DPF among the points indexed in $[\![1, i-1]\!]$; its cost is $O_{i,k,m-1}$. We thus have $O_{i,k,m} = O_{i,k,m-1}$, and (27) is an equality. Finally, (25) is proven by disjunction. □

Bellman equations of Proposition 13 can compute the optimal value $O_{i,k,m}$ by induction. A first method is a recursive implementation of the Bellman equations to compute the cost $O_{N,K,M}$ and store the intermediate computations $O_{i,k,m}$ in a memoized implementation. An iterative implementation is provided in Algorithm 5, using a defined order for the computations of elements $O_{i,k,m}$. An advantage of Algorithm 5 is that independent computations are highlighted for a parallel implementation. For both methods computing the optimal cost $O_{N,K,M}$, backtracking operations in the DP matrix with computed costs allow for recovery of the affectation of clusters and outliers in an optimal solution.

In Algorithm 5, note that some useless computations are not processed. When having to compute $O_{N,K,M}$, computations $O_{N,k,m}$ with $k+m < K+M$ are useless. $O_{N-1,K,M}$ will also not be called. Generally, triangular elements $O_{N-n,k,m}$ with $n+k+m < K+M$ are useless. The DP matrix $O_{n,k,m}$ is not fully constructed in Algorithm 5, removing such useless elements.

Algorithm 5: unified DP algorithm for K-M-$\oplus$-(α,γ)-BC2DPF

Input: - N points of $\mathbb{R}^2$, $E = \{x_1, \ldots, x_N\}$ such that for all $i \neq j$, $x_i \ \mathcal{I} \ x_j$;
- Parameters: $K \in \mathbb{N}^*$, $M \in \mathbb{N}$, $\oplus \in \{+, \max\}$, $\gamma \in \{0,1\}$ and $\alpha > 0$;

```
sort E following the order of Proposition 1
initialize matrix O with O_{i,k,m} := 0 for all m ∈ [[0; M]], k ∈ [[1; K − 1]],
i ∈ [[k; N − K + k]]

compute f_γ(C_{1,i})^α for all i ∈ [[1; N − K + 1]] and store in O_{i,1,0} := f_γ(C_{1,i})^α

for i := 2 to N
   compute and store f_γ(C_{i',i})^α for all i' ∈ [[1; i]]
   compute O_{i,k,0} := min_{j∈[[k,i]]} (O_{j−1,k−1,0} ⊕ f_γ(C_{j,i})^α) for all k ∈ [[2; min(K, i)]]
   for m = 1 to min(M, i − 2)
      compute O_{i,1,m} := min(O_{i−1,1,m−1}, f_γ(C_{1+m,i})^α)
      for k = 2 to min(K, i − m)
         compute O_{i,k,m} := min(O_{i−1,k,m−1}, min_{j∈[[k+m,i]]} (O_{j−1,k−1,m} ⊕ f_γ(C_{j,i})^α))
      end for
   end for
   delete the stored f_γ(C_{i',i})^α for all i' ∈ [[1; i]]
end for

initialize P = ∅, i_ = ī = N, m = M
for k = K to 1 with increment k ← k − 1
   compute ī := min{i ∈ [[i_ − m; i_]] | O_{i_,k,m} := O_{i_−i,k,m−i+i_}}
   m := m − ī + i_
   compute and store f_α(C_{i',ī}) for all i' ∈ [[1; ī]]
   find i_ ∈ [[1, ī]] such that i_ := argmin_{j∈[[k+m,i]]} (O_{j−1,k−1,m} ⊕ f_γ(C_{j,ī})^α)
   add C_{i_,ī} in P
   delete the stored f_α(C_{i',ī}) for all i' ∈ [[1; ī]]
end for

return O_{N,K,M} the optimal cost and the selected clusters P
```

Theorem 1. *Let $E = \{x_1, \ldots, x_N\}$ a subset of N points of $\mathbb{R}^2$, such that for all $i \neq j$, $x_i \mathcal{I} x_j$. When applied to the 2D PF E for $K \geqslant 2$, the K-M-$\oplus$-(α,γ)-BC2DPF problems are solvable to optimality in polynomial time using Algorithm 5, with a complexity in $O(KN^2(1+M))$ time and $O(KN(1+M))$ space.*

Proof. The validity of Algorithm 5 is proven by induction; each cell of the DP matrix $O_{i,k,m}$ is computed using only cells that were previously computed to optimality. Once the required cells are computed, a standard backtracking algorithm is applied to compute the clusters. Let us analyze the complexity. Let $K \geqslant 2$. The space complexity is in $O(KN(1+M))$, along with the size of the DP matrix, with the intermediate computations of cluster costs using, at most, O(N) memory space, only remembering such vectors due to the deleting operations. Let us analyze the time complexity. Sorting and indexing the elements of E (Proposition 1) has a time complexity in $O(N \log N)$. Once costs $f_\gamma(\mathcal{C}_{i',i})^\alpha$ are computed and stored, each cell of the DP matrix is computed, at most, in $O(N)$ time using Formulas (21)–(24). This induces a total complexity in $O(KN^2(1+M))$ time. The cluster costs are computed using N times Algorithm 3 and one time Algorithm 2; this has a time complexity in $O(N^2)$, which is negligible compared to the $O(KN^2(1+M))$ time computation of the cells of the DP matrix. The K backtracking operations requires a $O(N^2)$ time computation of the costs $f_\alpha(\mathcal{C}_{i',\bar{i}})$ for all $i' \in [\![1;\bar{i}]\!]$ and a given

i, M operations in $O(1)$ time to compute $\min\{i \in [\![\underline{i} - m; \underline{i}]\!] | O_{\underline{i},k,m} = O_{\underline{i}-i,k,m-i+\underline{i}}\}$ and $O(N)$ operations in $O(1)$ time to compute $\arg\min_{j\in[\![k+m,i]\!]}\left(O_{j-1,k-1,m} \oplus f_\gamma(C_{j,\bar{i}})^\alpha\right)$. Finally, the backtracking operations requires $O(KN^2)$ time, which is negligible compared to the previous computation in $O(KN^2(1+M))$ time. □

7. Specific Improvements

This section investigates how the complexity results of Theorem 2 may be improved, and how to speed up Algorithm 5, from a theoretical and a practical viewpoint.

7.1. Improving Time Complexity for Standard and Partial P-Center Problems

In Algorithm 5, the bottleneck for complexity are the computations $\min_{j\in[\![k+m,i]\!]}\left(O_{j-1,k-1,m} \oplus f_\gamma(C_{j,i})^\alpha\right)$, for $i \in [\![2, N]\!]$, $k \in [\![2, \min(K, i)]\!]$, $m \in [\![0, i-k]\!]$. When $\oplus = \max$, it is proven that such a minimization can be processed in $O(\log N)$ instead of $O(N)$ for the naive enumeration, leading to the general complexity results. This can improve the time complexity in the p-center cases.

Lemma 13. *Let $k \in [\![1, K]\!]$ and $j \in [\![1, N]\!]$. The application $m \in [\![0, M]\!] \longmapsto O_{j,k,m}$ is decreasing.*

Proof. Let $m \in\in [\![1, M]\!]$. For each $E' \subset E$, any feasible solution of k-$(m-1)$-$\oplus$-(α,γ)-BC2DPF in E' is a feasible solution of k-m-$\oplus$-(α,γ)-BC2DPF, with the partial versions defined by problems (11). An optimal solution of k-$m-1$-$\oplus$-(α,γ)-BC2DPF is feasible for k-$(m-1)$-$\oplus$-(α,γ)-BC2DPF, it implies $O_{j,k,m-1} \geqslant O_{j,k,m}$. □

Lemma 14. *Let $k \in [\![1, K]\!]$ and $m \in [\![0, M]\!]$. The application $j \in [\![1, N]\!] \longmapsto O_{j,k,m}$ is increasing.*

Proof. We yfirst note that the case $k = 1$ is implied by the Lemma 7, so that we can suppose in the following, that $k \geqslant 2$. Let $k \in [\![2, K]\!]$, $m \in [\![0, M]\!]$ and $j \in [\![2, N]\!]$. Let $\pi \in \Pi_K(E)$ be an optimal solution of k-m-$\oplus$-(α,γ)-BC2DPF among the points indexed in $[\![1, j]\!]$; its cost is $O_{j,k,m}$. Let $X \subset E$, the subset of the non-selected points, $|X| \leqslant M$, and $C_1, \ldots, C_k$ with the k subsets defining the costs, so that $X, C_1, \ldots, C_k$ is a partition of E and $\bigoplus_{k'=1}^{k} f_\gamma(C_{k'})^\alpha = O_{j,k,m}$. If $x_j \in X$, then $O_{j,k,m} = O_{j-1,k,m-1} \geqslant O_{j-1,k,m}$ using Lemma 13, which is the result. We suppose to end the proof that $x_j \notin X$ and re-index the clusters such that $x_j \in C_k$. We consider the clusters $C'_1, \ldots, C'_k = C_1, \ldots, C_{k-1}, C_k - x_k$. With X, a partition of $(x_l)_{l\in[\![1,j-1]\!]}$ is defined, with, at most, M outliers, so that it defines a feasible solution of the optimization problem, defining $O_{j-1,k,m}$ as a cost $OPT' \geqslant O_{j-1,k,m}$. Using Lemma 7, $OPT' \leqslant O_{j,k,m}$, so that $O_{j-1,k,m-1} \geqslant O_{j-1,k,m}$. □

Lemma 15. *Let $i \in [\![2, N]\!]$, $k \in [\![2, \min(K, i)]\!]$, $m \in [\![0, i-k]\!]$. Let $g_{i,k,m} : j \in [\![2, i]\!] \longmapsto \max(O_{j-1,k-1,m}, f_\gamma(C_{j,i})^\alpha)$. There is $l \in [\![2, i]\!]$, such that $g_{i,k}$ is decreasing for $j \in [\![2, l]\!]$, and then increases for $j \in [\![l+1, i]\!]$. For $j < l$, $g_{i,k} = f_\gamma(C_{j,i})^\alpha$ and for $j > l$, $g_{i,k} = O_{j-1,k-1,m}$.*

Proof. Similarly to the proof of Lemma 10, the following applications are monotone:
$j \in [\![1, i]\!] \longmapsto f_\gamma(C_{j,i})^\alpha$ decreases with Lemma 7,
$j \in [\![1, N]\!] \longmapsto O_{j,k,m}$ increases for all k with Lemma 14. □

Proposition 14. *Let $i \in [\![2, N]\!], k \in [\![2, K]\!], m \in [\![0, M]\!]$. Let $\gamma \in \{0, 1\}$. Once the values $O_{i',k-1,m'}$ in the DP matrix of Algorithm 2 are computed, Algorithm 6 computes $O_{i,k,m} = \min_{j\in[\![k+m,i]\!]} \max\left(O_{j-1,k-1,m}, f_\gamma(C_{j,i})^\alpha\right)$ calling $O(\log i)$ cost computations $f_\gamma(C_{j,i})$. This induces a time complexity in $O(\log^{1+\gamma} i)$ using straightforward computations of the cluster costs with Propositions 3 and 5.*

Algorithm 6: Dichotomic computation of $\min_{j \in [\![k+m,i]\!]} \max\left(O_{j-1,k-1,m}, f_\gamma(C_{j,i})^\alpha\right)$

input: indexes $i \in [\![2,N]\!]$, $k \in [\![2,\min(K,i)]\!]$, $m \in [\![0,i-k]\!]$, $\alpha > 0$ $\gamma \in \{0,1\}$;
a vector v containing $v_j := O_{j,k-1,m}$ for all $j \in [\![1,i-1]\!]$.

define $\underline{i} := k+m$, $\underline{v} = f_\gamma(C_{k+m,i})^\alpha$,
define $\overline{i} := i$, $\overline{v} := v_{i-1}$,
while $\overline{i} - \underline{i} \geqslant 2$
$i'' := \left\lfloor \frac{\underline{i}+\overline{i}}{2} \right\rfloor$
if $f_\gamma(C_{j,i})^\alpha < v_{i''}$ **then** set $\overline{i} := i''$ and $\overline{v} := v_{i''}$
else $\underline{i} := i''$ and $\underline{v} := f_\gamma(C_{j,i})^\alpha$
end while
return $\min(\underline{v}, \overline{v})$

Proof. Algorithm 6 is a dichotomic search based on Lemma 15, similarly to Algorithm 1, derived from Lemma 10. The complexity in Algorithm 6 is $O(\log i)$ cost computations $f_\gamma(C_{j,i})$. In the discrete case, such computations run in $O(\log i)$ time with Proposition 5, whereas it is $O(1)$ in the continuous case with Lemma 3. In both cases, the final time complexity is given by $O(\log^{1+\gamma} i)$. □

Computing $\min_{j \in [\![k+m,i]\!]} \max\left(O_{j-1,k-1,m}, f_\gamma(C_{j,i})^\alpha\right)$ in time $O(\log i)$ instead of $O(i)$ in the proof of Theorem 1 for p-center problem and variants, the complexity results are updated for these sub-problems.

Theorem 2. *Let $E = \{x_1, \ldots, x_N\}$ be a subset of N points of $\mathbb{R}^2$, such that for all $i \neq j$, $x_i \mathcal{I} x_j$. Whe napplied to the 2D PF E for $K \geqslant 2$, the K-M-max-(α,γ)-BC2DPF problems are solvable to optimality in polynomial time using Algorithm 4, with a complexity in $O(KN(1+M)\log N)$ time and $O(KN(1+M))$ space.*

Proof. The validity of Algorithm 5 using Algorithm 6 inside is implied by the validity of Algorithm 6, proven in Proposition 14. Updating the time complexity with Proposition 14, the new time complexity for continuous K-center problems is seen in $O(KN(1+M)\log N)$ time instead of $O(K(1+M)N^2)$, as previously. For the discrete versions, using Proposition 14 with computations of discrete cluster costs with Proposition 5 induces a time complexity in $O(KN(1+M)\log^2 N)$. The complexity is decreased to $O(KN(1+M)\log N)$, where the cluster costs are already computed and stored in Algorithm 5, and thus the computations of Algorithm 6 are seen in $O(1)$. tThisinduces the same complexity for discrete and continuous K-center variants. □

Remark 2. *For the standard discrete p-center, Theorem 2 improves the time complexity given in the preliminary paper [10], from $O(pN\log^2 N)$ to $O(pN\log N)$. Another improvement was given by Algorithm 1; the former computation of cluster costs has the same asymptotic complexity but requires two times more computations. tTis proportional factor is non negligible in practice.*

7.2. Improving Space Complexity for Standard P-Center Problems

For standard p-center problems, Algorithm 5 has a complexity in memory space in $O(KN)$, the size of the DP matrix. This section proves it is possible to reduce the space complexity into an $O(N)$ memory space.

One can compute the DP matrix for k-centers "line-by-line", with k increasing. This does not change the validity of the algorithm, with each computation using values that were previously computed to the optimal values. Two main differences occur compared to Algorithm 5. On one hand, the $k+1$-center values use only k-center computations, and the computations with $k' < k$ can be deleted once all the required k-center values are computed when having to compute only the K-center values, especially the optimal cost. On the other hand, the computations of cluster costs are not factorized, as in Algorithm 5; this does not make any difference in the continuous version, where Lemma 3 can to recompute cluster costs in $O(1)$ time when needed, whereas recomputing each cost induces the computations running in $O(\log N)$ for the discrete version with Algorithm 1.

The search order of operations slightly degrades the time complexity for the discrete variant, without inducing a change in the continuous variant. This allows only for computations of the optimal value; another difficulty is that the backtracking operations, as written in Algorithm 5, require storage of the whole stored values of the whole matrix. The issue is obtaining alternative backtracking algorithms that allow the computation of an optimal solution of the standard p-center problems using only the optimal value provided by the DP iterations, and with a complexity of, at most, $O(KN\log^\gamma N)$ time and $O(N)$ memory space. Algorithms 7 and 8 have such properties.

Algorithm 7: Backtracking algorithm using $O(N)$ memory space

input: - $\gamma \in \{0,1\}$ to specify the clustering measure;
- N points of a 2D PF, $E = \{z_1, \ldots, z_N\}$, sorted such that for all $i < j, z_i \prec z_j$;
- $K \in \mathbb{N}$ the number of clusters;
- OPT, the optimal cost of K-γ-CP2DPF;

output: $\mathcal{P}$ an optimal partition of K-γ-CP2DPF.

initialize $maxId := N$, $minId := N$, $\mathcal{P} = \varnothing$, a set of sub-intervals of $[\![1;N]\!]$.
for $k := K$ to 2 with increment $k \leftarrow k-1$
  set $minId := maxId$
  while $f_\gamma(\mathcal{C}_{minId-1,maxId})) \leqslant OPT$ **do** $minId := minId - 1$ **end while**
  add $[\![minId, maxId]\!]$ in $\mathcal{P}$
  $maxId := minId - 1$
end for
add $[\![1, maxId]\!]$ in $\mathcal{P}$
return $\mathcal{P}$

Algorithm 8: Backtracking algorithm using $O(N)$ memory space

input: - $\gamma \in \{0,1\}$ to specify the clustering measure;
- N points of a 2D PF, $E = \{z_1, \ldots, z_N\}$, sorted such that for all $i < j, z_i \prec z_j$;
- $K \in \mathbb{N}$ the number of clusters;
- OPT, the optimal cost of K-γ-CP2DPF;

output: $\mathcal{P}$ an optimal partition of K-γ-CP2DPF.

initialize $minId := 1$, $maxId := 1$, $\mathcal{P} := \varnothing$, a set of sub-intervals of $[\![1;N]\!]$.
for $k := 2$ to K with increment $k \leftarrow k+1$
  set $maxId := minId$
  while $f_\gamma(\mathcal{C}_{minId,maxId+1})) \leqslant OPT$ **do** $maxId := maxId + 1$ **end while**
  add $[\![minId, maxId]\!]$ in $\mathcal{P}$
  set $minId := maxId + 1$
end for
add $[\![minId, N]\!]$ in $\mathcal{P}$
return $\mathcal{P}$

Lemma 16. *Let $K \in \mathbb{N}, K \geqslant 2$. Let $E = \{z_1, \ldots, z_N\}$, sorted such that for all $i < j, z_i \prec z_j$. For the discrete and continuous K-center problems, the indexes given by Algorithm 7 are lower bounds of the indexes of any optimal solution. Denoting $[\![1,i_1]\!], [\![i_1+1,i_2]\!], \ldots, [\![i_{K-1}+1,N]\!]$, the indexes given by Algorithm 7, and $[\![1,i'_1]\!], [\![i'_1+1,i'_2]\!], \ldots, [\![i'_{K-1}+1,N]\!]$, the indexes of an optimal solution, we have, for all $k \in [\![1,K-1]\!], i_k \leqslant i'_k$*

Proof. This lemma is proven a decreasing induction on k, starting from $k = K - 1$. The case $k = K - 1$ is furnished by the first step of Algorithm 4, and $j \in [\![1, N]\!] \longmapsto f_\gamma(\mathcal{C}_{j,N})$ decreaswa with Lemma 7. WIth a given k, $i'_k \leqslant i_k$, $i_{k-1} \leqslant i'_{k-1}$ is implied by Lemma 2 and $d(z_{i_k}, z_{i_{k-1}-1}) > OPT$. □

Algorithm 8 is similar to Algorithm 7, with iterations increasing the indexes of the points of E. The validity is similarly proven, and this provides the upper bounds for the indexes of any optimal solution of K-center problems.

Lemma 17. *Let $K \in \mathbb{N}, K \geqslant 2$. Let $E = \{z_1, \dots, z_N\}$, sorted such that for all $i < j$, $z_i \prec z_j$. For K-center problems, the indexes given by Algorithm 8 are upper bounds of the indexes of any optimal solution. Denoting $[\![1, i_1]\!], [\![i_1 + 1, i_2]\!], \dots, [\![i_{K-1} + 1, N]\!]$, the indexes given by Algorithm 8, and $[\![1, i'_1]\!], [\![i'_1 + 1, i'_2]\!], \dots, [\![i'_{K-1} + 1, N]\!]$, the indexes of an optimal solution, we have, for all $k \in [\![1, K - 1]\!], i_k \geqslant i'_k$.*

Proposition 15. *Once the optimal cost of p-center problems are computed, Algorithms 7 and 8 compute an optimal partition in $O(N \log N)$ time using $O(1)$ additional memory space.*

Proof. We consider the proof for Algorithm 7, which is symmetrical for Algorithm 8. Let OPT be the optimal cost of K-center clustering with f. Let $[\![1, i_1]\!], [\![i_1 + 1, i_2]\!], \dots, [\![i_{K-1} + 1, N]\!]$ be the indexes given by Algorithm 7. Through this construction, all the clusters $\mathcal{C}$ defined by the indexes $[\![i_k + 1, i_{k+1}]\!]$ for all $k > 1$ verify $f_\gamma(\mathcal{C}) \leqslant OPT$. Let $\mathcal{C}_1$ be the cluster defined by $[\![1, i_1]\!]$; we have to prove that $f_\gamma(\mathcal{C}_1) \leqslant OPT$ to conclude the optimality of the clustering defined by Algorithm 4. For an optimal solution, let $[\![1, i'_1]\!], [\![i'_1 + 1, i'_2]\!], \dots, [\![i'_{K-1} + 1, N]\!]$ be the indexes defining this solution. Lemma 16 ensures that $i_1 \leqslant i'_1$, and thus Lemma 7 assures $f_\gamma(\mathcal{C}_{1,i_1}) \leqslant f_\gamma(\mathcal{C}_{1,i'_1}) \leqslant OPT$. Analyzing the complexity, Algorithm 7 calls for a maximum of $(K + N) \leqslant 2N$ times the clustering cost function, without requiring stored elements; the complexity is in $O(N \log^\gamma N)$ time. □

Remark 3. *Finding the biggest cluster with an extremity given and a bounded cost can be acheived by a dichotomic search. Rhis would induce a complexity in $O(K \log^{1+\gamma} N)$. To avoid the separate case $K = O(N)$ and $\gamma = 1$, Algorithms 7 and 8 provide a common algorithm running in $O(N \log N)$ time, which is enough for the following complexity results.*

The previous improvements, written in Algorithm 9, allow for new complexity results with a $O(N)$ memory space for K-centrer problems.

Algorithm 9: p-center clustering in a 2DPF with a O(N) memory space

Input:
- N points of $\mathbb{R}^2$, $E = \{x_1, \dots, x_N\}$ such that for all $i \neq j$, $x_i \mathcal{I} x_j$;
- $\gamma \in \{0, 1\}$ to specify the clustering measure;
- $K \in \mathbb{N}$ the number of clusters.

```
initialize matrix O with O_{i,k} := 0 for all i ∈ [[1; N]], k ∈ [[1; K − 1]]
sort E following the order of Proposition 1
compute and store O_{i,1} := f_γ(C_{1,i}) for all i ∈ [[1; N]] (with Algorithm 2 if γ = 1)
for k = 2 to K − 1
  for i = k + 1 to N − K + k
    compute and store O_{i,k} := min_{j∈[[2,i]]} max(O_{j−1,k−1}, f_γ(C_{j,i})) (Algorithm 6)
  end for
  delete the stored O_{i,k−1} for all i
end for
OPT := min_{j∈[[2,N]]} max(O_{j−1,K−1}, f_γ(C_{j,N})) with Algorithm 6
return OPT the optimal cost and a partition P given by backtracking Algorithm 7 or 8
```

Theorem 3. *Let $E = \{x_1, \ldots, x_N\}$ a subset of N points of $\mathbb{R}^2$, such that for all $i \neq j$, $x_i \mathcal{I} x_j$. When applied to the 2D PF E for $K \geqslant 2$, the standard continuous and discrete K-center problems, i.e., K-0-max-(α, γ)-BC2DPF, are solvable with a complexity in $O(KN \log^{1+\gamma} N)$ time and $O(N)$ space.*

Remark 4. *The continuous case improves the complexity obtained after Theorem 2, with the same time complexity and an improvement in the space complexity. For the discrete variant, improving the space complexity in $O(N)$ instead of $O(N)$ induces a very slight degradation of the time complexity, from $O(KN \log N)$ to $O(KN \log^2 N)$. Depending on the value of K, it may be preferable, with stronger constraints in memory space, to have this second version.*

7.3. Improving Space Complexity for Partial P-Center Problems?

This section tries to generalize the previous results for the partial K-center problems, i.e., K-M-max-(α, γ)-BC2DPF with $M > 0$. The key element is to obtain a backtracking algorithm that does not use the DP matrix. Algorithm 10 extends Algorithm 7 by considering all the possible cardinals of outliers between clusters k and $k+1$ for $k \in [\![0, K-1]\!]$ and the outliers after the last cluster. A feasible solution of the optimal cost should be feasible by iterating Algorithm 7 for at least one of these sub-cases.

Algorithm 10: Backtracking algorithm for K-M-max-(α, γ)-BC2DPF with $M > 0$

input: - a K-M-max-(α, γ)-BC2DPF problem
- N points of a 2D PF, $E = \{z_1, \ldots, z_N\}$, sorted such that for all $i < j$, $z_i \prec z_j$;
- OPT, the optimal cost of K-M-max-(α, γ)-BC2DPF problem;

output: $\mathcal{P}$ an optimal partition of K-M-max-(α, γ)-BC2DPF problem.

for each vector x of $K+1$ elements such that $\sum_{k=0}^{K} x[k] = M$
  initialize $maxId - x[K] := N$, $minId := N - x[K]$, $\mathcal{P} := \emptyset$, a set of sub-intervals of $[\![1; N]\!]$.
  for $k = K$ to 2 with increment $k \leftarrow k-1$
    set $minId := maxId$
    while $f_\gamma(\mathcal{C}_{minId-1,maxId}))^\alpha \leqslant OPT$ **do** $minId := minId - 1$ **end while**
    add $[\![minId, maxId]\!]$ in $\mathcal{P}$
    set $maxId := minId - 1 - x[K-1]$
  end for
  if $f_\gamma(\mathcal{C}_{1+x[0],maxId}))^\alpha \leqslant OPT$ **then** add $[\![1 + x[0], maxId]\!]$ in $\mathcal{P}$ and **return** $\mathcal{P}$
end for
return error "OPT is not a feasible cost for K-M-max-(α, γ)-BC2DPF "

It is crucial to analyze the time complexity induced by this enumeration. If the number of vectors x of $K+1$ elements is such such that $\sum_{k=0}^{K} x[k] = M$ is in $\Theta(K^M)$, then this complexity is not polynomial anymore. For $M = 1$, a time complexity in $O(KN \log N)$ would be induced, which is acceptable within the complexity of the computation of the DP matrix. Having $M \geqslant 2$ would dramatically degrade the time complexity. Hence, we extend the improvement results of space complexity only for $M = 1$, with Algorithm 11.

Theorem 4. *Let $E = \{x_1, \ldots, x_N\}$ a subset of N points of $\mathbb{R}^2$, such that for all $i \neq j$, $x_i \mathcal{I} x_j$. When applied to the 2D PF E for $K \geqslant 2$, partial K-center problems K-1-max-(α, γ)-BC2DPF, are solvable with a complexity in $O(KN \log^{1+\gamma} N)$ time and $O(N)$ space.*

7.4. Speeding-Up DP for Sum-Radii Problems

Similarly to Algorithm 6, this section tries to speed up the computations $\min_{j \in [\![k+m,i]\!]} \left(O_{j-1,k-1,m} + f_\gamma(\mathcal{C}_{j,i})^\alpha \right)$, which are the bottleneck for the time complexity in Algorithm 5. This section presents the stopping criterion to avoid useless computations in the $O(N)$ naive enumeration, but without providing proofs of time complexity improvements.

Algorithm 11: Partial p-center K-1-max-$(1,\gamma)$-BC2DPF with a O(N) memory space

Input:
- N points of $\mathbb{R}^2$, $E = \{x_1, \ldots, x_N\}$ such that for all $i \neq j$, $x_i \mathcal{I} x_j$;
- $\gamma \in \{0,1\}$ to specify the clustering measure;
- $K \in \mathbb{N}, K \geqslant 2$ the number of clusters.

initialize matrix O with $O_{i,k,m} := 0$ for all $i \in [\![1;N]\!], k \in [\![1;K-1]\!], m \in [\![0;1]\!]$
sort E following the order of Proposition 1
compute and store $O_{i,1,0} := f_\gamma(\mathcal{C}_{1,i})$ for all $i \in [\![1;N]\!]$ (with Algorithm 2 if $\gamma = 1$)
compute and store $f_\gamma(\mathcal{C}_{2,i})$ for all $i \in [\![2;N]\!]$ (with Algorithm 2 if $\gamma = 1$)
compute and store $O_{i,1,1} := \min(f_\gamma(\mathcal{C}_{2,i}), O_{i-1,1,0})$ for all $i \in [\![2;N]\!]$
for $k = 2$ to K
  for $i := k+1$ to $N - K + k$
    compute and store $O_{i,k,0} := \min_{j \in [\![2,i]\!]} \max(O_{j-1,k-1,0}, f_\gamma(\mathcal{C}_{j,i}))$ (Algorithm 6)
    compute and store $O_{i,k,1} := \min_{j \in [\![k+1,i]\!]} \max\left(O_{j-1,k-1,1}, f_\gamma(\mathcal{C}_{j,i})\right)$
    $O_{i,k,1} := \min(O_{i-1,k,0}, O_{i,k,1})$
  end for
  delete the stored $O_{i,k-1,m}$ for all i, m
end for
return $O_{N,K,1}$ the optimal cost and a partition $\mathcal{P}$ given by backtracking Algorithm 10

Proposition 16. *Let $m \in [\![0,M]\!]$, $i \in [\![1,N]\!]$ and $k \in [\![2,K]\!]$. Let β an upper bound for $O_{i,k,m}$. We suppose there exist $j_0 \in [\![1,i]\!]$, such that $f_\gamma(\mathcal{C}_{j_0,i})^\alpha \geqslant \beta$. Then, each optimal index j^*, such that $O_{i,k,m} = O_{j^*-1,k-1,m} + f_\alpha(\mathcal{C}_{j^*,i})$ necessarily fulfills $j^* > j_0$. In other words, $O_{i,k,m} = \min_{j \in [\![\max(k+m,j_0+1),i]\!]}\left(O_{j-1,k-1,m} + f_\gamma(\mathcal{C}_{j,i})^\alpha\right)$.*

Proof. With $f_\gamma(\mathcal{C}_{j_0,i})^\alpha \geqslant \beta$, Lemma 7 implies that for all $j < j_0$, $f_\gamma(\mathcal{C}_{j,i})^\alpha > f_\gamma(\mathcal{C}_{j_0,i})^\alpha \geqslant \beta$. Using $O_{i',k',m} \geqslant 0$ for all i', k' implies that for all $j < j_0$, $f_\gamma(\mathcal{C}_{j_0,i})^\alpha + O_{j_0-1,k-1,m} > \beta$, and the optimal index gives $O_{i,k,m} = \min_{j \in [\![k+m,i]\!]}\left(O_{j-1,k-1,m} + f_\gamma(\mathcal{C}_{j,i})^\alpha\right)$, which is superior to j_0. □

Proposition 16 can be applied to compute each value of the DP matrix using fewer computations than the naive enumeration. In the enumeration, β is updated to the best current value of $\left(O_{j-1,k-1,m} + f_\gamma(\mathcal{C}_{j,i})^\alpha\right)$. The index would be enumerated in a decreasing way, starting from $j = i$ until an index is found, such that $f_\gamma(\mathcal{C}_{j_0,i})^\alpha \geqslant \beta$, and no more enumeration is required with Proposition 16, ensuring that the partial enumeration is sufficient to find the wished-for minimal value. This is a practical improvement, but we do not furnish proof of complexity improvements, as it is likely that this would not change the worst case complexity.

8. Discussion

8.1. Importance of the 2D PF Hypothesis, Summarizing Complexity Results

Planar p-center problems were not studied previously in the PF case. The 2D PF hypothesis is crucial for the complexity results and the efficiency of the solving algorithms. Table 1 compares the available complexity results for 1D and 2D cases of some k-center variants.

The complexity for 2D PF cases is very similar to the 1D cases; the 2D PF extension does not induce major difficulties in terms of complexity results. 2D PF cases may induce significant differences compared to the general 2D cases. The p-center problems are NP-hard in a planar Euclidean space [17], since adding the PF hypothesis leads to the polynomial complexity of Theorem 1, which allows for an efficient, straightforward implementation of the algorithm. Two properties of 2D PF were crucial for these results: The 1D structure implied by Proposition 1, which allows an extension of DP algorithms [58,59], and Lemmas 3 and 6, which allow quick computations of cluster costs. Note that rectangular p-center problems have a better complexity using general planar results than using our Theorems 2 and 3. Our algorithms only use common properties for Chebyshev and Minkowski distances, whereas significant improvements are provided using specificities of Chebyshev distance.

Table 1. Comparison of the time complexity for 2D PF cases to the 1D and 2D cases.

Problem	1D Complexity		Our 2D PF Complexity		2D Complexity	
Cont. min-sum-K-radii	$O(N\log N)$	Proposition 12	$O(KN^2)$	Theorem 1	NP-hard	[40]
Cont. p-center	$O(N\log^3 N)$	[7]	$O(pN\log N)$	Theorems 2 and 3	NP-hard	[17]
Discr. p-center	$O(N)$	[33]	$O(pN\log N)$	Theorem 2	NP-hard	[17]
Cont. 1-center	$O(N)$	[20]	$O(N)$	Proposition 8	$O(N)$	[20]
Discr. 1-center	-	-	$O(N)$	Proposition 8	$O(N\log N)$	[29]
Cont. 2-center	-	-	$O(N\log N)$	Proposition 10	$O(N\log^2 N)$	[28]
Discr. 2-center	-	-	$O(N\log N)$	Proposition 10	$O(N^{4/3}\log^5 N)$	[32]
partial 1-center	-	-	$O(N\min(M,\log N))$	Proposition 9	$O(N^2\log N)$	[38]
Rect. 1-center	$O(N)$	[36]	$O(N)$	Proposition 2	$O(N)$	[36]
Rect. 2-center	$O(N)$	[36]	$O(N\log N)$	Proposition 10	$O(N)$	[36]
Cont. rect. p-center	$O(N)$	[36]	$O(pN\log N)$	Theorem 3	$O(N)$	[36]
Discr. rect. p-center	$O(N\log N)$	[36]	$O(pN\log N)$	Theorem 2	$O(N\log N)$	[36]

Note that our complexity results are given considering the complexity of the initial re-indexation with Proposition 1. This $O(N\log N)$ phase may be the bottleneck for the final complexity. Some papers mention results which consider that the data are already in the memory (avoiding an O(N) traversal for input data) and already sorted. In our applications, MOO methods such as epsilon-constraint provide already sorted points [3]. Using this means of calculating the complexity, our algorithms for continuous and discrete 2-center problems in a 2D PF would have, respectively, a complexity in $O(\log N)$ and $O(\log^2 N)$ time. A notable advantage of the specialized algorithm in a 2D PF instead of the general cases in 2D is the simple and easy to implement algorithms.

8.2. *Equivalent Optimal Solutions for P-Center Problems*

Lemmas 16 and 17 emphasize that many optimal solutions may exist; the lower and upper bounds may define a very large funnel. We also note that many optimal solutions can be nested, i.e., non-verifying the Proposition 2. For real-world applicationa, having well-balanced clusters is more natural, and often wished for. Algorithms 7 and 8 provide the most unbalanced solutions. One may balance the sizes of covering balls, or the number of points in the clusters. Both types of solutions may be given using simple and fast post-processing. For example, one may proceed with a steepest descent local search using two-center problem types for consecutive clusters in the current solution. For balancing the size of clusters, iterating two-center computations induces marginal computations in $O(\log^{1+\gamma} N)$ time for each iteration with Algorithm 6. Such complexity occurs once the points are re-indexed using Proposition 1; one such computation in $O(N\log N)$ allows for many neighborhood computations running in $O(\log^{1+\gamma} N)$ time, and the sorting time is amortized.

8.3. *Towards a Parallel Implementation*

Complexity issues are raised to speed-up the convergence of the algorithms in practice. An additional way to speed up the algorithms in practice is to consider implementation issues, especially parallel implementation properties in multi- or many-core environments. In Algorithm 5, the values of the DP matrix $O_{i,k,m}$ for a given $i \in [\![1;N]\!]$ requires only to compute the values $O_{j,k,m}$ for all $j < i$. Independent computations can thus be operated at the iteration i of Algorithm 5, once the cluster costs $f_\gamma(C_{i',i})^\alpha$ for all $i' \in [\![1;i]\!]$ have been computed, which is not the most time-consuming part when using Algorithms 2 and 3. This is a very useful property for a parallel implementation, requiring only $N-1$

synchronizations to process $O(KN^2(1+M))$ operations. Hence, a parallel implementation of Algorithm 5 is straightforward in a shared memory parallelization, using OpenMP for instance in C/C++, or higher-level programming languages such as Python, Julia or Chapel [60]. One may also consider an intensive parallelization in a many-core environment, such as General Purpose Graphical Processing Units (GPGPU). A difficulty when using this may be the large memory size that is required in Algorithm 5.

Section 7 variants, which construct the DP matrix faster, both for k-center and min-sum k-radii problems, are not compatible with an efficient GPGPU parallelization, and one would prefer the naive and fixed-size enumeration of Algorithm 5, even with its worse time complexity for the sequential algorithm. Comparing the sequential algorithm to the GPGPU parallelization, having many independent parallelized computations allows a huge proportional factor with GPGPU, which can compensate the worst asymptotic complexity for reasonable sized instances. Shared memory parallelization, such as OpenMP, is compatible with the improvements provided in Section 7. Contrary to Algorithm 5, Algorithms 9 and 11 compute the DP matrix with index k increasing, with $O(N)$ independent computation induced at each iteration. With such algorithms, there are only $K-2$ synchronizations required, instead of $N-1$ for Algorithm 5, which is a better property for parallelization. The $O(N)$ memory versions are also useful for GPGPU parallelization, where memory space is more constrained than when storing a DP matrix on the RAM.

Previously, the parallelization of the DP matrix construction was discussed, as this is the bottleneck in time complexity. The initial sorting algorithm can also be parallelized on GPGPU if needed; the sorting time is negligible in most cases. The backtracking algorithm is sequential to obtain clusters, but with a low complexity in general, so that a parallelization of this phase is not crucial. We note that there is only one case where the backtracking Algorithm has the same complexity as the construction of the DP matrix: the DP variant in $O(N)$ memory space proposed in Algorithm 11 with Algorithm 10 as a specific backtrack. In this specific case, the $O(K)$ tests with different positions of the chosen outlier are independent, which allows a specific parallelization for Algorithm 10.

8.4. Applications to Bi-Objective Meta-Heuristics

The initial motivation of this work was to support decision makers when an MOO approach without preference furnishes a large set of non-dominated solutions. In this application, the value of K is small, allowing for human analyses to offer some preferences. In this paper, the optimality is not required in the further developments. Our work can also be applied to a partial PF furnished by population meta-heuristics [5]. A posteriori, the complexity allows for the use of Algorithms 5, 9 and 11 inside MOO meta-heuristics. Archiving PF is a common issue of population meta-heuristics, facing multi-objective optimization problems [1,5]. A key issue is obtaining diversified points of the PF in the archive, to compute diversified solutions along the current PF.

Algorithms 5, 9 and 11 can be used to address this issue, embedded in MOO approaches, similarly to [49]. Archiving diversified solutions of Pareto sets has application for the diversification of genetic algorithms, to select diversified solutions for cross-over and mutation phases [61,62], but also for swarm particle optimization heuristics [63]. In these applications, clustering has to run quickly. The complexity results and the parallelization properties are useful in such applicationas.

For application to MOO meta-heuristics like evolutionary algorithms, the partial versions are particularly useful. Indeed, partial versions may detect outliers that are isolated from the other points. For such points, it is natural to process intensification operators to look for efficient solutions in the neighborhood, which will make the former outlier less isolated. Such a process is interesting for obtaining a better balanced distribution of the points along the PF, which is a crucial point when dealing with MOO meta-heuristics.

8.5. How to Choose K, M?

A crucial point in clustering applications the selection of an appropriate value of K. A too-small value of K can miss that instances which are well-captured with $K+1$ representative clusters. Real-world applications seek the best compromise between the minimization of K, and the minimization of the dissimilarity among the clusters. Similarly, with [11], the properties of DP can be used to achieve this goal. With the DP Algorithm 9, many couples $\{(k, O_{N,k})\}_k$ are computed, using the optimal k-center values with k clusters. Having defined a maximal value K', the complexity for computing these points is seen in $O(NK' \log^{1+\gamma} N)$. When searching for good values of k, the elbow technique, may be applied. Backtracking operations may be used for many solutions without changing the complexity. Rhe same ideas are applicable along the M index. In the previoulsy described context of MOO meta-heuristics, the sensitivity with the M parameter is more important than the sensitivity for the parameter K, where the number of archived points is known and fixed regarding other considerations, such as the allowed size of the population.

9. Conclusions and Perspectives

This paper examined the properties of p-center problems and variants in the special case of a discrete set of non-dominated points in a 2D space, using Euclidean, Minkowski or Chebyshev distances. A common characterization of optimal clusters is proven for the discrete and continuous variants of the p-center problems and variants. Thie can solve these problems to optimality with a unified DP algorithm of a polynomial complexity. Some complexity results for the 2D PF case improve the general ones in 2D. The presented algorithms are useful for MOO approaches. The complexity results, in $O(KN \log N)$ time for the standard K-center problems, and in $O(KN^2)$ time for the standard min-sum k-radii problems, are useful for application with a large PF. When applied to N points and able to ncover $M < N$ points, partial K-center and min-sum-K-radii variants are, respectively, solvable in $O(K(M+1)N \log N)$ and $O(K(M+1)N^2)$ time. Furthermore, the DP algorithms have interesting properties for efficient parallel implementation in a shared memory environment, such as OpenMP or using GPGPU. This allows their application for a very large PF with short solving times. For an application for MOO meta-heuristics such as evolutionary algorithms, the partial versions are useful for the detection of outliers where intensification phases around these isolated solutions may be processed in order to obtain a better distribution of the points along the PF.

Future perspectives include the extension of these results to other clustering algorithms. The weighted versions of p-center variants were not studied in this paper, which was motivated by MOO perspectives, and future perspectives shall consider extending our algorithms to weighted variants. Regarding MOO applications, extending the results to dimension 3 is a subject of interest for MOO problems with three objectives. However, clustering a 3D PF will be an NP-hard problem as soon as the general 2D cases are proven to be NP-hard. The perspective in such cases is to design specific approximation algorithms for a 3D PF.

Author Contributions: Conceptualization, N.D. and F.N.; Methodology, N.D. and F.N.; Validation, E.-G.T. and F.N.; Writing–original draft preparation, N.D.; Writing—review and editing, N.D.; Supervision, E.-G.T. and F.N. All authors have read and agreed to the published version of the manuscript.

Funding: This research received no external funding.

Institutional Review Board Statement: Not applicable.

Informed Consent Statement: Not applicable.

Data Availability Statement: Not applicable.

Conflicts of Interest: The authors declare no conflict of interest.

References

1. Peugeot, T.; Dupin, N.; Sembely, M.J.; Dubecq, C. MBSE, PLM, MIP and Robust Optimization for System of Systems Management, Application to SCCOA French Air Defense Program. In *Complex Systems Design & Management*; Springer: Berlin/Heidelberg, Germany, 2017; pp. 29–40. [CrossRef]
2. Dupin, N.; Talbi, E. Matheuristics to optimize refueling and maintenance planning of nuclear power plants. *J. Heuristics* **2020**, 1–43. [CrossRef]
3. Ehrgott, M.; Gandibleux, X. Multiobjective combinatorial optimization-theory, methodology, and applications. In *Multiple Criteria Optimization: State of the Art Annotated Bibliographic Surveys*; Springer: Berlin/Heidelberg, Germany, 2003; pp. 369–444.
4. Schuetze, O.; Hernandez, C.; Talbi, E.; Sun, J.; Naranjani, Y.; Xiong, F. Archivers for the representation of the set of approximate solutions for MOPs. *J. Heuristics* **2019**, *25*, 71–105. [CrossRef]
5. Talbi, E. *Metaheuristics: From Design to Implementation*; Wiley: Hoboken, NJ, USA, 2009; Volume 74.
6. Hsu, W.; Nemhauser, G. Easy and hard bottleneck location problems. *Discret. Appl. Math.* **1979**, *1*, 209–215. [CrossRef]
7. Megiddo, N.; Tamir, A. New results on the complexity of p-centre problems. *SIAM J. Comput.* **1983**, *12*, 751–758. [CrossRef]
8. Ravi, S.; Rosenkrantz, D.; Tayi, G. Heuristic and special case algorithms for dispersion problems. *Oper. Res.* **1994**, *42*, 299–310. [CrossRef]
9. Wang, D.; Kuo, Y. A study on two geometric location problems. *Inf. Process. Lett.* **1988**, *28*, 281–286. [CrossRef]
10. Dupin, N.; Nielsen, F.; Talbi, E. Clustering a 2d Pareto Front: P-center problems are solvable in polynomial time. In Proceedings of the International Conference on Optimization and Learning, Cádiz, Spain, 17–19 February 2020; pp. 179–191. [CrossRef]
11. Dupin, N.; Nielsen, F.; Talbi, E. k-medoids clustering is solvable in polynomial time for a 2d Pareto front. In Proceedings of the World Congress on Global Optimization, Metz, France, 8–10 July 2019; pp. 790–799. [CrossRef]
12. Borzsony, S.; Kossmann, D.; Stocker, K. The skyline operator. In Proceedings of the 17th International Conference on Data Engineering, Heidelberg, Germany, 2–6 April 2001; pp. 421–430.
13. Nielsen, F. Output-sensitive peeling of convex and maximal layers. *Inf. Process. Lett.* **1996**, *59*, 255–259. [CrossRef]
14. Arana-Jiménez, M.; Sánchez-Gil, C. On generating the set of nondominated solutions of a linear programming problem with parameterized fuzzy numbers. *J. Glob. Optim.* **2020**, *77*, 27–52. [CrossRef]
15. Daskin, M.; Owen, S. Two New Location Covering Problems: The Partial P-Center Problem and the Partial Set Covering Problem. *Geogr. Anal.* **1999**, *31*, 217–235. [CrossRef]
16. Calik, H.; Labbé, M.; Yaman, H. p-Center problems. In *Location Science*; Springer: Berlin/Heidelberg, Germany, 2015; pp. 79–92.
17. Megiddo, N.; Supowit, K. On the complexity of some common geometric location problems. *SIAM J. Comput.* **1984**, *13*, 182–196. [CrossRef]
18. Hochbaum, D. When are NP-hard location problems easy? *Ann. Oper. Res.* **1984**, *1*, 201–214. [CrossRef]
19. Hochbaum, D.; Shmoys, D. A best possible heuristic for the k-center problem. *Math. Oper. Res.* **1985**, *10*, 180–184. [CrossRef]
20. Gonzalez, T. Clustering to minimize the maximum intercluster distance. *Theor. Comput. Sci.* **1985**, *38*, 293–306. [CrossRef]
21. Daskin, M. *Network and Discrete Location: Models, Algorithms and Applications*; Wiley: Hoboken, NJ, USA, 1995.
22. Calik, H.; Tansel, B. Double bound method for solving the p-center location problem. *Comput. Oper. Res.* **2013**, *40*, 2991–2999. [CrossRef]
23. Elloumi, S.; Labbé, M.; Pochet, Y. A new formulation and resolution method for the p-center problem. *INFORMS J. Comput.* **2004**, *16*, 84–94. [CrossRef]
24. Callaghan, B.; Salhi, S.; Nagy, G. Speeding up the optimal method of Drezner for the p-centre problem in the plane. *Eur. J. Oper. Res.* **2017**, *257*, 722–734. [CrossRef]
25. Drezner, Z. The p-centre problem—heuristic and optimal algorithms. *J. Oper. Res. Soc.* **1984**, *35*, 741–748.
26. Hwang, R.; Lee, R.; Chang, R. The slab dividing approach to solve the Euclidean P-Center problem. *Algorithmica* **1993**, *9*, 1–22. [CrossRef]
27. Agarwal, P.; Procopiuc, C. Exact and approximation algorithms for clustering. *Algorithmica* **2002**, *33*, 201–226. [CrossRef]
28. Megiddo, N. Linear-time algorithms for linear programming in R3 and related problems. *SIAM J. Comput.* **1983**, *12*, 759–776. [CrossRef]
29. Brass, P.; Knauer, C.; Na, H.; Shin, C.; Vigneron, A. Computing k-centers on a line. *arXiv* **2009**, arXiv:0902.3282.
30. Sharir, M. A near-linear algorithm for the planar 2-center problem. *Discret. Comput. Geom.* **1997**, *18*, 125–134. [CrossRef]
31. Eppstein, D. Faster construction of planar two-centers. *SODA* **1997**, *97*, 131–138.
32. Agarwal, P.; Sharir, M.; Welzl, E. The discrete 2-center problem. *Discret. Comput. Geom.* **1998**, *20*, 287–305. [CrossRef]
33. Frederickson, G. Parametric search and locating supply centers in trees. In *Workshop on Algorithms and Data Structures*; Springer: Berlin/Heidelberg, Germany, 1991; pp. 299–319.
34. Karmakar, A.; Das, S.; Nandy, S.; Bhattacharya, B. Some variations on constrained minimum enclosing circle problem. *J. Comb. Optim.* **2013**, *25*, 176–190. [CrossRef]
35. Chen, D.; Li, J.; Wang, H. Efficient algorithms for the one-dimensional k-center problem. *Theor. Comput. Sci.* **2015**, *592*, 135–142. [CrossRef]
36. Drezner, Z. On the rectangular p-center problem. *Nav. Res. Logist. (NRL)* **1987**, *34*, 229–234. [CrossRef]
37. Katz, M.J.; Kedem, K.; Segal, M. Discrete rectilinear 2-center problems. *Comput. Geom.* **2000**, *15*, 203–214. [CrossRef]
38. Drezner, Z. On a modified one-center model. *Manag. Sci.* **1981**, *27*, 848–851. [CrossRef]

39. Hansen, P.; Jaumard, B. Cluster analysis and mathematical programming. *Math. Program.* **1997**, *79*, 191–215. [CrossRef]
40. Doddi, S.; Marathe, M.; Ravi, S.; Taylor, D.; Widmayer, P. Approximation algorithms for clustering to minimize the sum of diameters. *Nord. J. Comput.* **2000**, *7*, 185–203.
41. Gibson, M.; Kanade, G.; Krohn, E.; Pirwani, I.A.; Varadarajan, K. On metric clustering to minimize the sum of radii. *Algorithmica* **2010**, *57*, 484–498. [CrossRef]
42. Charikar, M.; Panigrahy, R. Clustering to minimize the sum of cluster diameters. *J. Comput. Syst. Sci.* **2004**, *68*, 417–441. [CrossRef]
43. Behsaz, B.; Salavatipour, M. On minimum sum of radii and diameters clustering. *Algorithmica* **2015**, *73*, 143–165. [CrossRef]
44. Mahajan, M.; Nimbhorkar, P.; Varadarajan, K. The planar k-means problem is NP-hard. *Theor. Comput. Sci.* **2012**, *442*, 13–21. [CrossRef]
45. Shang, Y. Generalized K-Core percolation in networks with community structure. *SIAM J. Appl. Math.* **2020**, *80*, 1272–1289. [CrossRef]
46. Tao, Y.; Ding, L.; Lin, X.; Pei, J. Distance-based representative skyline. In Proceedings of the 2009 IEEE 25th International Conference on Data Engineering, Shanghai, China, 29 March–2 April 2009; pp. 892–903.
47. Cabello, S. Faster Distance-Based Representative Skyline and *k*-Center Along Pareto Front in the Plane. *arXiv* **2020**, arXiv:2012.15381.
48. Sayın, S. Measuring the quality of discrete representations of efficient sets in multiple objective mathematical programming. *Math. Program.* **2000**, *87*, 543–560. [CrossRef]
49. Auger, A.; Bader, J.; Brockhoff, D.; Zitzler, E. Investigating and exploiting the bias of the weighted hypervolume to articulate user preferences. In Proceedings of the GECCO 2009, Montreal, QC, Canada, 8–12 July 2009; pp. 563–570.
50. Bringmann, K.; Cabello, S.; Emmerich, M. Maximum Volume Subset Selection for Anchored Boxes. In Proceedings of the 33rd International Symposium on Computational Geometry (SoCG 2017), Brisbane, Australia, 4–7 July 2017; Aronov, B., Katz, M.J., Eds.; Volume 77, pp. 22:1–22:15. [CrossRef]
51. Bringmann, K.; Friedrich, T.; Klitzke, P. Two-dimensional subset selection for hypervolume and epsilon-indicator. In Proceedings of the Annual Conference on Genetic and Evolutionary Computation, Vancouver, BC, Canada, 12–16 June 2014; pp. 589–596.
52. Kuhn, T.; Fonseca, C.; Paquete, L.; Ruzika, S.; Duarte, M.; Figueira, J. Hypervolume subset selection in two dimensions: Formulations and algorithms. *Evol. Comput.* **2016**, *24*, 411–425. [CrossRef]
53. Erkut, E. The discrete p-dispersion problem. *Eur. J. Oper. Res.* **1990**, *46*, 48–60. [CrossRef]
54. Hansen, P.; Moon, I. Dispersing facilities on a network. *Cahiers du GERAD* **1995**.
55. Dupin, N. Polynomial algorithms for p-dispersion problems in a 2d Pareto Front. *arXiv* **2020**, arXiv:2002.11830.
56. Dupin, N.; Nielsen, F.; Talbi, E. k-medoids and p-median clustering are solvable in polynomial time for a 2d Pareto front. *arXiv* **2018**, arXiv:1806.02098.
57. Dupin, N.; Nielsen, F.; Talbi, E. Dynamic Programming heuristic for k-means Clustering among a 2-dimensional Pareto Frontier. In Proceedings of the 7th International Conference on Metaheuristics and Nature Inspired Computing, Marrakech, Morocco, 27–31 October 2018.
58. Grønlund, A.; Larsen, K.; Mathiasen, A.; Nielsen, J.; Schneider, S.; Song, M. Fast exact k-means, k-medians and Bregman divergence clustering in 1d. *arXiv* **2017**, arXiv:1701.07204.
59. Wang, H.; Song, M. Ckmeans. 1d. dp: Optimal k-means clustering in one dimension by dynamic programming. *R J.* **2011**, *3*, 29. [CrossRef]
60. Gmys, J.; Carneiro, T.; Melab, N.; Talbi, E.; Tuyttens, D. A comparative study of high-productivity high-performance programming languages for parallel metaheuristics. *Swarm Evol. Comput.* **2020**, *57*, 100720. [CrossRef]
61. Zio, E.; Bazzo, R. A clustering procedure for reducing the number of representative solutions in the Pareto Front of multiobjective optimization problems. *Eur. J. Oper. Res.* **2011**, *210*, 624–634. [CrossRef]
62. Samorani, M.; Wang, Y.; Lv, Z.; Glover, F. Clustering-driven evolutionary algorithms: An application of path relinking to the quadratic unconstrained binary optimization problem. *J. Heuristics* **2019**, *25*, 629–642. [CrossRef]
63. Pulido, G.; Coello, C. Using clustering techniques to improve the performance of a multi-objective particle swarm optimizer. In Proceedings of the Genetic and Evolutionary Computation Conference, Seattle, WA, USA, 26–30 June 2004; pp. 225–237.

MDPI

Article

Adding Negative Learning to Ant Colony Optimization: A Comprehensive Study †

Teddy Nurcahyadi and Christian Blum *

Artificial Intelligence Research Institute (IIIA-CSIC), 08193 Bellaterra, Spain; teddy.nurcahyadi@iiia.csic.es
* Correspondence: christian.blum@iiia.csic.es

† This paper is an extended version of our paper published in ANTS 2020—12th International Conference on Swarm Intelligence, Barcelona, Spain, 26–28 October 2020.

Abstract: Ant colony optimization is a metaheuristic that is mainly used for solving hard combinatorial optimization problems. The distinctive feature of ant colony optimization is a learning mechanism that is based on learning from positive examples. This is also the case in other learning-based metaheuristics such as evolutionary algorithms and particle swarm optimization. Examples from nature, however, indicate that negative learning—in addition to positive learning—can beneficially be used for certain purposes. Several research papers have explored this topic over the last decades in the context of ant colony optimization, mostly with limited success. In this work we present and study an alternative mechanism making use of mathematical programming for the incorporation of negative learning in ant colony optimization. Moreover, we compare our proposal to some well-known existing negative learning approaches from the related literature. Our study considers two classical combinatorial optimization problems: the minimum dominating set problem and the multi dimensional knapsack problem. In both cases we are able to show that our approach significantly improves over standard ant colony optimization and over the competing negative learning mechanisms from the literature.

Keywords: ant colony optimization; mathematical programming; negative learning; minimum dominating set; multi-dimensional knapsack problem

Citation: Nurcahyadi, T.; Blum, C. Adding Negative Learning to Ant Colony Optimization: A Comprehensive Study. *Mathematics* **2021**, *9*, 361. https://doi.org/10.3390/math9040361

Academic Editor: Frank Werner
Received: 20 January 2021
Accepted: 9 February 2021
Published: 11 February 2021

1. Introduction

Metaheuristics [1,2] are approximate techniques for optimization. Each metaheuristic was originally introduced for a certain type of optimization problem, for example, function optimization or combinatorial optimization (CO). However, nowadays one can find a metaheuristic variant for different types of optimization problems. Ant colony optimization (ACO) [3,4] is a metaheuristic algorithm originally introduced for solving CO problems. The inspiration of ACO was the foraging behavior of natural ant colonies and, in particular, the way in which ant colonies find short paths between their ant hive and food sources. Any ACO algorithm works roughly as follows. At each iteration, a pre-defined number of artificial ants derive solutions to the considered optimization problem. This is done in a probabilistic way, making use of two types of information: (1) *greedy information* and (2) *pheromone values*. Then, some of these solutions—typically the best ones—are used to update the pheromone values. This is done with the aim of changing the probability distribution used for generating solutions such that high-quality solutions can be found. In other words, ACO is an optimization technique based on learning from positive examples, henceforth called *positive learning*. Most of the work on ACO algorithms from the literature focuses on solving CO problems, such as scheduling problems [5], routing and path-planning problems [6,7], problems related to transportation [8], and feature selection [9]. Several well-known ACO variants were introduced in the literature over the years, including the $\mathcal{MAX}$-$\mathcal{MIN}$ Ant System (MMAS) [10], Ant Colony System (ACS) [11], and the Rank-Based Ant System [12], just to name a few of the most important ones.

As already mentioned above, ACO is strongly based on positive learning, which also holds for most other metaheuristics based on learning. By means of positive learning the algorithm tries to identify those solution components that are necessary for assembling high-quality solutions. Nevertheless, there is evidence in nature that learning from negative examples, henceforth called *negative learning*, can play a significant role in biological self-organizing systems:

- Pharaoh ants (*Monomorium pharaonis*), for example, use negative trail pheromone a 'no entry' signals in order to mark unrewarding foraging paths [13,14].
- A different type of negative feedback caused by crowding at the food source was detected in colonies of *Lasius niger* [15]. This negative feedback enables the colony to maintain a flexible foraging system despite the strong positive feedback by the pheromone trails.
- Another example concerns the use of anti-pheromone hydrocarbons used by male tsetse flies. They play an important role in tsetse communications [16].
- Honeybees (*Apis mellifera ligustica*) were shown to mark flowers with scent and to strongly reject all recently visited flowers [17].

Based on these examples, Schoonderwoerd et al. [18] stated already in 1997 that it might be possible to improve ACOs' performance with an additional mechanism that tries to identify undesirable components with the help of a negative feedback mechanisms.

1.1. Existing Approaches

In fact, the ACO research community has made several attempts to design such a negative learning mechanism. Maniezzo [19] and Cordón et al. [20] were presumably the first ones to make use of an active decrease of pheromone values associated to solution components appearing in low-quality solutions. Montgomery and Randall [21] proposed three anti-pheromone strategies that were partially inspired by previous works that made use of several types of pheromones; see, for example, [22]. In their first approach, the pheromone values of those solution components that belong to the worst solution at each iteration are decreased. Their second approach—in addition to the standard pheromone—makes explicit use of negative pheromones. Each ant has a specific bias—different to the one of the other ants—towards each of the two types of pheromone. Finally, their third approach uses a certain number of ants at each iteration in order to explore the use of solution components with lower pheromone values, without introducing dedicated anti-pheromones. Unfortunately, the presented experimental evaluation did not allow clear conclusions about a potential advantage of any of the three strategies over standard ACO. Different extensions of the approaches from [21] were explored by Simons and Smith [23]. The authors admitted, however, that nearly all their approaches proved to be counter-intuitive. Their only idea that showed to be useful to some extent was to make use of a rather high amount of anti-pheromone at the early stages of the search process.

In [24], Rojas-Morales et al. presented an ACO variant for the multi dimensional knapsack problem based on opposite learning. The first algorithm phase serves for building anti-pheromone values with the intention to enable the algorithm during the second phase to avoid solution components that lead to low-quality solutions despite being locally attractive, due to a rather high heuristic value. Unfortunately, no consistent improvement over standard ACO could be observed in the results. In addition, earlier algorithm variants based on opposition-based learning were tested on four rather small TSP instances [25]. Another application to the TSP was proposed in Ramos et al. [26], where they proposed a method that uses a second-order coevolved compromise between positive and negative feedback. According to the authors, their method achieves better results than single positive feedback systems in the context of the TSP. Finally, the most successful strand of work on using negative learning in ACO deals with the application to constraint satisfaction problems (CSPs). Independently of each other, Ye et al. [27] and Masukane and Mizuno [28,29] proposed negative feedback strategies for ACO algorithms in the context of CSPs. Both approaches make use of negative pheromone values in addition to the standard pheromone values. Moreover, in both works the negative pheromone values are updated at each iteration with the worst solution(s) generated at that iteration. The difference is basically to be found in the way in which the negative pheromone values are used for generating new solutions. Finally, we would also like to mention a very recent negative learning approach from the field of multi-objective optimization [30].

1.2. Contribution and General Idea

When devising a negative feedback mechanism there are fundamentally two questions to be answered: (1) how to identify those solution components that should receive a negative feedback, and (2) how exactly to make use of the negative feedback. Concerning the first question, it can be observed that all the existing approaches mentioned in the previous section try to identify low-quality solution components on the basis of the solutions generated by the ACO algorithm itself. In contrast, the main idea of this article is to make use of an additional optimization technique for identifying these components. In particular, we test two possibilities in this work. The first one is the application of mathematical programming solvers—we used CPLEX—for solving opportunely defined sub-instances of the tackled problem instance. Second, we tested the use of an additional ACO algorithm that works independently of the main algorithm for solving the before-mentioned sub-instances.

We have tested this mechanism in a preliminary work [31] by applying it to the so-called capacitated minimum dominating set problem (CapMDS), with excellent results. In this extended work we first describe the mechanism in general terms in the context of subset selection problems, which is a large class of CO problems. Subsequently, we demonstrate its application to two classical NP-hard combinatorial optimization problems: the minimum dominating set (MDS) problem [32] and the multi dimensional knapsack problem (MDKP) [33]. Our results show that, even though positive learning remains to be the most important form of learning, the incorporation of negative learning improves the obtained results significantly for subsets of problem instances with certain characteristics. Moreover, for comparison purposes we implement several negative learning approaches introduced for ACO in the related literature. The obtained results show that our mechanism outperforms all of them with statistical significance.

2. Preliminaries and Problem Definitions

Even though the negative learning mechanism presented in this work is general and can be incorporated into ACO algorithms for any CO problem, for the sake of simplicity this study is conducted in the context of subset selection problems. This important class of CO problems can formally be defined as follows:

1. Set C is a finite set of n items.
2. Function $F : 2^C \mapsto \{\text{TRUE}, \text{FALSE}\}$ decides if a subset $S \subseteq C$ is a feasible solution. Henceforth, let $X \subseteq 2^C$ be the set of all feasible solutions.
3. The objective function $f : X \mapsto \mathbb{R}$ assigns a value to each feasible solution.

The optimization goal might be minimization or maximization. Numerous well-known CO problems can be stated in terms of a subset selection problem. A prominent example is the symmetric traveling salesman problem (TSP). Hereby, the edges E of the complete TSP graph $G = (V, E)$ correspond to item set C. Moreover, a subset $S \subseteq E$ is evaluated by function F as a feasible solution if and only if the edges from S define a Hamiltonian cycle in G. Finally, given a feasible solution S, the objective function value $f(S)$ of S is calculated as the sum of the distances of all edges from S. The optimization goal in the case of the TSP is minimization. In the following we explain both the MDS problem and the MDKP in terms of subset selection problems.

2.1. Minimum Dominating Set

The classical MDS problem—which is NP-hard—can be stated as follows. Given is an undirected graph $G = (C, E)$, with C being the set of vertices and E the set of edges. Given a vertex $c_i \in C$, $N(c_i) \subset C$ denotes the neighborhood of c_i in G. A subset $S \subseteq C$ is called a *dominating set* if and only if for each vertex $c_i \in C$ the following holds: (1) $c_i \in S$ or (2) there is at least one $c_j \in N(c_i)$ with $c_j \in S$. The MDS requires finding a feasible solution of minimum cardinality. This problem is obviously a subset selection problem in which C is the set of items, $F(S)$ for $S \subseteq C$ evaluates to TRUE if and only if S is a dominating set of G, and $f(S) := |S|$. A standard integer linear programming (ILP) model for the MDS problem can be stated as follows.

$$\text{minimize} \sum_{c_i \in C} x_i \tag{1}$$

subject to:

$$\sum_{c_j \in N(c_i)} x_j \geq 1 - x_i \quad \forall c_i \in C \tag{2}$$

$$x_i \in \{0,1\} \quad \forall c_i \in C \tag{3}$$

The model consists of a binary variable x_i for each vertex $c_i \in C$. The objective function counts the selected vertices, and the constraints (2) ensure that each vertex either belongs to the solution or has, at least, one neighbor that forms part of the solution. In the literature, there are many variants of the MDS problem. Examples include the minimum connected dominating set problem [34], the minimum total dominating set problem [35] and the minimum vertex weight dominating set problem [36]. The currently best metaheuristic approach for solving the MDS problem is a two-goal local search with inference rules from [37].

2.2. Multi Dimensional Knapsack Problem

The MDKP is also a classical NP-hard CO problem that is often used as a test case for new algorithmic proposals (see, for example, [38–40]). The problem can be stated as follows. Given is (1) a set $C=\{c_1,\ldots,c_n\}$ of n items and (2) a number of m resources. The availability of each resource k is limited by $\text{cap}_k > 0$, which is also called the capacity of resource k. Moreover, each item $c_i \in C$ consumes a fixed amount $r_{i,k} \geq 0$ from each resource $k = 1,\ldots,m$ (*resource consumption*). Additionally, each item $c_i \in C$ comes with a profit $p_i > 0$.

A candidate solution $S \subseteq C$ is a valid solution if and only if, concerning all resources, the total amount consumed by the items in S does not exceed the resource capacities. In other words, it is required that $\sum_{c_i \in S} r_{i,k} \leq \text{cap}_k$ for all $k = 1,\ldots,m$. Moreover, a valid solution S is labeled *non-extensible*, if no $c_i \in C \setminus S$ can be added to S without losing the property of being a valid solution. The problem requires to find a valid solution S of maximum total profit ($\sum_{c_i \in S} p_i$). The standard ILP for the MDKP is stated in the following.

$$\text{maximize} \sum_{c_i \in C} p_i \cdot x_i \tag{4}$$

subject to:

$$\sum_{c_i \in C} r_{i,k} \cdot x_i \leq \text{cap}_k \quad \forall k = 1,\ldots,m \tag{5}$$

$$x_i \in \{0,1\} \qquad \forall c_i \in C \tag{6}$$

This model is built on a binary variable x_i for each item $c_i \in C$. Constraints (5) are called the knapsack constraints. In general, the literature offers very successful exact solution techniques; see, for example, [41–43]. However, devising heuristic solvers still remains to be a challenge. Among numerous metaheuristic proposals for the MDKP problem, the currently best performing ones are the DQPSO algorithm from [44] and the TPTEA algorithm from [45].

3. MMAS: The Baseline Algorithm

Many of the negative learning approaches for ACO cited in Section 1.1 were introduced for different ACO variants. In order to ensure a fair comparison, we add both our own proposal as well as the approaches from the literature to the same standard ACO algorithm: $\mathcal{MAX}$-$\mathcal{MIN}$ Ant System (MMAS) in the hypercube framework [46], which is one of the most-used ACO versions from the last decade. In the following we first describe the standard MMAS algorithm in the hypercube framework for subset selection problems. This will be our baseline algorithm. Subsequently we describe the way in which the negative learning proposal from this paper and the chosen negative learning proposals from the literature are added to this baseline algorithm.

3.1. MMAS in the Hypercube Framework

The pheromone model $\mathcal{T}$ in the context of subset selection problems consists of a pheromone value $\tau_i \geq 0$ for each item $c_i \in C$, where C is the complete set of items. Remember that, in the context of the MDS, C is the set of vertices of the input graph, while C is the set of items in the case of the MDKP. The MMAS algorithm maintains three solutions throughout a run:

1. $S^{ib} \subseteq C$: the best solution generated at the current iteration, also called the *iteration-best* solution.
2. $S^{rb} \subseteq C$: the best solution generated since the last restart of the algorithm, also called the *restart-best* solution.
3. $S^{bsf} \subseteq C$: the *best-so-far* solution, that is, the best solution found since the start of the algorithm.

Moreover, the algorithm makes use of a Boolean control variable bs_update $\in$ {TRUE, FALSE} and the convergence factor $cf \in [0,1]$ for deciding on the pheromone update mechanism and on the question whether or not to restart the algorithm. At the start of the algorithm, solutions S^{bsf} and S^{rb} are initialized to NULL, the convergence factor is set to zero, bs_update is set to FALSE and the pheromone values are all initilized to 0.5 in function InitializePheromoneValues($\mathcal{T}$); see lines 2 and 3 of Algorithm 1. Then, at each iteration, n_a solutions are probabilistically generated in function Construct_Solution($\mathcal{T}$), based on pheromone information and on greedy information. The construction of solutions will be outlined in detail for both problems (MDS and MDKP) below. The generated solutions are stored in set $\mathcal{S}^{\text{iter}}$, and the best one from $\mathcal{S}^{\text{iter}}$ is stored as S^{ib}; see lines 5–10 of Algorithm 1. Then, the restart-best and best-so-far solutions—S^{rb} and S^{bsf}—are updated with S^{ib}, if appropriate; lines 11 and 12. Afterward, the pheromone update is conducted in function ApplyPheromoneUpdate($\mathcal{T}$, cf, bs_update, S^{ib},S^{rb},S^{bsf}) and the new value for the convergence factor cf is computed in function ComputeConvergenceFactor($\mathcal{T}$); lines 13 and 14. Finally, based on the values of cf and bs_update, the algorithm might be restarted. Such a restart consists in re-initializing all pheromone values, in setting the restart-best solution S^{rb} to NULL, and bs_update to TRUE. In the following, the function for the pheromone update and for the calculation of the convergence factor are outlined in detail.

ApplyPheromoneUpdate($\mathcal{T}$, cf, bs_update, S^{ib},S^{rb},S^{bsf}): the pheromone update that is described here is the same as in any other MMAS algorithm in the hypercube framework. First, the three solutions S^{ib}, S^{rb}, and S^{bsf} receive weights κ_{ib}, κ_{rb} and κ_{bsf}, respectively. A standard setting of these weights, depending on cf and bs_update, is provided in Table 1. It always holds that $\kappa_{ib} + \kappa_{rb} + \kappa_{bsf} = 1$. After having determined the solution weights, each pheromone value τ_i is updated as follows:

$$\tau_i := \tau_i + \rho \cdot (\xi_i - \tau_i) \quad , \tag{7}$$

where

$$\xi_i := \kappa_{ib} \cdot \Delta(S^{ib}, c_i) + \kappa_{rb} \cdot \Delta(S^{rb}, c_i) + \kappa_{bsf} \cdot \Delta(S^{bsf}, c_i) \tag{8}$$

Hereby, $\rho \in [0,1]$ is the so-called learning rate, and function $\Delta(S, c_i)$ evaluates to 1 if and only if item c_i forms part of solution S. Otherwise, the function evaluates to 0. Finally, after conducting this update, those pheromone values that exceed $\tau_{\max} = 0.999$ are set to $\tau_{\max}$, and those values that have dropped below $\tau_{\min} = 0.001$ are set to $\tau_{\min}$. Note that, in this way, a complete convergence of the algorithm is avoided. Finally, note that the learning mechanism represented by this pheromone update can clearly be labeled positive learning, because it makes use of the best solutions found for updating the pheromone values.

Table 1. Values for weights κ_{ib}, κ_{rb}, and κ_{bsf}. These values depend on the convergence factor *cf* and the Boolean control variable bs_update.

	bs_update = FALSE				bs_update = TRUE
	$cf < 0.4$	$cf \in [0.4, 0.6)$	$cf \in [0.6, 0.8)$	$cf \geq 0.8$	
κ_{ib}	1	2/3	1/3	0	0
κ_{rb}	0	1/3	2/3	1	0
κ_{bsf}	0	0	0	0	1

Algorithm 1 MMAS in the hypercube framework (baseline algorithm)

```
input: a problem instance with the complete item set C
S^bsf := NULL, S^rb := NULL, cf := 0, bs_update := FALSE
InitializePheromoneValues(T)
while termination conditions not met do
    S^iter := ∅
    for k = 1, ..., n_a do
        S^k := Construct_Solution(T)
        S^iter := S^iter ∪ {S^k}
    end for
    S^ib := best solution from S^iter
    if S^ib better than S^rb then S^rb := S^ib
    if S^ib better than S^bsf then S^bsf := S^ib
    ApplyPheromoneUpdate(T, cf, bs_update, S^ib, S^rb, S^bsf)
    cf := ComputeConvergenceFactor(T)
    if cf > 0.999 then
        if bs_update = TRUE then
            S^rb := NULL, and bs_update := FALSE
            InitializePheromoneValues(T)
        else
            bs_update := TRUE
        end if
    end if
end while
output: S^bsf, the best solution found by the algorithm
```

ComputeConvergenceFactor($\mathcal{T}$): Just like the pheromone update, the computation of the convergence factor is a standard procedure that works in the same way for all MMAS algorithms in the hypercube framework:

$$cf := 2\left(\left(\frac{\sum_{\tau_i \in \mathcal{T}} \max\{\tau_{\max} - \tau_i, \tau_i - \tau_{\min}\}}{|\mathcal{T}| \cdot (\tau_{\max} - \tau_{\min})}\right) - 0.5\right)$$

Accordingly, the value of *cf* is zero in the case when all pheromone values are set to 0.5. The other extreme case is represented by all pheromone values having either value $\tau_{\min}$ or $\tau_{\max}$. In this case, *cf* evaluates to one. Otherwise, *cf* has a value between 0 and 1. Herewith the description of all components of the baseline algorithm is completed.

3.1.1. Solution Construction for the MDS Problem

In the following we say that, if a vertex c_i is added to a solution S under construction, then c_i covers itself and all its neighbors, that is, all $c_j \in N(c_i)$. Moreover, given a set $S \subset C$—that is, a solution under construction—we denote by $N(c_i \mid S) \subseteq N(c_i)$ the set of uncovered neighbors of $c_i \in C$. The solution construction mechanism is shown in Algorithm 2. It starts with an empty solution $S = \emptyset$. Then, at each step, exactly one of the vertices of those that do not yet form part of S or that—with respect to S—have uncovered neighbors ($\overline{C}$) is chosen in function $\mathsf{ChooseFrom}(\overline{C})$ and added to S. The choice of a vertex in $\mathsf{ChooseFrom}(\overline{C})$ is done as follows. First, a probability $\mathbf{p}(c_i)$ is calculated for each $c_i \in \overline{C}$:

$$\mathbf{p}(c_i) := \frac{\eta_i \cdot \tau_i}{\sum_{c_k \in \overline{C}} \eta_k \cdot \tau_k} \tag{9}$$

Hereby, $\eta_i := |\overline{C}| + 1$ is the greedy information that we used. Then, a random number $r \in [0,1]$ is drawn. If $r \leq d_{\text{rate}}$, c_j (to be added to S) is selected such that $\mathbf{p}(c_j) \geq \mathbf{p}(c_i)$ for all $c_i \in \overline{C}$. Otherwise, c_j is chosen by roulette-wheel-selection based on the calculated probabilities. Note that d_{rate} is an important parameter of the algorithm.

Algorithm 2 MDS solution construction

input: a graph $G = (C, E)$
$S := \emptyset$
$\overline{C} := \{c_i \in C \mid c_i \notin S \text{ or } N(c_i \mid S) \neq \emptyset\}$
while $\overline{C} \neq \emptyset$ **do**
  $c_j := \mathsf{ChooseFrom}(\overline{C})$
  $S := S \cup \{c_j\}$
  $\overline{C} := \{c_i \in C \mid c_i \notin S \text{ or } N(c_i \mid S) \neq \emptyset\}$
end while
output: a valid solution S

3.1.2. Solution Construction for the MDKP

As in the MDS-case, the solution construction starts with an empty solution $S := \emptyset$, and at each construction step exactly one item c_j is selected from a set $\overline{C} \subseteq C$. The definition of $\overline{C}$ in the case of the MDKP is as follows. An item $c_k \in C$ forms part of $\overline{C}$ if and only if (1) $c_k \notin S$, and (2) $S \cup \{c_k\}$ is a valid solution. The probability $\mathbf{p}(c_i)$ for an item $c_i \in \overline{C}$ to be chosen at the current construction step is the same as in Equation (9), just that the definition of the greedy information changes. In particular, η_i is defined as follows:

$$\eta_i := \frac{p_i}{\sum_{k=1}^{m} r_{i,k}/\text{cap}_k} \qquad \forall\, c_i \in \overline{C}. \tag{10}$$

These greedy values are often called utility ratios in the related literature. Given the probabilities, the choice of an item $c_j \in \overline{C}$ is done exactly in the same way as outlined above in the case of the MDS problem.

4. Adding Negative Learning to MMAS

In the following we first describe our own proposal for adding negative learning to ACO. Subsequently, our implementations of some existing approaches from the literature are outlined.

4.1. Our Proposal

As mentioned in the introduction, for each negative learning mechanism there are two fundamental questions to be answered: (1) how is the negative information generated, maintained and updated, and (2) how is this information being used.

4.1.1. Information Maintenance

We maintain the information derived from negative learning by means of a second pheromone model $\mathcal{T}^{\text{neg}}$, which consists of a pheromone value τ_i^{neg} for each item $c_i \in C$. We henceforth refer to these values as the *negative pheromone values*. Whenever the pheromone values are (re-)initialized, the negative pheromone values are set to $\tau_{\min}$, which is in contrast to the standard pheromone values, which are set to 0.5 (see above).

4.1.2. Information Generation and Update

The generation of the information for negative learning is done by two new instructions, which are introduced between lines 9 and 10 of the baseline MMAS algorithm (Algorithm 1):

$$S^{\text{sub}} := \mathsf{SolveSubinstance}(\mathcal{S}^{\text{iter}}, cf) \quad (11)$$

$$\mathcal{S}^{\text{iter}} := \mathcal{S}^{\text{iter}} \cup \{S^{\text{sub}}\} \quad (12)$$

Function $\mathsf{SolveSubinstance}(\mathcal{S}^{\text{iter}}, cf)$ merges all solutions from $\mathcal{S}^{\text{iter}}$, resulting in a subset $C' \subseteq C$. Then an optimization algorithm is applied to find the best-possible solution that only consists of items from C'. In this work we have experimented with two options:

1. **Option 1:** Application of the ILP solver CPLEX 12.10. In the case of the MDS problem, the ILP model from Section 2.1 is used after adding an additional constraint $x_i = 0$ for all $c_i \in C \setminus C'$. In the case of the MDKP, we use the ILP model from Section 2.2 after replacing all occurrences of C with C'.
2. **Option 2:** Application of the baseline MMAS algorithm (Algorithm 1). In the case of both the MDS and the MDKP problem, this application of the baseline MMAS only considers items from C' for the construction of solutions. Moreover, this MMAS application uses its own pheromone values, parameter settings, etc. Finally, the best-so-far solution of this (inner) ACO is initialized with S^{ib}.

In both options, solution S^{sub}—which is returned by $\mathsf{SolveSubinstance}(\mathcal{S}^{\text{iter}}, cf)$—is the best solution between S^{ib} and the best solution found by the optimization algorithm (CPLEX, respectively baseline MMAS) in the allotted computation time. This computation time is calculated on the basis of a maximum computation time (t^{sub} CPU seconds) and the current value of the convergence factor, which is passed to function $\mathsf{SolveSubinstance}(\mathcal{S}^{\text{iter}}, cf)$ as a parameter. In particular, the allowed computation time (in seconds) is $(1 - cf)t^{\text{sub}} + 0.1cf$. This means that the available computation time for solving the sub-instance C' decreases with an increasing convergence factor value. The rationale behind this setting is that, when the convergence factor is low, the variance between solutions in $\mathcal{S}^{\text{iter}}$ is rather high and C' is therefore rather large, which means that more time is necessary to explore sub-instance C'.

The last action in function $\mathsf{SolveSubinstance}(\mathcal{S}^{\text{iter}}, cf)$ is the update of the negative pheromone values based on solution S^{sub}. This update only concerns the negative pheromone values of those components that form part of C'. The update formula is as follows:

$$\tau_i^{\text{neg}} := \tau_i^{\text{neg}} + \rho^{\text{neg}} \cdot (\xi_i^{\text{neg}} - \tau_i^{\text{neg}}) \quad , \quad (13)$$

where ρ^{neg} is the *negative learning rate* and $\xi_i^{\text{neg}} = 1$ if $c_i \notin S^{\text{sub}}$, resp. $\xi_i^{\text{neg}} = 0$ otherwise. In other words, the negative pheromone value of those components that do not form part of S^{sub} is increased.

4.1.3. Information Use

The negative pheromone values are used in the context of the construction of solutions. In particular, Equation (9) is replaced by the following one:

$$\mathbf{p}(c_i) := \frac{\eta_i \cdot \tau_i \cdot (1 - \tau_i^{\text{neg}})}{\sum_{c_k \in \overline{C}} \eta_k \cdot \tau_k \cdot (1 - \tau_k^{\text{neg}})} \quad (14)$$

In this way, those items that have accumulated a rather high negative pheromone value (because they have not appeared in the solutions derived by CPLEX, respectively the (inner) MMAS algorithm, to the sub-instances of previous iterations) have a decreased probability to be chosen for solutions in the current iteration. Note that a very similiar formula was used already in [27].

4.2. *Proposals from the Literature*

As mentioned before, the proposals from the literature were introduced in the context of several different ACO versions. In order to ensure a fair comparison, we reimplemented those proposals that we chose for comparison in the context of the baseline MMAS algorithm. In particular, we implemented four different approaches, which all share the following common feature. In addition to the iteration-best solution (S^{ib}), the restart-best solution (S^{rb}) and the best-so-far solution (S^{bsf}), these extensions of the baseline MMAS algorithm maintain the *iteration-worst* solution (S^{iw}), the *restart-worst* solution (S^{rw}) and the *worst-so-far* solution (S^{wsf}). As in the case of S^{rb} and S^{bsf}, solutions S^{rw} and S^{wsf} are initialized to NULL at the start of the algorithm. Then, the following three lines are introduced after line 12 of Algorithm 1:

$$S^{iw} := \text{worst solution from } \mathcal{S}^{\text{iter}}$$
$$\textbf{if } S^{iw} \text{ worse than } S^{rw} \textbf{ then } S^{rw} := S^{iw}$$
$$\textbf{if } S^{iw} \text{ worse than } S^{wsf} \textbf{ then } S^{wsf} := S^{iw}$$

The way in which these three additional solutions are used differs among the four implemented approaches.

4.2.1. Subtractive Anti-Pheromone

This idea is adopted from [21], but has already been used in similar form in [19,20]. Our implementation of this idea is as follows. After the standard pheromone update of the baseline MMAS algorithm (see line 13 of Algorithm 1), the following is done. First, a set B is generated by joining the items in solutions S^{iw}, S^{rw} and S^{wsf}, that is, $B := S^{iw} \cup S^{rw} \cup S^{wsf}$. Then, all those items in which the pheromone value receives an update from at least one of the solutions S^{ib}, S^{rb}, or S^{bsf} in the current iteration are removed from B. That is:

$$\textbf{if } \kappa_{ib} > 0 \textbf{ then } B := B \setminus S^{ib}$$
$$\textbf{if } \kappa_{rb} > 0 \textbf{ then } B := B \setminus S^{rb}$$
$$\textbf{if } \kappa_{bsf} > 0 \textbf{ then } B := B \setminus S^{bsf}$$

Afterward, the following additional update is applied:

$$\tau_i := \gamma \cdot \tau_i \quad \forall\, o_i \in B \tag{15}$$

In other words, the pheromone values of all those components that appear in "bad" solutions, but who do not form part of "good" solutions, are subject to a pheromone value decrease depending on the *reduction rate* γ. Finally, note that the solution construction procedure in this variant—which is henceforth labeled ACO-SAP—is exactly the same as in the baseline MMAS algorithm.

4.2.2. Explorer Ants

The explorer ants approach from [21]—henceforth labeled ACO-EA—is very similar to the previously presented ACO-SAP approach. The only difference is in the construction of solutions. This approach has an additional paramete: $p_{\text{exp}_\text{a}} \in [0, 1]$, the proportion of explorer ants. Given the number of ants (n_a) and p_{exp_a}, the number of explorer ants n_a^{exp} is calculated as follows:

$$n_a^{\text{exp}} := \max\{1, \lfloor p_{\text{exp}_\text{a}} \cdot n_a \rfloor\} \tag{16}$$

At each iteration, $n_a - n_a^{\exp}$ solution constructions are performed in the same way as in the baseline MMAS algorithm. The remaining $n_a^{\exp}$ solution constructions make use of the following formula (instead of Equation (9) for calculating the probabilities:

$$\mathbf{p}(c_i) := \frac{\eta_i \cdot (1 - \tau_i)}{\sum_{c_k \in \overline{C}} \eta_k \cdot (1 - \tau_k)} \tag{17}$$

In other words, explorer ants make use of the opposite of the pheromone values for constructing solutions.

4.2.3. Preferential Anti-Pheromone

Like our own negative learning proposal, the preferential anti-pheromone approach from [21] makes use of an additional set $\mathcal{T}^{\text{neg}}$ of pheromone values. Remember that $\mathcal{T}^{\text{neg}}$ contains a pheromone value τ_i^{neg} for each item $c_i \in C$. These negative pheromone values are initilized at the start of the algorithm as well as when the algorithm is restarted, to a value of 0.5. Moreover, after the update of the standard pheromone values in line 13 of the baseline MMAS algorithm, exactly the same update is conducted for the negative pheromone values:

$$\tau_i^{\text{neg}} := \tau_i^{\text{neg}} + \rho^{\text{neg}} \cdot (\xi_i^{\text{neg}} - \tau_i^{\text{neg}}) \quad , \tag{18}$$

where

$$\xi_i^{\text{neg}} := \kappa_{ib} \cdot \Delta(S^{iw}, c_i) + \kappa_{rb} \cdot \Delta(S^{rw}, c_i) + \kappa_{bsf} \cdot \Delta(S^{wsf}, c_i) \tag{19}$$

Hereby, $\rho^{\text{neg}} \in [0,1]$ is the negative learning rate, and function $\Delta(S, c_i)$ evaluates to 1 if and only if item c_i forms part of solution S. Moreover, values κ_{ib}, κ_{rb} and κ_{bsf} are the same as the ones used for the udpate of the standard pheromone values. This means that the learning of the negative pheromone values is dependent on the dynamics of the learning of the standard pheromone values.

The standard pheromone values and the negative pheromone values are used as follows for the construction of solutions. The probabilities for the a-th solution construction—where $a = 1, \ldots, n_a$—are determined as follows:

$$\mathbf{p}(c_i) := \frac{\eta_i \cdot (\lambda \tau_i + (1 - \lambda)\tau_i^{\text{neg}}))}{\sum_{c_k \in \overline{C}} \eta_k \cdot (\lambda \tau_k + (1 - \lambda)\tau_k^{\text{neg}}))} \quad , \tag{20}$$

where $\lambda := \frac{a-1}{n_a - 1}$. This means that $\lambda = 0$ for the first solution construction, which means that only the negative pheromones values are used. In the other extreme, it holds that $\lambda = 1$ for the n_a-th solution construction, that is, only the standard pheromone values are used. All other solution constructions combine both pheromone types at different rates. Note that this preferential anti-pheromone approach is henceforth labeled ACO-PAP.

4.2.4. Second-Order Swarm Intelligence

Our implementation of the second-order swarm intelligence approach from [26] works exactly like the ACO-PAP approach from the previous section for what concerns the definition and the update of the negative pheromone values. However, the way in which they are used is different. The item probabilities for the construction of solutions is calculated by the following formula:

$$\mathbf{p}(c_i) := \frac{\eta_i \cdot (\tau_i)^{\alpha} \cdot (\tau_i^{\text{neg}})^{(\alpha - 1)}}{\sum_{c_k \in \overline{C}} \eta_k \cdot (\tau_k)^{\alpha} \cdot (\tau_k^{\text{neg}})^{(\alpha - 1)}} \quad , \tag{21}$$

where $\alpha \in [0,1]$ is a parameter of the algorithm. Note that this approach is henceforth labeled ACO^{2o}.

4.3. Summary of the Tested Algorithms

In addition to the baseline MMAS algorithm (henceforth simply labeled ACO), and the four approaches from the literature (ACO-SAP, ACO-EA, ACO-PAP and ACO^{2o}) we test the following six versions of the negative learning mechanism proposed in this paper:

1. ACO-CPL$^{+}_{neg}$: The mechanism described in Section 4.1 using option 1 (CPLEX) for solving sub-instances.
2. ACO-CPL$_{neg}$: This algorithm is the same as ACO-CPL$^{+}_{neg}$, with the exception that Equation (12) is not performed. This means that the algorithm does make use of solution S^{sub} for additional positive learning. Studying this variant will show if, by solely adding negative learning, the algorithm improves over the baseline ACO.
3. ACO-CPL^{+}: This algorithm is the same as ACO-CPL$^{+}_{neg}$, apart from the fact that the update of the negative pheromone values is not performed. In this way, the algorithm only makes use of the additional positive learning mechanism obtained by adding solution S^{sub} to $\mathcal{S}^{iter}$.

The remaining three algorithm variants are ACO-ACO$^{+}_{neg}$, ACO-ACO$_{neg}$ and ACO-ACO^{+}. These algorithm variants are the same ones as ACO-CPL$^{+}_{neg}$, ACO-CPL$_{neg}$ and ACO-CPL^{+}, except that they make use of option 2 (baseline ACO algorithm) for solving the corresponding sub-instances at each iteration.

A summary of the parameters that arise in these 11 algorithms is provided in Table 2, together with a description of their function and the parameter value domains that were used for parameter tuning (which will be described in Section 5.2). Moreover, an overview on the parameters that are involved in each of the 11 algorithms is provided in Table 3.

Table 2. Summary of the parameters that arise in the considered algorithms, together with their description and the domains considered for parameter tuning.

Parameter	Description	Considered Domain
n_a	Number of solution constructions per iteration	$\{3, 5, 10, 20\}$
ρ	Learning rate	$\{0.1, 0.2, \ldots, 0.4, 0.5\}$
d_{rate}	Determinism rate for solution construction	$\{0.0, 0.1, \ldots, 0.8, 0.9\}$
ρ^{neg}	Negative learning rate	$\{0.1, 0.2, \ldots, 0.4, 0.5\}$
γ	Reduction rate for negative pheromone values	$\{0.1, 0.2, \ldots, 0.8, 0.9\}$
p_{exp_a}	Proportion of explorer ants	$\{0.1, 0.2, \ldots, 0.4, 0.5\}$
α	Exponent for the pheromone values	$\{0.01, \ldots, 0.99\}$
t^{sub}	Maximum computation time (seconds) for sub-instance solving	$\{1, 2, \ldots, 9, 10\}$
n_u^{sub}	Number of solution constructions in the inner application of the baseline ACO algorithm (option 2 for solving sub-instances)	$\{3, 5, 10, 20\}$
ρ^{sub}	Learning rate in the inner application of the baseline ACO algorithm (option 2 for solving sub-instances)	$\{0.1, 0.2, \ldots, 0.4, 0.5\}$
$d_{\text{rate}}^{\text{sub}}$	Determinism rate for solution construction in the inner application of the baseline ACO algorithm (option 2 for solving sub-instances)	$\{0.0, 0.1, \ldots, 0.8, 0.9\}$

Table 3. Summary of the parameters that arise in each algorithm.

Parameter	ACO	ACO-CPL$^{+}_{\text{neg}}$	ACO-CPL$_{\text{neg}}$	ACO-CPL^{+}	ACO-ACO$^{+}_{\text{neg}}$	ACO-ACO$_{\text{neg}}$	ACO-ACO^{+}	ACO-SAP	ACO-EA	ACO-PAP	ACO^{2o}
n_a	✓	✓	✓	✓	✓	✓	✓	✓	✓	✓	✓
ρ	✓	✓	✓	✓	✓	✓	✓	✓	✓	✓	✓
d_{rate}	✓	✓	✓	✓	✓	✓	✓	✓	✓	✓	✓
ρ^{neg}		✓	✓		✓	✓				✓	✓
γ								✓	✓		
p_{exp_a}									✓		
α											✓
t^{sub}		✓	✓	✓	✓	✓	✓				
n_a^{sub}					✓	✓	✓				
ρ^{sub}					✓	✓	✓				
$d_{\text{rate}}^{\text{sub}}$					✓	✓	✓				

5. Experimental Evaluation

The experiments concerning the MDS problem were performed on a cluster of machines with two Intel® Xeon® Silver 4210 CPUs with 10 cores of 2.20 GHz and 92 Gbytes of RAM. The MDKP experiments were conducted on a cluster of machines with Intel® Xeon® CPU 5670 CPUs with 12 cores (2.933 GHz) and at least 32 GB RAM. For solving the sub-instances in ACO-CPL$^{+}_{\text{neg}}$, ACO-CPL$_{\text{neg}}$ and ACO-CPL^{+} we used CPLEX 12.10 in one-threaded mode.

5.1. Problem Instances

Concerning the MDS problem, we generated a benchmark instance set with instances of different sizes (number of vertices $n \in \{5000, 10{,}000\}$), different densities (percentage of all possible edges $d \in \{0.1, 0.5, 1.0, 5.0\}$) and different graph types (random graphs and random geometric graphs). For each combination of n, d and graph type, 10 random instances were generated. This makes a total of 160 problem instances.

In the case of the MDKP we used a benchmark set of 90 problem instances with 500 items from the OR-Library (http://people.brunel.ac.uk/~mastjjb/jeb/info.html, accessed on 20 January 2021). This set consists of 30 instances with 5, 10, and 30 resources. Moreover, each of these three subsets contains 10 instances with resource tightness 0.25, 0.5, and 0.75. Roughly, the higher the value of the resource tightness, the more items can be placed in the knapsack. These 90 problem instances are generally known to be the most difficult ones available in the literature for heuristic solvers.

5.2. Algorithm Tuning

The scientific parameter tuning tool irace [47] was used for the purpose of parameter tuning. In particular we produced for each of the 11 algorithms (resp., algorithm versions) exactly one parameter value set for each problem (MDS problem vs. MDKP). For the purpose of tuning the algorithms for the MDS problem, we additionally generated for each combination of n, d (density), and graph type exactly one random instance. In other words, 16 problem instances were used for tuning, and the tuner was given a maximal budget of 2000 algorithm applications. In the context of tuning the algorithms for the MDKP, we randomly selected one of the 10 problem instances for each combination of "the number of resources" (5, 10, 30) and the instance tightness (0.25, 0.5, 0.75). Consequently, nine problem instances were used for tuning in the case of the MDKP. Remember that the parameter value domains considered for tuning are provided in Table 2. The parameter values that were determined by irace for the 11 algorithms and for the two problems are provided in Tables 4 (MDS problem) and 5 (MDKP).

Table 4. Parameter values for all algorithms for solving the minimum dominating set (MDS) problem.

Parameter	Algorithms										
	ACO	ACO-CPL$^{+}_{neg}$	ACO-CPL$_{neg}$	ACO-CPL^{+}	ACO-ACO$^{+}_{neg}$	ACO-ACO$_{neg}$	ACO-ACO^{+}	ACO-SAP	ACO-EA	ACO-PAP	ACO^{2o}
n_a	3	20	20	20	3	10	10	20	10	3	3
ρ	0.4	0.1	0.1	0.5	0.1	0.2	0.5	0.4	0.5	0.1	0.2
d_{rate}	0.9	0.9	0.8	0.9	0.9	0.9	0.9	0.9	0.9	0.9	0.9
ρ^{neg}	--	0.4	0.4	--	0.2	0.5	--	--	--	0.2	0.2
γ	--	--	--	--	--	--	--	0.6	0.6	--	--
p_{exp_a}	--	--	--	--	--	--	--	--	0.1	--	--
α	--	--	--	--	--	--	--	--	--	--	0.96
t^{sub}	--	8	7	6	6	7	8	--	--	--	--
n_a^{sub}	--	--	--	--	3	3	3	--	--	--	--
ρ^{sub}	--	--	--	--	0.3	0.5	0.4	--	--	--	--
d_{rate}^{sub}	--	--	--	--	0.7	0.7	0.5	--	--	--	--

Table 5. Parameter values for all algorithms for solving the multi dimensional knapsack problem (MDKP).

Parameter	Algorithms										
	ACO	ACO-CPL$^{+}_{neg}$	ACO-CPL$_{neg}$	ACO-CPL^{+}	ACO-ACO$^{+}_{neg}$	ACO-ACO$_{neg}$	ACO-ACO^{+}	ACO-SAP	ACO-EA	ACO-PAP	ACO^{2o}
n_a	20	10	20	20	10	10	20	20	20	3	10
ρ	0.3	0.4	0.1	0.4	0.1	0.1	0.3	0.1	0.2	0.1	0.3
d_{rate}	0.7	0.1	0.7	0.4	0.6	0.7	0.8	0.8	0.8	0.8	0.9
ρ^{neg}	--	0.2	0.5	--	0.4	0.5	--	--	--	0.1	0.2
γ	--	--	--	--	--	--	--	0.9	0.7	--	--
p_{exp_a}	--	--	--	--	--	--	--	--	0.3	--	--
α	--	--	--	--	--	--	--	--	--	--	0.95
t^{sub}	--	7	3	5	3	9	3	--	--	--	--
n_a^{sub}	--	--	--	--	5	10	10	--	--	--	--
ρ^{sub}	--	--	--	--	0.3	0.2	0.4	--	--	--	--
d_{rate}^{sub}	--	--	--	--	0.7	0.7	0.7	--	--	--	--

5.3. Results

Using the previously determined parameter values, each of the 11 considered algorithms was applied 30 times—that is, with 30 different random seeds—to each of the 160 MDS problem instances. Hereby, 500 CPU seconds were chosen as a time limit for the graphs with 5000 nodes, whereas 1000 CPU seconds were chosen as a time limit for each run concerning the graphs with 10,000 nodes. Moreover, each algorithm was applied 100 times to each of the 90 MDKP instances. This was done with a time limit of 500 s per run. Note that, in this way, the same computational resources were given to all 11 algorithms in the context of both tackled problems. The choice of 100 runs per instance in the case of the MDKP was done in order to produce results that are comparable to the best existing approaches from the literature, which were also applied 100 times to each problem instance.

Due to space restrictions we present a comparative analysis of the 11 algorithms in terms of *critical difference* (CD) plots [48] and so-called *heatmaps*. In order to produce the average ranks of all algorithms—both for the whole set of problem instances (per problem) as well as for instance subsets—the Friedman test was applied for the purpose of comparing the 11 approaches simultaneously. In this way we also obtained the rejection of the hypothesis that the 11 techniques perform equally. Subsequently, all pairwise algorithm comparisons were performed using the Nemenyi post-hoc test [49]. The obtained results are shown graphically (CD plots and heatmaps). The CD plots show the average algorithm ranks (horizontal axis) with respect to the considered (sub-)set of instances. In those cases in which the performances of two algorithms are below the critical difference threshold—based on a significance level of 0.05—the two algorithms are considered as statistically equivalent. This is indicated by bold horizontal bars joining the markers of the respective algorithm variants.

5.3.1. Results for the MDS Problem

Figure 1a shows the CD plot for the whole set of 160 MDS instances, while Figure 1b,c present more fine-grained results concerning random graphs (RGs) and random geometric graphs (RGGs), respectively. Furthermore, the heatmaps in Figure 2 show the average ranks of the 11 algorithms in an even more fine-grained way. The graphic shows exactly one heatmap for each algorithm. The ones of algorithms ACO-CPL$^{+}_{\text{neg}}$, ACO-CPL$_{\text{neg}}$ and ACO-CPL^{+} are shown in Figure 2a, the ones of algorithms ACO-ACO$^{+}_{\text{neg}}$, ACO-ACO$_{\text{neg}}$ and ACO-ACO^{+} in Figure 2b, and the ones of the remaining five algorithms in Figure 2c. The upper part of each heatmap shows the results for RGs, while the lower part concerns the results for RGGs. Each of these parts has two columns: the first one contains the results for the graphs with 5000 nodes, and the second one for the ones with 10,000 nodes. Moreover, each part has four rows, showing the results for the four considered graph densities. In general, the more yellow the cell of a heatmap, the better is the relative performance of the corresponding algorithm for the respective combination of features (graph type, graph size, and density).

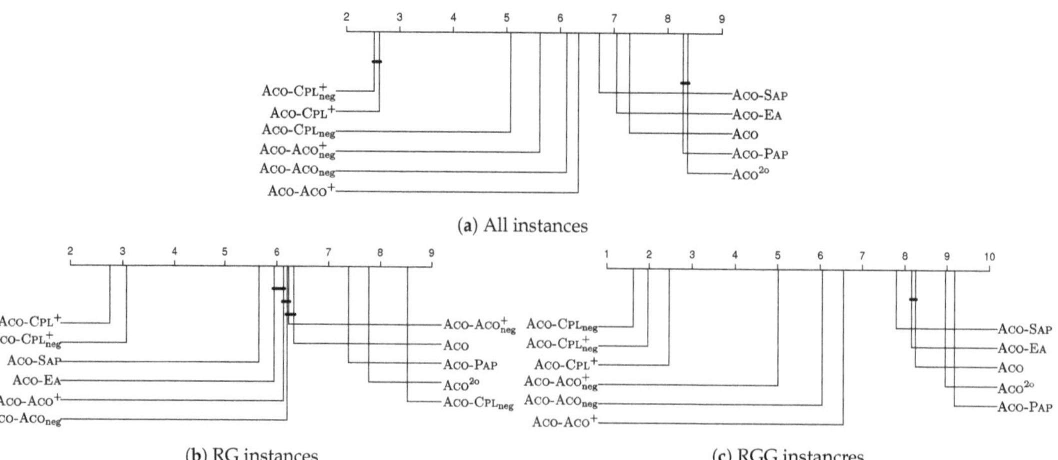

Figure 1. Criticial difference plots concerning the results for the MDS problem.

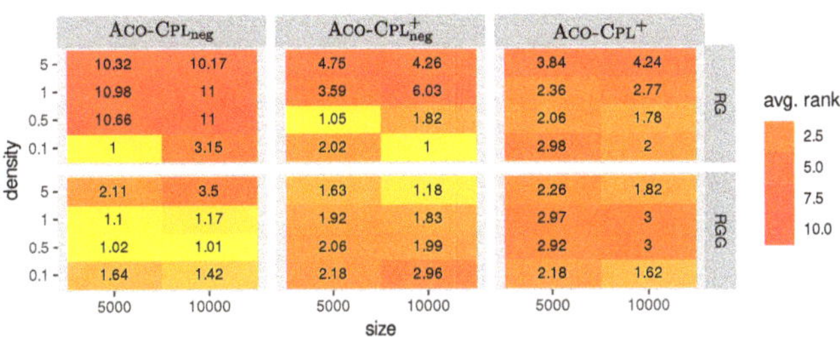

(**a**) Algorithms ACO-CPL$^+_{neg}$, ACO-CPL$_{neg}$, and ACO-CPL$^+$

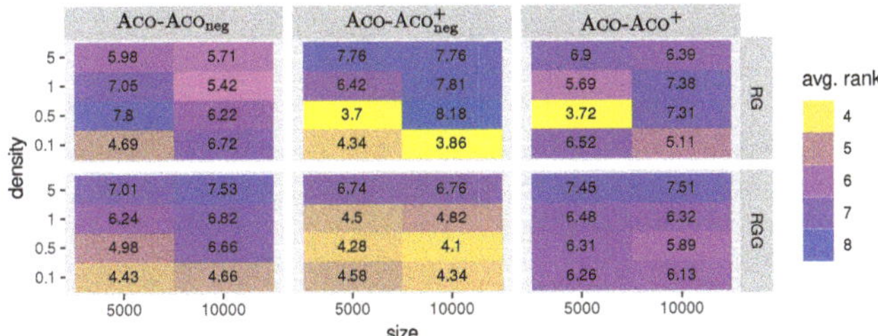

(**b**) Algorithms ACO-ACO$^+_{neg}$, ACO-ACO$_{neg}$, and ACO-ACO$^+$

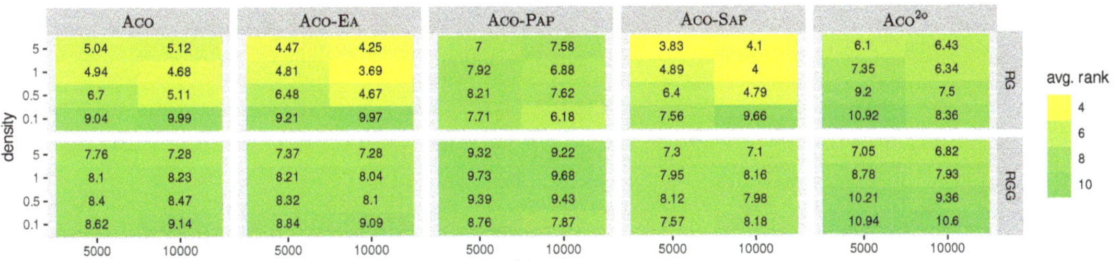

(**c**) Algorithms ACO, ACO-EA, ACO-PAP, ACO-PAP, and ACO^{2o}

Figure 2. Heatmaps concerning the results for the MDS problem.

The global CD plot from Figure 1a allows to make the following observations:

- All the six algorithm variants proposed in this paper significantly improve over the remaining five algorithm variants, that is, over the baseline MMAS (ACO) and over the four considered negative learning variants from the literature.
- The three algorithm variants that make use of CPLEX for generating the negative feedback (option 1) outperform the other three variants (making use of option 2) with statistical significance. This shows the importance of the way in which the negative feedback is generated. In fact, the more accurate the negative feedback, the better the global performance of the algorithm.
- Concerning the four negative learning mechanisms from the literature, it is shown that only ACO-SAP and ACO-EA are able to outperform the baseline MMAS algorithm. In contrast, ACO-PAP and ACO^{2o} perform significantly worse than the baseline MMAS algorithm.
- When comparing variants ACO-CPL$^+_{neg}$ and ACO-CPL$_{neg}$ with ACO-CPL$^+$, it can be observed that ACO-CPL$^+_{neg}$ has only a slight advantage over ACO-CPL$^+$ (which is not statistically significant). This means that, even though negative learning is useful, the additional positive feedback obtained by making use of solution S^{sub} for updating solutions S^{ib} and S^{rb} is very powerful.

- The comparison of the three algorithms making use of option 2 (ACO-ACO$^{+}_{\text{neg}}$, ACO-ACO$_{\text{neg}}$ and ACO-ACO^{+}) shows a significant difference to the comparison concerning the three algorithms using option 1: the two versions that make use of negative learning (ACO-ACO$^{+}_{\text{neg}}$ and ACO-ACO$_{\text{neg}}$) outperform the version without negative learning (ACO-ACO^{+}) with statistical significance. This can probably be explained by the lower quality of the positive feedback information, as solutions S^{sub} can be expected to be generally worse than solutions S^{sub} of the algorithm version using option 1.

When looking at the results in a more fine-grained way, the following can be observed:

- Interestingly, the graph type seems to have a big influence on the relative behavior of the algorithms. In the case of RGs, for example, ACO-CPL^{+} is the clear winner of the comparison with ACO-CPL$^{+}_{\text{neg}}$ in second place. However, the really interesting aspect is that ACO-CPL$_{\text{neg}}$ finishes last with statistical significance. This means that negative learning seems even to be harmful in the case of RGs. On the contrary, ACO-CPL$_{\text{neg}}$ is the clear winner of the competition in the context of RGGs, with ACO-CPL$^{+}_{\text{neg}}$ finishing in second place (with statistical significance), and ACO-CPL^{+} only in third place. This means that, in the case of RGGs, negative learning is much more important than the additional positive feedback provided by solution S^{sub}, which even seems harmful.
- Another interesting aspect is that, in the context of RGs, two negative learning versions from the literature (ACO-SAP and ACO-EA) clearly outperform our proposed negative learning variants using option 2.
- The heatmaps from Figure 2 also indicate some interesting tendencies. Negative learning in the context of our algorithm variants ACO-CPL$^{+}_{\text{neg}}$, ACO-CPL$_{\text{neg}}$, ACO-ACO$^{+}_{\text{neg}}$ and ACO-ACO$_{\text{neg}}$ seems to gain importance with an increasing sparsity of the graphs. On the other side, in the context of RGs, it is clearly shown that the relative quality of ACO-SAP and ACO-EA grows with increasing graph size (number of vertices) and with increasing density.

5.3.2. Results for the MDKP

Figure 3a shows the CD plot for the whole set of 90 MDKP instances, while Figure 3b–g present more fine-grained results concerning instances with different numbers of resources and with a varying instance tightness. Again, the heatmaps in Figure 4 complement this more fine-grained presentation of the results. The 11 algorithms are distributed in the same way as described in the context of the MDS problem into three heatmap graphics. Each heatmap (out of 11 heatmaps in total) has three rows: one for each number of resources (5, 10, 30). Moreover, each heatmap has three columns: one for each considered instance tightness (0.25, 0.5, 0.75). Interestingly, from a global point of view (Figure 3a) the relative difference between the algorithm performances is very similar to the one observed for the MDS problem. In particular, our negative learning variants using option 1 perform best. Again, ACO-CPL$^{+}_{\text{neg}}$ has a slight advantage over ACO-CPL^{+}, which is—like in the case of the MDS problem—not statistically significant. Basically there is only one major difference to the results for the MDS problem: ACO-SAP, one of the negative learning variants from the literature, outperforms ACO-ACO$_{\text{neg}}$ and ACO-ACO^{+}.

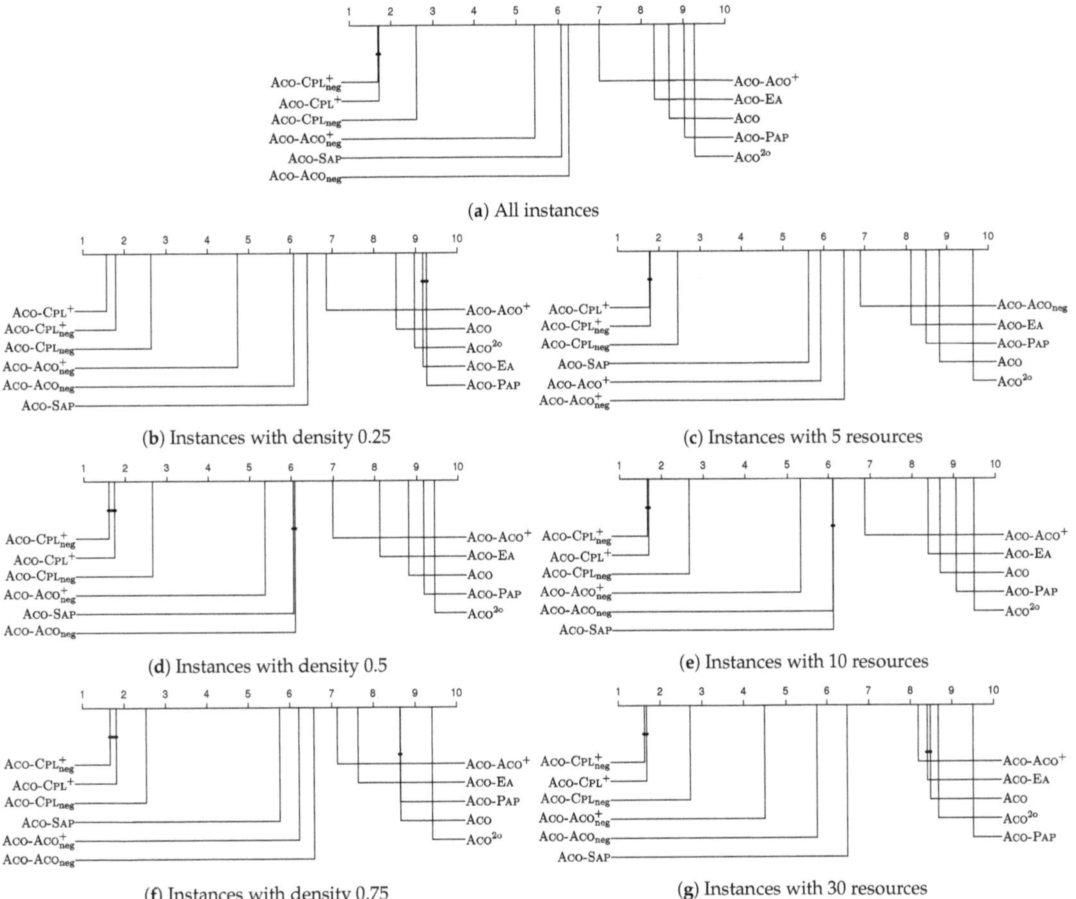

(**a**) All instances

(**b**) Instances with density 0.25

(**c**) Instances with 5 resources

(**d**) Instances with density 0.5

(**e**) Instances with 10 resources

(**f**) Instances with density 0.75

(**g**) Instances with 30 resources

Figure 3. Criticial difference plots for more fine-grained subdivisions of instances concerning the results for the MDKP problem.

When studying the results in a more fine-grained way, the following observations can be made:

- The negative learning component of our algorithm proposal seems to gain importance with a growing number of resources. This can especially be observed for algorithm variants ACO-CPL$^{+}_{neg}$, ACO-ACO$^{+}_{neg}$ and ACO-ACO$_{neg}$. However, there is an interesting difference between ACO-CPL$^{+}_{neg}$ and ACO-ACO$^{+}_{neg}$: while ACO-CPL$^{+}_{neg}$ improves with an increasing instance tightness, the opposite is the case for ACO-ACO$^{+}_{neg}$.
- Again, as in the case of the MDS problem, the relative performance of ACO-SAP, the best one of the negative learning variants chosen from the literature, is contrary to the relative performance of ACO-ACO$^{+}_{neg}$. In other words, the relative performance of ACO-SAP improves with a decreasing number of resources and with an increasing instance tightness.

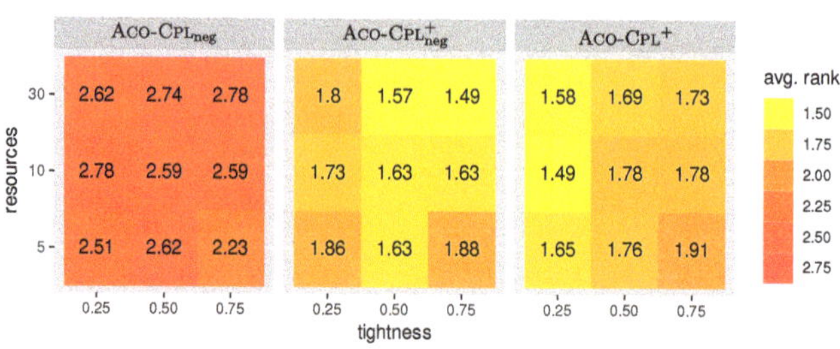

(**a**) Algorithms ACO-CPL$^{+}_{\text{neg}}$, ACO-CPL$_{\text{neg}}$, and ACO-CPL^{+}

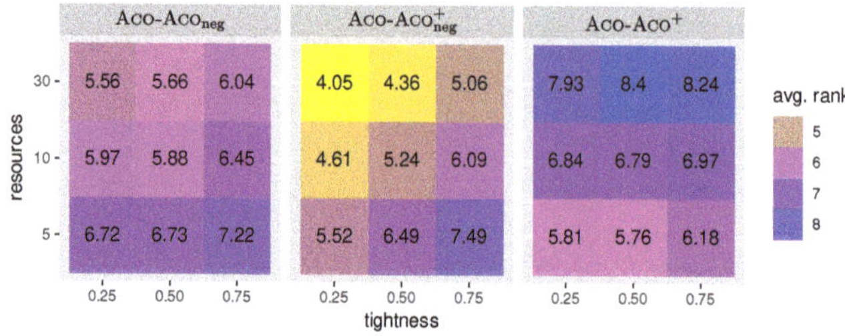

(**b**) Algorithms ACO-ACO$^{+}_{\text{neg}}$, ACO-ACO$_{\text{neg}}$, and ACO-ACO^{+}

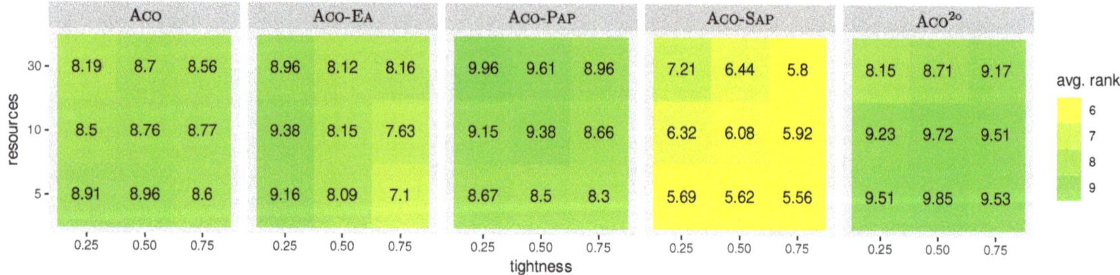

(**c**) Algorithms ACO, ACO-EA, ACO-PAP, ACO-PAP, and ACO^{2o}

Figure 4. Heatmaps concerning the results for the MDKP.

5.3.3. Comparison to the State-of-the-Art

Even though the objective of this study is not to outperform current state-of-the-art algorithms for the chosen problems, we are certainly interested to know how our globally best algorithm (ACO-CPL$^{+}_{\text{neg}}$) performs in comparison to the state-of-the-art.

In the case of the MDS problem we chose for this purpose one of the classical benchmark sets, which was also used in one of the latest published works [37]. This benchmark set is labeled UDG and consists of 120 graphs with numbers of vertices between 50 and 1000. For each of the six graph sizes, UDG contains graphs of two different densities. The benchmark set consists of 10 graphs per combination of graph size and graph density. Following the procedure from [37], we applied ACO-CPL$^{+}_{\text{neg}}$ 10 times with a time limit of 1000 CPU seconds for each application to each of the 120 instances of set UDG. Note that we did not specifically tune the parameters of ACO-CPL$^{+}_{\text{neg}}$. Instead, the same parameter values as in the previous section were used. The results are shown in a summarized way—as in [37]—in Table 6. In particular, each table row presents the results for the 10 instances of the respective instance family. For each of the six compared algorithms, the provided number is the average over the best solutions found for each of the 10 instances within 10 runs per instance. The best result per table row is indicated in bold face. Surprisingly, it can be observed that ACO-CPL$^+_{neg}$ matches the performance of the best two approaches. It is also worth mentioning that the five competitors of ACO-CPL$^+_{neg}$ in this table were all published since 2017 and are all based on local search. In particular, algorithm RLS$_o$ [50] was shown to outperform all existing ACO and hyper-heuristic algorithms, which were the state-of-the-art before this recent start of focused research efforts on sophisticated local search algorithms. Concerning computation time, in [37] it is stated that CC^{2}FS requires on average 0.21 s, FastMWDS requires 0.83 s, and FastDS requires 22.19 s to obtain the best solutions of each run. ACO-CPL$^+_{neg}$ is somewhat slower by requiring on average 36.14 s.

In the context of the MDKP, we compare ACO-CPL$^+_{neg}$ to the current state-of-the-art algorithms: a sophisticated particle swarm optimization algorithm (DQPSO) from [44], published in 2020, and a powerful evolutionary algorithm (TPTEA) from [45], published in 2018. As these two algorithms—in their original papers—were applied to the 90 benchmark problems used in this work, it was not required to conduct additional experiments with ACO-CPL$^+_{neg}$. A summarized comparison of the three algorithms is provided in Table 7. Each row contains average results for the 10 problem instances for each combination of the number of resources (5, 10, 30) and the instance tightness (0.25, 0.5, 0.75). In particular, we show averages concerning the best solutions found (table columns 3–5), the average solution quality obtained (table columns 6–8), and the average computation times required (table columns 9–11). As in the case of the MDS problem, we were surprised to see that ACO-CPL$^+_{neg}$ can actually compete with current state-of-the-art algorithms. The state-of-the-art results were even improved by ACO-CPL$^+_{neg}$ in some cases, especially for what concerns medium instance tightness for 5 and 10 resources, and low instance tightness for 30 resources. Moreover, the computation time of ACO-CPL$^+_{neg}$ is much lower than that of TPTEA, and comparable to the one required by DQPSO.

Table 6. MDS problem: summarized comparison to the state-of-the-art. Competitor names are accompanied by publication year and the reference.

Instance Family	CC^{2}FS	FastMWDS	RLS$_o$	ScBppw	FastDS	ACO-CPL$^+_{neg}$
	2017 [51]	2018 [52]	2018 [50]	2019 [53]	2020 [37]	
V50U150	**12.9**	**12.9**	**12.9**	**12.9**	**12.9**	**12.9**
V50U200	**9.4**	**9.4**	**9.4**	**9.4**	**9.4**	**9.4**
V100U150	**17.0**	**17.0**	**17.0**	17.3	**17.0**	**17.0**
V100U200	**10.4**	**10.4**	**10.4**	10.6	**10.4**	**10.4**
V250U150	**18.0**	**18.0**	**18.0**	19.0	**18.0**	**18.0**
V250U200	**10.8**	**10.8**	**10.8**	11.5	**10.8**	**10.8**
V500U150	**18.5**	**18.5**	18.6	20.1	**18.5**	**18.5**
V500U200	**11.2**	**11.2**	**11.2**	12.4	**11.2**	**11.2**
V800U150	**19.0**	**19.0**	19.1	20.9	**19.0**	**19.0**
V800U200	**11.7**	**11.7**	11.9	12.6	11.8	**11.7**
V1000U150	**19.1**	**19.1**	19.2	21.3	**19.1**	**19.1**
V1000U200	**12.0**	**12.0**	**12.0**	13.0	**12.0**	**12.0**

Table 7. MDKP: summarized comparison to the state-of-the-art.

# Resources	Tightness	Best			Average			Average Time		
		TPTEA	DQPSO	ACO-CPL$^{+}_{neg}$	TPTEA	DQPSO	ACO-CPL$^{+}_{neg}$	TPTEA	DQPSO	ACO-CPL$^{+}_{neg}$
5	0.25	**120,629.2**	120,627.7	120,628.6	120,612.70	**120,619.81**	120,611.94	3228.31	**117.53**	208.31
	0.5	219,511.6	219,511.9	**219,512.7**	219,505.29	219,505.79	**219,507.46**	2673.01	**79.51**	161.87
	0.75	**302,363.4**	302,362.8	302,363.0	**302,359.76**	302,358.98	302,356.40	2129.25	**66.05**	141.21
10	0.25	118,602.3	118,613.2	**118,613.5**	118,548.87	**118,574.88**	118,574.00	3639.40	**125.70**	232.00
	0.5	217,318.5	217,318.5	**217,321.9**	217,281.33	217,282.37	**217,283.24**	3811.47	**141.36**	184.32
	0.75	**302,601.4**	302,593.1	302,590.6	**302,583.25**	302,574.51	302,568.88	2950.25	**93.70**	174.51
30	0.25	115,571.0	115,518.0	**115,605.3**	115,494.70	115,421.82	**115,505.89**	3943.07	476.50	**231.05**
	0.5	**216,266.2**	216,195.3	216,236.2	**21,6200.55**	216,130.38	216,186.44	3542.84	407.97	**220.29**
	0.75	**302,445.1**	302,413.8	302,419.8	**302,414.08**	302,353.54	30,2374.68	3451.25	433.77	**216.52**

6. Discussion and Conclusions

Metaheuristics based on learning—such as ant colony optimization, particle swarm optimization and evolutionary algorithms—are generally based on learning from positive examples, that is, they are based on positive learning. However, examples from nature show that learning from negative examples can be very beneficial. In fact, there have been several attempts during the last two decades to find a way to beneficially add negative learning to ant colony optimization. However, hardly any of the respective papers were able to show that the proposed mechanism was really useful. This is with the exception of the strand of work on constraint satisfaction problems. The goal of this work was, therefore, to devise a new negative learning mechanism for ant colony optimization and to show its usefulness. The main idea of our mechanism is that the negative feedback should not be extracted from the main ant colony optimization algorithm itself. Instead, it should be produced by an additional algorithmic component. In fact, after devising a new negative learning framework, we have tested two algorithmic options for producing the negative information: (1) making use of the mathematical programming solver CPLEX, and (2) making use of the baseline ACO algorithm, but in terms of additional applications for solving sub-instances of the original problem instances.

All considered algorithm variants were applied to two NP-hard combinatorial optimization problems from the class of subset selection problems: the minimum dominating set problem and the multi dimensional knapsack problem. Moreover, four negative learning mechanisms from the literature were implemented on the basis of the chosen baseline ACO algorithm in order to be able to compare our proposals with existing approaches. The obtained results have shown, first of all, that the proposed negative learning mechanism—especially when using CPLEX for producing the negative feedback information—is superior to the existing approaches from the literature. Second, we have shown that, even though negative learning is not useful for all problem instances, it can be very useful for subsets of problem instances with certain characteristics. In the context of the minimum dominating set problem, for example, this concerns rather sparse graphs, while for the multi dimensional knapsack problem the proposed negative learning mechanism was especially useful for problem instances with rather many resources. From a global point of view, it was also shown that it is generally not harmful to add negative learning, because the globally best-performing algorithm variant makes use of negative learning. Finally, we were even able to show that our globally best-performing algorithm variant is able to compete with current state-of-the-art algorithms for both considered problems.

Future lines for additional work include the following aspects. First, we aim to apply the proposed mechanism also to problems of a very different nature. Examples include scheduling and vehicle routing problems. Second, we aim at experimenting with other alternatives for producing the negative feedback information.

Author Contributions: Methodology, T.N. and C.B.; programming, T.N; writing—original draft, T.N. and C.B.; writing—review and editing, T.N. and C.B. All authors have read and agreed to the published version of the manuscript.

Funding: This work was supported by project CI-SUSTAIN funded by the Spanish Ministry of Science and Innovation (PID2019-104156GB-I00).

Data Availability Statement: Both the problem instances used in this study and the detailed numerical results can be obtained from the corresponding author (C. Blum) on demand.

Acknowledgments: We acknowledge administrative and technical support by the Spanish National Research Council (CSIC).

Conflicts of Interest: The authors declare no conflict of interest. The funders had no role in the design of the study; in the collection, analyses, or interpretation of data; in the writing of the manuscript, or in the decision to publish the results.

References

1. Blum, C.; Roli, A. Metaheuristics in combinatorial optimization: Overview and conceptual comparison. *ACM Comput. Surv.* **2003**, *35*, 268–308. [CrossRef]
2. Gendreau, M.; Potvin, J.Y. *Handbook of Metaheuristics*, 3rd ed.; International Series in Operations Research & Management Science; Springer: Berlin, Germany, 2019.
3. Dorigo, M.; Stützle, T. Ant colony optimization: Overview and recent advances. In *Handbook of Metaheuristics*; Springer: Berlin, Germany, 2019; pp. 311–351.
4. Dorigo, M.; Stützle, T. *Ant Colony Optimization*; MIT Press: Cambridge, MA, USA, 2004.
5. Engin, O.; Güçlü, A. A new hybrid ant colony optimization algorithm for solving the no-wait flow shop scheduling problems. *Appl. Soft Comput.* **2018**, *72*, 166–176. [CrossRef]
6. Tirkolaee, E.B.; Alinaghian, M.; Hosseinabadi, A.A.R.; Sasi, M.B.; Sangaiah, A.K. An improved ant colony optimization for the multi-trip Capacitated Arc Routing Problem. *Comput. Electr. Eng.* **2019**, *77*, 457–470. [CrossRef]
7. Zhang, D.; You, X.; Liu, S.; Pan, H. Dynamic Multi-Role Adaptive Collaborative Ant Colony Optimization for Robot Path Planning. *IEEE Access* **2020**, *8*, 129958–129974. [CrossRef]
8. Jovanovic, R.; Tuba, M.; Voß, S. An efficient ant colony optimization algorithm for the blocks relocation problem. *Eur. J. Oper. Res.* **2019**, *274*, 78–90. [CrossRef]
9. Peng, H.; Ying, C.; Tan, S.; Hu, B.; Sun, Z. An improved feature selection algorithm based on ant colony optimization. *IEEE Access* **2018**, *6*, 69203–69209. [CrossRef]
10. Stützle, T.; Hoos, H.H. MAX–MIN ant system. *Future Gener. Comput. Syst.* **2000**, *16*, 889–914. [CrossRef]
11. Dorigo, M.; Gambardella, L.M. Ant colony system: A cooperative learning approach to the traveling salesman problem. *IEEE Trans. Evol. Comput.* **1997**, *1*, 53–66. [CrossRef]
12. Bullnheimer, B.; Hartl, R.R.; Strauss, C. A new rank-based version of the Ant System: A computational study. *Central Eur. J. Oper. Res.* **1999**, *7*, 25–38.
13. Robinson, E.J.H.; Jackson, D.E.; Holcombe, M.; Ratnieks, F.L.W. 'No entry' signal in ant foraging. *Nature* **2005**, *438*, 442. [CrossRef]
14. Robinson, E.J.H.; Jackson, D.E.; Holcombe, M.; Ratnieks, F.L.W. No entry signal in ant foraging (Hymenoptera: Formicidae): New insights from an agent-based model. *Myrmecol. News* **2007**, *10*, 120.
15. Grüter, C.; Schürch, R.; Czaczkes, T.J.; Taylor, K.; Durance, T.; Jones, S.M.; Ratnieks, F.L.W. Negative Feedback Enables Fast and Flexible Collective Decision-Making in Ants. *PLoS ONE* **2012**, *7*, e44501. [CrossRef]
16. Schlein, Y.; Galun, R.; Ben-Eliahu, M.N. Abstinons–Male-produced Deterrents of Mating in Flies. *J. Chem. Ecol.* **1981**, *7*, 285–290. [CrossRef]
17. Giurfa, M. The repellent scent-mark of the honeybee Apis mellifera tigustica and its role as communication cue during foraging. *Insectes Sociaux* **1993**, *40*, 59–67. [CrossRef]
18. Schoonderwoerd, R.; Holland, O.; Bruten, J.; Rothkrantz, L. Ant-Based Load Balancing in Telecommunications Networks. *Adapt. Behav.* **1997**, *5*, 169–207. [CrossRef]
19. Maniezzo, V. Exact and approximate nondeterministic tree-search procedures for the quadratic assignment problem. *INFORMS J. Comput.* **1999**, *11*, 358–369. [CrossRef]
20. Cordón, O.; Fernández de Viana, I.; Herrera, F.; Moreno, L. A New ACO Model Integrating Evolutionary Computation Concepts: The Best-Worst Ant System. In Proceedings of the ANTS 2000–Second International Workshop on Ant Algorithms, Brussels, Belgium, 8–9 September 2000; pp. 22–29.
21. Montgomery, J.; Randall, M. Anti-pheromone as a Tool for Better Exploration of Search Space. In Proceedings of the ANTS 2002–3rd International Workshop on Ant Algorithms, Brussels, Belgium, 12–14 September 2002; Lecture Notes in Computer Science; Dorigo, M., Di Caro, G., Sampels, M., Eds.; Springer: Berlin/Heidelberg, Germany, 2002; Volume 2463, pp. 100–110.
22. Iredi, S.; Merkle, D.; Middendorf, M. Bi-criterion optimization with multi colony ant algorithms. In Proceedings of the EMO 2001–International Conference on Evolutionary Multi-Criterion Optimization, Zurich, Switzerland, 7–9 March 2001; Lecture Notes in Computer Science; Zitzler, E., Deb, K., Thiele, L., Coello, C.A., Corne, D., Eds.; Springer: Berlin/Heidelberg, Germany, 2001; Volume 1993; pp. 359–372.
23. Simons, C.; Smith, J. Exploiting antipheromone in ant colony optimisation for interactive search-based software design and refactoring. In Proceedings of the GECCO 2016–Genetic and Evolutionary Computation Conference Companion, Denver, CO, USA, 20–24 July 2016; ACM: New York City, NY, USA, 2016; pp. 143–144.
24. Rojas-Morales, N.; Riff, M.C.; Coello Coello, C.A.; Montero, E. A Cooperative Opposite-Inspired Learning Strategy for Ant-Based Algorithms. In Proceedings of the ANTS 2018–11th International Conference on Swarm Intelligence, Rome, Italy, 29–31 October 2018; Lecture Notes in Computer Science; Dorigo, M., Birattari, M., Blum, C., Christensen, A.L., Reina, A., Trianni, V., Eds.; Springer International Publishing: Cham, Switzerland, 2018; Volume 11172, pp. 317–324.
25. Malisia, A.R.; Tizhoosh, H.R. Applying opposition-based ideas to the ant colony system. In Proceedings of the 2007 IEEE Swarm Intelligence Symposium, Honolulu, HI, USA, 1–5 April 2007; pp. 182–189.
26. Ramos, V.; Rodrigues, D.M.S.; Louçã, J. Second Order Swarm Intelligence. In Proceedings of the Proceedings of HAIS 2013–International Conference on Hybrid Artificial Intelligence Systems, Salamanca, Spain, 11–13 September 2013; Pan, J.S., Polycarpou, M.M., Woźniak, M., de Carvalho, A.C.P.L.F., Quintián, H., Corchado, E., Eds.; Springer: Berlin/Heidelberg, Germany, 2013; pp. 411–420.

27. Ye, K.; Zhang, C.; Ning, J.; Liu, X. Ant-colony algorithm with a strengthened negative-feedback mechanism for constraint-satisfaction problems. *Inf. Sci.* **2017**, *406–407*, 29–41. [CrossRef]
28. Masukane, T.; Mizuno, K. Solving Constraint Satisfaction Problems by Cunning Ants with multi-Pheromones. *Int. J. Mach. Learn. Comput.* **2018**, *8*, 361–366.
29. Masukane, T.; Mizuno, K. Refining a Pheromone Trail Graph by Negative Feedback for Constraint Satisfaction Problems. In Proceedings of the TAAI 2019–International Conference on Technologies and Applications of Artificial Intelligence, Kaohsiung City, Taiwan, 21–23 November 2019; pp. 1–6.
30. Ning, J.; Zhao, Q.; Sun, P.; Feng, Y. A multi-objective decomposition-based ant colony optimisation algorithm with negative pheromone. *J. Exp. Theor. Artif. Intell.* **2020**, in press. [CrossRef]
31. Nurcahyadi, T.; Blum, C. A New Approach for Making Use of Negative Learning in Ant Colony Optimization. In Proceedings of the ANTS 2020–12th International Conference on Swarm Intelligence, Barcelona, Spain, 26–28 October 2020; Lecture Notes in Computer Science; Dorigo, M., Stützle, T., Blesa, M.J., Blum, C., Hamann, H., Heinrich, M.K., Strobel, V., Eds.; Springer International Publishing: Cham, Switzerland, 2020; Volume 12421, pp. 16–28.
32. Garey, M.R.; Johnson, D.S. *Computers and Intractability*; Freeman: San Francisco, CA, USA, 1979; Volume 174.
33. Fréville, A. The multidimensional 0–1 knapsack problem: An overview. *Eur. J. Oper. Res.* **2004**, *155*, 1–21. [CrossRef]
34. Li, R.; Hu, S.; Liu, H.; Li, R.; Ouyang, D.; Yin, M. Multi-Start Local Search Algorithm for the Minimum Connected Dominating Set Problems. *Mathematics* **2019**, *7*, 1173. [CrossRef]
35. Yuan, F.; Li, C.; Gao, X.; Yin, M.; Wang, Y. A novel hybrid algorithm for minimum total dominating set problem. *Mathematics* **2019**, *7*, 222. [CrossRef]
36. Zhou, Y.; Li, J.; Liu, Y.; Lv, S.; Lai, Y.; Wang, J. Improved Memetic Algorithm for Solving the Minimum Weight Vertex Independent Dominating Set. *Mathematics* **2020**, *8*, 1155. [CrossRef]
37. Cai, S.; Hou, W.; Wang, Y.; Luo, C.; Lin, Q. Two-goal Local Search and Inference Rules for Minimum Dominating Set. In Proceedings of the Twenty-Ninth International Joint Conference on Artificial Intelligence, IJCAI-20, Yokohama, Japan, 11–17 July 2020; pp. 1467–1473. [CrossRef]
38. Chu, P.C.; Beasley, J.E. A genetic algorithm for the multidimensional knapsack problem. *Discret. Appl. Math.* **1994**, *49*, 189–212.
39. Wang, L.; Wang, S.Y.; Xu, Y. An effective hybrid EDA-based algorithm for solving multidiemnsional knapsack problems. *Expert Syst. Appl.* **2012**, *39*, 5593. [CrossRef]
40. Kong, X.; Gao, L.; Ouyang, H.; Li, S. Solving large-scale multidimensional knapsack problems with a new binary harmony search algorithm. *Comput. Oper. Res.* **2015**, *63*, 7–22. [CrossRef]
41. Vimont, Y.; Boussier, S.; Vasquez, M. Reduced costs propagation in an efficient implicit enumeration for the 01 multidimensional knapsack problem. *J. Comb. Optim.* **2008**, *15*, 165–178. [CrossRef]
42. Boussier, S.; Vasquez, M.; Vimont, Y.; Hanafi, S.; Michelon, P. A multi-level search strategy for the 0–1 multidimensional knapsack problem. *Discret. Appl. Math.* **2010**, *158*, 97–109. [CrossRef]
43. Mansini, R.; Speranza, M.G. Coral: An exact algorithm for the multidimensional knapsack problem. *INFORMS J. Comput.* **2012**, *24*, 399–415. [CrossRef]
44. Lai, X.; Hao, J.K.; Fu, Z.H.; Yue, D. Diversity-preserving quantum particle swarm optimization for the multidimensional knapsack problem. *Expert Syst. Appl.* **2020**, *149*, 113310. [CrossRef]
45. Lai, X.; Hao, J.K.; Glover, F.; Lü, Z. A two-phase tabu-evolutionary algorithm for the 0–1 multidimensional knapsack problem. *Inf. Sci.* **2018**, *436*, 282–301. [CrossRef]
46. Blum, C.; Dorigo, M. The hyper-cube framework for ant colony optimization. *IEEE Trans. Syst. Man Cybern. Part B* **2004**, *34*, 1161–1172. [CrossRef]
47. López-Ibánez, M.; Dubois-Lacoste, J.; Cáceres, L.P.; Birattari, M.; Stützle, T. The irace package: Iterated racing for automatic algorithm configuration. *Oper. Res. Perspect.* **2016**, *3*, 43–58. [CrossRef]
48. Calvo, B.; Santafé, G. scmamp: Statistical Comparison of Multiple Algorithms in Multiple Problems. *R J.* **2016**, *8*, 248–256. [CrossRef]
49. García, S.; Herrera, F. An Extension on "Statistical Comparisons of Classifiers over Multiple Data Sets" for all Pairwise Comparisons. *J. Mach. Learn. Res.* **2008**, *9*, 2677–2694.
50. Chalupa, D. An order-based algorithm for minimum dominating set with application in graph mining. *Inf. Sci.* **2018**, *426*, 101–116. [CrossRef]
51. Wang, Y.; Cai, S.; Yin, M. Local search for minimum weight dominating set with two-level configuration checking and frequency based scoring function. *J. Artif. Intell. Res.* **2017**, *58*, 267–295. [CrossRef]
52. Wang, Y.; Cai, S.; Chen, J.; Yin, M. A Fast Local Search Algorithm for Minimum Weight Dominating Set Problem on Massive Graphs. In Proceedings of the Twenty-Seventh International Joint Conference on Artificial Intelligence, IJCAI-18, Stockholm, Sweden, 13–19 July 2018; pp. 1514–1522.
53. Fan, Y.; Lai, Y.; Li, C.; Li, N.; Ma, Z.; Zhou, J.; Latecki, L.J.; Su, K. Efficient local search for minimum dominating sets in large graphs. In Proceedings of the International Conference on Database Systems for Advanced Applications, Chiang Mai, Thailand, 22–25 April 2019; Springer: Berlin, Germany, 2019; pp. 211–228.

MDPI

Article

Discrete Optimization: The Case of Generalized BCC Lattice

Gergely Kovács [1,†], Benedek Nagy [2,†], Gergely Stomfai [3,†], Neşet Deniz Turgay [2,*,†] and Béla Vizvári [4,†]

1 Department of Methodology of Applied Sciences, Edutus University, 2800 Tatabánya, Hungary; kovacs.gergely@edutus.hu
2 Department of Mathematics, Faculty of Arts and Sciences, Eastern Mediterranean University, Famagusta 99628, North Cyprus, Turkey; benedek.nagy@emu.edu.tr
3 ELTE Apáczai Csere János High School, 1053 Budapest, Hungary; 21asg@apaczai.elte.hu
4 Department of Industrial Engineering, Eastern Mediterranean University, Famagusta 99628, North Cyprus, Turkey; bela.vizvari@emu.edu.tr
* Correspondence: neset.turgay@emu.edu.tr
† These authors contributed equally to this work.

Abstract: Recently, operations research, especially linear integer-programming, is used in various grids to find optimal paths and, based on that, digital distance. The 4 and higher-dimensional body-centered-cubic grids is the nD ($n \geq 4$) equivalent of the 3D body-centered cubic grid, a well-known grid from solid state physics. These grids consist of integer points such that the parity of all coordinates are the same: either all coordinates are odd or even. A popular type digital distance, the chamfer distance, is used which is based on chamfer paths. There are two types of neighbors (closest same parity and closest different parity point-pairs), and the two weights for the steps between the neighbors are fixed. Finding the minimal path between two points is equivalent to an integer-programming problem. First, we solve its linear programming relaxation. The optimal path is found if this solution is integer-valued. Otherwise, the Gomory-cut is applied to obtain the integer-programming optimum. Using the special properties of the optimization problem, an optimal solution is determined for all cases of positive weights. The geometry of the paths are described by the Hilbert basis of the non-negative part of the kernel space of matrix of steps.

Keywords: integer programming; digital geometry; non-traditional grids; shortest chamfer paths; 4D grid; linear programming; optimization; digital distances; chamfer distances; weighted distances

Citation: Kovacs, G.; Nagy, B.; Stomfai, G.; Turgay, N.D.; Vizvari, B. Discrete Optimization: The Case of Generalized BCC Lattice. *Mathematics* **2021**, *9*, 208. https://doi.org/10.3390/math9030208

Academic Editor: Frank Werner
Received: 26 November 2020
Accepted: 18 January 2021
Published: 20 January 2021

1. Introduction

In digital geometry, by modeling the world on a grid, path-based distances are frequently used [1,2]. They allow to apply various algorithms of computer graphics and image processing and analysis, e.g., distance transformation [3]. Non traditional grids (related to various crystal structures) have various advantages over the traditional rectangular grids, e.g., having better packing density.

The face-centered cubic (FCC) and body-centered cubic (BCC) lattices are very important non-traditional grids appearing in nature. While the FCC grid is obtained from the cubic grid by adding a point to the center of each square face of the unit cubes, in the BCC grid, the additional points are in the centers of the bodies of the unit cells. Both FCC and BCC are point lattices (i.e., grid vectors point to grid points from any point of the grid). In this paper, we concentrate on the BCC grid and its higher dimensional generalizations. The BCC grid can be viewed as the union of the cubic lattices (the body-centers of the above-mentioned unit cubes form also a cubic lattice). The points located in edge-connected corners of a cube are called 2-neighbors, these points are from the same cubic lattice, while a corner and the body-center of a unit cell are 1-neighbors, as they are the closest neighbor point pairs in the 3D BCC lattice.

There is a topological paradox with the rectangular grids in every dimension, that can be highlighted as follows: considering a usual chessboard, the two diagonals contain different color squares, they go through on each other without a crossing, i.e., without a shared pixel. This is due to the fact that neighbor squares of a diagonal share only a corner point, and no side. One of the main advantages of the BCC grid is, that there are only face neighbor voxels, i.e., if two Voronoi bodies of the grid share at least one point on their boundary, then they share a full face (either a hexagon or a square). In this way, in the BCC grid, the topological paradox mentioned above cannot occur. The inner and outer part of the space is well-defined for any object built up by voxels of the BCC grid. Another important reason considering the BCC grid is its well applicability in graphical reconstruction.

The BCC lattice has been proven to be optimal for sampling spherically band-limited signals.

To perfectly reconstruct these signals from their discrete representations, around 30% fewer samples per unit volume have to be taken on a BCC grid than on an equivalent cubic grid. When the same number of samples is used with equivalent filters for resampling, a BCC-sampled volume representation ensures much higher quality of reconstruction than a cubic-sampled representation does [4–6].

Higher dimensional variants of FCC and BCC grids can also be defined and they have both theoretical and practical interest [7]. As their finite segments can be seen as graphs, they could also be used to build special architecture processor or computer networks. The BCC grid has the advantage that, independent of the dimension, exactly two types of neighborhoods are defined and thus, it allows relatively simple simulations (e.g., random walks) and computations. We note here that in the 4 dimensional extension of the BCC grid, the two types of neighbors have exactly the same Euclidean distance and in higher dimension, actually, the 2-neighbors of a point are closer than its 1-neighbor.

Concerning path-based (in grids, they are also called digital) distances, one of the simplest, but on the other hand, very practical and well applicable choices is to use chamfer distances [3]. These distances are, in fact, weighted distances based on various (positive) weights assigned to steps to various types of neighbors. These distances are studied in various grids [7–10] both with theoretical and practical analysis and also in connection with other fields including approximations of the Euclidean distance [11,12] and various other applications.

In this paper, similar to what we have used in other nontraditional grids (see, e.g., [8,13,14] for analogous results on the semi-regular Khalimsky grid and on the regular triangular grid), we use the tools for operation research to find shortest paths (optimal solution with the terminology of operational research) between any two points. Of course, shortest (also called minimal) paths can be found in various ways, for arbitrary graphs one may use Dijkstra algorithm [15]. However, the grids we use are much more structured than arbitrary graphs as we can refer to the vertices of the graph (points of the grid) by their coordinates. Our method gives explicit formulae for the shortest paths; in this way, our method is more efficient than the application of the Dijkstra algorithm for the grids. This efficiency is obtained by the mathematical analysis of the algorithms.

Since our approach, by using operational research techniques in the field of digital geometry is not common, in the next section we describe our grids and in Section 3 we recall briefly some basic concepts of the field operations research that we need for our work. Further, in Section 4 we represent the shortest path problem on the higher dimensional ($m > 3$) BCC grids as an integer programming problem. After filtering the potential bases in Section 5, the optimal solutions are determined in Section 6. Then, in Section 7, the Gomory cut is applied to ensure integer solution. We also give some details on the Hilbert bases of rational polyhedral cones in Section 8. Finally, the paper is concluded in Section 9.

2. The BCC Grid and Its Extensions to Higher Dimensions

Based on a usual description of the BCC grid, we can give the following definitions used throughout of this paper. Let us consider the $m > 1$ dimensional digital space $\mathbb{Z}^m$, especially, only the points that are represented by either only even or only odd coordinates:

$\mathbb{B}^m$ is a subset of $\mathbb{Z}^m$ such that it includes exactly the points $(x_1, \ldots, x_m)$ if and only if $x_1 \mod 2 \equiv x_2 \mod 2 \equiv \cdots \equiv x_m \mod 2$: $\mathbb{B}^m = \{x = (x_1, \ldots, x_m) \in \mathbb{Z}^m \mid$ the parity of all coordinates of x are the same$\}$.

Note that $\mathbb{B}^2$ is in fact the square grid with one of its unusual representations called 2D diamond grid/diagonal-square grid and $\mathbb{B}^3$ is the original BCC grid, however, in this paper we use mostly $\mathbb{B}^4$ and its higher dimensional generalizations.

These grids can be seen as the union of two m dimensional (hyper)cubic (also called rectangular) grids, they are referred as the even and odd (sub)lattices of the grid, based on the parity of the coordinates of their points.

There are two types of usually defined neighborhoods on $\mathbb{B}^m$: We say that the points $(x_1, \ldots, x_m)$ and $(y_1, \ldots, y_m)$ are 1-neighbors if and only if $|x_i - y_i| = 1$ for every $i \in \{1, \ldots, m\}$. These points of $\mathbb{B}^m$ are closest neighbors in Euclidean sense if $m \in \{2, 3\}$. Moreover, in each dimension, the 1-neighbor points are the closest point pairs containing points from both the even and odd sublattices. Two points $(x_1, \ldots, x_m)$ and $(y_1, \ldots, y_m)$ of the same sublattice are 2-neighbors if and only if there is exactly one $i \in \{1, \ldots, m\}$ such that $|x_i - y_i| = 2$ and the points agree on all other coordinate values, i.e., for every $j \in \{1, \ldots, m\}, i \neq j$ the equation $x_j = y_j$ holds. In case $m \in \{2, 3\}$, the 2-neighbor points are the second closest point pairs of $\mathbb{B}^m$. However, in dimension 4, the Euclidean distance of the two types of neighborhood relation coincide, while in higher dimensions ($m > 4$) the 2-neighbors are closer than the 1-neighbors in Euclidean sense.

Since we have two types of neighbors, we may use different weights for them in chamfer distances. The positive weights of the steps between 1- and 2-neighbors are denoted by w and u in this paper, respectively. We will use the term *neighbor* for both including 1- and 2-neighbors, and the term *step* as step from a point to one of its neighbors.

Then, the chamfer distance (also called weighted distance, and in this paper, we will refer to it simply with the term *distance*), of two points of the grid is defined as the weight of (one of the) smallest weighted path(s) between them, where a path is built up by steps to neighbor points.

Further, each of the mentioned grids is a point lattice, i.e., they are closed under addition of grid vectors. Based on that, when we look for a shortest path between two grid-points, w.l.o.g, we may assume that one of the points is the origin described by the m dimensional zero vector.

3. Theoretical Bases from Linear and Integer Programming

3.1. The Linear Programming Problem

The problem of linear programming is an optimization problem such that the mathematical form of both the objective function and the constraints are linear. The standard form of the problem is as follows:

$$\begin{aligned} \min f^T x \\ Gx = h \\ x \geq 0, \end{aligned}$$

where G is an $m \times n$ matrix, h is an m dimensional vector, f is an n dimensional vector of constants, and x is the n dimensional vector of variables. Notice that the non-negativity constraints are linear inequalities. In the practice, it is necessary to allow that some constraints are inequalities and some variables are "free", i.e., they may have positive, zero, and negative values as well. It is easy to see that any optimization problem with linear constraints and objective function can be transformed to the standard form by embedding the problem into a higher dimensional space.

3.2. The Simplex Method and Its Geometric Content

The linear programming and its first general algorithm, the simplex method was discovered by George Dantzig in 1947. However, it was published only in 1949 [16]. The simplex method is still in intensive use by professional solvers. It also gives a complete description of the convex polyhedral sets. This theory is summarized based on the book [17]. The author is well-known for the international operations research community as he obtained the EURO Gold Medal in 2003. However, this early book of him is not known for the international scientific community, although it is the best book written on the geometry of linear programming by the authors opinion.

A convex polyhedral set is the intersection of finitely many half-spaces. Notice that the ball is the intersection of infinitely many half-spaces. What is called 0-dimensional facet or corner point in the usual geometry, is called extreme point in this theory. A point is an extreme point of the polyhedral set if it is the only intersection point of the polyhedral set and a supporting hyperplane. Notice that all surface points of the ball are extreme points in this sense, as they are the intersection points of the ball and the tangent plane. The extreme points of the polyhedral set of the linear programming problem are the *basic feasible solutions*. For the sake of simplicity, assume that the rank of matrix G is m. Let B be a subset of the columns of G such that the vectors of B form a basis. The matrix formed from the elements of B is an $m \times m$ matrix. It is also denoted by B. Assume that matrix G and vector x are partitioned accordingly, i.e., $G = (B, N)$, and $x^T = (x_B^T, x_N^T)$. The basic solution of basis B is obtained if x satisfies the equation system and its x_N part is 0. It can be obtained as follows:

$$Gx = (B, N)\begin{pmatrix} x_B \\ x_N \end{pmatrix} = Bx_B + Nx_N = b.$$

Hence,

$$x_B = B^{-1}b - B^{-1}Nx_N. \tag{1}$$

Thus, the x_B part of the basic solution is $B^{-1}b$. The basic solution is *feasible* if $B^{-1}b \geq 0$.

The simplex method starts from an extreme point of the polyhedral set, i.e., from a basic feasible solution. The algorithm moves from here to a neighboring extreme point. Two extreme points are neighboring, if a one dimensional edge of the polyhedral set connects them. The value of the objecting function in the selected neighboring extreme point is at least as good as in the current extreme point. This procedure is repeated. It stops if the current extreme point is an optimal solution. There are important consequences. One is that if an optimal solution exists, then at least one basic feasible solution is optimal. The other one is as follows: Let the set of the extreme points of the polyhedral set be the set of vertices of an undirected graph. Two vertices are connected by an edge if and only if an edge of the polyhedral set connects them. This graph is a connected graph as (i) there is no restriction on the starting point and (ii) every other extreme point can be optimal, because its support plane determines a linear objective function such that only this extreme point is optimal solution.

The optimality condition of the simplex method is as follows. Let f_B be the vector consisting of the basic components of the objective function. The current basic feasible solution is optimal, if the inequality

$$f_B B^{-1} g_j - f_j \leq 0 \tag{2}$$

holds for all non-basic columns of G. In other words, (2) must hold for all columns g_j in N. If the inequality (2) is violated by a column g_j, then g_j can replace a vector of the basis such that the objective function value either improves or remains the same. The latter case will be excluded by the strict inequalities assumed among the data of the minimal path problem, see below. The replacement can also show that the problem is unbounded. However, this case is also excluded by the assumption of the step lengths.

3.3. The Gomory Cut

Gomory's method of integer programming is based on the observation that the basic variables are expressed by the other variables in the form of the equation system (1). Assume that one of these equations is

$$x_i = d_{i0} - \sum_{j \in K} d_{ij} x_j$$

where K is the index set of the non-basic variables. Assume further on, that d_{i0} is non-integer. It is the current value of the integer variable x_i. Let ϕ_j be the fractional part of the coefficient, i.e., the fractional part of d_{ij}

$$\phi_j = d_{ij} - \lfloor d_{ij} \rfloor, \quad j \in K \cup \{0\}.$$

Let us substitute these quantities into the equation. If the equation is rearranged such that all terms which are integers for sure, are in the left-hand side, the new form of the equation is obtained as follows:

$$x_i - \lfloor d_{i0} \rfloor + \sum_{j \in K} \lfloor d_{ij} \rfloor x_j = \phi_0 - \sum_{j \in K} \phi_j x_j.$$

Hence,

$$\phi_0 \equiv \sum_{j \in K} \phi_j x_j \pmod 1.$$

All the coefficients of this relation are between 0 and 1. Thus the two sides can be congruent only if the value of the right-hand side is in the set $\{ \phi_0, \phi_0 + 1, \phi_0 + 2, \dots \}$. Hence, the inequality

$$\sum_{j \in K} \phi_j x_j \geq \phi_0 \tag{3}$$

must be satisfied. The summation in (3) goes for the non-basic variables. The value of the non-basic variables is 0 in the basic solution. Thus, this inequality is not satisfied by the current basic feasible solution as the values of the variables in the sum are all 0. This inequality is the **Gomory cut**.

4. The Integer Programming Model and Its Linear Programming Relaxation

Many problems of combinatorial optimization are actually integer programming problems. The shortest path problem in a finite graph has also its integer programming version. A grid can be represented by an infinite graph. Therefore, the integer programming model is different.

The matrix of the steps of the 4-dimensional BCC grid are as follows:

$$\begin{array}{cccccccccccccccccccccccc}
a_1 & a_2 & a_3 & a_4 & a_5 & a_6 & a_7 & a_8 & d_9 & d_{10} & d_{11} & d_{12} & d_{13} & d_{14} & d_{15} & d_{16} & d_{17} & d_{18} & d_{19} & d_{20} & d_{21} & d_{22} & d_{23} & d_{24}
\end{array}$$

$$\begin{pmatrix}
2 & 0 & 0 & 0 & -2 & 0 & 0 & 0 & 1 & 1 & 1 & 1 & 1 & 1 & 1 & 1 & -1 & -1 & -1 & -1 & -1 & -1 & -1 & -1 \\
0 & 2 & 0 & 0 & 0 & -2 & 0 & 0 & 1 & 1 & 1 & 1 & -1 & -1 & -1 & -1 & 1 & 1 & 1 & 1 & -1 & -1 & -1 & -1 \\
0 & 0 & 2 & 0 & 0 & 0 & -2 & 0 & 1 & 1 & -1 & -1 & 1 & 1 & -1 & -1 & 1 & 1 & -1 & -1 & 1 & 1 & -1 & -1 \\
0 & 0 & 0 & 2 & 0 & 0 & 0 & -2 & 1 & -1 & 1 & -1 & 1 & -1 & 1 & -1 & 1 & -1 & 1 & -1 & 1 & -1 & 1 & -1
\end{pmatrix} \tag{4}$$

The matrix of the 4D BCC grid, the vectors $a_1, \dots, a_8$ represent steps between two neighbor points of the same sublattice (2-neighbors), while the vectors $d_9, \dots, d_{24}$ represent diagonal steps, i.e., steps between neighbor points of different sublattices (1-neighbors).

This matrix is denoted by A. It has the same role in the particular problem of the shortest path problem as matrix G in the general linear programming problem. The columns of A are denoted by $a_1, \dots, a_8, d_9, \dots, d_{24}$ The first 8 columns are steps within the same rectangular grid, i.e., in the even or in the odd sublattice (2-neighbors), and the last 16 columns are (diagonal) steps between the two rectangular grids (1-neighbors). As we have already mentioned, the weights of these steps are denoted by u and w. The number of columns is $2m + 2^m$ in the general m dimensional case having vectors $a_1, \ldots, a_{2m}, d_{2m+1}, \ldots, d_{2m+m^2}$, where $m \geq 2$ and is an integer. Thus, the size of the matrix is $m \times (2m + 2^m)$ in the general case.

It is supposed that the minimal path starts from the origin which is a point of the grid. The target point, i.e., the other end point of the path, is

$$\mathbf{b} = \begin{pmatrix} p_1 \\ p_2 \\ p_3 \\ p_4 \end{pmatrix}. \tag{5}$$

Then the optimization model of the minimal path is as follows:

$$\min u \sum_{j=1}^{8} x_j + w \sum_{j=9}^{24} x_j \tag{6}$$

$$\sum_{j=1}^{8} a_j x_j + \sum_{j=9}^{24} d_j x_j = \mathbf{b} \tag{7}$$

$$x_j \geq 0, \ j = 1, 2, \ldots, 24 \tag{8}$$

$$x_j \text{ is integer}, \ j = 1, 2, \ldots, 24 \tag{9}$$

5. Filtering the Potential Bases

The aim of this paper was to give an explicit formula for the minimal path in every possible case. The theory of linear optimization and its famous algorithm called simplex method are based on the analysis of linear bases. First problems (6)–(8) are solved by the simplex method. If the optimal solution is not integer, then Gomory cut is applied for completing the solution of problems (6)–(9). However, this case occurs only once as it is shown below. The matrix A has 24 columns of 4-dimension. Thus, there are

$$\binom{24}{4} = 10,626$$

potential candidates to be a basis and producing an optimal solution. The number of candidates increases in a fast way with the dimension. It is 850,668 in 5-dimensions. Obviously, many candidates are not bases as the columns are linearly dependent. If these candidates are filtered out, then still too many candidates remain. Thus, further filtering methods must be introduced.

The grid has a highly symmetric structure. It is enough to describe the minimal paths that the two end points of the path belong to a cone such that congruent cones cover the whole 4-dimensional space without overlapping. It is assumed that the path starts from the origin which is a point of the grid and goes to the point (5) where

$$p_1 > p_2 > p_3 > p_4 > 0. \tag{10}$$

The assumption is

$$p_1 > p_2 > \cdots > p_m > 0 \tag{11}$$

in the m-dimensional case.

The 4-dimensional case is discussed first. Assume that the optimal basis is B which consists of 4 columns of matrix A. Then, a solution of the linear equation system $B\mathbf{x} = \mathbf{b}$ is required with

$$\mathbf{x} = \begin{pmatrix} x_1 \\ x_2 \\ x_3 \\ x_4 \end{pmatrix} \geq \mathbf{0}.$$

Furthermore, it must also be an integer, but this constraint is checked in a second main step. First, the bases having columns from the last 16 columns of A, i.e., all elements of the matrix are either $+1$ or -1, are investigated. One example for such a matrix is

$$\begin{pmatrix} 1 & 1 & 1 & -1 \\ 1 & 1 & -1 & -1 \\ 1 & -1 & 1 & -1 \\ -1 & -1 & -1 & -1 \end{pmatrix}.$$

It is a basis because the columns are linearly independent and the determinant of the matrix is 8. However, this basis does not give a non-negative solution as all coefficients in the last row are negative. Similarly, it follows from the assumptions that if two equations are added or a higher index equation is subtracted from a lower index equation, then the right-hand side of the obtained new equation is positive. Hence, to obtain a non-negative solution, it is necessary that the obtained equation has at least one positive coefficient on the left-hand side. The value of any coefficient in the new row is either +2 or 0 or -2.

As a result of the high number of cases, we have used a computer program in Mathematica to filter out the cases not giving a non-negative solution. It investigates the four discussed conditions for every potential bases as follows. The described method can be generalized to the m dimensional case.

Theorem 1. *Let S be a selected set of m columns of matrix A. This subset is a basis and gives a non-negative solution for all right-hand sides satisfying the assumption $p_1 > p_2 > \cdots > p_m > 0$ only if*

1. *The determinant of the basis must be non-zero.*
2. *Every row must contain at least one positive element, i.e., at least one +1 or +2.*
3. *The sum of any two rows must contain at least one positive element, i.e., at least one +2.*
4. *(Assume that the rows of matrix A are indexed from 1 till m such that the top row has index 1 and the index of the last row is m.) If a higher index row is substituted from a lower index row, then the obtained new row must contain at least one positive element, i.e., at least one +2.*

Proof. No. 1 is obvious. No. 2 follows from the fact that the original right-hand sides are positive. Nos. 3 and 4 follow again from the fact that the right-hand side remains positive after the operation of the two equations. □

These four requirements produced a significant reduction of the candidates. Only 333 candidates remained from the initial

$$\begin{pmatrix} 24 \\ 4 \end{pmatrix} = 10,626$$

4×4 matrices. If only the last 16 columns are used, then 68 candidates remains out of 1820 ones. The latter remaining candidates are given in Appendix A Not all remaining candidates can provide with an optimal solution of the linear programming relaxation. This information is also provided in Appendix A.

6. The Optimal Solutions of the Linear Programming Relaxation

Assume that $m = 4$. The basis $\{a_1, a_2, a_3, a_4\}$ is always feasible if the coordinates of the target point are positive. As it was mentioned above, the simplex method can get to an optimal solution, if any, starting from any feasible solution. Therefore the analysis is started from this basis for the sake of convenience.

Lemma 1. *The basis $\{a_1, a_2, a_3, a_4\}$ is optimal if and only if*

$$2u - w \leq 0. \tag{12}$$

Proof. The inverse of the matrix of the basis is $\frac{1}{2}I_4$ where I_4 is the 4×4 unit matrix. Let a_j be a non-basic column of matrix A, i.e., $j \in \{5, 6, \ldots, 24\}$. The components of a_j $(5 \leq j \leq 8)$ or d_j $(9 \leq j \leq 24)$ are denoted by a_{1j}, a_{2j}, a_{3j}, and a_{4j}. The optimality condition (2) is

$$\frac{u}{2}\sum_{i=1}^{4} a_{ij} - w \leq 0. \tag{13}$$

Thus, (13) can be violated only if

$$\sum_{i=1}^{4} a_{ij} > 0. \tag{14}$$

Condition (14) is satisfied by the vectors d_9, d_{10}, d_{11}, d_{13}, and d_{17}. The left-hand side of (13) is $2u - w$ in case of vector d_9 and is $u - w$ for the four other vectors. As u, and w are positive, the condition $u - w > 0$ is stricter than $2u - w > 0$. It means that if $2u - w > 0$, then (13) is violated by the column of d_9, i.e., the basis is not optimal. If $2u - w \leq 0$, then (13) is true for all columns. □

It follows from the lemma that if $2u - w > 0 > u - w$, then the only possible change of the basis is that d_9 enters instead of a_4. If $2u - w > u - w \geq 0$, then d_9 still may enter the basis instead of a_4. If $u - w > 0$, the four other vectors may enter such that d_{10} enters instead of a_3 and each of d_{11}, d_{13}, and d_{17} enters instead of a_4. This kind of analysis is very long as there are many alternative optimal solutions.

In the proofs of Theorems 2 and 3, the term weakest condition is used. The optimality condition of a basis is that (2) holds for every non-basic vector. The particular form of (2) is (13) in the case of the basis $\{a_1, a_2, a_3, a_4\}$. The particular form is different for other bases. Notice that even (13) depends on the vector, i.e., the column of matrix A. The condition depends on the step length, i.e., on u and w which are positive. The term **weakest condition** refers to that condition which can be violated in the easiest way. For example, the weakest condition of (13) is when all four a_{ij} are equal to 1. It is the case when the value of the left-hand side is the greatest possible. Similarly, there is a greatest possible value of the particular form of (2) in the discussed cases in the proofs of Theorems 2 and 3.

Here is a theorem which gives an optimal solution in O(1) steps.

Theorem 2. *The basic feasible solution of one of the bases $\{a_1, a_2, a_3, a_4\}$, $\{a_1, a_2, a_3, d_9\}$, $\{a_1, a_2, d_9, d_{10}\}$, $\{a_1, d_9, d_{10}, d_{12}\}$, and $\{d_9, d_{10}, d_{12}, d_{16}\}$ is optimal.*

Proof. Lemma 1 states that the basis is optimal if $2u \leq w$. The simplex method is applied. As it was mentioned in the proof of the lemma, the weakest condition for entering the basis and improving the value of the objective function is $2u < w$. If this condition is satisfied, then d_9 enters instead of a_4 and the basis becomes $\{a_1, a_2, a_3, d_9\}$. The calculation of the optimality condition can be carried out based on formula (2). The weakest condition of entering the bases is $3u > 2w$. Thus, the basis $\{a_1, a_2, a_3, d_9\}$ is optimal if $2u \geq w \geq \frac{3}{2}u$. If w is just under $\frac{3}{2}u$, then a_{10} may enter the basis. The basis becomes $\{a_1, a_2, d_9, d_{10}\}$ and is optimal if $\frac{3}{2}u \geq w \geq u$. There are alternative optimal solutions in the next two simplex iterations. One option is selected in the statement in both steps. The basis $\{a_1, d_9, d_{10}, d_{12}\}$ is optimal if $u \geq w \geq \frac{1}{2}u$. Finally, the basis $\{d_9, d_{10}, d_{12}, d_{16}\}$ is optimal if $\frac{1}{2}u \geq w$. □

The potential optimal solutions and the objective function values are summarized in Table 1. The values of the not mentioned variables are 0.

With the exception of the first solution, all other solutions are integer valued as the components of the target points are either odd, or even. This fact implies that the Gomory cut must be applied only at basis $\{a_1, a_2, a_3, a_4\}$.

Table 1. The optimal solutions, i.e., distances between $(0,0,0,0)$ and (p_1, p_2, p_3, p_4).

Basis	Variables	The Value of the Objective Function	Optimality Condition
$\{a_1, a_2, a_3, a_4\}$	$x_1 = \frac{p_1}{2}, x_2 = \frac{p_2}{2}, x_3 = \frac{p_3}{2}, x_4 = \frac{p_4}{2}$	$u\frac{p_1+p_2+p_3+p_4}{2}$	$w \geq 2u$
$\{a_1, a_2, a_3, d_9\}$	$x_1 = \frac{p_1-p_4}{2}, x_2 = \frac{p_2-p_4}{2}, x_3 = \frac{p_3-p_4}{2}, x_9 = p_4$	$u\frac{p_1+p_2+p_3-3p_4}{2} + wp_4$	$2u \geq w \geq \frac{3}{2}u$
$\{a_1, a_2, d_9, d_{10}\}$	$x_1 = \frac{p_1-p_3}{2}, x_2 = \frac{p_2-p_3}{2}, x_9 = \frac{p_3+p_4}{2}, x_{10} = \frac{p_3-p_4}{2}$	$u\frac{p_1+p_2-2p_3}{2} + wp_3$	$\frac{3}{2}u \geq w \geq u$
$\{a_1, d_9, d_{10}, d_{12}\}$	$x_1 = \frac{p_1-p_2}{2}, x_9 = \frac{p_2+p_4}{2}, x_{10} = \frac{p_3-p_4}{2}, x_{12} = \frac{p_2-p_3}{2}$	$u\frac{p_1-p_2}{2} + wp_2$	$u \geq w \geq \frac{1}{2}u$
$\{d_9, d_{10}, d_{12}, d_{16}\}$	$x_9 = \frac{p_1+p_4}{2}, x_{10} = \frac{p_3-p_4}{2}, x_{12} = \frac{p_2-p_3}{2}, x_{16} = \frac{p_1-p_2}{2}$	wp_1	$\frac{1}{2}u \geq w$

Theorem 2 can be generalized to m-dimension. Some technical details must be discussed before the formalization and the proof of the theorem.

The first issue is how to generate the columns of matrix A in the m-dimensional case. As it was mentioned, the number of columns is $2m + 2^m$. The first $2m$ columns are the columns of two diagonal matrices. All elements in the main diagonal are 2, and -2 in the case of the first, and second matrices, respectively. A possible construction of the last 2^m columns is as follows. Let us write the integers from 0 to $2^m - 1$ in increasing order by m binary digits, each. Moreover, let us arrange the numbers vertically such that the top digit is the digit of 2^{m-1} and the lowest digit is the digit of 1. In the final step, every vector is mapped component-wise into a vector where each component is 1 or -1. A component 0 is mapped to 1 and a component 1 is mapped to -1. Hence, $d_{2m+1} = (1, 1, \ldots, 1)$ and $d_{2m+2^m} = (-1, -1, \ldots, -1)$. Hence, the next lemma follows immediately.

Lemma 2. *Let k be an integer such that $0 \leq k \leq m$. The components of vector d_{2m+2^k} are as follows: (a) the first $m - k$ components are 1, and (b) the last k components are -1.*

The matrix consisting of the columns d_{2m+2^r}, $r = 0, 1, \ldots, m-1$ is

$$\begin{pmatrix} 1 & 1 & 1 & \cdots & 1 \\ 1 & 1 & 1 & \cdots & -1 \\ & & & \cdots & \\ 1 & 1 & 1 & \cdots & -1 \\ 1 & 1 & -1 & \cdots & -1 \\ 1 & -1 & -1 & \cdots & -1 \end{pmatrix}. \tag{15}$$

Lemma 3. *The absolute value of the determinant of matrix (15) is 2^{m-1}.*

Proof. If row i is subtracted from row $i-1(i=2,\ldots,m)$, then the matrix

$$\begin{pmatrix} 0 & 0 & 0 & \cdots & 2 \\ 0 & 0 & 0 & \cdots & 0 \\ & & & \cdots & \\ 0 & 0 & 2 & \cdots & 0 \\ 0 & 2 & 0 & \cdots & 0 \\ 1 & -1 & -1 & \cdots & -1 \end{pmatrix} \tag{16}$$

is obtained. The determinants are equal. Furthermore, the absolute value of the determinant of matrix (16) is 2^{m-1}. □

Let k be an integer such that $0 \le k \le m$. $B(k)$ denotes the basis

$$\{a_i | i = 1,\ldots,m-k\} \cup \{d_{2m+2^r} | r = 0,1,\ldots,k-1\}$$

and its matrix as well. The matrix is as follows:

$$\left(\begin{array}{cccc|c|cccc} 2 & 0 & \cdots & 0 & 1 & 1 & 1 & \cdots & 1 \\ 0 & 2 & \cdots & 0 & 1 & 1 & 1 & \cdots & 1 \\ & & & & \cdots & & & & \\ 0 & 0 & \cdots & 2 & 1 & 1 & 1 & \cdots & 1 \\ \hline 0 & 0 & \cdots & 0 & 1 & 1 & 1 & \cdots & 1 \\ 0 & 0 & \cdots & 0 & 1 & 1 & 1 & \cdots & -1 \\ & & & & \cdots & & & & \\ 0 & 0 & \cdots & 0 & 1 & 1 & 1 & \cdots & -1 \\ 0 & 0 & \cdots & 0 & 1 & 1 & -1 & \cdots & -1 \\ 0 & 0 & \cdots & 0 & 1 & -1 & -1 & \cdots & -1 \end{array}\right). \tag{17}$$

Lemma 4. *Let k be an integer such that $0 \le k \le m$. The set of columns of matrix $B(k)$ form a basis of the m-dimensional Euclidean space.*

Proof. The determinant of the matrix consisting of the columns (17) is the product of two subdeterminants. One is a diagonal matrix where all diagonal elements are 2. The other one is a $k \times k$ matrix type (15). The determinants of both submatrices are different from zero. □

Lemma 5. *Let k be an integer such that $0 \le k \le m$. All components of the basic solution of basis*

$$\{a_i | i = 1,\ldots,m-k\} \cup \{d_{2m+2^r} | r = 0,1,\ldots,k \quad 1\}$$

is positive under the assumption of (11).

Proof. If $k=0$, then

$$x_i = \frac{p_i}{2} > 0,\ i = 1,\ldots,m. \tag{18}$$

The basic solution is determined by the equation

$$B(k)x = b. \tag{19}$$

Its solution is as follows:

$$x_i = \frac{p_i - p_{m-k+1}}{2} \quad i = 1,\ldots m-k, \tag{20}$$

$$x_{2m+1} = \frac{p_{m-k+1} + p_m}{2} \tag{21}$$

$$x_{2m+2^r} = \frac{p_{m-k+r} - p_{m-k+r+1}}{2}, \quad r = 1, \ldots, k-1 \tag{22}$$

Substituting the solution, it satisfies the equation system. □

Lemma 6. *The inverse of the matrix $B(1)$ is*

$$\left(\begin{array}{cccc|c} 0.5 & 0 & \cdots & 0 & -0.5 \\ 0 & 0.5 & \cdots & 0 & -0.5 \\ & & \cdots & & \\ 0 & 0 & \cdots & 0.5 & -0.5 \\ \hline 0 & 0 & \cdots & 0 & 1 \end{array}\right). \tag{23}$$

Let k be an integer such that $2 \le k \le m$. The inverse of $B(k)$ is

$$\left(\begin{array}{cccc|c|cccc} 0.5 & 0 & \cdots & 0 & -0.5 & 0 & 0 & \cdots & 0 \\ 0 & 0.5 & \cdots & 0 & -0.5 & 0 & 0 & \cdots & 0 \\ & & & & \cdots & & & & \\ 0 & 0 & \cdots & 0.5 & -0.5 & 0 & 0 & \cdots & 0 \\ \hline 0 & 0 & \cdots & 0 & 0.5 & 0 & 0 & \cdots & 0.5 \\ 0 & 0 & \cdots & 0 & 0 & 0 & 0 & \cdots & -0.5 \\ & & & & \cdots & & & & \\ 0 & 0 & \cdots & 0 & 0 & 0 & 0.5 & \cdots & 0 \\ 0 & 0 & \cdots & 0 & 0 & 0.5 & -0.5 & \cdots & 0 \\ 0 & 0 & \cdots & 0 & 0.5 & -0.5 & 0 & \cdots & 0 \end{array}\right). \tag{24}$$

Proof. The product of the two matrices is the unit matrix. □

Now it is possible to formalize the generalization of Theorem 2. The theorem gives a list of $m+1$ bases such that the basic solution of at least one basis is always optimal.

Theorem 3. *The basic solution of at least one of the bases (17) is always optimal.*

Proof. It follows from Lemmas 4 and 5 that each vector set of type (17) is a basis of the m-dimensional Euclidean space and its basic solution is feasible. The tool for investigating the optimality of these bases is formula (2).

Case 1, the basis $\{a_i | i = 1, \ldots, m\}$.

The inverse of the matrix of the basis is $\frac{1}{2} I_m$ where I_m is the $m \times m$ unit matrix. The basic part of the vector of the objective function, i.e., f_B, is the m-dimensional vector $(u, u, \ldots, u)$. Hence,

$$f_b B^{-1} = (u, u, \ldots, u) \frac{1}{2} I_m = \frac{u}{2}(1, 1, \ldots, 1). \tag{25}$$

A column d_{2m+r} $(1 \le r \le 2^m)$ may enter to the basis and can improve the solution if

$$\frac{u}{2}(1, 1, \ldots, 1) d_{2m+r} > w. \tag{26}$$

The weakest condition for entering the basis is obtained if the left-hand side is maximal. It is reached, if all components of d_{2m+r} is positive, i.e., at d_{2m+1}. Condition (26) is

$$\frac{um}{2} > w. \tag{27}$$

Notice, that vectors a_j $(m+1 \le j \le 2m)$ may not enter to the basis, because the basis becomes infeasible.

Case 2, the basis $\{a_i | i = 1, \ldots, m-k\} \cup \{d_{2m+2^r} | r = 0, 1, \ldots, k-1\}$ and $1 \leq k \leq m$. The same logic is applied as in Case 1. The first $m-k$ components of f_b are u and the last k components are w. It follows from (24) that

$$\begin{aligned} f_b B^{-1} &= (u, u, \ldots, u, w, w, \ldots, w) B(k)^{-1} \\ &= (\frac{u}{2}, \frac{u}{2}, \ldots, \frac{u}{2}, -\frac{(m-k)u}{2} + w, 0, \ldots, 0). \end{aligned} \tag{28}$$

The value of the general product $f_b B^{-1} g_j$ is $f_b B^{-1} a_j$ or $f_b B^{-1} d_j$, respectively, if $1 \leq j \leq 2m$ or $2m+1 \leq j \leq 2m+2^m$, respectively. Recall that the necessary and sufficient condition that a basic solution is optimal, is that condition (2) holds for all index j. Moreover, if j is the index of a basic variable, then the left-hand side of (2) is 0. The value of f_j is u if $1 \leq j \leq 2m$ and otherwise is w.

If $1 \leq j \leq m-k$, then $f_b B^{-1} a_j - f_j = 0$, because the vector is in the basis. If $j = m-k+1$, then $f_b B^{-1} a_j - f_j = -(m-k)u + 2w - u$. It is non-positive if and only if

$$w \leq \frac{(m-k+1)u}{2}. \tag{29}$$

If $m-k+2 \leq j \leq m$ or $2m-k+2 \leq j \leq 2m$, then $f_b B^{-1} a_j - f_j = -u < 0$. If $m+1 \leq j \leq 2m-k$, then $f_b B^{-1} a_j - f_j = -2u < 0$. Finally, if $j = 2m-k+1$, then $f_b B^{-1} a_j - f_j = u(m-k) - 2w - u$. It is non-positive if and only if

$$w \geq \frac{(m-k-1)u}{2}. \tag{30}$$

The components of the vector d_j are denoted by d_{ji}. Assume that $2m+1 \leq j \leq 2m+2^m$. In determining the maximal value of the left-hand side of (2), formula (28) is used. It follows immediately that the maximal value is achieved only if the first $m-k$ components of the column are 1. The last $k-1$ components are indifferent as they are multiplied by 0. Thus, the component $d_{j,m-k+1}$ is critical. The left-hand side of (2) as a function of this component is

$$\begin{aligned} &\frac{(m-k)u}{2} - d_{j,m-k+1}\left(\frac{(m-k)u}{2} - w\right) - w \\ &= \begin{cases} 0 & \text{if } d_{j,m-k+1} = 1 \\ u(m-k) - 2w & \text{if } d_{j,m-k+1} = -1 \end{cases} \end{aligned} \tag{31}$$

Thus, the left-hand side of (2) is nonpositive if $d_{j,m-k+1} = 1$ or $d_{j,m-k+1} = -1$ and

$$w \geq \frac{(m-k)u}{2}. \tag{32}$$

(32) is a stronger condition than (30). Hence, the basis is optimal if

$$\frac{(m-k+1)u}{2} \geq w \geq \frac{(m-k)u}{2}. \tag{33}$$

Notice that the intervals of the conditions (27) and (33) $k = 1, \ldots, m$ cover the non-negative half-line which implies the statement. □

Theorem 4. *If (33) holds for $1 \leq k \leq m$, then the length of the minimal path is*

$$\frac{u\left(\sum_{i=1}^{m-k} p_i - (m-k)p_{m-k+1}\right)}{2} + w p_{m-k+1}. \tag{34}$$

Proof. Notice that the optimal solution of the linear programming relaxation is integer valued. Formula (34) is obtained by substituting the optimal solution into the objective function. □

7. The Application of the Gomory Cut

The optimal solutions of the linear programming relaxation are given in (18)–(22). As all components of the target point are either odd or even, the optimal solutions given in (20)–(22) are integer. Non-integer optimal solution of the linear programming relaxation is obtained only, if the basis $\{a_i | i = 1, \dots, m\}$ is optimal and the target point is in different cubic lattice than the starting point. The linear programming optimal solution uses steps which do not change the cubic lattice. However, an integer feasible solution must contain an odd number from the last 2^m steps of matrix A. Hence, the inequality

$$\sum_{i=m+1}^{m+2^m} x_i \geq 1 \tag{35}$$

must be satisfied.

Lemma 7. *The Gomory cut is the inequality (35).*

Proof. The inverse of the basis is $\frac{1}{2} I_m$. Assume that the Gomory cut is generated from the first equation of (7). Thus, all integer parts are $\frac{1}{2}$. Hence, the form of (3) is

$$\sum_{i=m+1}^{m+2^m} \frac{1}{2} x_i \geq \frac{1}{2}. \tag{36}$$

It is equivalent to (35). □

Theorem 5. *If $\frac{um}{2} \leq w$ and $p_1, \dots, p_m$ are odd numbers, then an optimal solution is*

$$x_i = \frac{p_i - 1}{2}, \ i = 1, \dots, m, \ x_{2m+1} = 1, \ x_i = 0, \ i = m+1, \dots, 2m + 2^m, i \neq 2m+1 \tag{37}$$

and the optimal value is

$$u \sum_{i=1}^{m} \frac{p_i}{2} - \frac{mu}{2} + w. \tag{38}$$

Remark 1. *The first condition is the opposite of (27) and ensures that the basis $\{a_1, \dots, a_m\}$ is optimal.*

Proof. The inequality (35) is added to the problem as an equation with a non-negative slack variables s as follows:

$$\sum_{i=m+1}^{m+2^m} x_i - s = 1 \tag{39}$$

Thus, every column of matrix A is extended by one component. This component is 0 in the case of the first $2m$ columns and is 1 in the case of the last 2^m. Furthermore, a new column is added which belongs to variable s. This column is the negative $m+1$-st unit vector, as variable s appears only in the last equation. As it is shown below, the optimal basis of the extended problem is $\{1, \dots, m\} \cup \{2m+1\}$.

Notice that the products $B^{-1} g_j$ in formula (2) are the coordinates of vector g_j in the current basis. Thus, (2) can be checked if the coordinates of $B^{-1} g_j$ are known. The inverse of the extended matrix B is similar to (23).

Case 1. $m + 1 \leq j \leq 2m$. The coordinates are $(B^{-1} g_j)_j = -1$ and $(B^{-1} g_j)_i = 0, \ i \neq j$. Hence, the left-hand side of (2) is $-2u < 0$.

Case 2. $2m + 2 \leq j \leq 2m + 2^m$. The coordinates are

$$(B^{-1}g_j)_i = \begin{cases} 0 & \text{if } d_{ji} = 1,\ 1 \leq i \leq m \\ -1 & \text{if } d_{ji} = -1,\ 1 \leq i \leq m \\ 1 & \text{if } i = m+1, \end{cases}$$

where d_{ji} is the i-th component of vector d_j. Hence, the left-hand side of (2) is

$$u \sum_{i:d_{ji}=-1} 1 + w - w \leq -u < 0.$$

Case 3. $j = 2m + 2^m + 1$. It is the case of the unit vector of s. The coefficient of s in the objective function is 0. The coordinates are

$$(B^{-1}g_j)_i = \begin{cases} \frac{1}{2} & \text{if } 1 \leq i \leq m \\ -1 & \text{if } i = m+1. \end{cases}$$

Hence, the left-hand side of (2) is

$$\frac{um}{2} - w.$$

This value is nonpositive according to the conditions. Thus, the optimality condition is satisfied in all cases. □

Corollary 1. *If (33) holds for $1 \leq k \leq m$, then the length of the shortest path from the origin to the point $(p_1, p_2, \ldots p_m)^T$ is (34). If $\frac{um}{2} \leq w$ and $p_1, \ldots, p_m$ are even numbers, then the length of the shortest path is*

$$u \sum_{i=1}^{m} \frac{p_i}{2}. \tag{40}$$

If $\frac{um}{2} \leq w$ and $p_1, \ldots, p_m$ are odd numbers, then the length of the shortest path is (38).

8. The Hilbert Basis of the Nonnegative Part of the Kernel Space

Jeroslow reformulated an old paper of Hilbert [18] on the language of the contemporary mathematics and proved an important theorem [19]. Later Schrijver discovered the same theorem independently [20]. The Theorem 6 below contains the reformulated version of Hilbert's theorem (first statement) and the new theorem of Jeroslow and Schrijver.

Theorem 6. *Let C be a cone determined by finitely many linear inequalities such that the coefficients of the inequalities are rational numbers in the m dimensional space. Let*

$$\mathcal{S} = C \cap Z^m$$

be the set of the integer points in C.

- *There is a finite subset $\mathcal{T}$ of $\mathcal{S}$ such that for every vector $s \in \mathcal{S}$ there are non-negative integer weights α_t such that*

$$s = \sum_{t \in \mathcal{T}} \alpha_t t. \tag{41}$$

- *If C is a pointed cone, i.e., the origin is the extreme point of C, then there is only one set $\mathcal{T}$ with this property.*

Let $n = 2m + 2^m$. The kernel space of the general matrix A is

$$\{x \in R^n \mid Ax = 0\}. \tag{42}$$

Its non-negative part is the set

$$\mathcal{P} = \{x \in R^n \mid x \geq 0,\ Ax = 0\}. \tag{43}$$

$\mathcal{P}$ is a pointed cone. Let $\mathcal{H}$ be its Hilbert basis.

Although $\mathcal{H}$ seems an algebraic object, it has strong geometric and optimization meanings. Matrix A has the property in any dimension that if a vector a (or d) is its column, then the vector $-a$ (or $-d$) is a column of A as well. For the first look, the elements of $\mathcal{H}$ describe elementary circuits of the grid. For example, $a_1 + d_{17} + d_{24} = 0$ (see also Figure 1). Thus, starting from a point and making the steps of types 1, 17, and 24, the walk returns to the origin. However, there is another interpretation of the equation as follows:

$$a_1 + d_{17} + d_{24} = a_1 - d_{16} - d_9 = 0 \tag{44}$$

implying that

$$a_1 = d_{16} + d_9. \tag{45}$$

The meaning of the Equation (45) is that one step in the same rectangular sublattice is equivalent with two steps between the two sublattices (see also Figure 2). Thus, stepping in the same sublattice is better than the two steps between the sublattices if $u < 2w$.

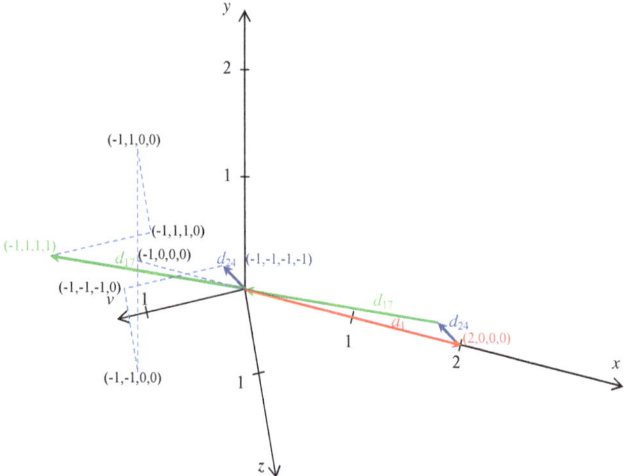

Figure 1. The sum of vectors a_1, d_{17}, and d_{24} is the four-dimensional zero vector. The coordinates of the construction are also shown.

If the opposite inequality is true, i.e., $u > 2w$, then it is better to step always between the sublattices. This example has further three similar cases as follows:

$$a_1 = d_{10} + d_{15},\ a_1 = d_{11} + d_{14},\ a_1 = d_{12} + d_{13}. \tag{46}$$

As we can see, a cycle in an undirected graph may have two different important interpretations as to walk a cycle arriving back to the initial point or to walk from a point to another in two different paths [21]. Many further composite steps can be obtained from Table 2. For example, it is possible to go from the origin to the point (2,2,0,0) by using steps a_1 and a_2 with length $2u$ or with $2w$ by using either steps d_9 and d_{12}, or d_{10} and d_{11} as it is shown in Figure 3. A cycle form of the same relation based on the above-mentioned diagonal steps can be seen in Figure 4.

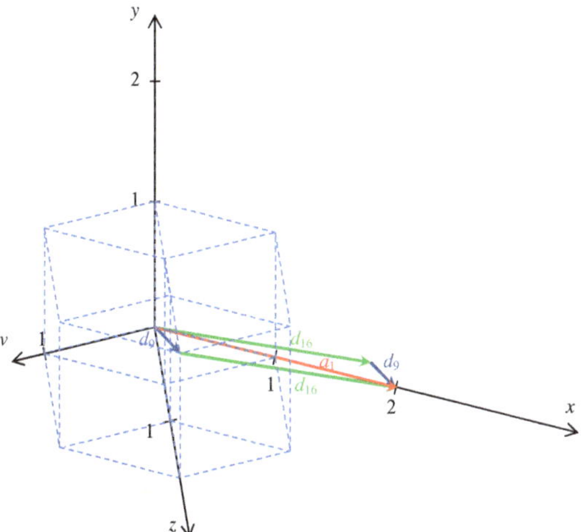

Figure 2. The sum of vectors d_9 and d_{16} is exactly the vector a_1 in the four-dimensional BCC grid. A unit hypercube is also shown.

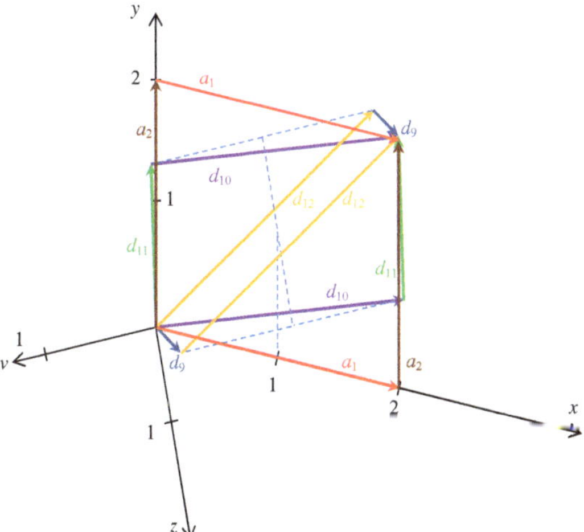

Figure 3. Alternative paths from $(0,0,0,0)$ to $(2,2,0,0)$, each is built up by two steps, in the four-dimensional BCC grid.

There is an iterative way to explore many elements of the Hilbert basis. The basic idea is that the known elements are excluded and the optimization problem looks for the next unknown element having the smallest l_1 norm. The model has variables x_j $j \in \{1, 2, \ldots, 24\}$, where the meaning of x_j is that step j is used how many times in the circuit as in (42) or (43). Thus, it must be an integer. There is a second set of variables denoted by y_j $j \in \{1, 2, \ldots, 24\}$. They are binary variables as follows:

$$y_j = \begin{cases} 1 & \text{if } x_j > 0 \\ 0 & \text{if } x_j = 0. \end{cases} \tag{47}$$

The meaning of y_j is if the step represented by the jth column of matrix A is included in the circuit or not.

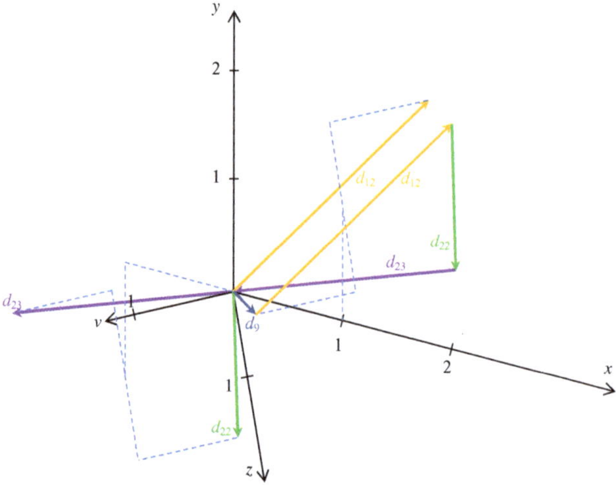

Figure 4. The sum of four diagonal vectors is the zero vector, each vector is also shown from the origin.

The objective function is the minimization of the number of used types of steps

$$\min \sum_{j=1}^{24} y_j. \tag{48}$$

Then the constraint being in the kernel space is similar to constraint (7) but the right-hand side must be the zero vector

$$\sum_{j=1}^{8} a_j x_j + \sum_{j=9}^{24} d_j x_j = 0. \tag{49}$$

The relation of the two types of variables is, that if $y_j = 1$ if and only if $x_j > 0$. Let M be a big positive number. The constraint to be claimed is

$$x_j \leq M y_j,\ j = 1, \ldots, 24. \tag{50}$$

Notice that (50) claims only that if x_j is positive, then $y_j = 1$. However, it is not possible in an optimal solution that $x_j = 0$ and $y_j = 1$. The zero vector must be excluded by claiming that the sum of the variables must be at least 1

$$\sum_{j=1}^{24} x_j \geq 1. \tag{51}$$

The variables must be non-negative integer and binary, respectively,

$$x_j \geq 0 \,\text{and integer};\ \ y_j = 0 \,\text{or}\, 1, j = 1, \ldots, 24. \tag{52}$$

The problems (48)–(52) are used in an iterative way. When a new member of the Hilbert basis is found, then it is excluded by the constraint that not all same types of steps can occur in a new element. Let $\bar{x}$ be the last optimal solution. Lest $S = \{j|\bar{x}_j > 0\}$. Then the inequality

$$\sum_{j \in S} y_j \leq |S| - 1 \tag{53}$$

is added to the problem as a new constraint. The next optimal solution must be different from $\bar{x}$.

Some trivially existing solution, i.e., if the sum of two columns is the zero vector, can be excluded immediately. Thus, the constraints

$$y_1 + y_5 \leq 1, \ldots, y_4 + y_8 \leq 1,\ y_9 + y_{24} \leq 1, \ldots, y_{16} + y_{17} \leq 1 \tag{54}$$

can be introduced.

The types obtained by iterative application of the model are summarized in Table 2.

Table 2. Types of the elements of the Hilbert basis.

Type	# of *a*'s	# of *d*'s	Example	# of Cases
1	2	0	$a_1 + a_5 = 0$	4
2	0	2	$d_9 + d_{24} = 0$	8
3	1	2	$a_1 + d_{17} + d_{24} = 0$	32
4	2	2	$a_1 + a_2 + d_{21} + d_{24} = 0$	48
5	0	4	$d_9 + d_{12} + d_{22} + d_{23} = 0$	24
6	3	2	$a_1 + a_2 + a_3 + d_{23} + d_{24} = 0$	32
7	1	4	$a_1 + d_9 + d_{20} + d_{22} + d_{23} = 0$	64
8	0	6	$2 \times d_9 + d_{16} + d_{20} + d_{22} + d_{23} = 0$	16
9	2	4	$2 \times a_1 + d_{17} + d_{20} + d_{22} + d_{23} = 0$	16
10	4	2	$a_5 + a_6 + a_7 + a_8 + 2 \times d_9 = 0$	16

A sequence of optimal bases of the linear relaxation is discussed in Theorem 3. The relation of the changes of the bases and the elements of the Hilbert basis is shown in Table 3. The relation can be obtained in the way shown in formula (44). Notice that the ratios of the numbers of a and d vectors are the same of the ratios of u and v where the optimal solution changes according to Theorem 2. The elements of the Hilbert basis can be obtained by taking the difference of the two optimal solutions.

Table 3. The relation of the elements of the Hilbert basis and the changes of basis.

Old Basis	New Basis	Hilbert Basis Element	Type	Ratio
a_1, a_2, a_3, a_4	a_1, a_2, a_3, d_9	$a_1, a_2, a_3, a_4, 2 \times d_{24}$	10	$\frac{4}{2} = 2$
a_1, a_2, a_3, d_9	a_1, a_2, d_9, d_{10}	$a_1, a_2, a_3, d_{23}, d_{24}$	6	$\frac{3}{2}$
a_1, a_2, d_9, d_{10}	a_1, d_9, d_{10}, d_{12}	a_1, a_2, d_{21}, d_{24}	4	$\frac{2}{2} = 1$
a_1, d_9, d_{10}, d_{12}	$d_9, d_{10}, d_{12}, d_{16}$	a_1, d_{17}, d_{24}	3	$\frac{1}{2}$

It is easy to see from formula (37) that when the optimal fractional solution is converted to optimal integer solution, then a type 10 member of the Hilbert basis has a key role. It is $a_1, a_2, a_3, a_4, 2 \times d_{24}$.

9. Conclusions

This paper continues the analysis of the BCC grids started in [22]. It generalizes the grids to 4 and higher dimensions. Minimal routes are determined from the origin to a target point. An integer linear programming model is applied. Its linear programming relaxation is solved by the simplex method. There is only one case when the optimal solution is fractional. The integrality is achieved by the Gomory method in this case. In 4D, Table 1 shows the direct formulae to compute distance based on Theorem 2 (and generalized to nD in Corollary 1). The non-negative cone of the kernel space of the matrix of the steps in the grid has an important role. The elements of its Hilbert basis describe the alternative routes, i.e., the geometry of the routes, and the changes of the bases during the simplex method.

Author Contributions: Formal analysis, G.K., G.S. and N.D.T.; visualization, B.N.; writing—original draft preparation, B.V.; writing–review and editing, G.K. and B.N. All authors have read and agreed to the published version of the manuscript.

Funding: This research received no external funding.

Conflicts of Interest: The authors declare no conflict of interest.

Appendix A

Table A1. The bases remained after the filtering of the 4-dimensional case containing no vectors from the first 8 vectors, i.e., containing only diagonal vectors. (The columns of matrix A are referred by their indices from 1 to 24.)

No.	Basis				Optimal	No.	Basis				Optimal
1	9	10	11	13	YES	2	9	10	11	14	YES
3	9	10	11	15	YES	4	9	10	11	16	YES
5	9	10	12	13	YES	6	9	10	12	14	YES
7	9	10	12	15	YES	8	9	10	12	16	YES
9	9	10	15	19	NO	10	9	10	15	20	NO
11	9	10	16	19	NO	12	9	10	16	20	NO
13	9	11	12	14	YES	14	9	11	14	18	NO
15	9	11	14	20	NO	16	9	11	16	18	NO
17	9	12	13	14	YES	18	9	12	13	18	NO
19	9	12	13	22	NO	20	9	12	14	15	YES
21	9	12	14	16	YES	22	9	12	14	17	NO
23	9	12	14	18	NO	24	9	12	14	20	NO
25	9	12	14	22	NO	26	9	12	14	23	NO
27	9	12	15	22	NO	28	9	12	16	18	NO
29	9	12	16	22	NO	30	9	14	15	20	NO
31	9	14	16	20	NO	32	9	16	18	19	NO
33	9	16	18	20	NO	34	9	16	20	22	NO
35	10	11	12	13	YES	36	10	11	13	14	YES
37	10	11	13	15	YES	38	10	11	13	16	YES
39	10	11	13	17	NO	40	10	11	13	18	NO
41	10	11	13	19	NO	42	10	11	13	21	NO
43	10	11	13	24	NO	44	10	11	14	17	NO
45	10	11	14	21	NO	46	10	11	15	17	NO
47	10	11	15	21	NO	48	10	11	16	21	NO
49	10	12	13	17	NO	50	10	12	13	19	NO
51	10	12	15	17	NO	52	10	13	15	19	NO
53	10	13	16	19	NO	54	10	15	17	19	NO
55	10	15	17	20	NO	56	10	15	19	21	NO
57	11	12	13	18	NO	58	11	12	14	17	NO
59	11	13	14	18	NO	60	11	13	16	18	NO
61	11	14	17	18	NO	62	11	14	17	20	NO
63	11	14	17	21	NO	64	12	13	14	17	NO
65	12	13	17	18	NO	66	12	13	17	22	NO
67	12	13	18	19	NO	68	12	14	15	17	NO

References

1. Klette, R.; Rosenfeld, A. *Digital Geometry—Geometric Methods for Digital Picture Analysis*; Morgan Kaufmann, Elsevier Science B.V.: Amsterdam, The Netherlands, 2004.
2. Rosenfeld, A.; Pfaltz, J.L. Distance functions on digital pictures. *Pattern Recognit.* **1968**, *1*, 33–61. [CrossRef]
3. Borgefors, G. Distance transformations in digital images. *Comput. Vision Graph. Image Process.* **1986**, *34*, 344–371. [CrossRef]
4. Csébfalvi, B. Prefiltered Gaussian Reconstruction for High-Quality Rendering of Volumetric Data sampled on a Body-Centered Cubic Grid. In Proceedings of the VIS 05—IEEE Visualization 2005, Minneapolis, MN, USA, 23–28 October 2005; pp. 311–318.
5. Csébfalvi, B. An Evaluation of Prefiltered B-Spline Reconstruction for Quasi-Interpolation on the Body-Centered Cubic Lattice. *IEEE Trans. Vis. Comput. Graph.* **2010**, *16*, 499–512. [CrossRef] [PubMed]
6. Vad, V.; Csébfalvi, B.; Rautek, P.; Gröller, M.E. Towards an Unbiased Comparison of CC, BCC, and FCC Lattices in Terms of Prealiasing. *Comput. Graph. Forum* **2014**, *33*, 81–90. [CrossRef]
7. Strand, R.; Nagy, B. Path-Based Distance Functions in n-Dimensional Generalizations of the Face- and Body-Centered Cubic Grids. *Discret. Appl. Math.* **2009**, *157*, 3386–3400. [CrossRef]

8. Kovács, G.; Nagy, B.; Vizvári, B. Weighted Distances and Digital Disks on the Khalimsky Grid—Disks with Holes and Islands. *J. Math. Imaging Vis.* **2017**, *59*, 2–22. [CrossRef]
9. Remy, E.; Thiel, E. Computing 3D Medial Axis for Chamfer Distances. In Proceedings of the DGCI 2000: Discrete Geometry for Computer Imagery—9th International Conference, Lecture Notes in Computer Science, Uppsala, Sweden, 13–15 December 2000; Volume 1953, pp. 418–430.
10. Sintorn, I.-M.; Borgefors, G. Weighted distance transforms in rectangular grids. In Proceedings of the ICIAP, Palermo, Italy, 26–28 September 2001; pp. 322–326.
11. Butt, M.A.; Maragos, P. Optimum design of chamfer distance transforms. *IEEE Trans. Image Process.* **1998**, *7*, 1477–1484. [CrossRef] [PubMed]
12. Celebi, M.E.; Celiker, F.; Kingravi, H.A. On Euclidean norm approximations. *Pattern Recognit.* **2011**, *44*, 278–283. [CrossRef]
13. Kovács, G.; Nagy, B.; Vizvári, B. On disks of the triangular grid: An application of optimization theory in discrete geometry. *Discret. Appl. Math.* **2020**, *282*, 136–151. [CrossRef]
14. Kovács, G.; Nagy, B.; Vizvári, B. Chamfer distances on the isometric grid: A structural description of minimal distances based on linear programming approach. *J. Comb. Optim.* **2019**, *38*, 867–886. [CrossRef]
15. Ahuja, R.K.; Magnanti, T.L.; Orlin, J.B. *Network Flows: Theory, Algorithms and Applications*; Prentice Hall: Upper Saddle River, NJ, USA, 1993.
16. Dantzig, G.B. Programming in a Linear Structure, Report of the September 9, 1948 meeting in Madison. *Econometrica* **1949**, *17*, 73–74.
17. Prékopa, A. *Lineáris Programozás I*; Hungarian, Linear Programming I; Bolyai János Matematikai Társulat (János Bolyai Mathematical Society): Budapest, Hungary, 1968.
18. Hilbert, D. Über die Theorie der algebrischen Formen. *Math. Ann.* **1890**, *36*, 473–534. [CrossRef]
19. Jeroslow, R.G. Some basis theorems for integral monoids. *Math. Oper. Res.* **1978**, *3*, 145–154.
20. Schrijver, A. On total dual integrality. *Linear Algebra Its Appl.* **1981**, *38*, 27–32. [CrossRef]
21. Nagy, B. Union-Freeness, Deterministic Union-Freeness and Union-Complexity. In Proceedings of the DCFS 2019: Descriptional Complexity of Formal Systems—21st IFIP WG 1.02, International Conference, Lecture Notes in Computer Science, Kosice, Slovakia, 17–19 July 2019; Volume 111612, pp. 46–56.
22. Kovács, G.; Nagy, B.; Stomfai, G.; Turgay, N.D.; Vizvári, B. On Chamfer Distances on the Square and Body-Centered Cubic Grids—An Operational Research Approach. 2021. Unpublished work.

MDPI
Grosspeteranlage 5
4052 Basel
Switzerland
www.mdpi.com

Mathematics Editorial Office
E-mail: mathematics@mdpi.com
www.mdpi.com/journal/mathematics